Molecular Evolution of the Major Histocompatibility Complex

NATO ASI Series

Advanced Science Institutes Series

A series presenting the results of activities sponsored by the NATO Science Committee, which aims at the dissemination of advanced scientific and technological knowledge, with a view to strengthening links between scientific communities.

The Series is published by an international board of publishers in conjunction with the NATO Scientific Affairs Division

A Life Sciences B Physics	Plenum Publishing Corporation London and New York
C Mathematical and Physical Sciences D Behavioural and Social Sciences E Applied Sciences	Kluwer Academic Publishers Dordrecht, Boston and London
F Computer and Systems Sciences G Ecological Sciences H Cell Biology I Global Environmental Change	Springer-Verlag Berlin Heidelberg New York London Paris Tokyo Hong Kong Barcelona Budapest

NATO-PCO DATABASE

The electronic index to the NATO ASI Series provides full bibliographical references (with keywords and/or abstracts) to more than 30000 contributions from international scientists published in all sections of the NATO ASI Series. Access to the NATO-PCO DATABASE compiled by the NATO Publication Coordination Office is possible in two ways:

- via online FILE 128 (NATO-PCO DATABASE) hosted by ESRIN, Via Galileo Galilei, I-00044 Frascati, Italy.
- via CD-ROM "NATO-PCO DATABASE" with user-friendly retrieval software in English, French and German (© WTV GmbH and DATAWARE Technologies Inc. 1989).

The CD-ROM can be ordered through any member of the Board of Publishers or through NATO-PCO, Overijse, Belgium.

Series H: Cell Biology Vol. 59

Molecular Evolution of the Major Histocompatibility Complex

Edited by

Jan Klein

Max-Planck-Institut für Biologie
Abt. Immungenetik
Correnssstr. 42
W-7400 Tübingen
FRG

and

Dagmar Klein

Dept. of Microbiology and Immunology
University of Miami
School of Medicine
Miami, Florida
USA

Springer-Verlag
Berlin Heidelberg New York London Paris Tokyo
Hong Kong Barcelona Budapest
Published in cooperation with NATO Scientific Affairs Division

Proceedings of the NATO Advanced Research Workshop on MHC Evolution held at the Sheraton Royal Resort and Racquet Club, Key Biscayne, Florida (USA), April 28 – May 1, 1991.

ISBN-13:978-3-642-84624-3 e-ISBN-13:978-3-642-84622-9
DOI: 10.1007/978-3-642-84622-9

Softcover reprint of the hardcover 1st edition 1991

Typesetting: Camera-ready copy provided by the editors
31/3140-543210 – Printed on acid-free paper

Preface

From molecules to populations and back

In biology, the most vigorous organisms often ensue from a union of two disparate, pure lines. In science, too, laws of hybrid vigor seem to operate at the interface between two disciplines, an interface that often proves to be fertile ground for germinating concepts and new outlooks. The fringes of research into the major histocompatibility complex (*Mhc*) have provided such an interface several times in the past and the encounters have invigorated fields such as transplantation biology, cellular immunology, and immunogenetics. In the last few years, a new interface has been emerging between *Mhc* and evolutionary genetics, and particularly the branch of evolutionary genetics dealing with molecular evolution. *Mhc* research relies upon molecular evolutionary genetics, with its grand superstructure of mathematical formulations, to come to grips with the events leading to and maintaining the *Mhc* polymorphism. Without the armament of rigorous statistical procedures developed by evolutionary geneticists, the intricate relationships among *Mhc* genes cannot be resolved. It will undoubtedly be a molecular geneticist who is the final arbiter in the dispute concerning the nature of the selection pressure molding the *Mhc* genes. And it is doubtful whether the true function of *Mhc* can ever be comprehended without the vantage point afforded by the elucidation of its evolutionary history. We might learn the minutiae about the involvement of *Mhc* in the interactions of cells and their receptors by studying the mouse and the human immune systems and ignoring all other species, but we might never understand the purpose of these interactiosn if we do not descend the evolutionary ladder right down to the beginning of the *Mhc*. Only a die-hard ignoramus can fail to appreciate the importance of the historical perspective in immunology.

On the other hand, evolutionary genetics, too, has much to gain from including *Mhc* among its favorite paradigms. Not only can the *Mhc* make major inroads into some of the perennial problems of evolutionary biology such as the mode of speciation; it can also create grounds for entirely new concepts such as those emanating from the analysis of the trans-specific character of the *Mhc* polymorphism. Thus, *Mhc* research and evolutionary genetics are equal partners who can greatly benefit from knowing each other intimately.

To facilitate the formation of partnership, *Mhc* researchers and evolutionary geneticists were brought together on a "blind date" at the Sheraton Royal Biscayne Hotel, Key Biscayne, Florida for a meeting that lasted from April 28 until May 1, 1991. The present book is based on this Conference. It does not reproduce it faithfully (the participants' arms were not twisted to contribute to the Proceedings, two evenings of discussion were not recorded, authors were permitted to include material not presented at the meeting, and they were encouraged to speculate as freely as they desired), but it does, we hope, reflect the *spirit* of the Conference. It was the editors' intention to make it more than a memento of a pleasant get-together; we hope to enkindle the interest of those who might stand outside this galvanizing interface.

We thank the North Atlantic Treaty Organization, without whose generous financial support the Conference could not have been held. We would like to express our appreciation to Drs. Herbert Tichy, Vladimir Vincek, Masanori Kasahara, and J. Wayne Streilein, who helped us at various stages in the organization of the meeting. The real *esprit de corps* of the meeting and the moving force behind these Proceedings was, however, Ms. Lynne Yakes, to whom we, as always, remain indebted.

Jan Klein
Tübingen

Dagmar Klein
Key Biscayne

List of Participants

Conference on Major Histocompatibility Complex Evolution
Key Biscayne, Florida, April 28 - May 2, 1991

Alper, Chester	Department of Pediatrics, Harvard Medical School, Boston, Massachusetts, USA
Andersson, Göran	University of Uppsala, Biomedical Center, Department of Medical Genetics, Uppsala, Sweden
Arnaiz-Villena, Antonio	Immunologia, Hospital 12 de Octubre, Madrid, Spain
Auffray, Charles	Institut d'Embryologie, Nogent Sur Marne, France
Bartl, Simona	Department of Pathology and Howard Hughes Medical Institute, Stanford University School of Medicine, Stanford, CA and Hopkins Marine Station of Stanford University, Pacific Grove, CA, USA
Begovich, Ann B.	Department of Human Genetics, Cetus Corporation, Emeryville, CA, USA
Bontrop, Ronald E.	TNO Primate Center, Rijswijk, The Netherlands
Brunsberg, Ulrica	Department of Animal Breeding and Genetics, Uppsala, Sweden
Burrous, Mary R.	The Rockefeller University, Laboratory of Neurobiology, New York, NY, USA
Chen, Z.W.	Harvard Medical School, New England Regional Primate Research Center, Southborough, MA, USA
Crew, Mark D.	Department of Pathology, UCLA School of Medicine, Center for Health Sciences, Los Angeles, CA, USA
David, Chella S.	Department of Immunology, Mayo Clinic, Rochester, MN, USA
Dawkins, Roger L.	Department of Clinical Immunology, Royal Perth Hospital, Perth, Western Australia
Degli-Esposti, Mariapia	Department of Clinical Immunology, Royal Perth Hospital, Perth, Western Australia
Elferink, B.G.	Department of Immunohaematology and Blood Bank, University Hospital Leiden, Leiden, The Netherlands
Ellis, Shirley	Institute of Molecular Medicine, Molecular Immunology Group, John Radcliffe Hospital, Headington, Oxford, UK
Erlich, Henry	Human Genetics Department, Cetus Corporation, Emeryville, CA, USA
Fischer Lindahl, Kirsten	Howard Hughes Medical Institute, University of Texas Southwestern, Dallas, TX, USA
Flajnik, Martin	Department of Microbiology & Immunology, University of Miami School of Medicine, Miami, FL, USA
Garber, Ted	Texas A & M University, Department of Veterinary Pathobiology, College Station, TX, USA

Gaur, Lakshmi	Department of Microbiology & Immunology, Bowman Gray School of Medicine, Winston-Salem, NC, USA
Geliebter, Jan	Rockefeller University, New York, NY, USA
Grahovac, Blazenka	Max-Planck-Institut für Biologie, Abteilung Immungenetik, Tübingen, Germany
Greenberg, Andrew	Department of Microbiology & Immunology, University of Miami School of Medicine, Miami, FL, USA
Guardiola, John	International Institute of Genetics and Physics, Napoli, Italy
Gyllensten, Ulf	Medical Genetics, Biomedical Center, Uppsala, Sweden
Hashimoto, Keeichiro	Institute for Comprehensive Medical Science, Fujita Health University, Toyoake, Aichi, Japan
Hauptfeld, Vera	Washington University School of Medicine, Department of Genetics, St. Louis, Missouri, USA
Heise, Eugene R.	Bowman Gray School of Medicine, Wake Forest University, Winston-Salem, NC, USA
Hill, Adrian	University of Oxford, Molecular Immunology Group, Institute of Molecular Medicine, John Radcliffe Hospital, Headington, Oxford, England
Howard, Jonathan C.	Agricultural and Food Research Council, Institute of Animal Physiology, Babraham, Cambridge, UK
Hughes, Austin L.	Department of Biology, College of Science, Erwin W. Mueller Laboratory, The Pennsylvania State University, University Park, PA, USA
Jones, Patricia P.	Dept. of Biological Sciences, Stanford University, Stanford, CA, USA
Kasahara, Masanori	Department of Microbiology & Immunology, University of Miami School of Medicine, Miami, FL, USA
Kaufman, Jim	Basel Institute for Immunology, Basel, Switzerland
Kawaguchi, Hiroshi	Yokohama City University School of Medicine, Department of Dermatology, Urafune-cho, Minami-ku, Yokohama 236, Japan
Kenter, Marcel	Department of Blood Bank & Immunohaematology, University Hospital Leiden, Leiden, The Netherlands
Kesari, Krishna	Rockefeller University, New York, NY 10021, USA
Klein, Dagmar	Department of Microbiology & Immunology, University of Miami School of Medicine, Miami, FL, USA
Klein, Jan	Max-Planck-Institut für Biologie, Abteilung Immungenetik, Tübingen, Germany and Department of Microbiology & Immunology, University of Miami School of Medicine, Miami, FL, USA
Klitz, William	Department of Integrative Biology, University of California, Berkeley, CA, USA
Lapierre, Lynne A.	Box 150, The Rockefeller University, New York, NY, USA
Lardy, M.M.	Dept. Transplantation Immunology, CLB, Amsterdam, The Netherlands

Lawlor, David	Department of Cell Biology, Sherman Fairchild Bldg., Stanford University, Stanford, CA, USA
Martinko, John	Department of Microbiology, Southern Illinois University, Carbondale, IL, USA
Mayer, Werner	Max-Planck-Institut für Biologie, Abteilung Immungenetik, Tübingen, Germany
McDevitt, Hugh O.	Department of Microbiology and Immunology, Stanford, CA, USA
Nei, Masatoshi	Department of Microbiology and Institute of Molecular Evolutionary Genetics, Mueller Laboratory, Pennsylvania State University, University Park, PA, USA
Neyt, Christine	Rockefeller University, New York, NY 10021, USA
O'hUigin, Colm	Max-Planck-Institut für Biologie, Abteilung Immungenetik, Tübingen, Germany
Ono, Hideki	Max-Planck-Institut für Biologie, Abteilung Immungenetik, Tübingen, Germany
Passmore, Howard C.	Rutgers State University of New Jersey, Department of Biological Sciences, Nelson Biological Laboratory, Piscataway, NJ, USA
Pease, Larry R.	Department of Immunology, Mayo Clinic, Rochester, MN, USA
Pontarotti, P.	Centre d'Immunologie de Marseille-Luminy, INSERM, Parc Scientifique de Luminy, Marseille, France
Potts, Wayne K.	Department of Pathology, University of Florida, Gainesville, FL, USA
Radman, Miro	Institut Jacques Monod, Centre National de la Recherche Scientifique, Université de Paris VII, Paris, France
Ripley, Lynn S.	Department of Microbiology and Molecular Genetics, New Jersey Medical School, University of Medicine and Dentistry of New Jersey, Newark, NJ, USA
Ritte, Uzi	The Hebrew University of Jerusalem, Department of Genetics, Jerusalem, Israel
Salomon, Hugh	University of California, Berkeley, CA, USA
Salomonsen, Jan	Basel Institute for Immunology, Basel, Switzerland
Sant'Angelo, Derek	State University of New Jersey, Nelson Biological Laboratories, Piscataway, NJ, USA
Satta, Yoko	National Institute of Genetics, Department of Population Genetics, Mishima, Shizuoka-ken, Japan
Schönbach, Christian	Max-Planck-Institut für Biologie, Abteilung Immungenetik, Tübingen, Germany
Schultz, Jane	University of Pittsburgh, Pittsburgh, PA, USA
She, Jin-Xiong	Department of Pathology, University of Florida, Gainesville, FL, USA
Shum, Benny	University of Miami, Department of Microbiology and Immunology, Miami, FL, USA
Snoek, M.	Division of Molecular Genetics, The Netherlands Cancer Institute, Amsterdam, The Netherlands
Sullivan, John S.	Blood Transfusion Service, Sydney, NSW, Australia
Takahata, Naoyuki	National Institute of Genetics, Institute of Population Genetics, Mishima, Japan

Tichy, Herbert	Max-Planck-Institut für Biologie, Abteilung Immungenetik, Tübingen, Germany
Vincek, Vladimir	Department of Microbiology & Immunology, University of Miami School of Medicine, Miami, FL, USA
Wakeland, Edward K.	Department of Pathology, University of Florida, Gainesville, FL, USA
Wang, C.-R.	Howard Hughes Medical Institute, Department of Microbiology and Biochemistry, University of Texas Southwestern Medical Center, Dallas, TX USA
Watkins, David I.	New England Regional Primate Research Center, Southborough, MA, USA
Wiley, Donald C.	Department of Biochemistry and Molecular Biology, Harvard University and Howard Hughes Medical Institute, Harvard University, Cambridge, Massachusetts, USA
Ye, Ying	Department of Pathology, University of Florida, Gainesville, FL, USA

Contents

Title, Authors **Page**

ORGANIZATION AND EVOLUTION OF THE MHC CHROMOSOMAL REGION: AN OVERVIEW

Corine Vernet, G. Chimini, J. Boretto, P. Le Bouteiller, and P. Pontarotti
Centre d'Immunologie de Marseille-Luminy, Marseille Cedex 9, France

Two years ago we wrote a review : Organization and evolution of the MHC region (Chimini and Pontarotti 1991). In this review, we presented the organization of this region in human, compared different genomic organizations in the human population, and ended up with the comparison of the man and the mouse organization. The main conclusion of this review was that class II and class III regions have kept their identities between these two mammalian orders and that was not the case for the class I region. Paradoxically (at least for us two years ago) The region around class I genes seems to be almost monomorphic, in the other end large scale polymorphisms were shown for the class II class III region. During these last two years progress was made in the MHC map mainly for the MHC class I region. Novel coding sequences were found in the MHC of both human and mice. In the light of these new results we will revise here some of our conclusions.

Organization of the MHC chomosomal region

Figure 1 shows a precise and complete map of the MHC region. For the class II and the class III regions several new coding sequences were evidenced these last years (Sargent et al. 1989; Spies et al. 1989; Kendall et al . 1990; Trowsdale et al.1990; Spies et al. 1990). We will not go into details here, but we would like to mention the presence of the peptide transporter called in the human case RING4 or PSF (Trowsdale et al.1990; Spies et al.1990; Spies and Demars 1991). This may argue for teleological clustering of genes in the genome.

The class I region is shown more precisely in Figure 2. *HLA-F*, one of the most telomeric class I genes (Koller et al. 1989) is located 250 kb from *HLA-A*. *HLA-H* and *HLA-G* are located between *HLA-A* and *HLA-F*, other class I pseudogenes are located within the map. It should be noted that all these genes are oriented the same way with the 5' end of the gene oriented to the telomere. *HLA-E* is located 550 kb centromeric to *HLA-A*, but in this case the 5' part of the HLA-E gene is located to the centromere thus in inverse orientation than the class I cluster around *HLA-A*. Still no linking is available between B, C and E , but the distance is at least 700 kb. Phenomenon other than inequal crossing over have to be considered to explain the organization of this region. Several coding sequences are evidenced in the region, also no idea of the function is available yet as they are conserved in the mouse they could be useful to understand the evolution of the region (unpublished data).

NATO ASI Series, Vol. H 59
Molecular Evolution of the
Major Histocompatibility Complex
Edited by J. Klein and D. Klein

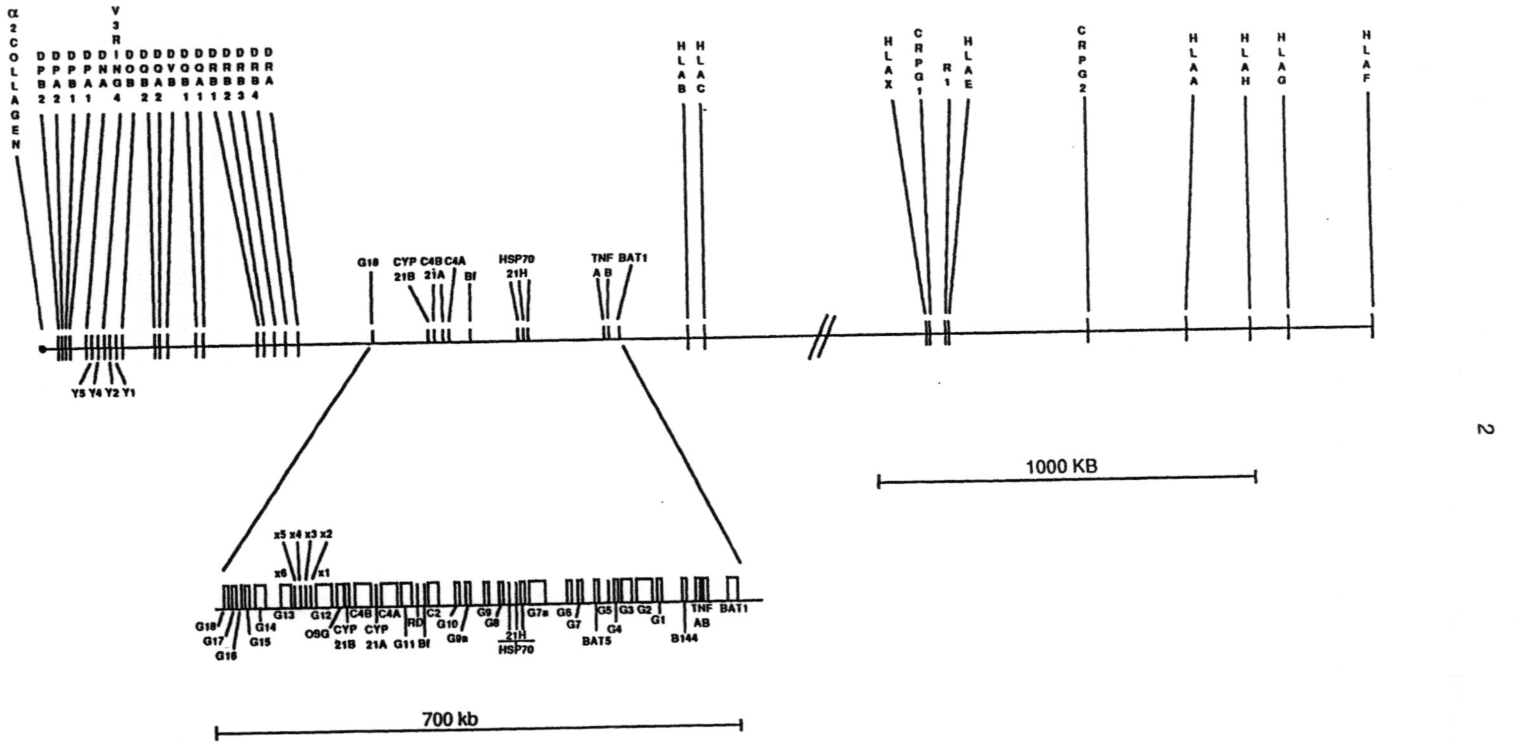

Fig. 1. Map of the *Mhc* region. The novel genes identified recently in the class II are from Sargent et al. 1989 and Kendall et al. 1990 (G series and B144); from Spies et al. 1989 and Levy-Strauss et al. 1988 (RD gene); from Sargent et al. 1989 (Hsp 70). For the class II region they are from Trowsdale et al. 1990 and Spies et al. 1990 (RING4 or PSF and Y1, Y2, Y4, and Y5 genes) and from Hanson 1989 (α2 collagenase). The map of the class I region is the result of unpublished observations (RI, CRPGI, and CRPGII).

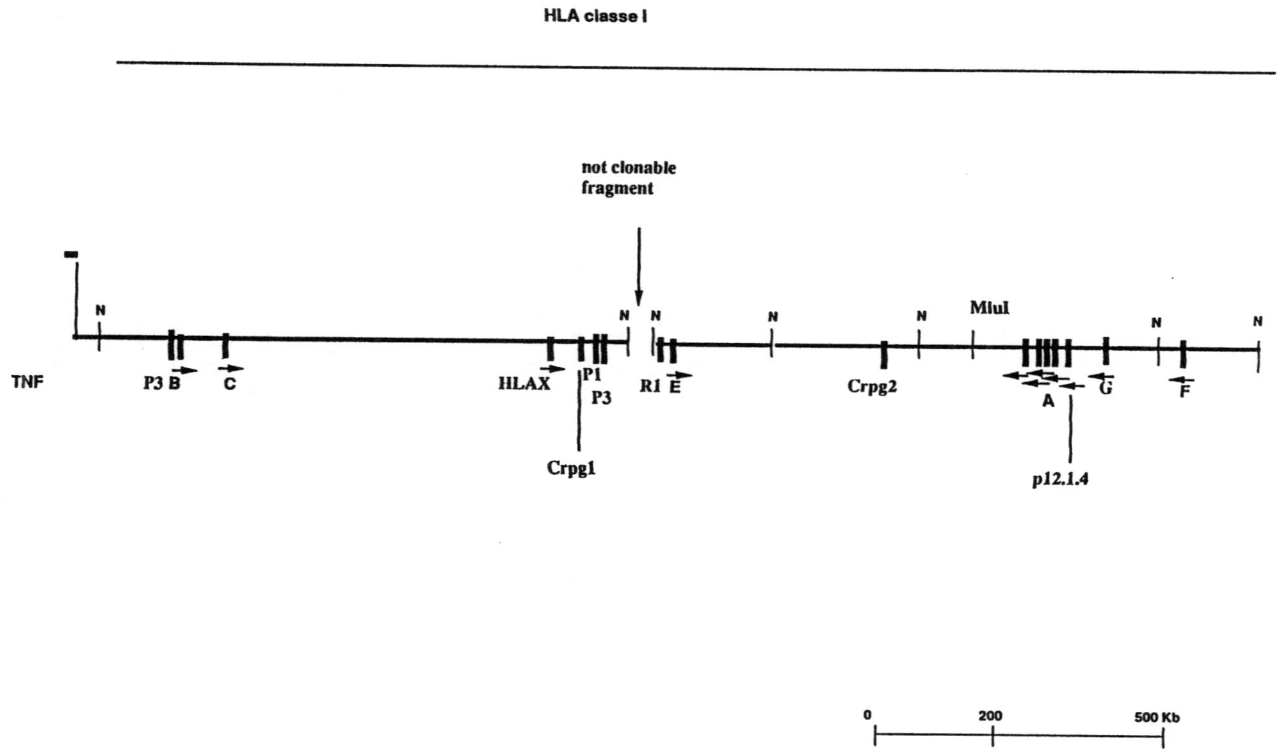

Fig. 2. Restriction map of human class I region relative to *Not* I (from Dunham et al. 1987 and unpublished observations).

Large scale genomic variation.

Intraspecies and interspecies comparison, at genomic level were mainly investigated by sequence comparison (kb level) or by chromosomal comparison (cytogenetic and synthenic group analysis about 20 000 kb range). Large scale cloning (with cosmid or yeast artificial chromosome vectors) and large scale genomic mapping (with pulsed field gel electrophoresis) provide the possibility of investigating variation at the megabase level. Because more and more probes are available in the MHC, this region can be used as paradigm for such a study; insertions, deletions, inversions, and translocations and other can be shown in this way. We will investigate first intraspecific large scale variation, and then interspecific variation.

Intraspecific variation.

We will review the MHC region starting with the class I region. This region contains several homologous genes and most of them are pseudogenes. Unequal crossing over events can explain the presence of so many homologous genes. If this phenomenon is very active a variation in the number of genes from one individual to another could be expected. PFG analysis of hybridization pattern variability in the human population provides a rapid way to check this hypothesis. The analysis of the macrorestriction fragment length polymorphism patterns from a panel of human homozygous typing cell lines showed little differences for fragments detected either by general class I probes or *A*, *B*, and *C* locus specific probes. (Chimini et al. 1988 and unpublished results). For instance, the *Mlu* I fragment containing the *HLA-A* gene, which is now known (see above) to contain 10 class I genes showed little polymorphisms. This implies that expansion contraction events due to unequal crossing over events in the HLA class I region are unfrequently found in the human population . It should be noted that this result has not as we said previously (Chimini and Pontarotti. 1991) to be contrast with what was found for the laboratory mice for two reasons: first, the laboratory mice are products of crosses of different mice subspecies and specifically from *Mus musculus* and *Mus domesticus*. Second,the cluster of class I gene which underwent the most expansion and contraction is the Tl cluster. This cluster, as we will see later, is absent in the human genome. However two sizes differences were detected. The *Mlu* I fragment detected by the *HLA-A* locus specific probe has a smaller size (Chimini et al. 1988; 40 kb less) in the case of the *A24* haplotype and in the case of *A31* haplotype (not published) a gain in size of 60kb was found. We were able to map, using probes isolated from the YAC clone and cosmid, this insertion and this deletion. The deletion maps centromeric to the *HLA-A* gene, the insertion maps telomeric to *HLA-A* gene. The specific organization of *A24* and *A31* chromosomal region should be helpful in understanding the phylogenetic tree of the *HLA-A* alleles.

Class II region

As for the class I region, the organization of the chromosomal fragment containing class II genes has been investigated by PFG patterns analysis in different haplotypes by Tokunaga and co-workers (1988). They reported variability of PFG profiles according to specific supratypes. (i.e. sets of MHC alleles preferentially associated-extended haplotypes), and provided evidence for supratypic specific insertion and deletion of DNA segments in the class II region. It should be noted, however, that most of the deletions they found are located distally to the *DRα* gene, and therefore more appropriately mapped to the class III region. Unfortunately they do not extend their study to the *DR4* haplotype, which contains a greater number of *DRβ* loci (Böhme et al. 1985; Bell et al. 1987). A comparison can however be made with the PFG map reported by Hardy et al. (1986) which is carried out on a *DR4* typed cell line. The two maps are consistent except for the finding of a 100 kb insertion between *DQ* and *DR* in the *DR4* haplotype. The presence of a fourth *DRβ* gene in *DR4* haplotype (Spies et al. 1985; Böhme et al 1985) is not sufficient to account for the 100 kb insertion. Thus other non-class II sequences are probably present in this inserted (or deleted) DNA fragment.

The overall chromosomal arrangement of the class II region in unrelated subjects is thus maintained. A few supratypic associated differences are reported, which could explain on a molecular basis the preferential association of alleles. The differences in molecular organization could in fact prevent recombination between different supratypes. Indeed Gregersen et al. (1988) observed the absence of a recombinant haplotype supporting recombination between the *DR4/DR7* and *DR1/DR2* supertypic groups. They also report that haplotypes from different supratypes tend to undergo recombination at different locations. These differences in site of recombination are probably a reflection of differences in the molecular organization of each supratype.

Class III region

Two genes encode *C4* and correspond to *C4A* and *C4B* loci. They are 22 kb in size and are nearly identical except for an extra 6-7 kb intron near the 5' end of the *C4A* gene and some alleles of *C4B* (Carroll et al. 1984). Flanking each *C4* gene the *21OHA* and *21OHB* loci are found. A single duplication event involving an ancestral DNA segment of 30 kb probably resulted in the duplication of *C4* and *21OH* genes. A further duplication of either *C4A* or *C4B* loci, resulting in three *C4* genes per chromosome (1.2% of frequency) was suggested based on results from genetic studies at the protein level (Raum et al. 1984; Uring-Lambert et al.1984). Furthermore genetic studies show that C4A or C4B protein is absent from the serum of some individuals and that the deficiency maps to the *C4* locus.(O'Neill et al 1982). Absence of *C4* is defined as the result of a null allele (*C4AQO,C4BQO*).

These results were confirmed at the DNA level by Carroll and co-workers (1984), who described at least in one haplotype three *C4* genes resulting from duplication of a 30 kb DNA fragment containing *C4* and *21OH* genes. They further showed by Southern analysis that haplotype *C4AQ0* may result from a deletion of a 30 kb fragment containing C4A and 21OHB (2 cases out of a sample of 18) and haplotype *C4BQ0* corresponds once again to a 30 kb deletion this time involving C4B and 21OH A (1 out of 6; Carroll et al. 1985). Unequal crossing over events could explain the ability of *C4* and *21OH* to undergo expansion and contraction. Besides this well documented contraction: expansion event within the class III region PFG analysis demonstrates other size variations in the region, i.e. deletions or insertions of DNA fragments of variable size just distally to the *DR*α gene (Togunaga et al. 1988).

Interspecific comparison

The conservation of a given sequence throughout evolution is a valuable hint to its physiological relevance since it identifies protein coding sequences and regulatory regions, which for their functional constraints undergo selective pressure, whereas noncoding and nonregulatory regions diverge more rapidly (Hayashida and Miyata 1983).

Furthermore, the interspecific conservation of syntenic linkage groups (genetic loci which are linked on the same region of a given chromosome) maintaining the same overall organization among the coding and regulatory sequences may reveal a functional constraint against spreading. Interactions among the contained genes have to be present so that maintenance of intact chromosomal segments is favoured by selection. Common regulatory mechanisms either *cis* and *trans* acting may control the expression of the whole cluster of genes. Control regions may directly mediate the binding of transacting factors to regulatory promoters or enhancers immediately surrounding the genes of interest. This type of position dependent control is reported for instance for the globin gene cluster where it determines stage specific expression LAR (**L**ocus **A**ctivation **R**egion; see for example Forrester et al. 1990). Coevolution and maintenance of linkage group may also reflect constraint to proximity for correct coordinate expression of paired proteins (as an example IA α and β chains in the mouse class II system (Braunstein and Germain 1987). We will compare here MHC organization in the two best known systems: mouse and man. It should be noted here that rodents belong to an outgroup compared to the one including primates carnivores lagomorphs and artiodactylas, and that more similarities would be revealed if we compared for example primates and artiodactylas.

Class I

In man, *HLA-A, -B*, and probably *-C* loci code for classical transplantation antigens as in mouse *H-2KD* and *-L* loci. Their immunological function is well documented in both systems. No direct functional relationship between the two species is found for the other class I related sequences. A structural comparison is also difficult as human Class I sequences show the same degree of homology either to *K, D*, or *L* or to any member of Qa family of genes. This finding can be explained, first by the occurrence of duplication events from a common ancestor after splitting of specific ancestors. Second, if duplication occurred before splitting, gene conversion events could lead to the interspecific homogenization of class I sequences and therefore to the actual loss of specific identity. Alternatively all these genes were present in the common ancestor but deletions occurred in the different lineages. If we compare the maps of the two species (Fig. 3), several major differences are evident.

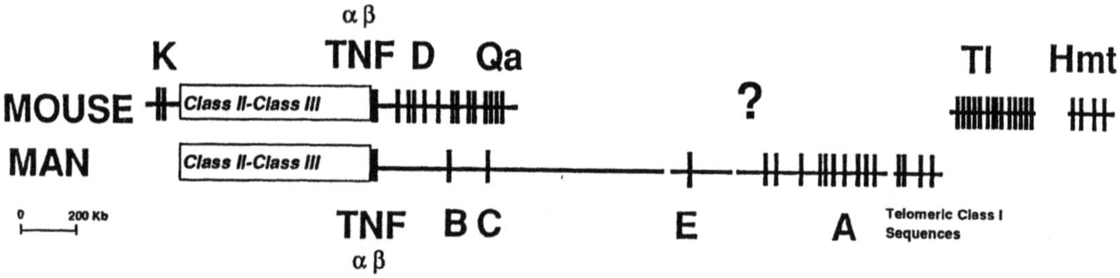

Fig. 3. Molecular organization of the class I region: mouse/man comparative map. Maps are arbitrarily aligned on *TNF* genes. Sizes of class II and class III region are not drawn to scale. The distance from the *Qa* to *Tl* cluster and *Hmt* is unknown.

In the mouse, two different groups of genes are described. Group I corresponds to sequences displaying an high level of similarity (*K, D, L*, and *Qa* cluster of genes), while Group II is composed of a variable number of members, highly dissimilar among them and with respect to Group I genes (Rogers 1985; Brorson et al. 1989). Genes belonging to Group II are located in the *Tl* and the *Hmt* region (Rogers 1985; Brorson et al. 1989) telomeric to the *Qa* region, *Tl* and *Hmt* gene clusters are absent from the MHC region in man.

In man all class I genes display a high degree of homology; no divergent genes, that could be classified as belonging to group II are evidenced. Only group I genes are present. In the mouse group I genes are splitted in two regions, several hundreds of kb apart, Figure 3 shows that the overall organization of the mouse MHC is, from the centromere to the telomere: class I *H-2K* (*H-2K2* and *H-2K*), class II, class III, class I *H-2D* and *Qa* genes, whereas for man it is: class II, class III, class I. No evidence for an equivalent split of class I genes is found in man. An explanation for this discrepancy is given by Bodmer (1981) who suggests that the *H-2K* locus arose by translocation. Abe and co-workers (1988) detected five new genes (*KE* genes) in the *H-2K* region. These genes are present in the human genome and are found centromerically to *HLA-DP* genes (J. Trowsdale personal communication) at the same position where they are found in the mouse. Thus translocation of *H-2K2* and *-K* into the preexisting *KE* region can be postulated.

Most human class I genes are clustered around *HLA-A* in about a 300 kb fragment . Howewer, *HLA-E* is found at 550 kb centromeric to *HLA-A* in an inversed orientation and *HLA-B* and *HLA-C* genes are at least 700 kb centromeric to the *HLA-E* gene.On the whole, however, the human class I region seems larger than the murine one, where except for *H-2K* and *H-2K*, *Tl*, and *Hmt* regions all class I genes are clustered from the *H-2D* locus within 400 kb. In conclusion we have not found orthologies between man and mouse class I genes and the organization of the corresponding region between the two species is different. Hughes has argued (Hughes 1991) that the non-orthology between class I genes can be explained by independent gene duplication and not by concerted evolution. It is obvious that if repeated duplication and deletion occur, the general organization between two independent species would be different. Thus the different class I region organization between man and the mouse argues too that the lack of orthology is due to independent gene duplication.

Class II

Figure 4 shows a comparative map of class II organization in mouse and man (Steinmetz et al .1986; Monaco et al. 1990; Spies et al 1990; Trowsdale et al.1990). More class II genes, generated by duplication (in the human lineage or a deletion in the mouse lineage) are present in the human than in mouse, and the distances between the sequences are larger in human. Conservation of the region is however striking: most murine genes have a human counterpart with a high degree of interspecific sequence homology. Their transcriptional orientation is maintained invariant in the two species as well as the order of homologous genes on the chromosome. Thus in the case of the class II region we found orthologies between the genes and about the same chromosomal organization; this is an inverse situation of that in the class I region. It should be noted, too, that the non-class I genes are found in the same location in the two species.

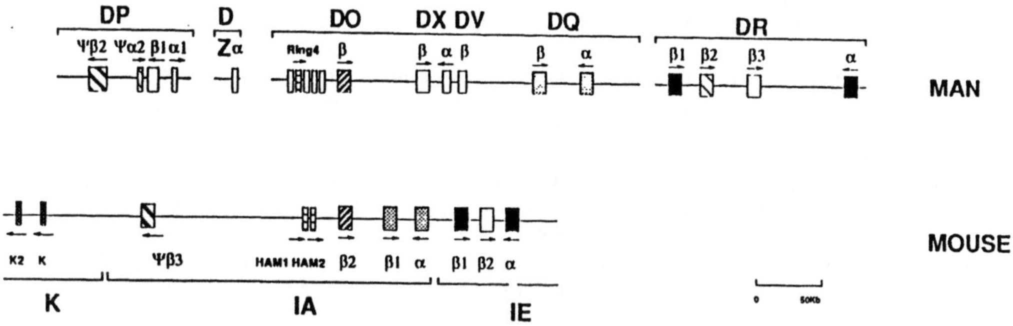

Fig. 4. Molecular organization of class II region; mouse/man comparative map (see Steinmetz et al. 1986; Monaco et al. 1990; Deverson et al. 1990; Trowsdale et al. 1991). Maps are arbitrarily aligned on the homologous pair DO^b-IAb2. Interspecific gene homologies are shown by the same background pattern. Empty boxes represent genes which do not have a counterpart in the other species. Lightly hatched boxes indicate human pseudogenes without murine counterpart. Distances separating the human subregions are not drawn to scale. Arrows indicate transcriptional orientations. HAM I is orthologous to RING4 (Monaco et al.1990; Deverson et al. 1990). HAM II is paralogous to RING 4, according to a recent publication (Spies and DeMars 1991), the human ortholog of HAMII should be found in the same region.

Class III

We will not develop this paragraph because Dr. M. Snoek will develop this point. We would just like to underline that a nearly perfect orthology has been found.

In conclusion, class I regions and genes have lost their identities when comparisons are made between man and mouse. Howewer the organization of this region is relatively stable within the human population. Class II and class III regions and genes have kept their identities between man and mouse, but more intraspecifics large scale variations than for the class I region are found. This apparent paradox can be explained by the fact that large scale polymorphisms for the class II and class III regions might be transpecifics, thus these organizations are stable. In the case of the class I region the organization is maybe too recent. Large scale mapping in the case of the apes of the class I and the class II class III regions will prove or rule out this hypothesis.

References

Abe, K., Wei, J.F., Wei, F.S., Hsu, Y.C., Uehara, H., Artzt, K., and Bennett, D.: Searching for coding sequences in the mammalian genome: the H-2K region of the mouse MHC is replete with genes expressed in embryos. *EMBO 7*:3441-3449, 1988

Bell, J.I., Denney, D.Jr., MacMurray, A., Foster, L., Watling, D., and McDevitt, H.O.: Molecular mapping of class II polymorphisms in the human major histocompatibility complex. I. DR beta. *J Immunol 139*: 562-573, 1987

Braustein, N.S. and Germain, R.N.: Allele-specific control of Ia molecule surface expression and conformation:Implications for a general model of Ia structure-function relationships. *Proc Natl Acad Sci USA 84*:2921-2925, 1987

Brorson, KA., Richards, S., Hunt III, S,W., Cheroutre, H., Fisher Lindahl, K., and Hood, L.: Analysis of a new class I gene mapping to the *Hmt* region of the mouse. *Immunogenetics 30*:273-283, 1989

Böhme, J., Andersson, M., Anderson, G., Möller, E., Peterson, P.A., and Rask, L.: HLA-DR beta genes vary in number between different DR specificities, whereas the number of DQ beta genes is constant. *J Immunol 135*:2149-2155, 1985

Carroll, M.C., Belt, T., Palsdottir, A., and Porter, R.R.: Structure and organization of the C4 genes. *Philos. Trans R Soc Lond (Biol). 306*:379-388, 1984

Chimini, G. and Pontarotti, P.: Organization and evolution of the chromosomal region containing the human major histocompatibility complex. *Immunogenetics of the Major Histocompatibility Complex I*:65-99, 1991

Chimini, G., Pontarotti, P., NGuyen, C., Toubert, A., Boretto, J., and Jordan, B.R.: The chromosome region containing the higly polymorphic HLA class I genes displays limited large scale variability in the human population. *EMBO J 7*: 396-400, 1991

Deverson, E.V., Gow, I.R., Coadwell, J.V., Monaco, J.J., Butcher, G.W., and Howard, J.C.: MHC class II region encoding proteins related to the multidrug resistance family of transmembranes transporters. *Nature 348*: 738-741, 1990

Dunham, J., Sargent, C.A., Trowsdale, J., and Campbell, R.D.: Molecular map of the human major histocompatibility complex by pulsed field gel electrophoresis. *Proc Natl Acad Sci USA 84*:7237-7241, 1987

Forrester, W.C., Epner, E., Driscoll, M.C., Enver, T., Brice, M., Papayannopoulou, T., and Groudine, M.: A deletion of the human ß-globin locus activation region causes a major alteration in chromatine structure and replication across the entire ß-globin locus. *Genes Develop 4*:1637-1649, 1991

Gregersen, P.K., Kao, H., Nunez-Roldan, A., Hurley, C.K., Karr, R.W., and Silver, J.: Recombination sites in the HLA class II region are haplotypes dependent. *J Immunol 141*:1365-1368, 1991

Hanson, I., Gorman, P., Lui, V.C.H., Cheah, K.S., Salomon, E., and Trowsdale, J.: The human α2 (XI) collagen gene (COL11A2) maps to the centromeric border of the major histocompatibility complex on chromosome 6. *Genomics 5*:925-931, 1989

Hardy, D.A., Bell, J.I., Long, E.O., Lindesten, T., and McDevitt, H.O.: Mapping of the class II region of the human major histocompatibility complex by pulsed-field gel electrophoresis. *Nature 323*:453-455, 1986

Kendall, E., Sargent, C.A., and Campbell, R.D.: Human major histocompatibility complex contains a new cluster of genes between the HLA-D and the complement C4 loci. *Nucleic Acids Res 18*: 7251-7257, 1991

Koller, B,H., Geraghty, D,E., DeMars, R., Duvick, L., Rich, S.S., and Orr, H.T.: Chromosomal organization of the human major histocompatibility complex class I gene family. *J Exp Med 169*: 469-480, 1989

Lévi-Strauss, M., Carroll, M.C., Steinmetz, M., and Meo, T.: A previousely undetected MHC gene with an unusual periodic structure. *Science 240*: 201-204, 1988

Monaco, J.J., Cho, S., and Attaya, M.: Transport protein genes in the MHC: possible implications for antigen processing. *Science 250*:1723-1726, 1990

Raum, D., Surgenor, T., Awdeh, Z., Marcus, D., Blumenthal, M., Yunis, E.J., and Alper, C.A.: An unusual "morphologic" variant of BF S. *Am J Hum Genet 36*:346-351, 1984

Rogers, J.H.: Mouse histocompatibility-related genes are not conserved in other mammals. *EMBO J 4*: 749-753, 1985

Sargent, C.A., Dunham, I., and Campbell, D.C.: Identification of multiple HTF-island associated genes in the human major histocompatibility complex class III region. *EMBO J 8*: 2305-2312, 1989a

Sargent, C.A., Dunham, I., Trowsdale, J., and Campbell, R.D.: Human major histocompatibility complex contains genes for the major heat shock protein HSP70. *Proc Natl Acad Sci USA 86*:1968-1972, 1989b

Spies, T, Bresnahan, M., Bahram, S., Arnold, D., Blanck, G., Mellins, E., Pious, D., and DeMars, R.: A gene in the human histocompatibility complex class II region controlling the class I antigen presentation pathway. *Nature 348*:744-747, 1990

Spies, T. and DeMars, R.: Restored expression of major histocompatibility class I molecules by gene transfert of a putative peptide transporter. *Nature 351*, 323-324, 1991

Spies, T., Sorrentino, R., Boss, J.M., Okada, K., and Strominger, J.L.: Structural organization of the DR subregion of the major histocompatibility complex. *Proc Natl Acad Sci USA 82*: 5165-5169, 1985

Steinmetz, M., Stephan, D., and Fisher Lindahl, K.: Gene organization and recombinational hotspots in the murine major histocompatibility complex. *Cell 44*:895-904, 1986

Togunaga, K., Saueracker, G., Kay, P.H, Anand, R., and Dawkins, R.L.: Extensive deletions and insertions in different MHC supratypes detected by pulsed field gel electrophoresis. *J Exp Med 168*: 933-940, 1988

Trowsdale, J., Hanson, I., Mockridge, I., Beck, S., Townsend, A., and Kelly, A.: Sequences encoded in the class II region of the MHC related to the 'ABC' superfamily of transporters. *Nature 348*:742-744, 1991

RECONSTRUCTION OF PHYLOGENETIC TREES AND EVOLUTION OF MAJOR HISTOCOMPATIBILITY COMPLEX GENES

Masatoshi Nei and Andrey Rzhetsky

Department of Biology and Institute of Molecular Evolutionary Genetics, The Pennsylvania State University, 328 Mueller Laboratory, University Park, PA 16802, USA

There are many different methods of phylogenetic reconstruction for DNA sequence data, and their advantages and disadvantages are discussed by considering various factors such as the constancy of evolutionary rate, extent of sequence divergence, variation in evolutionary rate over the sequence, number of nucleotides examined, number of sequences used, etc. It is shown that the bootstrap method of testing the stability of the branching pattern of a tree is not really accurate in evaluating the probability level of the clustering of a group of sequences and that it sometimes leads to an erroneous conclusion particularly when it is applied to the maximum parsimony method. Nevertheless, it is a useful method for obtaining a rough idea of the stability of the branching pattern of a tree. Application of the neighbor-joining and maximum parsimony bootstrap methods to Gyllensten et al.'s sequence data for the DQB locus alleles from primates confirmed the trans-species polymorphism between humans and gorillas, but the trans-species polymorphism between humans and chimpanzees was ambiguous. However, sequence data for the DQA locus alleles confirmed the trans-species polymorphism among all of the humans, chimpanzees and gorillas.

In the study of the evolution of the major histocompatibility complex (MHC) loci phylogenetic analysis of DNA sequences has become an important tool. Only by using this method can one study the evolutionary relationships of alleles from different loci of the complex as well as those of alleles from different species. Therefore, it is now routine for MHC researchers to reconstruct a phylogenetic tree. It should be noted, however, that the reconstruction of phylogenetic trees is not a simple problem and that all statistical methods of phylogenetic reconstruction depend on various assumptions that do not necessarily hold with real data. It is therefore important to use a method that is appropriate for the data set used and to be cautious about the interpretation of the results obtained.

NATO ASI Series, Vol. H 59
Molecular Evolution of the
Major Histocompatibility Complex
Edited by J. Klein and D. Klein

In this paper we would like first to review the advantages and disadvantages of several tree-making methods that are currently used for molecular data and then to reanalyze some of the phylogenetic trees for MHC genes in the literature. It is not our intention to explain the procedure of each tree-making method in detail. The reader who is interested in this subject may consult Nei (1987; 1991) and Felsenstein (1988).

Tree-making methods that are often used for molecular data

1. Average distance method (UPGMA)

The original idea of this method was presented by Sokal and Michener (1958), but their procedure is different from the one currently in use. A formal presentation of this method is given in Sneath and Sokal (1973) and Nei (1975). The principle of this method is to cluster DNA sequences according to the degrees of sequence similarities. This method requires the assumption of constant rate of evolution for all evolutionary lineages. Computer simulations have shown that this method is quite efficient in obtaining the correct tree when the rate of evolution is constant and all the DNA sequences are distantly related to each other (Tateno et al. 1982; Sourdis and Krimbas 1987). It is also useful for obtaining a rough idea of the similarities of different groups of sequences. In general, however, this method does not produce reliable phylogenetic trees for DNA sequence data.

2. Fitch and Margoliash's (1967) method

This is one of the first tree-making methods that were used for molecular data. It is known that this method reconstructs the correct tree topology (branching pattern) when there is no backward and no parallel substitutions (condition of an additive tree). For these reasons, this method is still widely used among molecular biologists. However, computer simulation studies (e.g., Tateno et al. 1982; Saitou and Imanishi 1989) have shown that when the underlying assumption of this method is violated it is not a very efficient method.

3. Neighbor-joining (NJ) method

The principle of this method (Saitou and Nei 1987) is to obtain a tree by minimizing the sum of all branch lengths. Unlike the minimum evolution method by Cavalli-Sforza and Edwards (1967), this method does not examine all possible topologies. Rather, comparison of different topologies is built in the algorithm, and the final tree with both topology and branch lengths is automatically produced. This method is also known to produce the correct tree when the condition of an additive tree is satisfied. Computer simulations (Saitou and Nei 1987; Sourdis and Nei 1988; Saitou and Imanishi 1989) have shown that this method is more efficient than most distance matrix methods and is often better than the maximum parsimony method.

4. Maximum parsimony (MP) method

In this method, the DNA sequences of ancestral species are inferred from those of extant species by considering a particular tree topology, and the minimum number of evolutionary changes that are required to explain all the observed nucleotide differences among the sequences is computed. This number is obtained for all possible or all likely topologies, and the topology that shows the smallest number of evolutionary changes is chosen as the final tree. This method is mainly for finding the topology of a tree, and branch lengths can be estimated only under certain assumptions (Fitch 1971). When the number of substitutions per site between different sequences is small and the number of nucleotides examined is large (>1000), this method is very powerful in obtaining the correct tree (Sourdis and Nei 1988). However, when the rate of evolution varies extensively with evolutionary lineage, this method often produces an erroneous tree and is generally less efficient than the NJ and maximum likelihood methods. Furthermore, when a large number of sequences are used and the number of nucleotides examined is small, as is often the case with data obtained by the polymerase chain reaction method, the MP method produces a large number of equally parsimonious trees. Therefore, it is difficult to choose the best possible tree.

5. Maximum likelihood (ML) method

This is the most elaborate method so far invented (Felsenstein 1981). If the pattern of nucleotide substitution is simple and all nucleotide sites follow the same substitution pattern, this is a very efficient method (Saitou and Imanishi 1989). However, the actual pattern of nucleotide substitution is quite complicated and usually varies from nucleotide site to nucleotide site. In this case it is not clear how efficient this method is. Furthermore, this method requires an enormously long computer time compared with other methods unless the number of sequences used is very small.

Accuracy of the tree obtained

When a tree is obtained by a tree-making method, one is naturally interested in the accuracy of the tree, i.e., how good the tree is compared with other alternative trees. There are two types of test in this case. One is to test the topological difference between two trees, whereas the other is to examine the accuracy of the branching pattern of a tree obtained. The former test is a very difficult problem, and most statistical methods so far proposed depend on assumptions which do not necessarily hold in reality (see Nei 1991). Therefore, one has to be cautious in using these methods.

The test of the accuracy of the branching pattern of a tree obtained is done by examining whether a particular cluster of sequences in the tree is reliable or not. There are two ways of testing this reliability. One is to compute the standard error of the length of the branch that unites all the sequences in the cluster (e.g., branch a in Figure 1), and the other is to use the bootstrap method (Efron 1982). For a UPGMA tree, the standard error of a branch length can be computed by Nei et al.'s (1985) method, and the computation of the standard error is relatively simple. However, the standard error for a tree obtained by other tree-making methods is not easy to compute, unless the number of sequences used is four or five (Li 1989).

The bootstrap method was first used by Felsenstein (1985) to test the reliabilities of sequence clusters in a maximum parsimony (MP) tree. The principle of this method is very simple. If there are n phylogenetically informative sites, one can resample n sites with replacement and reconstruct a MP tree using this new set of data. If one repeats this process many times (say 500 times), he or she can examine how often a particular cluster (say, sequences HB1.1, HB1.2, and HB1.3 in Figure 2) appears in the replicate trees. If the frequency of occurrence of this cluster is 95 percent or higher, one may conclude that the cluster of the sequences is statistically significant. This test, of course, can be used for any other tree-making method. Indeed, it is much simpler to use this method for the NJ method than for the MP method. Dr. Thomas Whittam at The Pennsylvania State University has already developed a computer program for the NJ bootstrap method.

This test is intuitively appealing, but the theoretical basis of the test remains unclear. M. Nei and S. Kumar (unpublished) investigated this problem for the NJ and MP methods by using computer simulation. In this investigation we set up a model tree for six species that is

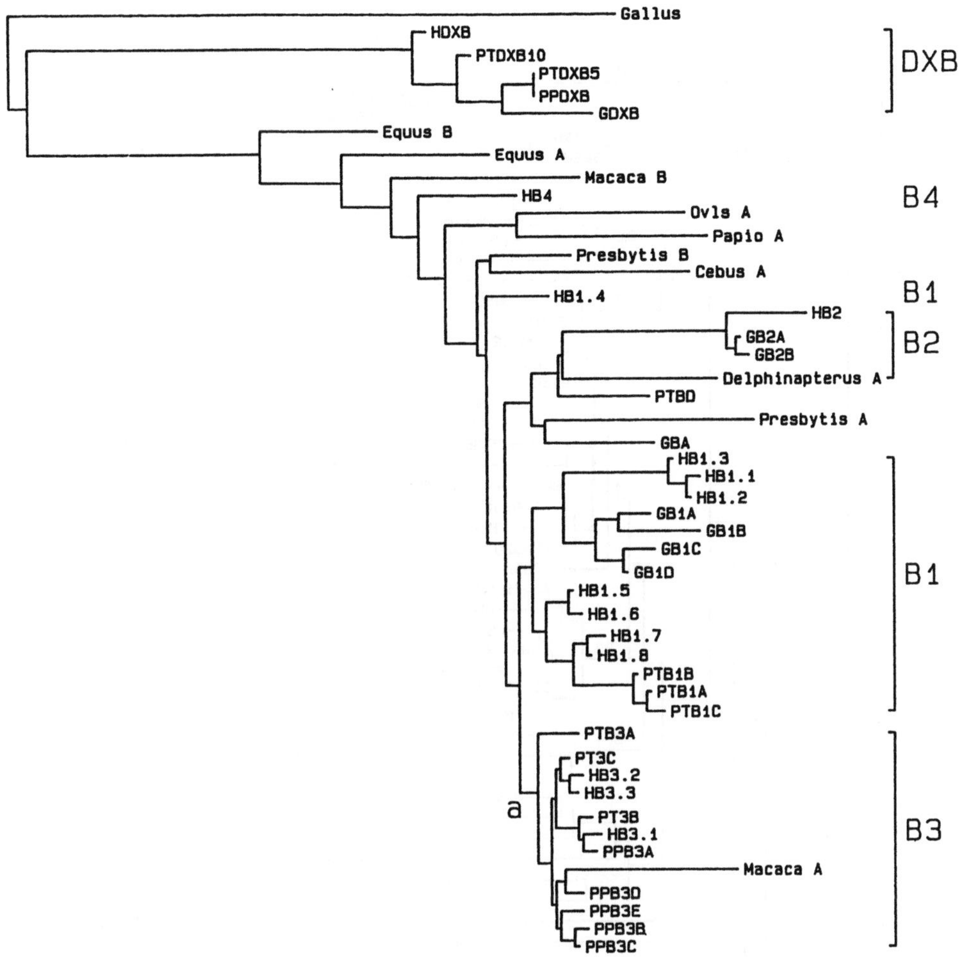

Fig. 1. Phylogenetic tree for alleles at the class II *DQB* locus from primate and several non-primate mammalian species. This is a reproduction (with permission) of Gyllensten et al.'s (1990) tree constructed by the NJ method for amino acid sequence data. 57 amino acid residues in the polymorphic second exon were used. H: Human. PT: Common chimpanzee (*Pan troglodytes*). PP: Pygmy chimpanzees (*Pan paniscus*). G: Gorillas. Papio: Baboon (*Papio leucophaeus*). Macaca: Rhesus monkey (*Macaca mulatta*). Presbytis: Langurs (*Presbytis entellus*). Cebus: Capuchin monkey (*Cebus capuchinus*). Equus: Horse. Ovis: Sheep. Delphinapterus: Beluga. Gallus: Chicken. B1, B2, B3, and B4 represent different allelic groups in humans and apes. There is one B1 group allele in man (HB1.4), which did not cluster with other B1 alleles.

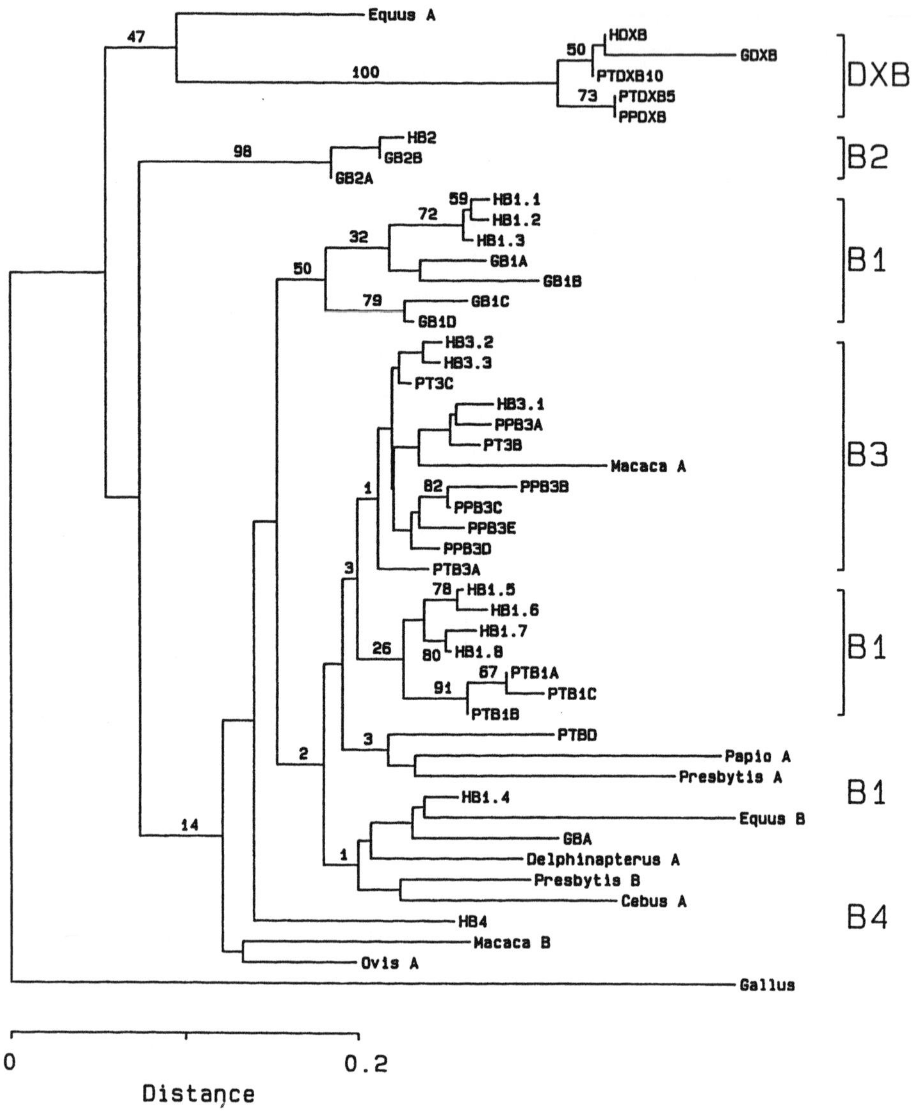

Fig. 2. Neighbor-joining tree constructed by our computer program for the same set of amino acid sequence data as those used by Gyllensten et al. (1990). B1 group alleles are now split into three subgroups. However, any of the subgroup is not statistically confirmed according to the bootstrap test. [This test was conducted by using Tom Whittam's (1991) NJBOOT program.] The number for an internal branch represents the bootstrap confidence limit. It is given only for important branches. DXB alleles do not belong to the DQB locus.

identical with model tree C of Saitou and Imanishi (1989) and generated simulated DNA sequences following the model tree (a=0.05 and 300 or 600 nucleotides used). We then reconstructed a tree for the sequence data using the NJ and MP methods. When we repeated this process many times (100 replications), the NJ method produced the correct tree (topology) in most replications, but the MP method produced many incorrect trees. This result is in agreement with that obtained by Saitou and Imanishi (1989). We then conducted the bootstrap test for each of the correct and incorrect trees. Our results indicated that although most of the NJ trees obtained were correct most sequence clusters in the trees showed a 95 percent or lower bootstrap confidence limit (BCL) when 300 nucleotides were used. This suggests that even if the BCL is lower than 95 percent the branching pattern of a NJ tree is generally correct.

When the MP method was used, however, contrary results were occasionally obtained. That is, when an incorrect cluster of sequences was obtained, the cluster appeared in more than 95 percent of the bootstrap replications. This indicates that the bootstrap test applied to MP trees can establish an incorrect tree as though it were statistically confirmed.

Nevertheless, our investigation has shown that when the rate of evolution is more or less constant the bootstrap test is a useful method to obtain a rough idea of the accuracy of the tree obtained, though the BCL does not really prove or disprove the accuracy of a sequence cluster involved. The BCL for a cluster is usually higher when the topology is correct than when it is incorrect.

Phylogenetic trees for alleles at the DQB locus in primates

Although immunologists are now routinely constructing phylogenetic trees, they tend to use a computer program available uncritically and not to try to spend much time in the interpretation of the results obtained. Furthermore, they do not always describe detailed aspects of the computation, so it is often unclear which option of a computer program was used. We therefore examined several phylogenetic trees that were recently published. Here we report the results of one of our such investigations.

Gyllensten et al. (1990) constructed a phylogenetic tree for 48 alleles at the class II DQB locus from several primate species and a few non-primate mammals. The neighbor-joining tree reconstructed is reproduced in Figure 1. Their parsimony tree was somewhat different from the NJ tree, but each of the allelic groups B1, B2, and B4 formed one cluster except that HB1.4 was separate from other B1 alleles. However, since the number of amino acids used in this study was quite small (57) and some of the sequences were closely related, we reanalyzed the same set of data to examine the reliability of the tree with the bootstrap method.

The authors stated that the NJ tree is based on the genetic distance between the amino acid sequences. We therefore used the standard distance for amino acid sequences, i.e., $d = -\log_e(1-p)$, where p is the proportion of different amino acids between two sequences. To our surprise, however, the tree obtained was quite different from that of Gyllensten et al. (Fig. 2). The most conspicuous difference is the splitting of B1 alleles into three subgroups. Of course, the first B1 group (HB1.1, HB1.2, HB1.3, GB1A, GB1B, GB1C, and GB1D) has a BCL of only 50 percent. This suggests that the division into the three subgroups is not statistically significant.

To examine whether our use of distance d generated a tree which is different from that of Gyllensten et al.'s, we tried the p distance as well as Fitch and Margoliash's (1967) amino acid distance. [Ulf Gyllensten (pers. comm.) informed me that they used the p distance.] Both of these distances, however, produced essentially the same results as those in Figure 2. We also constructed the maximum parsimony trees for the same set of data using David Swofford's PAUP program (version 3.0). This method generated numerous MP trees, so that we printed only the first five trees. (This method requires an enormous computer time when the number of sequences is large.) In the first five MP trees, B1 alleles always produced three clusters similar to our NJ tree. At present, we are not sure what caused the difference between our tree and Gyllensten et al.'s.

One problem with the amino acid sequence data is that the number of informative sites is small. We therefore constructed a NJ tree from nucleotide sequence data. This tree shows a splitting pattern somewhat different from that of amino acid sequence data (Fig. 3). B1 alleles now form four different clusters rather than three, and allele B1.4 is clustered with Cebus A and Presb B instead of Equus B and GBA. However, most of the clusters are not statistically significant, so that the difference between the topologies of the trees in Figure 2 and 3 could be due to sampling errors. This suggests that the number of nucleotides examined (171) in this study is too small to determine the evolutionary history of alleles examined.

Swofford's PAUP computer package includes a program for the bootstrap test for a MP

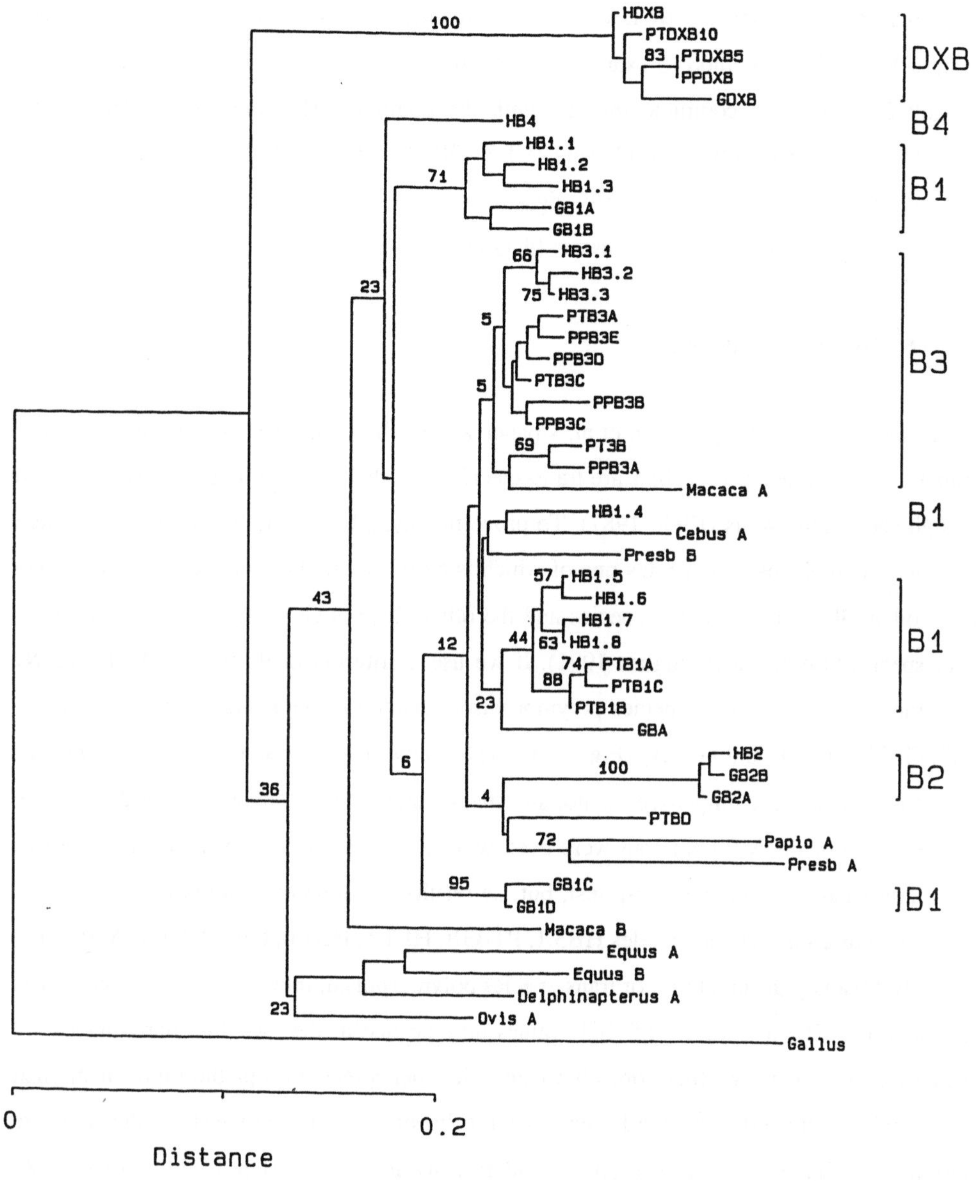

Fig. 3. Neighbor-joining tree constructed by our computer program for DNA sequence data. Notations are the same as those in Figure 2. Some genus names are abreviated.

tree. This program gives the BCL value for each internal branch that leads to a sequence cluster. We used this method to examine the reliability of the clusters in the parsimony tree. (It took 55.5 hours to complete the test with the computer Macintosh IIci!) The results obtained showed that only three clusters (HB2, GB2A, GB2B; HDX1, PTDX10, PTDX5, GDX; GB1C, GB1D) were statistically significant at the 95 percent level, and no statistical resolution was obtained for other clusters (data not shown).

Trans-species polymorphism

One of the purposes of Gyllensten et al.'s paper was to show that the polymorphic alleles in humans, chimpanzees, and gorillas are trans-specific, i.e., the same allelic pairs have persisted through speciation events (Klein 1987). To prove this hypothesis, one has to show that there are at least two alleles in a species one of which is closer to an allele in another species than to the other allele in the sample species and the other allele is closer to another allele in the other species (see Nei and Hughes 1991). If we use a criterion of 95 percent BCL, the NJ bootstrap confirms the trans-species polymorphism between humans and gorillas (e.g., HB2, GB2, GB2B, HB1.1, and GB1A) (Fig. 2). However, any of the trans-species polymorphisms between humans and chimpanzees is not statistically significant in either Figure 2 or Figure 3. The results of the MP bootstrap were even worse, as mentioned above. This may be due to the crude nature of the bootstrap test, but the results are quite discouraging.

Figure 2 suggests that alleles HB3.1, PTB3B, HB1.5, HB1.6, HB1.8, PTB1A, PTB1B, and PTB1C satisfy the condition of trans-species polymorphism, though the NJ bootstrap does not support it. The low values of BCL values may be due to the fact that many alleles are included in this tree. We therefore eliminated all other alleles except the eight mentioned above and constructed a new NJ tree. The results of the bootstrap test of this tree are presented in Figure 4(A). Alleles HB3.1 and PTB3B are now clustered with a BCL of 88 percent. However, alleles HB1.5 and HB1.6 made a separate cluster, and the cluster of HB1.8, PTB1A, PTB1B, and PTB1C showed a BCL of only 47 percent. Therefore, these results do not really prove trans-species polymorphism at the statistical level. However, when we analyzed DNA sequence data for the same set of alleles, trans-species polymorphism was established at the 98 percent probability level [Fig. 4(B)]. Since DNA sequences are more informative than amino acid sequences in this case, this will probably support the idea of

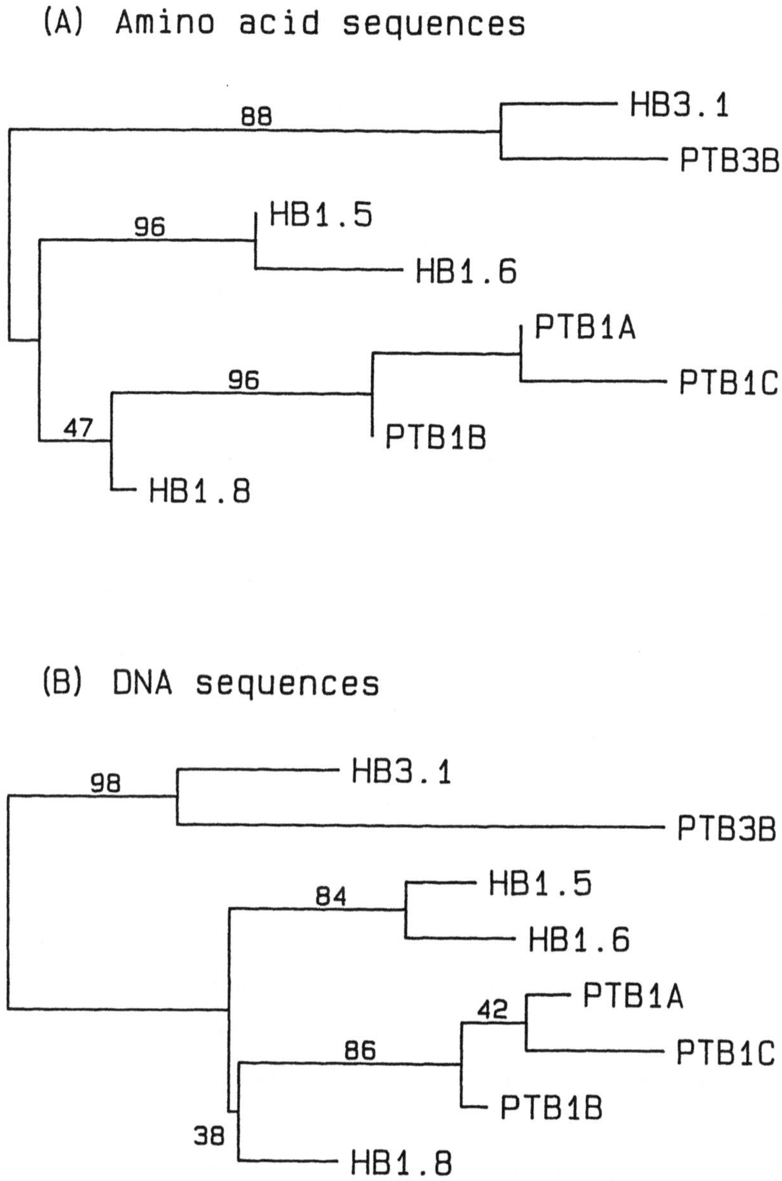

Fig. 4. Bootstrap tests for the eight alleles which show transspecific polymorphism in Figure 2. (A) Amino acid sequence data. (B) DNA sequence data. The number for each internal branch represents the bootstrap confidence limit (BCL).

trans-species polymorphism. Nevertheless, we should be cautious about these results because they are not consistent with those of Figures 2 and 3.

Gyllensten and Erlich (1989) also studied the trans-species polymorphism at the DQA locus among humans and apes. In this analysis they used only the MP method. Since they presented amino acid sequence data, we constructed a NJ tree from them and conducted the bootstrap test. The results are presented in Figure 5. The topology of the tree in this figure is similar to that of Gyllensten and Erlich (1989), though the splitting pattern of the major groups of clusters is different. The clusters of A1, A2, A3, and A4 alleles are the same as those of Gyllensten and Erlich. Furthermore, the A1, A3, and A4 clusters are statistically significant. Therefore, these results provide strong evidence of trans-species polymorphism.

Discussion

Trans-species polymorphism is one of the unique features in the evolution of MHC loci. If we note that different alleles from a species are often highly diverged at the nucleotide level (Klein and Figueroa 1986) and the same pairs of allelic lineages are often shared by closely related species (e.g. McConnell et al. 1988), there seems to be no question about the validity of this idea. However, it is not very clear how long such a long-term polymorphism has persisted in the population.

Theoretically, it is a simple matter to study trans-species polymorphism. If one can show that the same pair of allelic lineages are shared by two species, it suggests trans-species polymorphism. In practice, however, the phylogenetic tree reconstructed may not represent the correct tree. This happens quite often when the number of nucleotides examined is small or when the lengths of internal branches of a tree are short. Since the DNA sequences studied by the PCR technique are often short, one should be cautious about the trees obtained by this method. As mentioned earlier, the theoretical basis of the bootstrap test is not well established. However, when it is applied to the NJ method, the test seems to be conservative. It is therefore encouraged that this test be used whenever the number of nucleotides examined is small or when the lengths of internal branches are short.

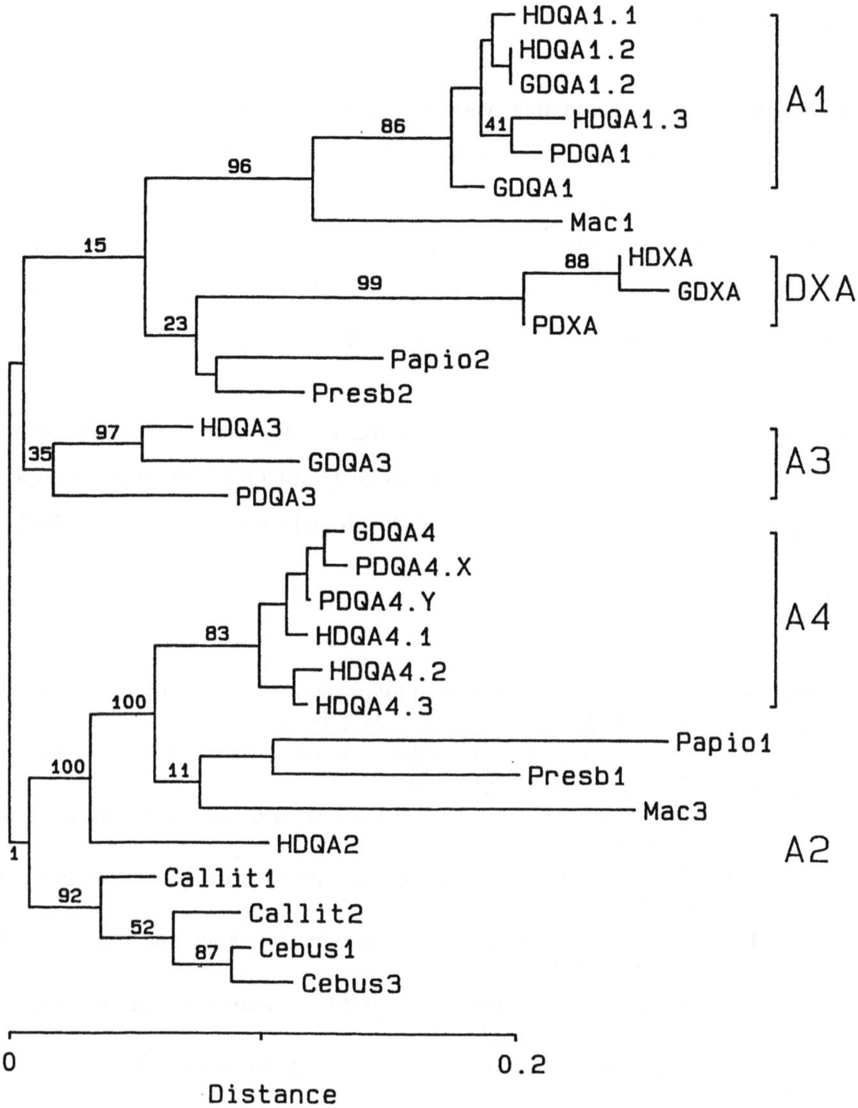

Fig. 5. Neighbor-joining tree for alleles at the DQA locus from several primate species. A1, A2, A3, and A4 group alleles make separate clusters as in the case of Gyllensten and Erlich's tree. H: Human. P: Chimpanzee. G: Gorilla. Callit: Marmoset. Other notations (abbreviations) are the same as those in Figure 1. The bootstrap test establishes the trans-specific polymorphism of A1 and A4 alleles in humans, chimpanzees, and gorillas.

Computer programs

Computer programs for computing various distances for DNA sequences (numbers of nucleotide substitutions per site, synonymous substitutions per synonymous site, nonsynonymous substitutions per nonsynonymous site, etc.) and for constructing (NJDRAW) and testing (NJBOOT) a neighbor-joining tree are available upon request from the Institute of Molecular Evolutionary Genetics. Please send IBM compatible 3.5 or 5.25 inch floppy diskettes. Each program requires one 360 Kb diskette.

Acknowledgment. We thank Tatsuya Ota for making the computer-drawn figures. Tom Whittam provided us with a NJ bootstrap program (NJBOOT). This study is supported by research grants from the National Institute of Health and the National Science Foundation.

References

Cavalli-Sforza, L.L. and Edwards, A.W.F: Phylogenetic analysis: Models and estimation procedures. *Amer J Hum Genet 19*: 233-257, 1967

Efron, B.: The jackknife, the bootstrap and other resampling plans. Society for Industrial and Applied Mathematics, Philadelphia 1982

Felsenstein, J.: Evolutionary trees from DNA sequences: A maximum likelihood approach. *J Mol Evol 17*: 368-376, 1981

Felsenstein, J.: Confidence limits on phylogenies: an approach using the bootstrap. *Evolution 39*: 783-791, 1985

Felsenstein, J.: Phylogenies from molecular sequences: Inference and reliability. *Ann Rev Genet 22*: 521-565, 1988

Fitch, W.M.: Toward defining the course of evolution: minimum change for a specific tree topology. *Syst Zool 20*: 406-416, 1971

Fitch, W.M. and Margoliash, E.: Construction of phylogenetic trees. *Science 155*: 279-284, 1967

Gyllensten, U.B. and Erlich, H.A.: Ancient roots for polymorphism at the HLA-DQα locus in primates. *Proc Natl Acad Sci 86*: 9986-9990, 1989

Gyllensten, U.B., Lashkari, D., and Erlich, H.A.: Allelic diversification at the class II *DQB* locus of the mammalian major histocompatibility complex. *Proc Natl Acad Sci 87*: 1835-1839, 1990

Klein, J.: Origin of major histocompatibility complex polymorphism: The trans-species hypothesis. *Hum Immunol 19*: 155-162, 1987

Li, W.-H.: A statistical test of phylogenies estimated from sequence data. *Mol Biol Evol 6*: 424-435, 1989

McConnell, T.L., Talbot, W.S., McIndoe, R.A., and Wakland, E.K.: The origin of MHC class II gene polymorphism within the genus *Mus*. *Nature 332*: 651-654, 1988

Nei, M.: *Molecular Population Genetics And Evolution*, North-Holland, Amsterdam and New

York 1975

Nei, M.: *Molecular Evolutionary Genetics*. Columbia University Press, New York 1987

Nei, M.: Relative efficiencies of different tree-making methods for molecular data. *In* M.M. Miyamoto and J.L. Cracraft (eds.): *Recent advances in Phylogenetic studies of DNA sequences*. Oxford University Press. Oxford and New York (submitted 1991)

Nei, M. and Hughes, A.L.: Polymorphism and evolution of the major histocompatibility complex loci in mammals. *In* R.K. Selander, A.G. Clark, and T.S. Whittam (eds.): *Evolution at the Molecular Level*, pp. 222-247. Sinauer Associates, Sunderland, MA 1991

Nei, M., Stephens, J.C., and Saitou, N.: Methods for computing the standard error of branching points in an evolutionary tree and their application to molecular data from humans and apes. *Mol Biol Evol 2*: 66-85, 1985

Saitou, N. and Imanishi, T.: Relative efficiencies of the Fitch-Margoliash, maximum-parsimony, maximum-likelihood, minimum-evolution, and neighbor-joining methods of phylogenetic tree construction in obtaining the correct tree. *Mol Biol Evol 6*: 514-525, 1989

Saitou, N. and Nei, M.: The neighbor-joining method: A new method for reconstructing phylogenetic trees. *Mol Biol Evol 4*: 406-425, 1987

Sneath, P.H.A. and Sokal, R.R.: Numerical Taxonomy. Freeman, San Francisco 1973

Sokal, R.R. and Michener, C.D.: A statistical method for evaluating systematic relationships. *University of Kansas Sci Bull 28*: 1409-1438, 1958

Sourdis, J. and Krimbas, C.: Accuracy of phylogenetic trees estimated from DNA sequence data. *Mol Biol Evol 4*: 159-166, 1987

Sourdis, J. and Nei, M.: Relative efficiencies of the maximum parsimony and distance-matrix methods in obtaining the correct phylogenetic tree. *Mol Biol Evol 5*: 298-311, 1988

Tateno, Y., Nei, M., and Tajima, F.: Accuracy of Estimated phylogenetic trees from molecular data: I. Distantly related species. *J Mol Evol 18*: 387-404, 1982

TRANS-SPECIES POLYMORPHISM OF HLA MOLECULES, FOUNDER PRINCIPLE, AND HUMAN EVOLUTION

Naoyuki Takahata

Department of Population Genetics, National Institute of Genetics, Mishima 411, Japan

There are three main hypotheses for the origin of Homo sapiens, called the candelabra, multiregional and Noah's Ark. Their essential difference is in the role and extent of gene migration which might have occurred during the Middle and Upper Pleistocene. The candelabra assumes no migration and thus parallel evolution of H. sapiens in several regional localities while the Noah's Ark assumes the complete replacement of Homo erectus populations in the Old World, around 200,000 years ago, by anatomically modern H. sapiens from Africa. The multiregional hypothesis allows continuous but presumably infrequent gene migration throughout the Pleistocene. It had been difficult to distinguish these hypotheses based solely upon morphological characters, because of limited amounts of fossil records and little knowledge about the evolutionary mechanisms. This situation in the study of human evolution has recently been changed dramatically by the introduction of DNA technology. It seems now possible to examine more deeply the history of organisms and the population dynamics including speciation processes. It is argued that although the data of mitochondrial (mt) DNA sequences sampled from human populations contradict the candelabra hypothesis, they can be consistent with the other alternatives and do not provide further information about what might have happened during the Middle and Upper Pleistocene. In contrast, HLA polymorphisms exhibit trans-species modes of evolution and therefore have a great potential in examining the human evolution in much longer time scales. The trans-species polymorphisms clearly contradict the extreme form of the Noah's Ark as well as the founder principle as being important in speciation. Combined with DNA sequence data of HLA, population genetics theory predicts that the long-term effective size of the human population is of the order of 10^5. This figure is about ten times larger than that inferred from the allele frequency data of other proteins and enzymes, suggesting that the population had been structured, not to a small extent, until the Late Pleistocene.

NATO ASI Series, Vol. H 59
Molecular Evolution of the
Major Histocompatibility Complex
Edited by J. Klein and D. Klein

Introduction

There are several alternative hypotheses for the origin of *Homo sapiens* (e.g. Smith and Spencer 1987; Lewin 1988; Mellars and Stringer 1989), but they all agree that the first demonstrable migration of *Homo erectus* took place from Africa 1.0-1.5 million years ago. What has been debated is whether all living populations had a recent origin in the Late Pleistocene, some hundred thousand years ago, or whether they evolved in many different regions from local archaic populations of *H. erectus*. The debate is concerned also with racial differentiation. There are two extreme hypotheses, the candelabra and the Noah's Ark (Howells 1976). The candelabra assumes no migration and parallel evolution of modern *H. sapiens* in several regional localities at the same time. The Noah's Ark, on the other hand, assumes the complete replacement of populations in the Old World by anatomically modern *H. sapiens* from Africa. In the former view, racial differentiation is very old while in the latter it is relatively young. There can still be many possibilities between the two extremes. One such is a modified version of the candelabra, called the multiregional hypothesis (Wolpoff 1987, 1989). It allows gene flow among different populations, although it is not clear from his writing to what extent gene flow is presumed. In any case, the existing hypotheses for the origin of modern *H. sapiens* differ essentially in the role and extent of migration which might have occurred during the Middle and Upper Pleistocene. The problem thus appears to be the one that must be studied more rigorously from the viewpoint of population genetics.

In the first and second parts of this paper, I review some theoretical results on population structure and bottlenecks and re-examine the mitochondrial (mt) DNA data of Cann, Stoneking and Wilson (1987). The data are clearly incompatible with the candelabra hypothesis, but they do not provide any information about the history of human populations in the Middle and Upper Pleistocene. In the third part, I review a theory of allelic genealogy (Takahata and Nei 1990; Takahata 1990) which I believe is relevant to the HLA data (Klein 1986). The theory and trans-species modes of HLA polymorphisms (Klein 1980) suggest that the human population has never experienced severe bottlenecks since it first diverged from African apes. Thus, the Noah's Ark may be rejected if it assumes a very small number of founding individuals (Klein, Gutknecht and Fischer 1990). By the same token, the founder principle of Mayr (1977) often invoked for speciation cannot be applied in its extreme form to human evolution (Takahata 1990). Some results of computer simulation for a model of the interplay among mutation, balancing selection, genetic drift and gene flow in a geographically structured population are also presented to account for the extraordinary polymorphism observed at classical HLA loci.

Population structure and mitochondrial DNA

Because the human population appears to be to some extent structured (Nei and Roychoudhury 1982; Lewin 1988), theory needs to take account of this fact. Here I use a simple version of Wright's (1931) island model which assumes that the entire population consists of a finite number of demes or subpopulations (Maruyama 1970). Use of this model is based on the conservative assumption that if anything is unlikely in the island model, it is more unlikely in other highly structured population models. I regard the island model of population structure as relevant to the multiregional hypothesis. In its simple form, the hypothesis envisions a stable structure of population subdivision throughout the Pleistocene (Fig. 1). This time period amounts to t_e = 50,000-75,000 generations if the generation time of humans is 20 years. The Noah's Ark (see Stringer 1990 for review) or the mitochondrial Eve hypothesis contends that the out-of-Africa of *H. sapiens* took place around 200,000 years ago and swept out the preexisting *H. erectus* populations. This amounts to t_s = 10,000 generations. Note that the deepest branch in the mtDNA ancestry constructed by Cann et al. (1987) is about this long. Neutrality (Kimura 1968) for the existing mtDNA variation is assumed.

The population considered here consists of L demes each of which has an effective size N (Wright 1931; Maruyama 1970) so that there are NL diploid individuals in total. I first present some mathematical formulas about gene genealogy at an autosomal locus (Takahata 1991a) and then modify them for mtDNA. The per-generation migration rate is the same in both sexes, denoted by m, and when emigration occurs from one deme, the L–1 remaining demes are equally likely to receive immigrants. The average fraction of immigrants in a recipient deme from a donor is m/(L–1) every generation. Assume that n_i genes are sampled from the ith deme, but n_i may be 0 for some demes because they were not sampled. The total numbers of demes arbitrarily chosen and genes randomly sampled from demes are r ($r \le L$) and $n = n_1 + n_2 + \cdots + n_L$, respectively.

When migration is limited ($0 < 4Nm \le 0.1$), it is most likely that orthologous genes within each deme coalesce to or are descended from a common ancestor within the deme: coalescence and migration processes are decoupled. Coalescence time T at which the most recent ancestor is traced back is a random variable. It was shown in Takahata (1991a) that the probability density of T being t generations can be approximated by a linear combination of

$$\frac{j(j-1)m}{L-1}\exp\{-\frac{j(j-1)mt}{L-1}\}\ ,\quad j = 2, 3, ..., r. \qquad [1]$$

The mean and variance of T become

$$E\{T\} = \frac{L-1}{m}(1-\frac{1}{r}) \quad \text{and} \quad V\{T\} = (\frac{L-1}{m})^2 \sum_{j=2}^{r} \frac{1}{\{j(j-1)\}^2}\ .$$

In the above and subsequently, E{ } and V{ } stand for mean and variance. It is clear that not only migration rate (m) but also the number (r) of demes from which a sample is taken is an

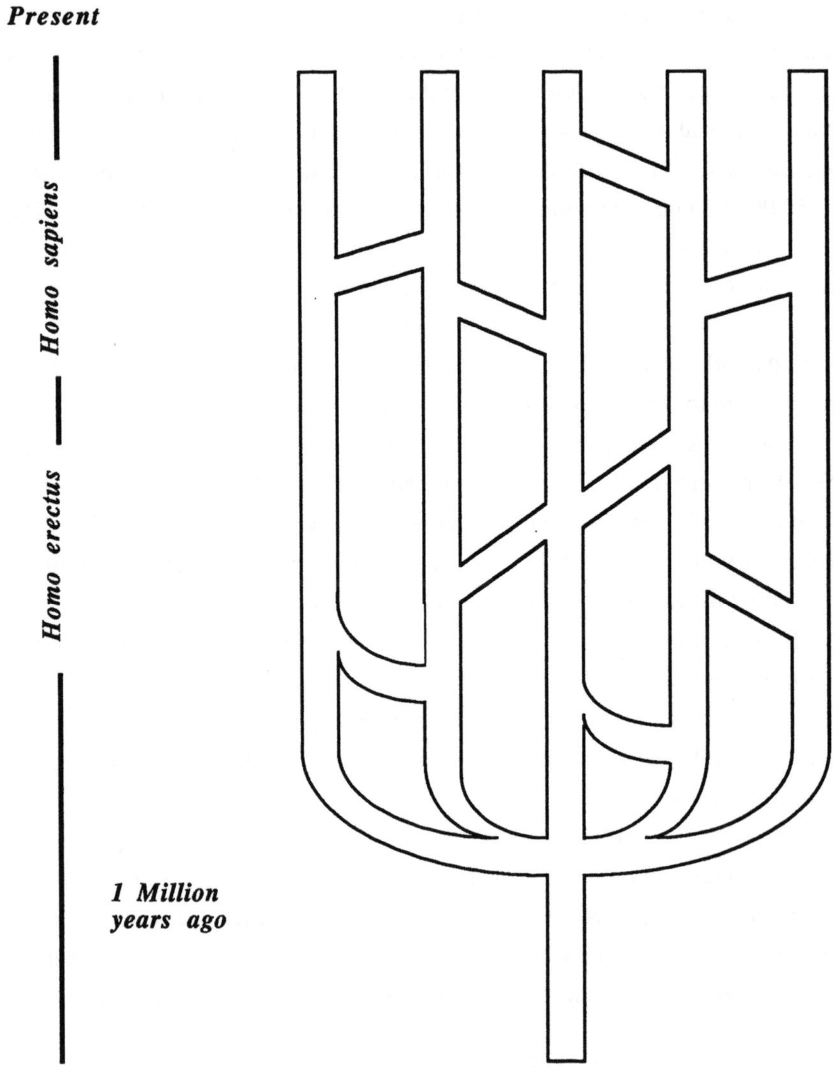

Fig. 1. A model of population structure for the multiregional hypothesis which assumes continuous gene flow throughout the Pleistocene.

important parameter to determine coalescence time T, i.e., the longest branch length of ancestry (see also Takahata 1988).

If on the other hand $4Nm \geq 10$, there must be a number of gene migrations before sampled genes coalesce to a common ancestor. In this case, the ancestral lineages are expected to be distributed at random over the demes even before any pair of genes were descended from a common ancestor. The probability density of T can be approximated by a linear combination of

$$\frac{j(j-1)}{4NL}\exp\{-\frac{j(j-1)t}{4NL}\}\ , \quad j = 2, 3, ..., n \qquad [2]$$

(Takahata 1991a). With high migration rates, it does not matter how the initial sample is taken from demes and the initial sample size n does not greatly change the distribution of T. These can be seen from the mean and variance, which are given by

$$E\{T\} = 4NL(1-\frac{1}{n}) \quad \text{and} \quad V\{T\} = (4NL)^2\sum_{j=2}^{n}\frac{1}{\{j(j-1)\}^2}\ .$$

These expressions are not dependent on r, which is in sharp contrast to the case of low migration limit. Note that $m/(L-1)$ in Eq. [1] and 4NL in Eq. [2] play mathematically the same role in determining the ancestry of genes.

For mtDNA, it is necessary to modify the above results because of the differences in inheritance and ploidy. The modification is, however, straightforward by interpreting m as female migration rate and by replacing 2N by N_f (the number of breeding females in a generation). These changes are based on the maternal transmission and haploidy of mtDNA (Takahata 1985).

When there is no migration ($m = 0$), Eq. [1] cannot be used. In this situation, genealogy of genes sampled from different localities is expected to reflect the shape of candelabra. The exhaustive sampling of mtDNA from world-wide human populations, however, failed to find such a shape and no pairs of gene lineages have lasted for 1 million years (Cann et al. 1987; Horai 1991). Therefore the mtDNA data are certainly inconsistent with the candelabra hypothesis.

For the multiregional hypothesis to be able to account for such mtDNA data, one of the following conditions A and B must be met. Condition A is based on the assumption of $N_fm < 1$. In this case, the ancestry of mtDNA or T is mainly determined by rare migration events. It is then necessary that Eq. [1] does not predict branch lengths longer than $t_s = 10^4$. This requirement can be expressed roughly by

$$\int_{t_s}^{\infty}\frac{j(j-1)m}{L-1}\exp\{-\frac{j(j-1)mt}{L-1}\}dt = \exp\{-\frac{j(j-1)mt_s}{L-1}\} << 1$$

for any j = 2, 3, ..., r. Thus, the following inequalities must hold

$$N_fL < \frac{L}{m} < t_s = 10^4 \ . \qquad [3]$$

On the other hand, condition B is based on the assumption of $N_fm > 1$. In this case, the ancestry is mainly determined by intensity of random genetic drift within demes. For the longest branch length or T to be shorter than $t_s = 10^4$, it is necessary that drift cannot be very weak (large N_f), or more quantitatively

$$\int_{t_s}^{\infty} \frac{j(j-1)}{2N_fL} \exp\{-\frac{j(j-1)t}{2N_fL}\}dt = \exp\{-\frac{j(j-1)t_s}{2N_fL}\} << 1$$

must be met for any j = 2, 3, ..., n. Thus, the following inequalities are required

$$\frac{L}{m} < N_fL < t_s = 10^4 \ . \qquad [4]$$

Eqs. [3] and [4] indicate that the total number of breeding females (N_fL) must be limited to

$$N_fL < 10^4 \qquad [5]$$

and migration cannot be particularly infrequent, or

$$\frac{m}{L} > 10^{-4} \, . \qquad [6]$$

These are necessary conditions for the island model of population structure to be compatible with the mtDNA ancestry. Again it is clear that condition [6] surely rejects the candelabra hypothesis which assumes m = 0 throughout the Pleistocene. In the absence of information on the number of demes (L), however, it is difficult to estimate the absolute value of m and judge whether or not conditions [5] and [6] meet the multiregional hypothesis. There is another serious problem, which is related to the short ancestry of mtDNA and may be looked at in the following way.

In order to estimate N_fm (the number of female migrants per generation), Stoneking et al. (1990) used the F_{ST} statistic (measuring the genetic variation among different demes) and applied it to the New Guinea population. Their F_{ST} value was 0.31, leading to $N_fm = 1.1$ under the present model of population structure. A very similar estimate (Nm = 2.3) was obtained by using protein polymorphisms (Nei and Roychoudhury 1982) under the assumption that the sex ratio is one and migration is sexually unbiased. According to Wright's criteria, these estimates imply that there has been enough migration that the human population has been nearly panmictic (randomly mating, or not locally differentiated genetically). One may therefore assert that this extent of gene flow is inconsistent with the multiregional hypothesis.

A problem in this argument is that F_{ST} equilibrates rather rapidly (Nei, Chakravarti and Tateno 1977; Crow and Aoki 1984; Takahata and Nei 1984). When mutation is negligibly small relative to migration and L is large, this can be shown from the explicit non-equilibrium solution of $F_{ST}(t)$:

$$F_{ST}(t) = \frac{1 - \exp\{-(2m + 1/N_f)t\}}{1 + 2N_f m} \qquad [7]$$

with $F_{ST}(0) = 0$ initially (Takahata, unpublished). From $N_f m = 1.1$, Eqs. [5] and [7], it is clear that $t = 10^4$ generations are sufficiently long for F_{ST} to reach its equilibrium value. In other words, F_{ST} does not give any information about the population structure in the remote past, similar to the mtDNA ancestry. Hence, the estimated levels of gene flow are largely irrelevant to the multiregional hypothesis if the main concern is with the human population structure during the Middle and Upper Pleistocene. Nevertheless, the estimated value of $N_f m = 1.1$ or $Nm = 2.3$ suggests that racial differentiation, if it actually occurred during the Late Pleistocene (see Nei and Livshits 1990 for review), must invoke local selection which could overcome the moderate level of gene flow revealed by protein and mtDNA polymorphisms.

Bottleneck and Noah's Ark

In the preceding section, I have shown that based on a model of stable population structure and neutrality, the data of mtDNA contradict the candelabra but cannot test the multiregional hypothesis if the recent human population has satisfied the conditions of $N_f L < 10^4$ and $m/L > 10^{-4}$. An obvious alternative exists, however. It assumes that $N_f L$ may be larger than 10^4 during the Late Pleistocene but there was a bottleneck immediately before this period. There are some theoretical studies of gene genealogy based on this assumption (M. Slatkin and H. C. Harpending, personal communications). If the bottleneck was so severe as to truncate the mtDNA ancestry around 200,000 years ago, such a model obviously becomes consistent with the data, which in fact is the interpretation of the Noah's Ark hypothesis.

To model a bottleneck, it is important to specify both the reduced population size (N_b) and its duration period (t_b) since it is the ratio of t_b to N_b that determines the ancestry of orthologous genes (Takahata 1989). If N_b is very small, the effects are substantial even when t_b is short. Conversely, when N_b is moderately large but t_b is long, the bottleneck has also substantial effects. A rough but useful inequality to determine the effects is whether or not

$$\exp\{-\frac{t_b}{N_b}\} << 1$$

holds true. If the mitochondrial Eve hypothesis suggests that $N_b = 2$ or if the Noah's Ark presumes that $N_b \approx 100$ immediately before 200,000 years, then $t_b \approx 10$ or 100 is sufficiently long in order to truncate the mtDNA ancestry to be less than 200,000 years. In a more moderate form, the Noah's Ark may assume $N_b \approx 1000$. In this case, $t_b > 1000$ is necessary. However, N_b cannot be too large compared with the deepest branch length of the mtDNA ancestry: otherwise it could have been longer when $N_fL > 10^4$. To model the Noah's Ark, it is therefore assumed that

$$N_b < 10^3 \quad \text{and} \quad t_b > N_b . \qquad [8]$$

As mentioned, if $N_fL < 10^4$ and $m/L > 10^{-4}$ during the Late Pleistocene, the mtDNA data are irrelevant either to the multiregional hypothesis or to the number of the founding individuals which might emerge in Africa 200,000 years ago. The same data are also compatible with the Noah's Ark model if it is specified by [8]. Under the condition of [8], it does not matter whether $N_fL > 10^4$ or $< 10^4$. But if $N_fL > 10^4$ turns out be the case, then the mtDNA data require a fairly severe bottleneck that took place around 200,000 years ago. In the absence of such information about the size of the recent human population, it is inevitable to conclude that both the multiregional and Noah's Ark models are consistent with the mtDNA data (cf. Stringer 1990). Detailed examination of nuclear loci should be very informative in this respect. Although there are only few loci thus far subjected to intensive scrutiny as for mtDNA, there is one exceptional genetic system. This system is the major histocompatibility complex (Mhc), epitomized as HLA in humans.

Trans-species modes of HLA polymorphism and the founder principle

If genes evolve according to a symmetric model of strong balancing selection, relatively weak mutation and drift, it is possible to construct a theory of allelic genealogy, the relationships among different allelic lines (Takahata and Nei 1990; Takahata 1990). Since some forms of balancing selection are responsible for evolution of HLA (Hedrick and Thomson 1983; Hughes and Nei 1988; Nei and Hughes 1991; Hedrick, Whittam and Parham 1991; references therein), it is interesting to compare the theory with the unusual polymorphisms observed at classical HLA loci (Klein 1986).

In this section, s and u stand for the selective advantage of rare alleles or that of heterozygotes over homozygotes and the mutation rate, respectively. Suppose a panmictic population of size NL in which n_c common allelic lines are maintained. It was shown that divergence time T of different alleles can be approximated by a linear combination of exponential functions (Takahata 1990)

$$\frac{j(j-1)}{4NLf_s}\exp\{-\frac{j(j-1)t}{4NLf_s}\}, \qquad j = 2, 3, ..., n_c \tag{9}$$

where

$$f_s \approx \frac{\sqrt{s}}{u\sqrt{2NL}}[\ln\{\frac{s}{8\pi NLu^2}\}]^{-3/2} . \tag{10}$$

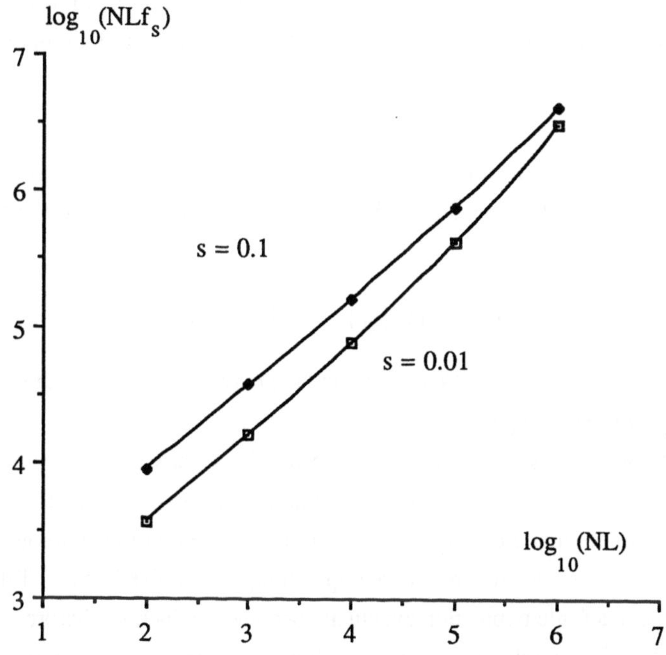

Fig. 2. Effective population size NLf_s at a locus subjected to balancing selection. s = selective advantage of rare alleles or that of heterozygotes over homozygotes. Mutation rate u is assumed to be 4.5×10^{-6} per antigen recognition site per generation. The population is panmictic, consisting of NL diploid breeding individuals.

In short, the allelic genealogy is mathematically the same as the neutral gene genealogy, except for the time scale (cf. Eqs. [1] and [2]). The scaling factor in the allelic genealogy is f_s, which can be much larger than 1 for small NLu and large NLs (Fig. 2).

The mean and variance of T are given by

$$E\{T\} = 4NLf_s(1-\frac{1}{n_c}) \quad \text{and} \quad V\{T\} = (4NLf_s)^2\sum_{j=2}^{n_c}\frac{1}{\{j(j-1)\}^2} \;,$$

respectively, and the effective number of alleles $n_e = E\{n_c\}$ becomes

$$E\{n_c\} = \sqrt{\frac{4NLs}{\ln\{s/(8\pi NLu^2)\}}}$$

(Kimura and Crow 1964). The reciprocal of $E\{n_c\}$ is the expected homozygosity so that

$$H = 1 - \frac{1}{E\{n_c\}}$$

is the expected heterozygosity. Note that the mean divergence time between two different allelic lines is $E\{T^*\} = 2NLf_s$ generations from $E\{T\}$ with $n_c = 2$ and that of the most distantly related lines is given by $E\{T\}$. Thus, the ratio R of $E\{T\}$ to $E\{T^*\}$ simply becomes

$$R = 2H \qquad [11]$$

irrespective of the scaling factor f_s. The average number of selected mutations that can accumulate in an allelic line during $E\{T\}$ and $E\{T^*\}$ generations, denoted by $E\{D\}$ and $E\{D^*\}$, becomes

$$E\{D\} \approx 2HE\{n_c\} \qquad \text{and} \qquad E\{D^*\} \approx E\{n_c\}, \qquad [12]$$

and thus this ratio is also 2H. Eqs. [11] and [12] may be used to test this balancing selection model in light of DNA sequence data.

One important aspect of balancing selection is to increase the effective population size in a similar way that population subdivision plays. The balanced allelic lines are relatively immune to extinction and can persist for a number of generations without complete fixation. This immunity is not absolute in a finite population, however (Klein and Takahata 1990). Any particular line in a finite population eventually becomes extinct, but before going to extinction a balanced allelic line tends to descend new lines successfully. The average number of such successive descendants in an allelic line is expressed as in Eq. [12]. An allelic line segregating in the population thus accumulates $E\{n_c\}$ selected mutations or experiences $E\{n_c\}$ allelic turnovers on average, at each of which occurrences a single copy starts to form a new descendant line in the population. The average distance between a pair of segregating lines is

$2E\{T^*\}$ and therefore $2E\{n_c\}$ in terms of the number of selected mutations. This distance is generally much longer than that for a pair of neutral genes, consistent with extremely high levels of HLA polymorphisms and the trans-species mode of evolution. Because of the long persistence time of polymorphic alleles, neutral mutations may also occur so as to further differentiate each selected allele. Such linked neutral mutations can contribute to increasing the number of alleles substantially, although neutral alleles belonging to the same selected line are closely related (Takahata 1990).

Eqs. [1], [2] and [9] are mathematically the same and fundamental to the present discussion, so it is useful to examine them more carefully. If those equations are rewritten in a single form of

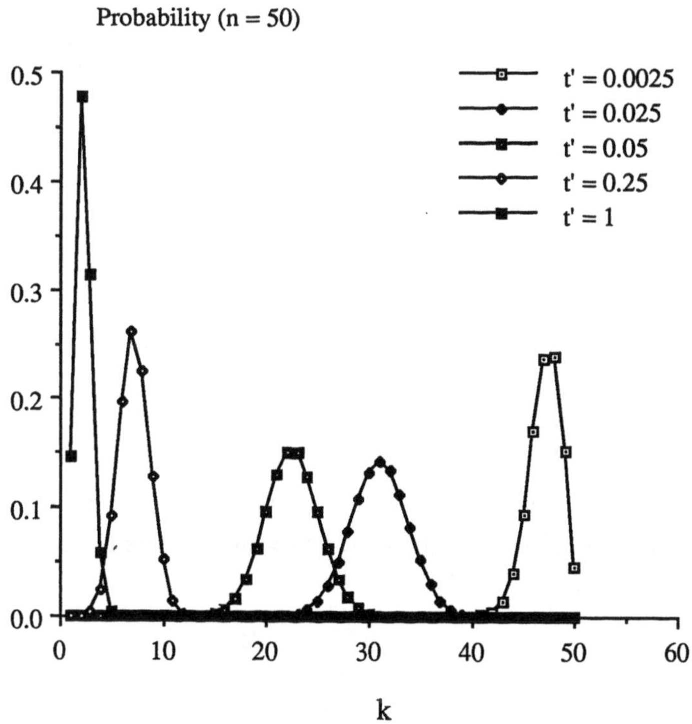

Fig. 3. The probability $g_{nk}(t')$ that there are k ancestral lineages t' units of time ago for a sample of n genes or balanced allelic lines randomly drawn from a panmictic population. The time units are different for neutral genes in the island model with low and high migration rates as well as for balanced alleles.

$$f_j(t) = \frac{j(j-1)}{4NL\lambda}\exp\{-\frac{j(j-1)t}{4NL\lambda}\}, \qquad [13]$$

then $\lambda = (L-1)/(4NLm)$ for Eq. [1], $\lambda = 1$ for Eq. [2] and $\lambda = f_s$ for Eq. [9]. An appropriate linear combination of Eq. [13] for j = 2, 3,..., n (or r, n_c) is given by Eqs. [4] and [5] in Takahata and Nei (1985). From these, they also derived the probability $g_{nk}(t)$ that there were k (= 1, 2, ..., n) genes or allelic lines t generations ago for a sample of size n (see also Tavaré 1984; Watterson 1984).

Fig. 3 shows various values of $g_{nk}(t')$ with n = 50 in which time t' is measured in units of $2NL\lambda$ generations. An important feature is that the time scale of ancestral relationships of genes or alleles is of the order of $2NL\lambda$ generations. For instance, when t' becomes 0.25, the likely values of k are smaller than 10. For t' = 1, they become smaller than 5 and for t' = 10, k = 1 is almost certain. Table 1 shows the most likely value of k for given t' and n which was computed from Tavaré's (1984) expression

$$g_{nk}(t') = \sum_{j=k}^{n} \frac{(2j-1)(-1)^{j-k}k_{(j-1)}n_{[j]}}{k!(j-k)!n_{(j)}}\exp\{-\frac{j(j-1)t'}{2}\} \qquad [14a]$$

for $2 \le k \le n$, and

$$g_{n1}(t') = 1 - \sum_{j=2}^{n} \frac{(2j-1)(-1)^{j}n_{[j]}}{n_{(j)}}\exp\{-\frac{j(j-1)t'}{2}\} \qquad [14b]$$

for k = 1, where $a_{(j)} = a(a+1)\cdots(a+j-1)$ and $a_{[j]} = a(a-1)\cdots(a-j+1)$ for $j \ge 1$, and $a_{(0)} = a_{[0]} = 1$.

Another feature worth mentioning is that for given time spans, t_1 and t_2, $g_{nk}(t)$ satisfies

$$g_{nk}(t_1+t_2) = \sum_{j=k}^{n} g_{nj}(t_1)g_{jk}(t_2)$$

(Takahata 1989). In particular, if population size changes abruptly during these time spans (denoted by N_1L_1 and N_2L_2), the $g_{nk}(t)$ value immediately before $t_1 + t_2$ generations can be computed by first evaluating

$$t' = \frac{t_1}{2N_1L_1\lambda_1} + \frac{t_2}{2N_2L_2\lambda_2}$$

and then substituting this t' for the formula of $g_{nk}(t')$ in Eq. [14]. Recall that λ is generally a function of N, and note that an obvious extension for more than two changes in population size can be expressed as

$$t' = \sum_{i=1} \frac{t_i}{2N_iL_i\lambda_i} . \qquad [15]$$

Clearly, a bottleneck (small N_iL_i) can increase the t' value so that $g_{nk}(t')$ becomes concentrated on smaller k values (Fig. 3). As an example, suppose that there was such a population size change that $N_1L_1 = 10^6$, $N_2L_2 = 10^3$, $t_1 = 10^4$, and $t_2 = 10^3$. If neutral mutations are concerned and $N_im > 1$ for t_i generations (i = 1, 2), then $\lambda_i = 1$ from Eq. [2]. In this case, t' = 0.505 so that there are only a few ancestral lineages that can go back beyond 11,000 generations or 220,000 years. For mtDNA, the effective size is one quarter as large, so t' = 2.2, implying that it is most likely that only one mitochondrial lineage existed before 11,000 generations. This confirms the previous argument for the multiregional or Noah's Ark model.

Table 1. The maximum likelihood estimates of the number (k) of distinct genes or alleles for a given sample of size n and time $t' = t/(2NL\lambda)$: t is measured in units of generations.

t'	n = 10	n = 20	n = 30	n = 40	n = 50
0.01	10	19	26	33	40
0.02	10	17	23	28	34
0.03	9	16	21	25	28
0.04	9	15	19	22	25
0.05	8	14	17	20	21
0.06	8	13	16	18	20
0.07	8	12	15	17	17
0.08	7	11	14	16	17
0.09	7	11	13	15	16
0.10	7	11	12	14	15
0.15	6	8	9	10	11
0.20	5	7	8	8	9
0.25	5	6	7	7	7
0.30	4	5	6	6	6
0.35	4	5	5	5	5
0.40	4	4	5	5	5
0.45	3	4	4	4	4
0.50	3	4	4	4	4
0.55	3	3	4	4	4
0.60	3	3	3	3	3
0.65	3	3	3	3	3
0.70	2	3	3	3	3
0.75	2	3	3	3	3
0.80	2	2	3	3	3
0.85	2	2	2	2	3
0.90	2	2	2	2	2
0.95	2	2	2	2	2
1.00	2	2	2	2	2
2.00	1	1	1	1	1

Of main concern here is whether or not the above bottleneck model is also consistent with the trans-species mode of HLA polymorphisms. Because some forms of balancing selection are responsible for HLA polymorphisms, λ must be chosen as f_s. For s = 0.01~ 0.1 and u = 4.5×10^{-6}, $N_1L_1\lambda_1 \approx 5 \times 10^6$ and $N_2L_2\lambda_2 \approx 5 \times 10^4$ (Fig. 2). As before, $t_1 = 10^4$ and $t_2 = 10^3$. In this case, $t' \approx 0.011$ and the distribution of $g_{nk}(t')$ with n = 50 has a peak around k = 40 (Table 1). Thus, it is possible that many alleles could be passed on through this rather mild bottleneck. However, if the bottleneck becomes as strong as $N_2 = 100$ and $t_2 = 10^3$, then $t' \approx 0.1$ and the theory predicts that only 15 allelic lines must have been transmitted.

It is essential to estimate the number of ancestral allelic lines (k) for given t from actual data. For this purpose, it may be necessary to calibrate the evolutionary rate of molecules. In the case of HLA, there is a difficulty that results from the very nature due to trans-species modes of evolution and from the usual saturation problem of DNA sequence changes (Kimura 1983; Nei 1987). If closely related species such as humans and chimpanzees are used, the species divergence does not coincide with interspecific allelic divergences. Without inferring properly the extent of polymorphisms that already existed in the common ancestral species, it is therefore difficult to calibrate the evolutionary rate. On the other hand, if distantly related species are used, an obvious problem is how to infer correctly multiple substitutions per nucleotide site under the strongly biased GC content (70-93%) at the third codon positions of HLA class I and II loci (Hayashida and Miyata 1983; Hughes and Nei 1988; Takahata 1991b).

Methods to avoid the above difficulty must be developed. One such may be to use the minimum divergence (D_m) between Mhc alleles that are sampled from various species with known divergence times. The rate is then calibrated from D_m divided by twice the species divergence time. Of course it is still an overestimate, since interspecific allelic divergences must have occurred prior to the species splitting. Nevertheless, it will provide an accurate figure if various species comparisons become available. And more importantly, this approach of using the minimum divergence, without knowing the rate, allows to estimate the minimum number (k) of distinct lines around a species splitting and therefore the minimum number of breeding individuals involved in this speciation event. The larger the number of alleles, the larger number of individuals required. A preliminary study shows that for n = 27 alleles at the DRB locus, there are at least k = 12 distinct alleles at the time of human and chimpanzee divergence, about 5 million years (or 2.5×10^5 generations) ago (Satta et al., in preparation). In order for k to be 12 with appreciable probability, t' in $g_{nk}(t')$ must be as small as 0.1 (Table 1). The ratio of 2.5×10^5 to t' = 0.1 is 2.5×10^6, implying that the value of $2NL\lambda = 2NLf_s$ over 5 million years must be at least this large. Fig. 2 then converts NLf_s to NL being of the order of 10^5. If the population size have fluctuated after the species divergence, this estimate of NL may be regarded as the long-term effective size (N_e) in the Wright's sense,

$$\frac{1}{N_e} = \frac{1}{\sum\limits_{i=1} t_i} \sum_{i=1} \frac{t_i}{N_i L_i \lambda_i} \qquad [16]$$

(cf. Eq. [15] and see Crow and Kimura 1970). The result of $N_e \approx 10^5$ indicates that any extreme form of bottlenecks or founder events could not occur in the human lineage throughout these 5 million years. Since there were several speciation events in the course of the human evolution, once per 1 million years on average (Lewin 1988), it must therefore be concluded that such speciation occurred without any severe reduction in population size. Clearly, trans-species modes of HLA polymorphisms contradict the extreme form of founder principle and the Noah's Ark.

A similar approach can be taken for the founding population of modern humans. In this case, it is necessary to assume the clock-like evolution of HLA, because genetic changes among alleles must be translated into their actual separation times. Once this is done, it is possible to estimate k and apply the above method to set an lower bound of breeding individuals at such a particular stage of human evolution.

In a more general situation, the numbers of distinct alleles may be inferred at various evolutionary times. Suppose that for a sample of size n there were k_1, k_2, k_3, and k_4 alleles t_1, t_1+t_2, $t_1+t_2+t_3$, and $t_1+t_2+t_3+t_4$ generations ago, respectively. After converting these t into t', this sequential change can be expressed by a likelihood of

$$L(t') = g_{nk_1}(t_1')g_{k_1k_2}(t_2')g_{k_2k_3}(t_3')g_{k_3k_4}(t_4'),$$

or

$$L(t') = g_{nk_1}(b_1t')g_{k_1k_2}(b_2t')g_{k_2k_3}(b_3t')g_{k_3k_4}(b_4t') \qquad [17]$$

where $t' = t_1'$ ($b_1 = 1$) and $b_i = t_i'/t'$ ($i = 2, 3, 4$). Since b_is are usually known, the problem is simply to determine the value of t'. This can be done by the maximum likelihood method, which chooses the value of t' so as to maximize the likelihood in Eq. [17] or the natural logarithm of L(t'). For instance, if $b_2 = 2$, $b_3 = 9$, $b_4 = 13$, $n = 27$, $k_1 = 21$, $k_2 = 17$, $k_3 = 12$, and $k_4 = 8$, then the maximum value of logL(t') becomes –8.196 at $t' = 0.0095$. The likelihood ratio test shows that the 90% confidence interval of t' ranges from 0.006 to 0.015. Thus, the value of t' is likely to be about 0.01, or the harmonic mean of NLf_s is about 50 t_1. If $t_1 = 2.5 \times 10^4$ generations, then $NLf_s = 1.25 \times 10^6$ which in turn suggests that $N_e \approx 10^5$ from Fig. 2.

The above selection model of HLA has assumed a random mating population, based on which the harmonic mean of NLf_s was estimated as about 10^6 over 5 million years and the harmonic mean of the actual number of breeding individuals (N_e) was estimated as the order of 10^5. A problem is that this estimate of N_e is about ten times larger than that based on protein polymorphisms (Nei and Graur 1984; Nei 1987). It was noted that if $N_e = 10^4$, $s = 0.1$, $Nm >$ 1 (random mating) and $u = 4.5 \times 10^{-6}$ for these 5 million years, it is not easy even for balancing selection models to account for the observed large number of alleles at some of HLA loci (Takahata 1990). One may argue that intragenic recombination or exon shuffling might have been involved in generating new alleles. However, as discussed in Klein and Takahata (1990), some linkage disequilibria are subject also to trans-species modes of evolution. In particular, the DR region is known to be generally frozen in terms of recombination (Fan et al. 1989). Despite such apparent absence of recombination, there are at least 27 alleles at the DRB1 locus in the human population (Klein et al. 1990). Therefore, as far as this locus is concerned, it is unlikely that intragenic recombination has played significant roles in producing new alleles (see Klein and Schönbach 1991 for a more updated review on this matter).

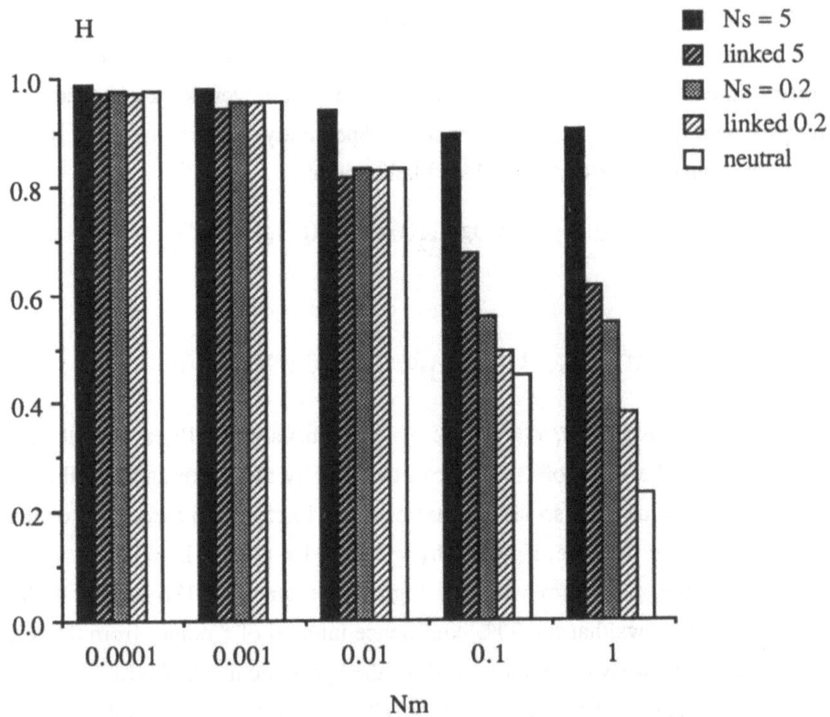

Fig. 4. Simulation results for the mean heterozygosity (H) in the island model of population structure at equilibrium (simulation results of 100 replications). N = 10, L = 50, v (neutral mutation rate linked to a selected site) = 0.0001, u (mutation rate at a selectively balanced site) = 0.0001. The values of H for a linked neutral locus are indicated by linked 5 and linked 0.2 when Ns = 5 or 0.2 at selected site. Open bars indicate the cases where both sites are neutral.

Do HLA polymorphisms really indicate that the human population has been structured over a long period of time? If so, to what extent? These questions do not appear to be trivial, because although balancing selection and population subdivision can individually increase genetic variation, their effects are not additive. If a population is isolated into many small demes, balancing selection becomes less efficient, reducing genetic variation: nevertheless, such inefficiency may be offset by increased neutral variations due to relatively strong isolation. This can be shown as follows. Because the expected heterozygosity for neutral genes is

$$H = 1 - \frac{1}{1+4NLu+\frac{(L-1)^2}{L-1+mL/u}},$$

it approaches $1 - 1/\{L(1+4Nu)\} \geq 1 - 1/L$ as m becomes smaller than u (Maruyama 1970; Takahata 1983). Thus, there exists tremendous genetic variation in a highly isolated population

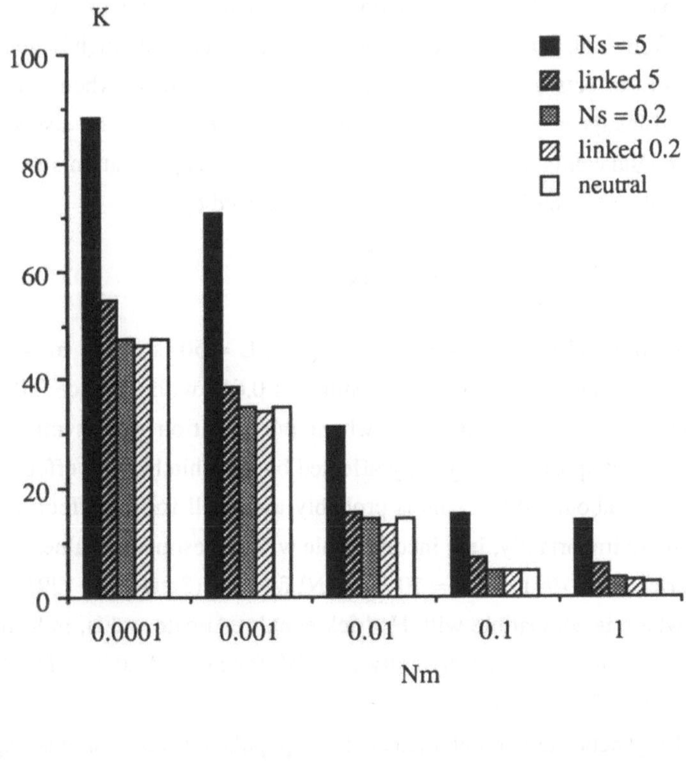

Fig. 5. The expected number (K) of alleles in the whole population. The parameter values and notations are the same as those of Fig. 4.

even without balancing selection. It should be noted, however, that the condition of $m < u$ appears to be unrealistic to the modern human population and that for more realistic values of m such as $m > Lu$, H is close to the equilibrium value of $4NLu/(1+4NLu)$ that is expected in a panmictic population (Kimura and Crow 1964).

In any case, there seems to be no theoretical study about this problem and apparently there is a difficulty in its analytic treatment. I therefore conducted a computer simulation to observe joint effects of population subdivision and balancing selection on heterozygosity (H) and the number of alleles (K). To confirm the previous claim that balancing selection has an effect similar to population subdivision for linked neutral mutations at rate v, the simulated locus is assumed to have two completely linked sites, selected and neutral, both following the infinite allele model of Kimura and Crow (1964) which assumes that every mutation produces a new allele.

Simulation results may be summarized as follows. (1) When $Ns \geq 0.2$ and $NLs \geq 10$, there are substantial increases in H and K at the selected site irrespective of the value of Nm. The effect of selection on H is less conspicuous for $Nm \leq 0.01$ because of its high value due to isolation. On the other hand, the value of K for $Nm = 0.1$ is essentially the same as that of $E\{n_c\}$ in the panmictic case. This is in sharp contrast to the neutral case (Wright 1931) and the difference results from long persistence times of balanced alleles. Although in general the K is fairly sensitive to L, effects of population subdivision become clear when Nm becomes as small as 0.01. (2) Linkage effects of selected mutation on a neutral site are also substantial, as mentioned earlier. Balancing selection acts in the same way as population subdivision does. In a panmictic population, the effect on H can be approximated by

$$H = \frac{4NLf_s v}{1+4NLf_s v} .$$

For instance, the above H value is 0.655 for $N = 10$, $L = 50$, $s = 0.5$, $m = 0.1$, $u = v = 0.0001$, while the corresponding simulation result was 0.614 with variance 0.0338 (Fig. 4). However, at the level of nucleotide sites where the mutation rate is much lower, the heterozygosity may not appear to be greatly affected by such hitchhiking effect. If $v = 10^{-8}$ and $NLf_s = 10^5$, H is about 0.4%. This is probably too small for the effect to be detected statistically, but more importantly, it is incompatible with the estimated values at the HLA-A and -B loci (Hedrick et al. 1991). If $v = 10^{-8}$ and $NLf_s = 10^6$ (Satta et al. 1991), however, H becomes 3.8%, which is compatible with Hedrick et al.'s estimate and is, indeed, more than ten times larger than the nucleotide diversity of 0.4% for mtDNA, 0.2% for ß-globin and growth hormone, and 0.3% for insulin.

In short, for genetic variation at a selected site, population subdivision is important only when migration is fairly limited ($Nm < 0.1$), although the critical value of Nm depends on intensity of balancing selection within demes as well as the number of demes.

Conclusion

The mtDNA sequence and protein allele frequency data both are by and large consistent with the neutral theory (Kimura 1968, 1983) and predicted that the human population has not been particularly structured geographically and genetically. Important estimates thereof are $NL \approx 10^4$ and $Nm \approx 2$ (Nei 1987). If this estimate of Nm is applied to the modern human population in the Late Pleistocene, racial differentiation must have been driven by positive selection. However, these data do not rule out the possibilities of the presence of loci subjected to balancing selection and of the geographic structure of the ancient human population. In fact, HLA polymorphisms suggest that the long-term effective population size might have been of the order of 10^5 and the human population might have been structured as highly as $Nm = 0.01$. The disparity between these two inferences may be reconciled by noting that balanced alleles are documents of tens of million years in the hominid evolution while neutral alleles are those that are likely less than one million years. It may not therefore be unreasonable to conclude that the dispersal rate in humans has increased during the Late Pleistocene and therefore *decreased* the global effective size.

It is still conceivable that some mild bottlenecks might have occurred in such a way that most of neutral variants were lost by drift but balanced variants were passed on. The dynamics of neutral alleles is sensitive to changes in population size and thus bottlenecks, whereas that of balanced alleles is relatively insensitive unless the size becomes extremely small (Fig. 2). It is worth mentioning, however, that the unimodal distribution of the pairwise distances in Cann et al. (Fig. 1, 1987) cannot be taken as direct evidence for a bottleneck if the mtDNA sequences are not a random sample and/or it is not shown statistically that such a distribution is unlikely in a panmictic population. In any event, trans-species transmission of HLA polymorphism (Klein 1986) will further demonstrate how small the human population could be in its almost entire evolutionary history since it first branched off from African apes.

I thank Drs. P. W. Hedrick, J. Klein, M. Nei and Y. Satta for stimulating discussions and useful comments on an early version of this paper. This work is supported in part by a grant from the Japanese Ministry of Education, Science and Culture.

References

Cann, R.L., Stoneking, M., and Wilson, A.C.: Mitochondrial DNA and human evolution. *Nature 325*: 31-36, 1987

Crow, J.F. and Aoki, K.: Group selection for a polygenic behavioral trait: estimating the degree of population subdivision. *Proc Natl Acad Sci USA 81*: 6073-6077, 1984

Crow, J.F. and Kimura, M.: *An Introduction to Population Genetics Theory*. Harper & Row, New York 1970

Fan, W., Kasahara, M., Gutknecht, J., Klein, D., Mayer, W. E., Jonker, M., and Klein, J.: Shared class II MHC polymorphism between humans and chimpanzees. *J Hum Immunol 26*: 107-121, 1989.

Hayashida, H. and Miyata, T.: Unusual evolutionary conservation and frequent DNA segment exchange in class I genes of the major histocompatibility complex. *Proc Natl Acad Sci USA 80*: 2671-2675, 1983

Hedrick, P.W. and Thomson, G.: Evidence for balancing selection at HLA. *Genetics 104*: 449-456, 1983

Hedrick, P.W., Whittam, T.S., and Parham, P.: Heterozygosity at individual amino acid sites: Extremely high levels for HLA-A and B genes. *Proc Natl Acad Sci USA*, in press 1991

Horai, S.: Molecular phylogeny and evolution of human mitochondrial DNA. *In* M. Kimura and N. Takahata (eds.): *New Aspects of the Genetics of Molecular Evolution,* pp. 135-152, Japan Sci Soc Press, Springer-Verlag, Tokyo, Berlin 1991

Howells, W.W.: Explaining modern man: Evolutionists versus migrationists. *J Hum Evol 5*: 477-495, 1976

Hughes, A.L. and Nei, M.: Pattern of nucleotide substitution at major histocompatibility complex class I loci reveals overdominant selection. *Nature 335*: 167-170, 1988

Kimura, M.: Evolutionary rate at the molecular level. *Nature 217*: 624-626, 1968

Kimura, M.: *The Neutral Theory of Molecular Evolution.* Cambridge University Press, Cambridge, 1983

Kimura, M. and Crow, J.F.: The number of alleles that can be maintained in a finite population . *Genetics 49*: 725-738, 1964

Klein, J.: Generation of diversity at MHC loci: Implications for T cell receptor repertories. *In* M. Fougereau and J. Dausset (eds.): *Immunology 80,* pp. 239-253, Academic Press, London 1980

Klein, J.: *Natural History of the Major Histocompatibility Complex.* John Wiley & Sons, New York 1986

Klein, J., Gutknecht, J., and Fischer, N.: The major histocompatibility complex and human evolution. *Trends Genet. 6*: 7-11, 1990

Klein, J. and Takahata, N.: The major histocompatibility complex and the quest for origins. *Immunol Rev 113*: 5-25, 1990

Klein, J. and Schönbach, C.: Origins of *Mhc* diversity. *In* B.G. Solheim, E. Möller, and S. Ferrone (eds.): *The HLA System. Basic Concepts and Clinical Importance in Transplantation Immunology*. Springer-Verlag, Berlin 1991

Lewin, R.: *In the Age of Mankind.* Smithsonian Book, Washington D.C. 1988

Maruyama, T.: Effective number of alleles in a subdivided population. *Theor Pop Biol 1*: 273-306, 1970

Mayr, E.: *Population, Species, and Evolution.* Harvard University Press, Cambridge, MA 1977

Mellars, P. and Stringer, C.: *The Human Revolution: Behavioral and Biological Perspectives on the Origin of Modern Humans.* Princeton University Press, Princeton, New Jersey 1989

Nei, M.: *Molecular Evolutionary Genetics.* Columbia University Press, New York, 1987

Nei, M., Chakravarti, A., and Tateno, Y.: Mean and variance of F_{ST} in a finite number of incompletely isolated populations. *Theor Pop Biol 11*: 291-306, 1977

Nei, M. and Roychoudhury, A.K.: Genetic relationship and evolution of human races. *Evol Biol 14*: 1-59, 1982

Nei, M. and Graur, D.: Extent of protein polymorphism and the neutral mutation theory. *Evol Biol 17*: 73-118, 1984
Nei, M. and Livshits, G.: Evolutionary relationships of Europeans, Asians, and Africans at the molecular level. *In* N.Takahata and J.F. Crow (eds.): *Population Biology of Genes and Molecules*, pp. 251-265. Baifukan, Tokyo 1990
Nei, M. and Hughes, A.L.: Polymorphism and evolution of the major histocompatibility complex loci in mammals. *In* R.K. Selander, A.G. Clark, and T.S. Whittam (eds.): *Evolution at the Molecular Level*, pp. 222-247, Sinauer Associates, Sunderland, Massachusetts 1991
Smith, F.H. and Spencer, F.: *The Origins of Modern Humans*. Alan R. Liss, New York 1987
Stoneking, M., Jorde, L.B., Bhatia, K., and Wilson, A.C.: Geographic variation of human mitochondrial DNA from Papua New Guinea. *Genetics 124*: 717-733, 1990
Stringer, C.: The emergence of modern humans. *Sci Amer 263*: 98-104, 1990
Takahata, N.: Gene identity and genetic differentiation of populations in the finite island model. *Genetics 104*: 497-521, 1983
Takahata, N.: Population genetics of extranuclear genomes: A model and review. *In* T. Ohta and K. Aoki (eds.): *Population Genetics and Molecular Evolution*, pp.195-212, Japan Sci Soc Press, Springer-Verlag, Tokyo, Berlin 1985
Takahata, N.: The coalescent in two partially isolated diffusion populations. *Genet Res Camb. 52*: 213-222, 1988
Takahata, N.: Gene genealogy in three related populations: Consistency probability between gene and population trees. *Genetics 122*: 957-966, 1989
Takahata, N.: A simple genealogical structure of strongly balanced allelic lines and trans-species evolution of polymorphism. *Proc Natl Acad Sci USA 87*: 2419-2423, 1990
Takahata, N.: Genealogy of neutral genes and spreading of selected mutations in a geographically structured population. Submitted, 1991a
Takahata, N.: Overdispersed molecular clock at the major histocompatibility complex loci. *Phil Trans R Soc Lond B. 243*: 13-18, 1991b
Takahata, N. and Nei, M.: F_{ST} and G_{ST} statistics in the finite island model. *Genetics 107*: 501-504, 1984
Takahata, N. and Nei, M.: Gene genealogy and variance of interpopulational nucleotide differences. *Genetics 110*: 325-344, 1985
Takahata, N. and Nei, M.: Allelic genealogy under overdominant and frequency-dependent selection and polymorphism of major histocompatibility complex loci. *Genetics 124*: 967-978, 1990
Tavaré, S.: Lines-of-descent and genealogical processes, and their applications in population genetics models. *Theor Pop Biol 26*: 119-164, 1984
Watterson, G. A.: Lines-of-descent and the coalescent. *Theor Pop Biol 26*: 77-92, 1984
Wolpoff, M. H., Zhi, W.X., and Thorne, A.G.: Modern *Homo sapiens* origins: A general theory of hominid evolution involving the fossil evidence from east Asia. *In* F.H. Smith and F. Spencer (eds.).: *The Origin of Modern Humans*, pp. 411-483, Alan R. Liss, New York 1987
Wolpoff, M.H.: Multiregional evolution: the fossil alternative to Eden. *In* P. Mellars and C. Stringer (eds.): *The Human Revolution*, pp. 62-108, Princeton University Press, New Jersey 1989
Wright, S.: Evolution in Mendelian populations. *Genetics 16*: 97-159, 1931

CALIBRATING EVOLUTIONARY RATES AT MAJOR HISTOCOMPATIBILITY COMPLEX LOCI

Yoko Satta*, Naoyuki Takahata*, Christian Schönbach°, Jutta Gutknecht°, and Jan Klein°+
*Department of Population Genetics, National Institute of Genetics, Mishima 411, Japan
°Max-Planck-Institut für Biologie, Abteilung Immungenetik, 7400 Tübingen, Germany
+Department of Microbiology and Immunology, University of Miami School of Medicine, Miami, FL 33101, USA

Unlike alleles at many other loci, major histocompatibility complex (Mhc) locus alleles often differ by nucleotide substitutions at more than one site, often as many as 88 sites. The substitutions accumulate gradually during evolution by the same process that leads to the divergence of genes in two biological species. The difference between the inter- and intraspecific variation is that in the former, substitutions become fixed in the population (reach a frequency of 1.0), whereas in the latter, they reach polymorphic frequencies (≥ 0.01, < 1.0). Since accumulation of interspecific differences is believed by many geneticists to proceed with a clock-like regularity within certain taxonomic groups, there is no a priori reason why the same should not be true for the accumulation of polymorphic differences. Here we demonstrate the validity of this assumption by comparing alleles at the Mhc-DRB and Mhc-DQB loci of different primate species. We then estimate the evolutionary rates at the DRB and DQB loci; the overall rates of these loci are 0.97 ± 0.17 and 1.2 ± 0.39 (site/billion years), respectively. However, the rate of the sites (both synonymous and nonsynonymous) encoding the peptide (antigen)-binding region (PBR) is 4 to 7 times higher than in the rest of the gene. As previously suggested, the enhanced nonsynonymous rate at the PBR is most likely due to balancing selection, but the PBR as a whole may be a hot spot of nucleotide substitutions.

Trans-species polymorphism and rate calibration method

The fundamental observation on which we base our considerations is that much of *Mhc* polymorphism predates speciation (Klein 1980; Arden and Klein 1982; Klein 1986; Klein and Figueroa 1986; Figueroa et al. 1988; Mayer et al. 1988; Lawlor et al. 1988; McConnell et al. 1988; Fan et al. 1989; Gyllensten and Erlich 1989). To explain, let us assume the existence of four *Mhc* alleles, *A1* and *A2* in the species H, as well as *A3* and *A4* in the species O (see left-hand panel of Figure 1). What one often finds is that the genetic distance (roughly the number of nucleotide differences divided by the total number of nucleotides compared) between *A1* and *A3* or *A2* and *A4* is smaller than that between *A1* and *A2* or *A3*

NATO ASI Series, Vol. H 59
Molecular Evolution of the
Major Histocompatibility Complex
Edited by J. Klein and D. Klein

and *A4*. This observation can be explained by postulating that *A1* separated from *A2* not during the tenure of species H, but before H and O separated from a common ancestor Z. The same then applies to alleles *A3* and *A4*. Hence, in this case the time of interspecific gene divergence differs from the time of species divergence t_S. If we designate this difference t_d, we see that t_d assumes different values for different gene pairs. Some genes separated long before species separation and hence have a large t_d, while others may have diverged close to the time of species divergence and thus have t_d close to zero (see Takahata and Nei 1985 on the subject of neutral genes).

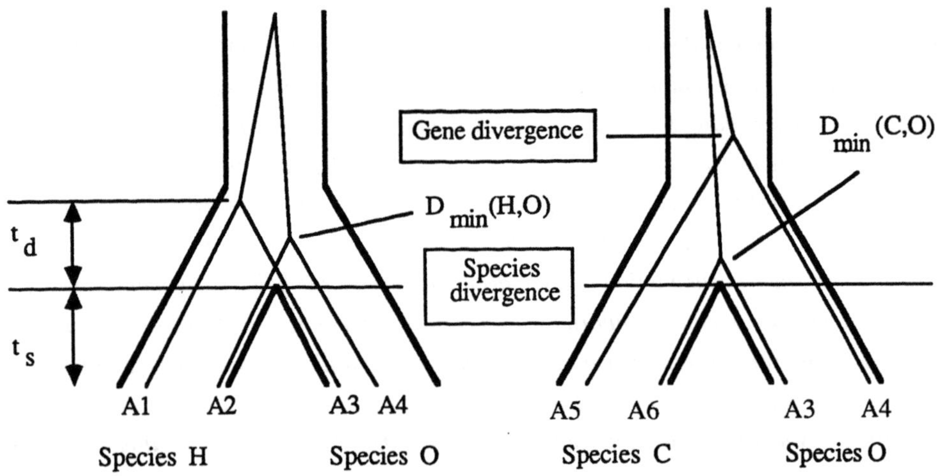

Fig. 1. The principle of the minimum-minimum distance method. Minimum t_d (time between gene and species divergence) is sought by comparing the minimum distance between species H and O ($D_{min(H,O)}$) to the minimum distance between species C and O ($D_{min(C,O)}$). The minimum-minimum distance is indicated as $D_{m.m(H,C,O)} = D_{min(C,O)}$.

If t_d is large, we cannot know *when* different genes actually diverged, because we have no way of knowing how long the genes diverged from each other before species divergence. If we were able, however, to identify genes which diverged close to the time of species divergence, we could use their divergence times to calibrate the evolutionary rate of genes even when they show trans-species transmission of polymorphisms. The task is thus to find genes with minimum t_d.

The principle of the approach we used to obtain allelic divergence rates is depicted schematically in Figure 1. We consider three species, H, O, and C, H having diverged from O at approximately the same time as C diverged from O. For each species, we have a collection of alleles at an *Mhc* locus whose sequences we know. We compare each allele of one species with all alleles of the second and third species and determine nucleotide differences. We then correct the differences for multiple hits by an appropriate formula

(Jukes and Cantor 1969) and determine genetic distances for all the gene pairs. We take the minimum distance values from the (H-O) and from the (C-O) comparison, and then the minimum value of these two. We plot this minimum-minimum distance against the time of separation of the three species to obtain a line whose slope gives us the evolutionary rate and which we then use to calibrate the time of divergence of the allelic lineages.

This method departs from the traditional methods used. Usually, it is assumed that t_s is much longer than t_d even when t_d *per se* might be long. This assumption would be valid for large t_s, in which case the main concern is the accurate estimation of the actual number of substitutions from nucleotide differences. Extensive analysis of available sequence information has revealed, however, that when we compare genes sampled from distantly related species, some have evolved at very different rates in different lineages (Wu and Li 1985; Li et al. 1985). The rate constancy or molecular clock hypothesis appears to hold true only within but not between taxonomic groups (Britten 1986). In what follows, we therefore examine the validity of this hypothesis restricting ourselves to the hominoid and the Old World simian lineages. Since polymorphisms that existed in ancestral species may be substantial when closely related species are compared, it was necessary to assess the extent

Table 1. Intraspecific variation (on the diagonal) and net nucleotide differences (above the diagonal) of *Mhc-DRB* sequences. The sampled number of alleles from each species is as follows: human (42), chimpanzee (3), gorilla (4), orang-utan (1), macaque (5), tamarin (2), and cattle (1). Only the sequences coding for the b1 domain were compared.

	Human	Chimp	Gorilla	Orang	Macaque	Tamarin	Cattle
Human	0.086	0.001	0.004	0.048	-0.001	0.095	0.117
Chimp		0.109	0.005	0.006	-0.003	0.079	0.106
Gorilla			0.101	0.052	0.006	0.108	0.113
Orang				N.D.	0.028	0.177	N.D.
Macaque					0.137	0.083	0.079
Tamarin						0.007	0.244
Cattle							N.D.

of the ancestral polymorphism. One method proposed for doing this (Nei and Li 1979; Nei 1987) is based on the assumption that the extent of polymorphisms is kept constant throughout evolutionary time so that it is by and large the same in the extant as in the ancestral species. In effect, this method tries to compute the net nucleotide substitutions that accumulated after species divergence. For primate *Mhc* genes, however, we found the method unreliable because the estimated value of the net nucleotide substitutions fluctuated to a large extent or even became negative (Table 1). This is to be expected if alleles diverged long before species divergence, which is in fact the case for *Mhc* alleles exhibiting trans-species polymorphism.

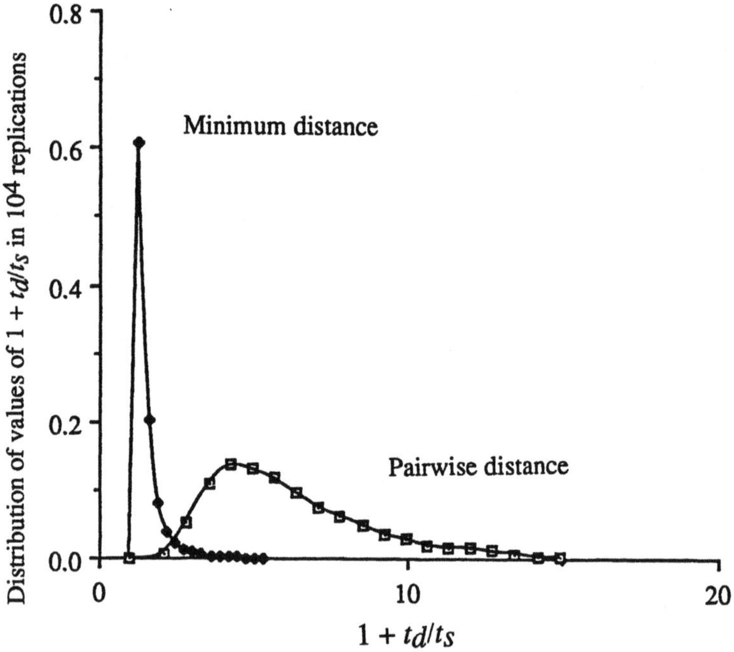

Fig. 2. Computer-generated distribution of the ratio of interspecific allelic divergence time, $t_s + t_d$, to species divergence time, t_s, in units of million years (myr; abscissa). The interspecific allelic divergence times were defined by the minimum distance method (denoted by *minimum distance*), and the pairwise method (denoted by *pairwise distance*). The value of t_s is set as 6 myr, the number of alleles sampled is 27 from one species and 3 from the other, and the population parameter values are chosen so as to imitate the situation of human and chimpanzee.

Simulation

In order to examine the accuracy of the minimum distance method, we conducted a computer simulation. We were particularly interested in the distribution of the ratio of $t_s + t_d$ to t_s. The ratio depends on the value of t_s and the number of genes sampled from each of two species, but if it is close to 1, our contention that the species divergence time can be used to calibrate the evolutionary rate is justified. To efficiently generate allelic ancestries for two species that diverged t_s years ago, we used the theory of allelic genealogy (Takahata 1990) which is based on the model of symmetric balancing selection (Hughes and Nei 1988; Takahata and Nei 1990). The expected allelic genealogy is similar to the family relationships of neutral genes (Hudson 1983; Tavaré 1984), but the time scale of allelic genealogy is much longer because of balancing selection.

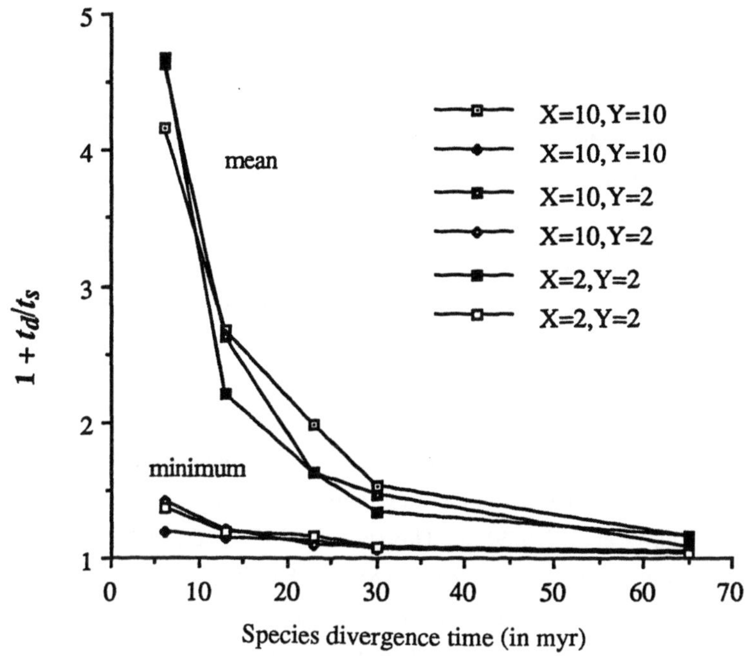

Fig. 3. The average value of the ratio 1 + t_d/t_s (obtained from 10^4 repeats of each simulation. The number of alleles sampled from two species is specified by the values of X and Y. Note that the minimum distance (*minimum*) is always much smaller than the mean pairwise distance (*mean*) and close to the species divergence time t_s ranging from 6 to 65 myr.

Figure 2 shows that the distribution of 1 + t_d/t_s (indicated by *minimum*) ranges from 1 to 5.3, and, more importantly, that it concentrates around 1.2. We conclude therefore that although our method may still underestimate allelic divergence times, the extent of such a bias is very small. In contrast, the distribution of the mean allelic divergence time in all interspecific pairwise comparisons ranges broadly from 1 to 15 in units of t_s, indicating the fact of trans-species polymorphism. Thus the mean allelic divergence time cannot be used to calibrate the evolutionary rate accurately. Figure 3 shows how sample size influences the relationship between species divergence time and the 1 + t_d/t_s ratio. Again the difference between the estimates by the two methods is conspicuous, particularly when closely related species are compared. Although both mean and minimum allelic divergence times are not affected to a great extent by the sample size in the range we examined, it is desirable to have many alleles from each species: the more alleles compared, the higher the probability of finding a small value of t_d.

Results and Discussion

Figure 4 shows the results of the application of the minimum distance method to the sample of human (*Homo sapiens*, Marsh and Bodmer 1990), chimpanzee (*Pan troglodytes*, Fan et al. 1989), orang-utan (*Pongo pygmaeus*, Schönbach, Ç. and Klein, J. unpublished data), pigtail macaque (*Macaca nemestrina*, Zhu et al. 1991), cotton-top tamarin (*Saguinus oedipus*, Grahovac et al. 1991), dog (*Canis familiaris*, Sarmiento et al. 1990), and domestic cattle (*Bos taurus*, Groenen 1990) *DRB* alleles. The numbers of *DRB* genes sampled from the above species are 27, 3, 1, 5, 2, 1, and 1, respectively. The pairwise comparisons of these 40 alleles provide species separation times of 6 myr (human-chimp), 13 myr (human-orang-utan and chimp-orang-utan), 23 myr (human-macaque, chimp-macaque, orang-utan-macaque), 30 myr (human-tamarin, chimp-tamarin, orang-utan-tamarin, and macaque-tamarin), and 65 myr (human-cattle, human-dog, chimp-cattle, etc.); the species separation times are from Hasegawa et al. (1989) and Hasegawa and Kishino (1991), but they are still controversial (Miyamoto et al. 1988; Cowen 1990); the horizontal bars in Figure 4 indicate this uncertainty. The calculated minimum-minimum distance was that between human *DRB1*1101* and macaque *DRB1*01* alleles (point 4 in Figure 4) and this value gave the minimum-minimum substitution rate of 0.95 per site per billion years (Table 2). The same analysis was carried out using synonymous and nonsynonymous substitutions separately with similar results to those obtained by analyzing total substitutions (Table 2).

Similar comparisons were also made for a sample of 49 *DQB1* genes from the following species (the number of genes is given in parentheses): human (*Homo sapiens*, 18, Marsh and Bodmer 1991; Gyllensten et al. 1990), chimpanzee (*Pan troglodytes*, 7, Gyllensten et al. 1990), pygmy chimpanzee (*P. paniscus*, 5, Gyllensten et al. 1990), lowland gorilla (*Gorilla gorilla*, 8, Gyllensten et al. 1990), orang-utan (*Pongo pygmaeus*, 1, L. Gaur, E. Heise, and J. Gutknecht, unpublished data), gibbon (*Hylobates lar*, 1, Gaur et al. 1991), crab-eating macaque (*Macaca fascicularis*, 1, Gaur et al. 1991), baboon (*Papio leucophaeus*, 1, Gyllensten et al. 1990), langur (*Presbytis entellus*, 2, Gyllensten et al. 1990), white-throated capuchin (*Cebus capucinus*, 1, Gyllensten et al. 1990), white whale (*Delphinapterus leucas*, 1, Gyllensten et al. 1990), horse (*Equus caballus*, 2, Gyllensten et al. 1990), sheep (*Ovis aries*, 1, Gyllensten et al. 1990). In contrast to the *DRB* genes, which included most of the coding region, the *DQB* sequences were only 171 nucleotides long and were comprised only of the region coding for the bulk of the first domain of the class II b chain. The results obtained using these two samples were nevertheless comparable (Table 2).

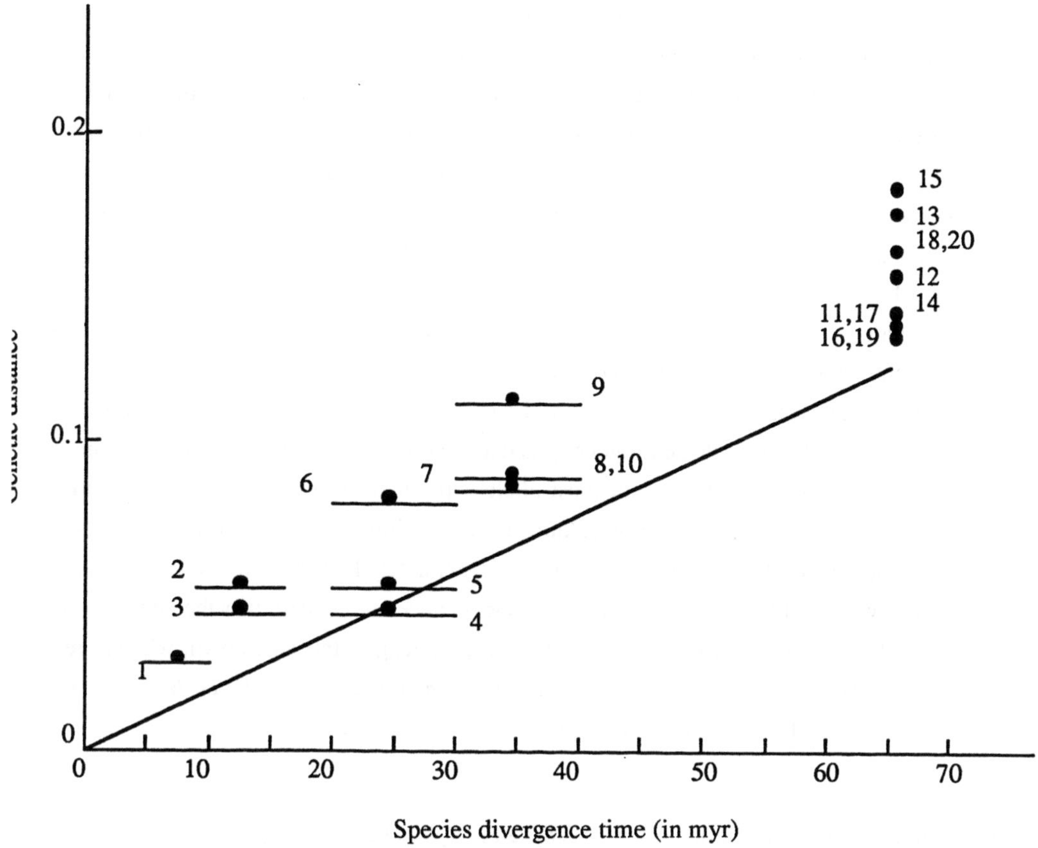

Fig. 4. Estimation of the minimum-minimum substitution rate for a group of *Mhc-DRB* alleles. Each filled circle (1 to 20) is the minimum distance from one interspecific comparison. The slope of the line is the minimum-minimum substitution rate. Comparisons : 1, human *vs*. chimp; 2, chimp *vs*. orang-utan; 3, human *vs*. orang-utan; 4, human *vs*. macaque; 5, chimp *vs*. macaque; 6, orang-utan *vs*. macaque; 7, human *vs*. tamarin; 8, chimp *vs*. tamarin; 9, macaque *vs*. tamarin ; 10, orang-utan *vs*. tamarin; 11, human *vs*.cattle; 12, macaque *vs*. cattle; 13, chimp *vs*. cattle; 14, orang-utan *vs* cattle; 15, tamarin *vs*.cattle; 16, macaque *vs*. dog; 17, human *vs*. dog; 18, chimp *vs*. dog; 19, orang-utan *vs*. dog; 20, tamarin *vs*. dog. References for the sequences are given in the text. Horizontal bars crossing solid circles indicate the range in the estimates of species divergence time (myr) that is still in debate.

Two issues are critically relevant to this analysis. The first is the assumption that the substitution rate is constant at the *Mhc-DRB* locus over the period considered here (23 myr, the divergence time between humans and macaque). Figure 4 shows that the values of minimum gene divergence increase approximately proportionally to the species divergence time. This observation indicates that during 23 myr, nucleotide substitutions have not yet reached a saturation level at which recurrent substitutions would lead to lowering of the rate and hence to levelling off of the curve. To test whether the substitution rate may not have slowed down or accelerated in some lineages, we compared the number of substitutions in each lineage against those in the cattle and dog. Had some lines evolved at different rates, these differences would become apparent in the comparison with rates involving the cattle and dog sequences. In reality, however, the rates in different lineages were similar to each other (the number of observed substitutions was 84 to 107, with a mean of 93.6 and variance of 30.1). Even so, one possible objection can be raised against this analysis: the compared lineages share a branch and substitutions in this shared part could, theoretically, mask rate heterogeneity in other portions of the dendrogram. For this reason we carried out the relative rate test (Wilson et al. 1977; Wu and Li 1985; Kimura 1987) using cattle sequences as a reference (Table 3). Rate heterogeneity is detected only in seven pairs of alleles among the total of 338 comparisons so that most alleles appear to have accumulated nucleotide changes in a clock-like manner. For unknown reasons, however, the substitution rates of hominoid and Old World simians appear to be approximately half those in the cattle lineage (see also Hayashida and Miyata 1983). Nonetheless, since our concern was the constancy of rates after the divergence of hominoid and Catarrhini lineages, this slowdown, whether real or not, does not affect our calculations. Hence we conclude that the *Mhc-DRB* genes have evolved at a constant rate in the last 23 myr.

Table 2. Estimated substitution rates for *DRB* and *DQB* genes (x 10—3/site/myr). PBR, peptide-binding region.

Substitutions	*DRB* including PBR	*DRB* excluding PBR	*DQB* including PBR
Total	0.97 ± 0.17	0.67 ± 0.15	1.2 ± 0.39
Synonymous	0.96 ± 0.35	0.80 ± 0.33	-
Nonsynonymous	0.70 ± 0.10	0.52 ± 0.15	-

The second issue we must consider is the divergence of alleles after speciation. If identical genes were transmitted into the recently emerging species at the time of divergence, then the number of postspeciation substitutions could be small, and lead to a biased estimate of the substitution rate. However, the minimum-minimum distance is as large as 30 for a pair of *DRB1* genes from the human and macaque lineages, and the last 23 myr is a sufficiently long time for a number of substitutions to occur. Indeed, we have assumed that this distance is generated mainly after speciation. If on the other hand the rate is so low that only a few postspeciation substitutions could occur, we must assume -- absurdly -- that the remaining 20 or so substitutions predated the time of human and macaque divergence. In this case, the required time of allelic divergence in the ancestral species would become unreasonably long. Thus, it appears that our assumption that most of the substitutions we observed for the pair of alleles accumulated after the human and macaque divergence is not violated and the estimation bias, if present, must be small.

Table 3. Relative rate test by which an allele X shows significantly higher substitution rates than allele Y ($P < 0.1\%$).

Pairs of alleles from species X and Y	Differences (X-Y) and sampling error
*Hosa - DRB4*0101 vs. Mane - DRB4*04*	0.034 ± 0.010
*Saoe - DRB*01 vs. Hosa -DRB1*0401*	0.042 ± 0.012
*Saoe - DRB*01 vs. Hosa - DRB1*0404*	0.044 ± 0.011
*Saoe - DRB*02 vs. Hosa - DRB1*0401*	0.038 ± 0.011
*Saoe - DRB*02 vs. Hosa - DRB1*0404*	0.040 ± 0.011
*Saoe - DRB*01 vs. Mane - DRB3*03*	0.039 ± 0.012
*Saoe - DRB*01 vs. Mane - DRB4*04*	0.042 ± 0.012

*The numbers of alleles compared were human, 27; chimpanzee, 3; orangutan, 1; macaque, 5; and tamarin, 2. The total number of pairs examined was 338 of which only seven showed rate heterogeneity. The reference species was the cattle. Abbreviations: *Popy*, orang-utan; *Hosa*, human; *Saoe,* tamarin; and *Mane*, macaque.

The functional *Mhc* genes, or more specifically the parts coding for the peptide (antigen)-binding region (PBR) of the molecule, are believed to be subjected to balancing selection (Hughes and Nei 1988, 1989). To assess the effect selection has on the substitution rate of *Mhc* alleles, we compared the synonymous and nonsynonymous substitution rates, as well as rates of sequences with or without the putative PBR-encoding part (Table 2; because the *DQB* sequences were so short, we did not include them in these comparisons). We found the nonsynonymous rates to be lower than the synonymous ones, although the ratio of nonsynonymous to synonymous rates (approximately 0.7) is higher for *Mhc* genes than for other genes (less than 0.4). This observation is consistent with the hypothesis of Hughes and Nei (1988) that natural selection is responsible for the increased nonsynonymous rates of *Mhc* genes. Indeed, the nonsynonymous rates of sequences with PBR are significantly higher than those of sequences from which the PBR has been excluded, indicating that nonsynonymous substitutions in PBR affect the rate. From Table 2 and the number (16) of putative codons involved in the class II PBR, we can compute the nonsynonymous rate at the PBR to be seven times higher than that in the rest of the gene. Likewise we can estimate the synonymous rate at the PBR to be increased four times compared to the rest of the gene. Thus, since not only nonsynonymous but also synonymous rates are higher in the PBR, the region is a hot spot of both types of substitutions.

We also estimated the overall evolutionary rate of the ß1 domain-encoding segment as 1.5 x 10^{-9} per site per year, which is 1.5 times higher than that for the whole *DRB1* gene (Table 2). If we apply this rate to intraspecific variation in the ß1 domain-encoding segment (Table 1), we can estimate the mean divergence time of *DRB1* alleles as 29 myr for human, 36 myr for the chimpanzee, 34 myr for the gorilla, 46 myr for the macaque, and 2.3 myr for the tamarin. This confirms that many polymorphic *Mhc* alleles in primate species diverged long ago (Klein 1986; Mayer et al. 1988; McConnell et al. 1988; Lawlor et al. 1988; Gyllensten and Erlich 1989). An application of such extraordinary polymorphisms to the study of the human population will be presented elsewhere.

Acknowledgments. We thank Ms. Lynne Yakes for editorial assistance. This work was supported by grant AI 23667 from the National Institutes of Health, Bethesda, Maryland, USA and a grant from the Ministry of Education, Science and Culture, Japan.

References

Arden, B. and Klein, J.: Biochemical comparison of major histocompatibility complex molecules from different subspecies of *Mus musculus*. Evidence for *trans*-specific evolution of alleles. *Proc Natl Acad Sci USA 79*: 2342-2346, 1982

Britten, R. J.: Rates of DNA sequence evolution differ between taxonomic groups. *Science 231*:1393-1398, 1986

Cowen, R.: *History of Life*. Blackwell, Cambridge 1990

Fan, W., Kasahara, M., Gutknecht, J., Klein, D., Mayer, W. E., Jonker, M., and Klein, J.: Shared class II polymorphisms between human and chimpanzees. *Hum Immunol 26*:107-121, 1989

Figueroa, F., Günther, E., and Klein, J.: MHC polymorphism predating speciation. *Nature 335*: 265-267, 1988

Grahovac, B., Mayer, W., Vincek, V., Figueroa, F., O'hUigin, C., Tichy, H., and Klein, J.: Major histocompatibility complex *DRB* genes of a New World monkey, the cotton-top tamarin (*Saguinus oedipus*). *Mol Biol Evol*, in press 1991

Groenen, M.A.M., van der Poel, J.J., Dijkhof, R.J.M., and Giphart, M.J.: The nucleotide sequence of bovine MHC class II *DQB* and *DRB* genes. *Immunogenetics 31:* 37-44, 1990

Gyllensten, U.B. and Erlich, H. A.: Ancient roots for polymorphism at the HLA-DQα locus in primates. *Proc Natl Acad Sci USA 86*: 9986-9990, 1989

Gyllensten, U.B., Lashkari, D., and Erlich, H.A.: Allelic diversification at the class II *DQB* locus of the mammalian major histocompatibility complex. *Proc. Natl. Acad. Sci. USA 87*: 1835-1839, 1990

Hasegawa, M., Kishino, H., and Yano, T.: Estimation of branching dates among primates by molecular clocks of nuclear DNA which slowed down in Hominoidea. *J Hum Evol 18*: 461-476, 1989

Hasegawa, M. and Kishino, H.: DNA sequence analysis and evolution of Hominoidea. *In* M. Kimura and N. Takahata (eds.): *New Aspects of the Genetics of Molecular Evolution,* pp. 303-317, Springer Verlag, Tokyo/Berlin 1991

Hayashida, H. and Miyata, T.: Unusual evolutionary conservation and frequent DNA segment exchange in class I genes of the major histocompatibility complex. *Proc Natl Acad Sci USA 80*: 2671-2675, 1983

Hudson, R. R.: Testing the constant-rate of neutral allele model with protein sequence data. *Evolution 37*: 203-217, 1983

Hughes, A.L. and Nei, M.: Pattern of nucleotide substitution at major histocompatibility complex class I loci reveals overdominant selection. *Nature 335*: 167-170, 1988

Hughes, A.L. and Nei, M.: Nucleotide substitution at major histocompatibility complex class II loci: Evidence for overdominant selection. *Proc Natl Acad Sci USA 86*: 958-962, 1989

Jukes, T.H. and Cantor, C.H.: Evolution of protein molecules. *In* H. N. Munro (ed): *Mammalian Protein Metabolism*, pp. 21-132, Academic Press, New York 1969

Kimura, M.: Molecular evolutionary clock and the neutral theory. *J Mol Evol 26*: 24-33, 1987

Klein, J.: Generation of diversity at *MHC* loci: Implications for T cell receptor repertoires. *In* M. Fougereau and J. Dausset (eds.): *Immunology 80*: *Progress in Immunology IV*, pp. 239-253, Academic Press, New York 1980

Klein, J.: *Natural History of the Major Histocompatibility Complex*, Wiley, New York 1986

Klein, J. and Figueroa, F.: Evolution of the major histocompatibility complex. *CRC Crit. Rev. Immunol. 6*: 295-386, 1986

Lawlor, D.A., Zenmour, J., Ennis, P.P., and Parham, P.: HLA-A and B polymorphisms predated the divergence of humans and chimpanzees. *Nature 335*: 268-271,1988

Li, W-H., Luo, C.C., and Wu, C.I.: Evolution of DNA sequences. *In* R.J. MacIntyre (ed.): *Molecular Evolutionary Genetics*, pp. 1-94 Plenum, New York 1985

Marsh, S.G.E. and Bodmer, J.G.: HLA-DRB nucleotide sequences. *Immunogenetics 31*: 141-144, 1990

Marsh, S.G.E. and Bodmer, J.G.: HLA class II nucleotide sequences. *Immunogenetics 33*:321-334, 1991

Mayer, W.E., Jonker, M., Klein, D., Ivanyi, P., van Seventer, G., and Klein, J.: Nucleotide sequences of chimpanzee MHC class I alleles: evidence for *trans*-species mode of evolution. *EMBO J. 7*: 2765-2774, 1988

McConnell, T.J., Talbot, W.S., McIndoe, R.A., and Wakeland, E.K.: The origin of MHC class II gene polymorphism within the genus *Mus. Nature 332*: 651-654, 1988

Miyamoto, M.M., Koop, B.F., Slightom, J.L., Goodman, M., and Tennant, M.R.: Molecular systematics of higher primates: Genealogical relations and classification. *Proc Natl Acad Sci USA 85*: 7627-7631, 1988

Nei, M.: *Molecular Evolutionary Genetics*, Columbia University Press, New York 1987

Nei, M. and Li, W.-H.: Mathematical model for studying genetic variation in terms of restriction endonucléases. *Proc Natl Acad Sci USA 76*: 5269-5273, 1979

Sarmiento, U.M., Sarmiento, J.I., and Storb, R.: Allelic variation in the *DR* subregion of the canine major histocompatibility complex. *Immunogenetics 32*: 13-19, 1990

Takahata, N.: A simple genealogical structure of strongly balanced allelic lines and trans-species evolution of polymorphism. *Proc Natl Acad Sci USA 87*: 2419-2423, 1990

Takahata, N. and Nei, M.: Gene genealogy and variance of interpopulational nucleotide differences. *Genetics 110*: 325-344, 1985

Takahata, N. and Nei, M.: Allelic genealogy under overdominant and frequency-dependent selection and polymorphism of major histocompatibility complex loci. *Genetics 124*: 967-978, 1990

Tavaré, S.: Lines of descent and genealogical processes, and their applications in population genetics models. *Theor Popul Biol 26*: 119-164, 1984

Wilson, A.C., Carlson, S.S., and White, T.J.: Biochemical evolution. *Annu Rev Biochem 46*: 573-639, 1977

Wu, C.-I. and Li, W.H.: Evidence for higher rates of nucleotide substitution in rodents than in man. *Proc Natl Acad Sci USA 82*: 1741-1745, 1985

Zhu, Z., Vincek, V., Figueroa, F., Schönbach, C., and Klein, J.: *Mhc-DRB* genes of the pigtail macaque (*Macaca nemestrina*): Implications for the evolution of human *DRB* genes. *Mol Biol Evol*, in press 1991

CONCERTED MUTAGENESIS: ITS POTENTIAL IMPACT ON INTERPRETATION OF EVOLUTIONARY RELATIONSHIPS

Lynn S. Ripley
Department of Microbiology and Molecular Genetics, UMDNJ-New Jersey Medical School, Newark, NJ 07013, USA

Spontaneous mutations are produced occasionally by mechanisms that produce more than one base substitution in a single mutational event. The mechanisms responsible for these concerted substitutions often depend on templated, but misaligned DNA synthesis. Thus, concerted mutations have two unusual properties: (1) they simultaneously produce multiple substitutions and (2) the specificity of these substitutions is directly determined by a DNA sequence elsewhere in the gene. If the molecular evolution of DNA sequences includes concerted mutations, classical interpretations of lineage relationships predicted solely on the characteristics of single substitutions may be compromised. Comparisons of DNA sequences from rodent HLA class II β genes are presented from the perspective of the ability of concerted mutations to account for the appearance of short DNA motifs in many different alleles. The motifs are found to be related to nearby DNA sequences in a manner suggesting that concerted mutagenesis provides a parsimonious explanation for the sequences of the motifs. Similar motifs appear to arise independently, and the proposed mutational mechanisms often permit motif shuffling. Thus, concerted mutagenesis may also explain the distributions of the motifs among alleles. Comparisons among the rodent sequences and with a human orthologue suggest that concerted mutagenesis offers a reasonable mechanistic explanation for convergence. Improved understanding of lineage relationships that take into account concerted mutagenesis may be able to provide a clearer picture of the mechanisms responsible for the unusual allelic diversity of many genes of the major histocompatibility complex.

The molecular evolution of DNA sequences is likely to be shaped by many factors. The patterns of DNA sequence variation in evolutionary time reflect the accumulation of mutations, their reassortment by recombination and their selection in populations. Our understanding of the evolutionary patterns of sequence changes and the molecular events responsible is incomplete. Advances in sequencing technologies are now providing a truly extensive catalogue of evolutionary DNA

NATO ASI Series, Vol. H 59
Molecular Evolution of the
Major Histocompatibility Complex
Edited by J. Klein and D. Klein

sequence changes which can be expected to identify more clearly the DNA patterns of intra- and inter-species variation.

It is generally accepted that a substantial fraction of the DNA sequence divergence in protein coding regions is due to be the accumulation of neutral, single base substitutions (Kimura, 1977). These mutations are hypothesized to accumulate because they are rarely subject to negative Darwinian selection and are occasionally stochastically fixed in a population. The operation of positive Darwinian selection as a directive force in shaping the specific accumulation of DNA sequence changes is believed to be more rare.

The patterns of DNA sequence polymorphisms in humans are consistent with the view that they often reflect the accumulation of neutral, single base changes (Bowcock et al. 1991). However, positive selection also appears to be a factor, although its quantitative contribution is presently unknown. Comparisons of polymorplic sequence changes in genes subject to alternative degrees of selection might provide a much clearer view of the contributions of neutral and selective forces to evolutionary change. Such comparisons may sometimes be able to reveal special characteristics of the operative selective forces (Kreitman, 1991).

There is modest precedent for postulating that the characteristics of spontaneous mutation may partially shape the evolution of DNA sequences. The patterns of DNA evolution of a particular gene will be influenced by the phenotypes that the spontaneous mutations produce because the patterns of observed variation are filtered by the operation of negative selection. For example, frameshifts would rarely contribute to the molecular evolution of translated sequences because of their high propensity for inactivating the gene. The relative contribution by small additions or deletions ("frameshifts") to the molecular evolution of non-coding DNA sequences might be considerably more frequent (e.g. Tautz et al. 1986). In contrast, neutrally accumulated base substitutions may contribute often to DNA sequences that code for required proteins because a high fraction of base substitutions have no detectable phenotypic effect (e.g. Gordon et al. 1988).

The accumulation of neutral mutations reflects the mutation rate. Thus, to the degree that mutants are equivalently neutral, accumulated mutations should also reflect the specificity of spontaneous base substitution mutation. For example, at sites at which both transition and transversion base substitutions are phenotypically silent, the transition:transversion ratio should be greater than 1:2, reflecting the higher spontaneous frequency of transitions. Indeed, this pattern is generally found (e.g. Li et al. 1985). An experimental demonstration of the genomic impact of increased frequency and changed specificity of spontaneous mutation has been carried out in *E. coli*. In the case investigated, a large, specific increase in the spontaneous A:T -> C:G mutation rate quickly produced an increase in the G:C-base pair content of the DNA (Cox and Yanofsky, 1967).

Less explored, is the possibility that the specificity of spontaneous mutation might influence the characteristics of positively selected mutations. The rate of accumulation of positively selected mutations is governed by the intensity of selection. Thus, mutation frequencies are not a useful guide to identifying the spontaneous mutations most likely to be selected. Instead, mutations conferring high selectability should predominate. High selectability in protein-coding regions is anticipated to be conferred by the creation of amino acid changes. Often the amino acid changes appear to be created by multiple, nearby DNA base substitutions. Do these substitutions accumulate one-by-one? Most analyses of evolutionary relationships implicitly assume so; each substitution is counted as an individual mutation. However, several classes of spontaneous mutations can introduce multiple base substitutions into a DNA sequence in a single mutational event (Ripley, 1990). Might these concerted DNA base substitutions contribute to evolutionarily observed amino acid sequence changes?

The concerted production of multiple base changes in a single mutational event was first detected among mutant frameshift sequences deduced from the amino acid sequences of mutants in the iso-1-cytochrome c gene of yeast (Stewart and Sherman, 1974). It was not until DNA sequencing and studies of DNA structure had progressed that a mutational mechanism accounting for specific concerted sequence changes could be described (Ripley, 1982). Initially, the substitutions had been mysterious, appearing to bear no relationship to one another. However, it is now clear that the substituted sequence is produced by templated polymerization of DNA from sequences lying at a distance from the mutant site (Ripley, 1990).

Although concerted frameshift-base substitution changes are not the most frequent class of spontaneous mutations, they appear to represent approximately 10% of spontaneous frameshifts in bacteriophage T4 and in yeast (Ripley, 1982; Ripley et al. 1986; Shinedling et al. 1987; Hampsey and Sherman, 1988). Improved understanding of the mechanisms responsible for some of these mutations clearly suggest that there is no inherent need for these mutations to include frameshifts; the production of concerted substitutions without accompanying frameshifts is expected.

From the perspective of evolutionary selection, concerted sequence changes could have several advantages over the accumulation of single base substitutions for creating selectable protein products. First, by producing more than one substitution in a single codon, the concerted mutation could bypass an amino acid change subject to negative selection. Second, by producing substitutions in neighboring codons, concerted mutations could produce novel amino acid motifs in a single step, again bypassing intermediate mutations that might be subject to negative selection. Third, even when no negative selection exists against intermediate sequence changes, the frequency of a concerted mutation coupled to positive selection for the product might lead to the accumulation of sequence changes in populations at higher frequencies than the accumulated occurrence and fixing of neutral single base substitutions.

Only a few modest attempts have been made to look for contributions of concerted mutational events to the molecular evolution of DNA sequences (Ripley and Glickman, 1983: Glickman and Ripley, 1984; Golding and Glickman, 1985, Klein and Schonbach, 1991). Although considerable evidence that is consistent with contributions by concerted mutations has been found, further studies that explore the impact that concerted mutations may have on evolution or its interpretation in particular DNA sequences have not been carried out.

The genes of the mammalian major histocompatibility complex (MHC) appear to offer an ideal possibility for examining the impact of concerted mutagenesis on molecular evolution. The MHC loci from many organisms have been sequenced and some of these show an extraordinary degree of polymorphism. At least some MHC allelic variation appears to be ancient and is maintained in multiple comtemporary species. Many of the alleles within a species differ by large numbers of nucleotide differences. The most extensive diversity is found in portions of *HLA* proteins believed to bind to antigens and comparisons between diversity in the antigen binding regions and adjacent regions suggested the significant participation of positive Darwinian selection in the evolution of the sequences (Hughes and Nei 1988, 1989).

This article briefly reviews the presently defined mechanisms of concerted mutagenesis. Simple models describing the production of concerted mutations during DNA polymerization reactions are presented. Because the models for concerted mutation predict the specific DNA sequence change produced, searches can be made among mutant sequences for examples consistent with the models. An example of how a mutational model can be used to search for base substitutions that may arise by a templated mechanism is presented, using *E. coli* mutant sequences. Finally, a subset of the database for mammalian MHC class II allelic polymorphisms is examined. This search revealed sets of DNA sequence motifs which are highly consistent with the hypothesis that they arose by templated, concerted mutagenesis. The data suggest that concerted mutagenesis may provide an excellent rationale for some characteristic of MHC allelic sequences that are poorly explained by alternative models.

The Characteristics of Concerted Mutagenesis

The most studied classes of concerted mutagenesis produce frameshifts, but the mechanisms responsible do not demand a frameshift as a final product of the event (Ripley, 1990). Concerted mutations are produced in single mutational events. However, at the DNA sequence level, the mutations resemble combinations of base substitutions and/or deletions and/or additions of adjacent or nearby nucleotides. We have regularly referred to these concerted changes as "complex" mutations, thereby avoiding names that describe the changes in terms of molecular events usually associated with other mutational mechanisms. Several different mechanisms produce complex, concerted sequence changes.

So far, the best studied class of concerted mutations are those in which concerted changes are brought about by templated DNA metabolism mediated by misaligned DNA intermediates. It is these mutations that will be considered here. However, it should be appreciated that apparently untemplated, concerted mutations may occur. It is not easy to distinguish untemplated processes from failure to identify the template. However, one hallmark of templated mutations is that the same DNA sequence changes are repeatedly produced (Ripley, 1982). Often, the DNA templates are nearby and can then be readily identified by inspection (de Boer and Ripley, 1984, Papanicolaou and Ripley, 1989).

Templated concerted mutations are chosen as an important focus for evaluating the potential role of concerted mutagenesis in evolution for two reasons. (1) Because templated concerted mutations can occur repeatedly with the same outcome, instances of separate occurrences in different lineages could present a particular challenge to evolutionary analysis. (2) The identification of a potential template for the concerted changes lends support to the concerted mutagenesis hypothesis. As will be seen, this support is particularly suggestive where correlations have been found between a sequence change in the suspected template and a complementary change at the site of concerted mutagenesis.

Misaligned DNA substrates as intermediates in mutagenesis

All of the concerted mutational mechanisms to be described are ectopically templated. That is, the DNA undergoes some misalignment during polymerization which ultimately leads to the wrong sequence in the right place or the right sequence in the wrong place, depending on your point of view. There is no requirement to postulate misincorporation or non-template-directed incorporation of nucleotides by the DNA polymerase to account for the mutant sequences.

The first mutations to be linked to DNA misalignments were frameshifts. Streisinger and co-workers (1966) proposed a misalignment hypothesis that accounted for the correlation between the DNA context of the most frequent frameshifts and the sequences of the frameshifts. The proposed misalignments could account for the fact that frameshifts observed in monotonic runs of A:T or G:C pairs were the deletion or addition of 1 bp (the repeat motif) in the run and for the fact that frameshifts in repeats that were longer than 1 bp were also additions or deletion of the repeat. The example in Figure 1 illustrates how a DNA misalignment during DNA polymerization could produce additions or deletions. The misalignment of DNA from the normal site of polymerization to a second locally complementary site, provides a substrate for templated polymerization. The product is identical to the

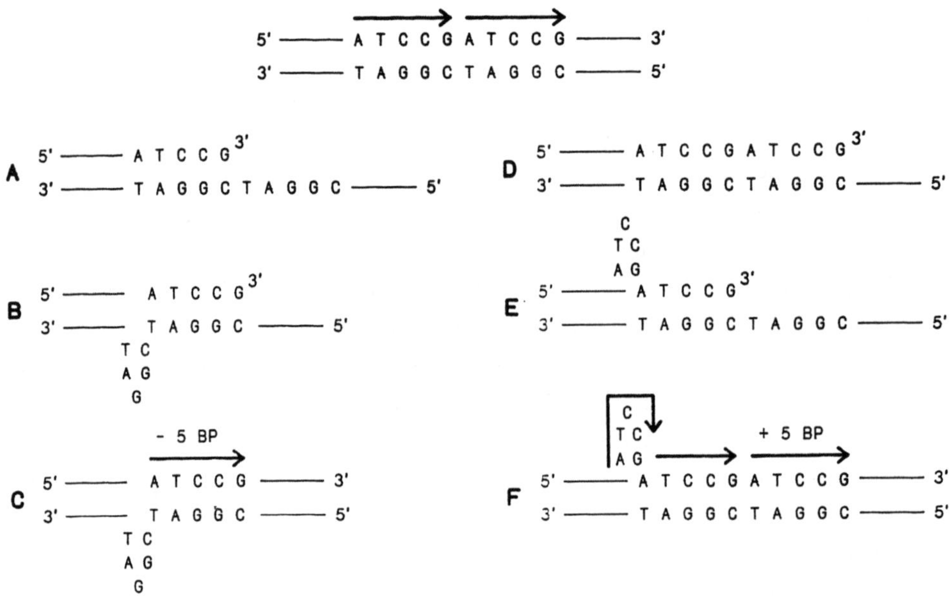

Figure 1. Misalignment in a direct tandem repeat sequence can account for deletions or duplications of one copy of the repeat.

The original tandem repeat of the sequence ATCCG is shown at the top. The pathway A->C describes how a misalignment of the newly synthesized strand during the course of polymerization could lead to the loss of one copy of the repeat (deletion of 5 base pairs). The pathway D->F describes how a different misalignment of the newly synthesized strand during the course of polymerization could lead to the duplication of one copy of the repeat (addition of 5 base pairs). These quite different mutational results are dictated by the relative direction of misalignment of the two DNA strands. Production of a loop-out in the template leads to a deletion, while production of a loop-out in the primer leads to a duplication. Although, as illustrated, one strand consistently serves as a template for the two pathways, a pair of related intermediates can account for the same mutations during DNA synthesis when the other strand is the template.

sequences observed in mutants. Evidence consistent with these misalignment models has accumulated (Farabaugh et al. 1978; Pribnow et al. 1981; Albertini et al. 1982; Singer and Westlye, 1988; Papanicolaou and Ripley, 1989). However, some internal contradictions suggested that there was room to accomodate additional mechanisms to explain some mutations in repeats (Glickman and Ripley, 1984; Streisinger and Owen, 1985, Ripley et al. 1986). Advances in defining the mechanisms of some mutagen-induced frameshifts clearly demonstrate that mechanisms other than the misalignments proposed by Streisinger can produce frameshifts in repeats (Burnouf et al. 1989; Masurekar et al. 1991).

As the exploration of frameshift mutagenesis expanded due to the ability to sequence larger numbers of mutants, increasing numbers of frameshift sequences could not be accounted for by the original Streisinger hypothesis. However, some of these sequences were found to be accounted for by related hypotheses that invoked different DNA misalignments.

Mutational mechanisms involving quasipalindromic misalignments

Misalignments permitted by the self-complementarity of palindromes were developed to account for DNA sequence changes, called complex framesifts, which were combinations of changed sequence and numbers of base pairs (Ripley, 1982; de Boer and Ripley, 1984). The DNA sequence changes could be accounted for by a DNA misalignment involving the self-complementary portions of the palindrome, followed by DNA polymerization in the misaligned configuration and realignment to the original position. Figure 2 shows how this model can account for a DNA sequence from bacteriophage T4. The quasipalindromic misalignment model is distinguished from the Streisinger misalignment model not only by the details of the DNA misalignment, but also by the requirement for realignment of the DNA back to the original site.

The model presented in Figure 2 postulates an intrastrand DNA misalignment to initiate the mutation. Plausible alternative interstrand misalignments can also explain this mutagenesis (Ripley, 1982; Ripley and Glickman, 1983). During *in vitro* DNA polymerization by *E. coli* polymerase I, both kinds of misalignments were detected, but frequent mutations were consistent with intrastrand misalignment and only rarely did mutant sequences require the interstrand misalignment hypotheses (Papanicolaou and Ripley, 1989). Regardless of the structural details of the misalignments involved, the possibility that palindromic misalignments might account for an observed DNA sequence change can be identified by searching the local DNA sequence for a second sequence which is exactly palindromic to the mutational change. The palindromic relationship between the mutant sequence and its template is shown by arrows in Figure 2. The hypothetical template should be partially complementary to the initial DNA sequence at the mutant site. This complementarity defines the pairing that can mediate the primer misalignment before misaligned polymerization.

It is not yet possible to predict which DNA sequences may be most frequently subject to palindromic misalignment mutagenesis. Factors other than the extent of the complementarity that initiates the misalignment are involved and discontinuities during DNA polymerization correlate to increased mutagenesis (Papanicolaou and

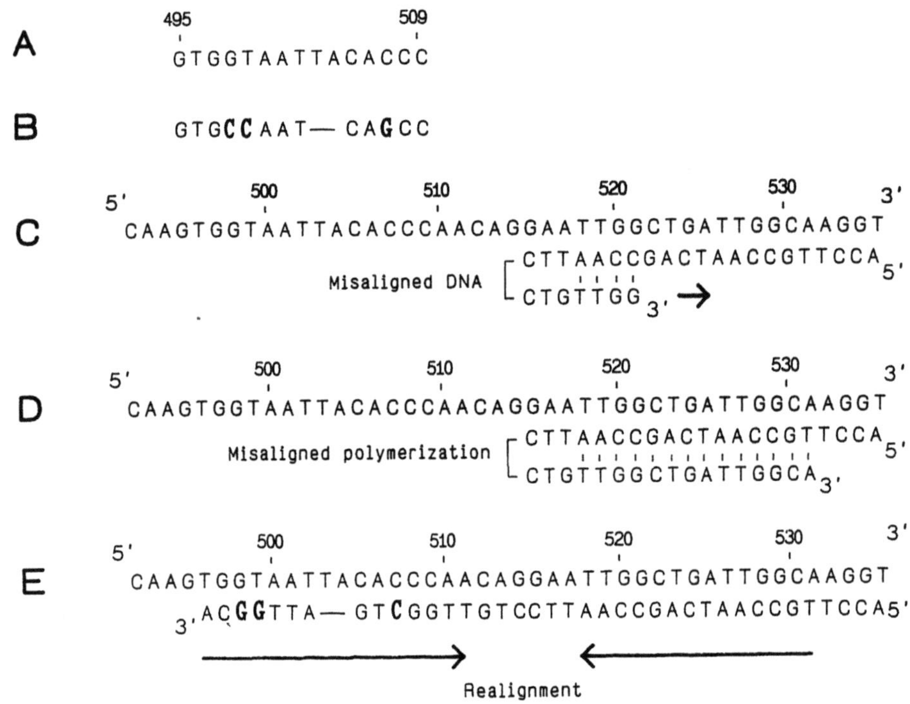

Figure 2. Palindromic DNA misalignments can account for concerted base changes and frameshifts.

The top sequence illustrated in A represents the starting sequence in the *rIIB* gene of bacteriophage T4. B illustrates a mutant sequence arising spontaneously (de Boer and Ripley, 1984). The mutant was selected for its frameshift phenotype; the base substitutions are phenotypically silent. The concerted nature of the mutagenesis leads to the recovery of the base substitutions. For example, these base substitutions have never been associated with any other example of spontaneous or induced frameshifts in the gene. The three substitutions, if independent, would be expected to occur in one sequences at a frequency of about 10^{-24}. Nonetheless, the base substitute and the frameshift predicted by this specific misalignment have been isolated several times. C, D and E illustrate how Misalignment, DNA synthesis and Realignment account for the concerted DNA sequence changes. The arrows in E illustrate the palindrome created by the mutation.

Ripley, 1991). Some, but not all DNA polymerases produce these mutations at high frequencies *in vitro* and different polymerases exhibit different specificities (Papanicolaou and Ripley, 1989, 1991; Kunkel et al. 1989).

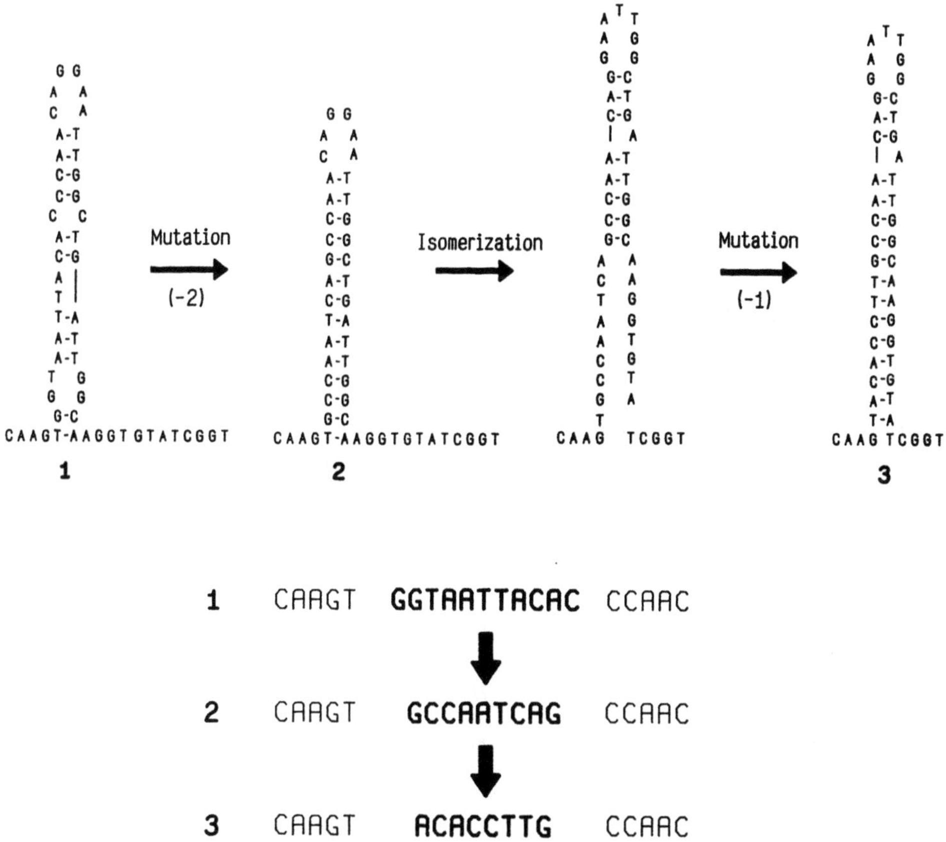

Figure 3. Sequential quasipalindromic mutations in T4 DNA sequences.

1, 2 and 3 indicate the initial, first mutant and second mutant sequences, respectively. The sequence at the concerted mutational sites are shown in palindromic formats at the top, which demonstrate the template for concerted mutation. When the concerted changes are shown in conventional format at the bottom, it is clear that the separation of these mutations by a single mutational step would not be apparent without consideration of the larger DNA context. Notice that mutation 3 wipes out all evidence of mutation 2. Moreover, sequence 2 could have been produced from sequence 3 as well as from sequence 1. Similarly, sequence 3 could have been produced from sequence 1 as well as from sequence 2. However, neither sequence 2 nor sequence 3 can be concertedly mutated to sequence 1 because there is no template for sequence 1.

Some DNA sequences frequently misalign at alternative sites *in vitro* (Papanicolaou and Ripley, 1991) and *in vivo* (Ripley, unpublished). For example, the mutant sequence shown in Figure 2 was subject to additional selection for more frameshifts. Among these new frameshifts was another quasipalindromically templated complex mutation, shown in Figure 3. An intriguing consequence of this final frameshift mutation is that it wipes out the first frameshift mutation that was its direct progenitor. If these mutations were encountered in an evolutionary context, their descent pattern could not be deduced. Certainly, the fact that the sequences are separated by only single mutations would not be a classical interpretation.

Mutational mechanisms depending on misalignments between quasidirect repeats

Other concerted sequence changes can be attributed to misalignments involving partially homologous, directly repeated sequences. In some cases in T4, the template may be quite far away. Figure 4 describes an example. Base substitutions (and a frameshift) are created at an upstream position consistent with the use of the downstream sequence as a template. Details of the interaction are not yet well-studied, but the templated nature of the mutation is strongly supported by the fact that we have isolated a large number (>50) of examples having this precise sequence change (Ripley et al. 1986; Ripley unpublished).

```
       Initial sequence
                                        (256 bp)
C G T C G T G| T T |T T G A A A A A T—//—C G T C G T G|C T G|T T G A A A A A T

C G T C G T G|C T G|T T G A A A A A T—//—C G T C G T G|C T G|T T G A A A A A T

      Concerted mutation                                  Template
```

Figure 4. Concerted mutagenesis directed by a quasidirect repeat in T4 DNA.

The initial upstream TT sequence is converted in a single step to a CTG sequence by mutation. The template for the change appears to be a sequence 256 bp downstream which shares considerable flanking complementarity.

The initial misalignments involved in quasidirect repeat mutagenesis are not distinguishable from the misalignments proposed by Streisinger. The distinction is that the DNA synthesized in the misaligned position does not remain there, but realigns back to the initial position. The requirement for realignment back to the initial position is also a requirement for the concerted mutations involving quasipaldindromic misalignments described above. *A priori*, it seems somewhat surprising that mutations involving two misalignment events are detectably frequent. It appears that DNA misalignments represent an important challenge to DNA fidelity.

Analysis of deletion (Glickman and Ripley, 1984) and frameshift (Schaaper et al. 1986) specificities in the *E. coli lacI* gene suggested that some mutant sequences were explained by misalignments mediated by quasidirect repeats. A search for evidence that mutagenesis templated by quasidirect repeats may be responsible for some base substitution mutations in the same gene is presented later. Just as in the case of quasipalindromes, the search uses the specificity of observed mutations to identify potential templates. The templates are predicted by the mutagenesis model to be exactly complementary to the mutant sequence, consistent with template function, and to be partially complementary to the initial DNA sequence with which it must misalign.

Rearranging frameshift intermediates to produce concerted base substitutions

In vitro studies identified hotspots for base substitutions that were adjacent to other hotspots for frameshifts (Kunkel and Alexander, 1986). Although the base substitutions and frameshifts do not arise concertedly, their high frequencies, adjacent locations and specificities suggested that misaligned pairing might offer an opportunity to convert the intermediates of frameshift mutagenesis to base substitutions or the intermediates of base substitution mutagenesis to frameshifts (Ripley, 1990). A general model for converting frameshift intermediates to base substitutions is shown in Figure 5. The essence of the dislocation model (the left half of Figure 5) is that a frameshift misalignment, represented by the dash in A, permits the incorporation of a cytosine. However, as shown in B, the return of the elongating strand to its original, aligned position and further elongation creates the base substitution. The net result, as shown in C is the T->G transversion.

In vitro studies at two sites clearly support the dislocation model (Kunkel and Soni, 1988), although no hypothesis yet describes why these events are frequent at some sites but not at other apparently similar sites. There is no theoretical requirement that dislocation be initiated from a frameshift that lies in a repeat (Ripley, 1990). The critical factor is that by definition the transient frameshift-heterozygote is a misaligned duplex because it contains an extra base in one strand compared to the other.

A literal extension of the dislocation model provides a way to think about how a frameshift mutation coupled to dislocation could be responsible for concerted base substitutions at adjacent sites. This model requires only that the misaligned synthesis postulated in the dislocation process be extended for an additional number of bases. A hypothetical mutation is illustrated in the right half of Figure 5. Notice that the number and identity of the mutated bases is defined entirely by the original position of the frameshift and the distance extended before realignment and further extension. Inspection of the final sequence change can identify the template but can not distinguish whether the initiating frameshift was a 1 base addition to the left or a 1 base deletion to the right of the concerted change. Dislocation involving +1 frameshifts has been recently suggested (Bebenek et al. 1989).

Searches for sequence changes that might be accounted for by dislocation are found by examining the local sequence. The identity of the substituted bases is defined by the template offered by the bases neighboring the mutation site. Substitutions that appear to be non-contiguous may in fact be part of the same mutational event when there are repeated bases in the sequence (Figure 5). The initiating frameshift might sometimes be larger than +1 or -1 base. For example +2 and -2 bp frameshifts occur at about 1/10 the frequency of 1 bp frameshifts both in vitro and *in vitro* (Ripley et al, 1986; de Boer and Ripley, 1988). However, in specialized sequences 2 bp frameshifts can be abundant (Freund et al. 1989).

It is in total now possible to describe three general mechanisms by which DNA misalignments, by serving as substrates for DNA polymerases, can lead to the production of single or concerted substitutions as well as complex frameshifts. Because the specific changes produced are so varied and are entirely determined by the local DNA context, it is possible that studies of mutagenesis may often underestimate the role of misalignment-mediated mutagenesis.

Do Misalignments Contribute to Base Substitution Mutagenesis in *E. coli*?

As pointed out above, spontaneous deletions and frameshifts in the *lacI* gene of *E. coli* sometimes have sequences that suggest their production by misalignments mediated by quasidirect repeats. Thus, this mutational mechanism may be an excellent prospect for uncovering heretofore unrecognized contributions of misalignment-mediated mutagenesis to spontaneous base substitution mutagenesis. Examples consistent with the possibility that quasidirect repeats might produce mutant sequences in *lacI* have previously been pointed out (Shinedling et al. 1987). However, no systematic examination of the hypothesis has been reported. It appears that the methyl-directed mismatch repair system of *E. coli* is highly effective in removing polymerization errors (Schaaper and Dunn, 1987). Thus, it may be that a large proportion of spontaneous mutations in this organism are those that escape mismatch repair. Perhaps mutagenesis mediated by quasidirect repeat DNA misalignments falls in this category.

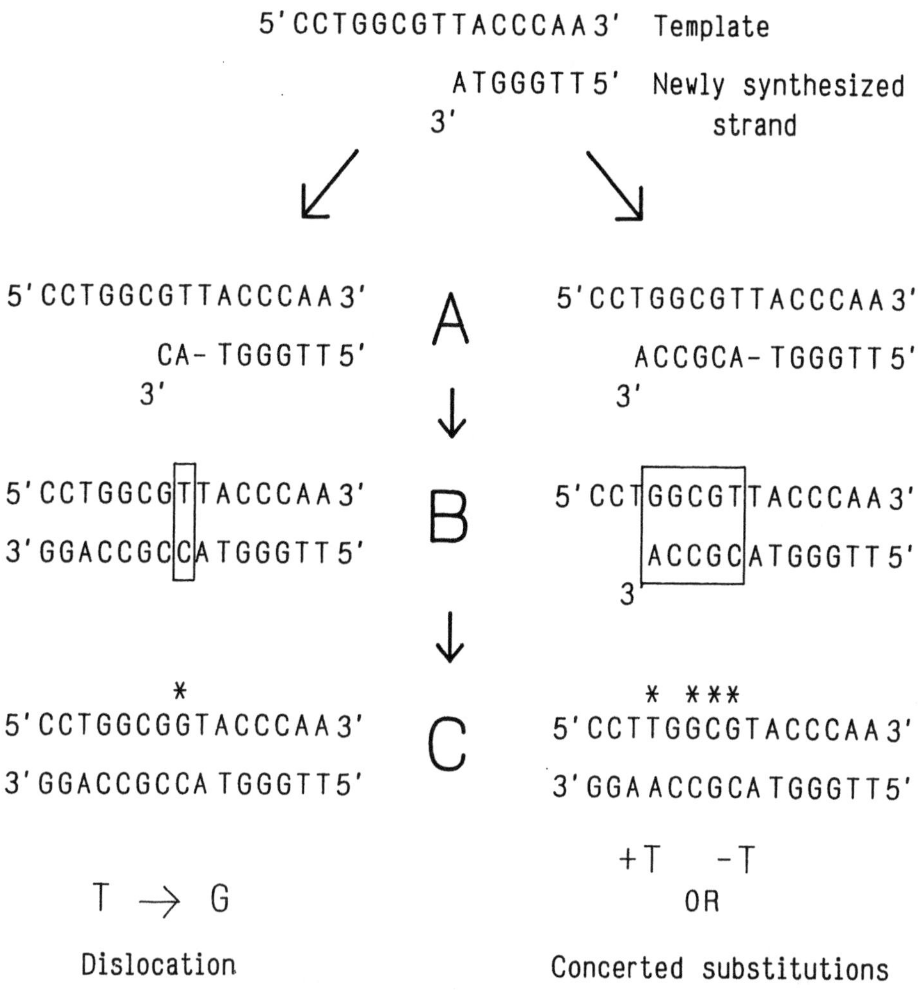

Figure 5. Dislocation mutagenesis and concerted substitutions.

At the left is illustrated the production of a T->G transversion mediated by DNA polymerization from a misaligned configuration involving a frameshift intermediate. Step A shows that the frameshift has deleted an A in the newly synthesized strand and that the polymerase has incorporated a templated C. In step B, realignment of the frameshift misalignment produces an intermediate which now has a base substitution rather than a frameshift. Resolution of the heterozygote to fully mutant form is illustrated in step C. * marks the substitution. At the right, the theoretical potential for dislocation to produce concerted substitutions is illustrated. The only difference is that DNA polymerization continues further before resolution of the frameshift misalignment as shown in step A. The resolution in step B now produces multiple changes which, when resolved in step C, produce multiple nearby changes in the sequence. * marks the detected substitutions.

If misalignments are creating some of the real, detected mutations, these should coincide with quasidirect repeat sequences more frequently than do random sequences. The survey presented below compared the ability of local quasidirect repeats to account for observed base substitutions in four separate studies of mutagenesis in the *E. coli lacI* system. These results were compared to the ability of the quasidirect repeats to account for randomly created sequence changes.

Random mutations were generated as described in Table 1 and compared to the sequences of identified *lacI* mutants (Gordon et al. 1988) to define their coincidence with real mutations. The 75 random mutations were selected in three groups of 25.

Table 1. The base pair substitutions in *E. coli lacI* sequences consistent with templated production by a nearby quasidirect repeat.

	Mutants	Total number	Extended[a] Misalignments
1.	Random mutants[b]	75	15 (20%)
	a. Undetected[c]	66	11 (16%)
	b. Detected	9	4 (44%)
2.	Spontaneous[d]	14	6 (43%)
3.	Spontaneous (*ung*$^-$)[e]	24	5 (20%)
4.	Spontaneous (*pol*A1)[f]	24	7 (29%)
5.	NMAM-induced[g]		
	a. G:C transitions	17	1 (6%)
	b. Remaining substitutions	13	6 (46%)

a Misalignments were initially sought out by computer, demanding a 5/6 bp match to the sequence at the mutant site. Such matches within 100 bp in either direction of the mutation were further examined for extended homologies, which were allowed to contain additional mismatches. Matches which could not be extended or which lie outside of the surrounding 200 bp were not considered in this analysis.

b Random mutations were generated using a random number to define the position of the sequence change. In a second step, the change itself was defined by random use of numerals 1-4 for A,C,G, and T, respectively. Each site was assigned 2 numerals. The first numeral was random, the second was random, except that it could not be identical to the first. When the base in the non-transcribed strand was the same as the base defined by the first numeral, the second numeral (base) was used.

c 9 of the random mutations coincide with DNA sequence changes that in fact have already been detected among *lacI* mutant sequences (Gordon et al. 1988).

d Mutant sequences from Schaaper et al. 1986.

e Mutant sequences from Fix and Glickman, 1987.

f Mutant sequences from Fix et al. 1987.

g Mutant sequences from Horsfall and Glickman, 1989.

None of the groups differed from the others in the frequency of correlation to quasidirect repeats. In one of the groups, only mutations in the N-terminal portion of the gene were considered. This was done because some of the mutational assays only evaluated sequences in this portion of the gene. However, because this group did not differ from the others, it is likely that the frequency of quasidirect repeats is not greatly different in this portion of the gene than it is in the entire gene.

Potential templates were identified by the existance of a 5 bp sequence (identified by computer search) that was perfectly complementary to the mutation. To be considered a template for the mutation the template was arbitrarily required to lie within 100 bp of the mutation, but no additional DNA, outside the genetic target of detectable mutations, was considered in instances where the mutation lay close to an end of the gene. Each potential template was also required to exhibit several additional bases of complementarity on at least one side of the 5 bp sequence. Such extension often included interruptions of perfect homology.

The results are summarized in Table 1. Overall, random mutations were associated with quasidirect repeats approximately 20% of the time. In contrast, spontaneous base pair substitutions were associated with quasidirect repeats 43% of the time. Although the sample size is small, but this increase is consistent with the suggestion that some spontaneous base substitutions are templated by quasidirect repeats. Further support was found when the sequences of mutants arising after perturbation of spontaneous mutagenesis by genes that participate in *E. coli* DNA metabolism were examined.

E. coli encodes a uracil-N-glycosylase enzyme that removes deaminated cytosine (uracil) from *E. coli* DNA. Mutations that inactivate this gene (*ung*$^-$) specifically increase G:C -> A:T transition mutations. This specificity is consistent with the incorporation of adenine opposite the uracil at the sites that formerly contained cytosine (Fix and Glickman, 1987). Clearly the increased mutations in the mutant should not be related to mutagenesis mediated by quasidirect repeats. Indeed, there is no correlation above random between base substitutions and quasidirect repeats in the spontaneous spectrum from the uracil-N-glycosylase mutant.

Inactivation of *E. coli* PolI (*polA1*) was also found to change the spontaneous mutagenesis specificity (Fix et al. 1987). The intermediate level of substitutions correlating to quasidirect repeat templates suggests that mutations increased by the *polA1* defect may include both classes.

Finally, mutations induced by N-nitroso-N-methyl-N-α-acetoxymethylamine (NMAM) were examined (Horsfall and Glickman, 1989). The induced base substitutions fall into two categories: (1) G:C -> A:T transitions which increase as a fraction of the spectrum with increasing dose of mutagen and (2) all other base substitutions which decrease as a fraction of the spectrum with dose. The absolute frequencies of both classes increase with dose. The two categories differ both in specificity and in dose response strongly suggesting that they arise by different mechanisms. The G:C site transitions in category 1 do not correlate to potential quasidirect repeat templates, consistent which a mutational mechanism in which the mutagen-induced base damage led to mispairing. However, the remaining substitutions in category 2 show a correlation well over random to the existence of templates which could account for some of these changes.

Altogether, the survey demonstrates that base substitutions arising by mechanism that are unlikely to involve misalignments (e.g. Table 1, 3 and 5a) share with random mutations a low correlation with potential quasidirect repeat templates. In contrast spontaneous base substitutions (Table 1, 2) and the miscellaneous category of NMAM-induced substitutions (Table 1, 5b) show a higher than random correlation with quasidirect repeats consistent with the view that templated mutagenesis directed by quasidirect repeats may contribute to some base substitution pathways in *E. coli.* These data cannot demonstrate the degree to which the substitutions that can be explained by the quasidirect repeat hypothesis actually arise by that mechanism, but represent an upper estimate for that fraction. The infrequent occurrence of multi-base substitutions in any of the spectra reported suggest that multi-base substitutions mediated by quasidirect repeats arise at lower frequencies than do single base substitutions. Whether such a difference could be due to decreased production of the mutation and/or to increased repair of the heterozygotes is unknown. Nonetheless, misalignments and templated mutations may contribute more widely to base substitution mutagenesis than has been previously suspected.

Concerted Mutagenesis and the Evolution of MHC

Comparisons of nucleotide substitutions at major histocompatability complex class II genes of humans and mice suggest that neutral mutations rarely contribute to allelic differences in the portions of the gene believed to be mediate antigen binding (Hughes and Nei, 1989). The characterization of both class I and class II MHC loci suggest that the rate of amino acid substitution in the antigen binding region is enhanced by positive Darwinian selection. However, the nature of this selection is somewhat controversial. It is not at all clear what drives the selection (Klein, 1987). It is argued that the selection can not be based solely on simple selection for advantageous mutations because such selection does not account for the large number of alleles and the long persistence of specific DNA motifs at these loci (Nei,

1987). These factors might be accounted for by overdominant selection (heterozygote advantage) (Hughes and Nei, 1988, 1989). Neither the nature nor extent of heterozygote advantage is known and some natural populations have not yet been shown to display heterozygosity for alleles that in mice and man show a great deal of heterozygosity (Klein, 1987; Nei and Hughes, 1991).

Inheritance of multiple alleles at speciation has been postulated to explain the observation that some alleles of distantly related species are more similar to one another than they are to other alleles within a species (Klein, 1987; Figueroa et al. 1988; Lawlor et al. 1988). Analyses of individual animals indicate that MHC loci are not subject to exceptionally high rates of mutation. Indeed, silent substitutions appear to have been fixed at lower rates in MHC loci than they have been in the globin gene family (Hughes and Nei, 1988).

It has been suggested that diversity is increased by allelic recombination in humans (Gorski and Mach, 1986, Wu et al. 1986; Gyllensten et al. 1991). In rodent studies (McConnell et al. 1988), recombination was tracked using a retroposon insertion in the intron between exons B1 and B2 of group I-αβ alleles in *Mus* species. Association of the retroposon sequence with alternative alleles was found, suggesting the possibility that recombination was responsible for exon shuffling.

Because much of the diversity in these *Mus* alleles appears to lie in the B2 exon, sequencing of this exon for a large number of alleles was carried out. The sequence analysis of allelic diversity in the B2 exon led to the suggestion that the role of recombination in generating allelic diversity should be extended to include intraexonic shuffling of many short DNA motifs (She et al. 1990). This analysis did not attempt to estimate recombination frequencies, but suggested that the accumulation of diversity by recombination might not be rapid (She et al. 1990).

The analysis by She et al. (1990) assumes that the short DNA sequence motifs are the result of accumulated base substitutions. Working under the constraints of this hypothesis the attractiveness of the recombinational transfer mechanism is sharply increased because the accumulation of precisely identical, but independent base substitutions in independent lineages is quantitatively unlikely. These accumulations are particularly unlikely when no alleles exhibiting partial accumulations are found and/or when some of the base differences in the motif are silent. Thus, if the motifs originally arise by the accumulation of substitutions, recombination of those motifs offers a parsimonious explanation for their repeated occurrence.

However, as we have seen, extended DNA sequence motifs may arise by concerted mutation and the frequencies of some concerted mutations are significant (Ripley et al. 1986). If some allelic sequence motifs are the product of concerted substitutions, rather than accumulated single substitutions, concerted mutagenesis may provide a parsimonious and quantitatively attractive explanation for the appearance of these motifs in otherwise different allelic contexts. An examination of rodent exon B2 sequences from the perspective of concerted mutagenesis is presented below.

```
      1                                2
Allele                          Sequence
------       -----------------------------------
             54                                88
             |                                  |
 dom6        CGCATACGGCTCGTGACCAGATACATCTACAAC
 (10)         R  I  R  L  V  T  R  Y  I  Y  N

 dom5        CGCATACGGCTTGTGACCAGATACATCTACAAC
 (3)          R  I  R  L  V  T  R  Y  I  Y  N

 spc4        CGCATACGGCTCGTGACCAGATGCATCTACAAC
 (1)          R  I  R  L  V  T  R  C  I  Y  N

 mol1        CGCATACGGCTCGTGATCAGATACATCTACAAC
 (4)          R  I  R  L  V  I  R  Y  I  Y  N

  k          CGCATACGGCTTGTGATCAGATACATCTACAAC
 (1)          R  I  R  L  V  I  R  Y  I  Y  N

 dom9        GCGATACGGCTCGTGATCAGATACATCTACAAC
 (1)          A  I  R  L  V  I  R  Y  I  Y  N

 dom13       CGCATACGGCTCGTGAGCAGATACATCTACAAC
 (1)          R  I  R  L  V  S  R  Y  I  Y  N

 dom12       CGCATACGGCTCGTGAGCAGAAACATCTACAAC
 (1)          R  I  R  L  V  S  R  N  I  Y  N

 car2        CGCATACGGCTCGTGGCCAGAAACATCTACAAC
 (1)          R  I  R  L  V  A  R  N  I  Y  N

 dom14       CGCATACGATCTGTGACCAGATACATCTACAAC
 (1)          R  I  R  S  V  T  R  Y  I  Y  N

 dom1        CGCATACGATCTGTGAACAGATACATCTACAAC
 (9)          R  I  R  S  V  N  R  Y  I  Y  N

 dom4        CGCATACGATCTGTGAACAGATTCATCTACAAC
 (1)          R  I  R  S  V  N  R  F  I  Y  N

 dom11       CGCATACGATCTGTGGACAGATACATCTACAAC
 (5)          R  I  R  S  V  D  R  Y  I  Y  N

 mus4        CGCATACGAGCTGTGGACAGATACATCTACAAC
 (1)          R  I  R  A  V  D  R  Y  I  Y  N

 dom7        CGCATACGTTCTGTGACCAGAAACATCTACAAC
 (1)          R  I  R  S  V  T  R  N  I  Y  N

 crv1        CGCATACGAGGTCTGACCAGATACGTCTACAAC
 (1)          R  I  R  G  L  T  R  Y  V  Y  N

 cok1        CGCATACGACATGTGACCAGATACATCTACAAC
 (1)          R  I  R  H  V  T  R  Y  I  Y  N
```

```
ptx1     CGCATACGGCATGTGAGCAGTTACATCTACAAC
(1)       R  I  R  H  V  S  S  Y  I  Y  N

car1     CGCATACGTATAGTGACCAGATACATCTACAAC
(1)       R  I  R  I  V  T  R  Y  I  Y  N

 b       CGCATACGATATGTGACCAGATACATCTACAAC
(3)       R  I  R  Y  V  T  R  Y  I  Y  N

spt1     CGCATACGATATGTGACCAGAAACATCTACAAC
(1)       R  I  R  Y  V  T  R  N  I  Y  N
```

Figure 6. DNA and amino acid sequences from the BS2 segment of exon 2.

[1]An example of one allele is given for each sequence. The total number of alleles in the sample that are identical to the illustrated sequence is indicated in ( ).
[2]Sequences are all from She et al. (1990). No differences in any *Mus* sequence was found between 41 and 54 or between 88 and 90. Rat alleles are not shown; see Figure 8.

The conserved motifs of exon B2 and concerted mutagenesis

A substantial fraction of the variability in class II gene alleles is found in a portion of the second exon of the β gene (exon B2) sequence that is believed to correspond to part of the antigen-binding cleft (Brown et al. 1988). Although the actual polypeptides that bind to all of these alleles are unknown, it has been demonstrated that alternative alleles preferentially bind alternative groups of polypeptides (Sette et al. 1989). Altogether, these characteristics provide a rationale for predicting that the highly variant regions of exon B2 could be subject to positive selection on the basis of antigen binding.

The analysis by She et al. divided the 273 base pair sequence of exon B2 into five segments. In each segment five to eleven alternative sequence motifs were defined. In many instances these motifs were found in more than one of the 46 alleles studied. In other instances, the motifs were unique multiple base changes. Sometimes alternative motifs in a segment differ more from one another than they do from the consensus. Some motifs are shared by alleles from species believed to be separated by a minimum of 10 Myr. Because such a time frame would have encompassed multiple speciations, these motifs are argued to represent trans-species evolution. However, because some multiple base motifs are unique or occassional, more recent occurrences of multiple base change motifs appear to be possible.

The search for concerted mutations focused on one segment (BS2) of the exon (She et al. 1990). The BS2 segment lies between bases 41 and 90 in the exon (Figure 6) and is believed to correspond to a portion of a β-pleated sheet structure lying at the base of the antigen binding cleft (Brown et al. 1988). BS2 was chosen for the search because of its extensive diversity and because it is at neither terminus of the exon. Complete analysis of the 5' and 3' most segments was not undertaken because of the unavailability of the neighboring intron sequences for many alleles.

The ATCT motif

The ATCT motif at bp 62-65 was found 16 times, often in association with different combinations of additional substitutions at 69, 70, and 75 (Figure 6 ***dom14***, ***dom1***, ***dom4*** and ***dom11***). The ***dom14*** allele, illustrated in Figure 7, contains only the ATCT motif. The initial A of the motif is silent. All rodent alleles in the sample have an arginine codon (CGX) at this site. The remaining three changes produce a TCT serine condon. Although TCX is the codon for serine, no silent third position changes are found in any of the 16 alleles.

These four substitutions are adjacent transitions when compared to the consensus, which for these 4 nucleotides is represented by 25 alleles. The high proportion of variation that is associated with this single motif within the exon becomes apparent when the remainder of the ***dom14*** allele is compared to the consensus sequence. The ***dom14*** allele differs from the consensus in the entire exon by only four additional substitutions (excluding a highly conserved deletion motif in the alpha-helix region). None of the other substitutions are adjacent, and none are silent; all are transversions that produce an amino acid substitution.

The substitution of four consecutive bases at positions 62-65 in the ***dom14*** sequence can be templated by a concerted mutational mechanism involving a quasipalindromic sequence. A nearby template is provided by the sequence at 72-75 (Figure 7). Each of the sixteen different rodent alleles having the ATCT motif at position 62 to 65 also has the AGAT template. This data does not address how many independent mutational events may be represented by the 16 alleles in question, but it clearly indicates the possibility that the number may be larger than one.

The sequence of the ***dom7*** allele (Figure 7) suggests that the number of concerted mutations is larger than one. In ***dom7***, the motif is TTCT rather than ATCT. The A->T difference between the ATCT and TTCT motifs is silent. However, it is not necessary to hypothesize that this silent transversion was simply accumulated. In fact, the TTCT motif is exactly the change predicted by the sequence lying in the proposed template (position 72 and 75). In the ***dom7*** allele, a transversion at 75 (producing a substitution of threonine for tyrosine) has occurred. This transversion provided the template for the silent transversion observed at 62.

Although other explanations for the sequences of ***dom7*** and ***dom14*** are possible, concerted mutagenesis clearly provides a highly parsimonious hypothesis. The explanation offered for the ***dom7*** sequence implies that in this allele, concerted mutation occurred after the transversion mutation at 75. This ordered interpretation may ultimately be able to provide a powerful approach to deducing lineage relationships that take into account the characteristics of concerted mutagenesis.

	55 70 87
	|R I R L V |T R Y I Y N|
Rodent consensus[1]	CGCATACGGCTCGTGACCAGATACATCTACAAC
	R I R **S** V T R Y I Y N
M. m. domesticus dom 14	CGCATACG**ATCT**GTGACCAGATACATCTACAAC
	⟶ ⟵
	R I R **S** V T R **N** I Y N
M. m. domesticus dom 7	CGCATACG**TTCT**GTGACCAGA**A**ACATCTACAAC
	⟶ ⟵
	R I R **G** **L** T R Y **V** Y N
M. cervicolor popaeus cru 1	CGCATACG**AGGTC**TGACCAGATAC**G**TCTACAAC
	⟶ ⟵
	R I R **S** V **N** R Y I Y N
M. m. domesticus dom 1[2]	CGCATACG**ATCT**GTGA**A**CAGATACATCTACAAC
	⟶ ⟵
	R I R **S** V **D** R Y I Y N
M. m. domesticus dom 11[3]	CGCATACG**ATCT**GTG**GA**CAGATACATCTACAAC
	⟶ ⟵

Figure 7. Concerted production of ATCT and related motifs in *Mus* alleles.

[1]The illustrated concensus is the concensus from the alleles examined by She et al. (1990). DNA sequence changes and amino acid sequence changes when compared to this consensus are highlighted. Arrows mark the palindromic relationship between proposed template and concerted sequence changes.

[2]As shown in Figure 3 this ***dom1*** sequence motif is shared by 8 additional alleles (***q, p, dom2, dom3, dom8, mus1, mus6, spt3***).

[3]As shown in Figure 3 this ***dom11*** sequence motif is shared by 4 additional alleles (***s, f, cas2, spc3***).

Thus, lineage predictions might now place ***dom7*** into any closely related lineage, without concern that the lineage include the ATCT or TTCT motif, because the TTCT change may represent only a single concerted mutation.

Other motifs overlapping in position with the ATCT motif include that seen in the ***crv1*** allele from ***M. cervicolor popaeus***, a Thai mouse (Figure 7). Mutations in this motif differ from the consensus at 5 consecutive positions and from ***dom14*** by 4 consecutive changes. ***crv1*** has an AGGTC motif rather than the ATCT(G) motif of ***dom14***. The AGGTC can be produced from the ***dom14*** motif by three base substitutions (T->G at 63, C->G at 64 and G->C at 66). However, a concerted mutation, templated by a quasipalindrome can produce all three substitutions. The template as shown in Figure 7 includes bp 68-72. Thus, the AGGTC motif could be produced from the consensus by two mutations, or from a ***dom14***-like allele in a single mutation. Thus, just as we have observed for mutation in bacteriophage T4 (Figure 3), consecutive or alternative concerted mutations offer the opportunity for extensive diversity from only two mutational events. The sequences of additional alleles from ***M. cervicolor popaeus*** might be quite interesting in probing the number of independent concerted mutations that may occur in this region.

Motifs shorter that those of four or five base pairs may be produced by concerted mutations. Nine alleles that have the ATCT motif also have a C->A transversion at 70. Four more alleles have this change and an A->G transition at 69. Because these single or double substitutions produce amino acid substitutions, there is no clear quantitative or selective advantage to evoking a templated mechanism to explain the changes. However, this portion of the BS2 region clearly yields diverse sequences by alternative misalignments, and thus the possibility for explaining the substitutions at 69 and 70 by templated mechanisms was also explored.

The ***dom1*** allele (Figure 7) is an example of an allele having both the C->A transversion at 70 and the ATCT motif at 62-65. Notice that the transversion at 70, extends the quasipalindromic relationship between the template and the concertedly produced ATCT motif. If the mutation at 70 preceded the concerted production of the ATCT motif, its presence would increase the number of paired bases in the misaligned DNA that mediates the mutation. If many ATCT motifs arise by independent mutational events, this change might be frequently associated with the ATCT change because it increased the frequency of concerted mutation. Thus, those alleles having the ATCT change would be likely to have the C->A transversion. Alternatively, the C->A transversion might arise after concerted mutagenesis. Indeed, a second instance of misalignment-mediated base substitution mutagenesis could be proposed, in which the smaller palindrome now produces a DNA intermediate which extends the palindrome at 70 using as a template the bases at 66 and 67. [The orientation of this change demands an interstrand (strand-switch) misalignment (Ripley, 1982; Papanicolaou and Ripley, 1989) rather than the intrastrand misalignment presented in Figure 2.]

The addition of a G transition at 69 (***dom11***) to the A change at 70 (***dom1***) might represent an accumulated sequence change, perhaps driven by the asparagine to aspartate change. However, it is worth noting that a concerted GA change could be produced as a single dislocation mutation. Starting with the consensus sequence, the double substitution mutation could be initiated by the deletion of a G opposite C at either 70 or 71, followed by templated synthesis to position 72, followed by realignment and further synthesis, just like the mutation described in Figure 5.

The concerted appearance of an AC -> GA change at the 69-70 site could alternatively be directed by a quasidirect repeat template. The template lies at position 118 and is GTGgaC, where the lower case bases template the substitutions. This template sequence is not a characteristic of the consensus, but is found in several alleles, namely ***ptx1***, ***dom10*** and ***mus4***. The change to GA at 69-70 is found in ***mus4*** and thus offers a potential explanation for the mutation in ***mus4*** by a mechanism distinct from those offered for other alleles. The apparently widespread but potentially independent occurrence of GA motifs might be accounted for by the concerted production of GA sequences which are then prone to acquiring the ATCT motif because of the enhanced palindromic pairing provided by the A as described above.

Concerted mutagenesis and rat allele motifs

Rat sequences appear to differ substantially from mouse sequences in this portion of the B2 exon (She et al. 1990). However, many of the differences are accounted for by closely linked base pair substitution mutations that can be explained by concerted mutational mechanisms.

Two rat alleles, ***rr1*** (Figure 8) and ***rr2*** (not shown), display a TATC motif at 63-66. When compared to the rodent consensus (dominated by ***Mus*** alleles) the 4 substitutions change the amino acid sequence from leucine-valine to tyrosine-leucine. The rat sequence can be templated from a nearby quasipalindrome. The misalignment is closely related to that hypothesized to produce the ATCT motif in mice (Figure 8). Thus, the misalignment responsible for the ATCT motif could also be produced from the ***rr1*** sequence, and identification of other rat alleles with that motif would come as no surprise. Indeed, because the mutation proposed to account for TATC would remove all traces of ATCT, it is even possible that ATCT might have been a direct precursor of TATC.

The *rn2* rat allele carries an AGT motif rather than ATCT or TATC (Figure 8). This motif, like the TTCT motif of *dom7* (Figure 7), is complementary to a mutation in the upstream template sequence. A likely template can be seen in the *rr3* sequence ACACT (75-79) which has not yet accumulated the G change at 77 characteristic of *rn2*. Of some interest is the observation that the AGT motif encodes serine just as does the ATCT motif. Perhaps whatever selection leads to accumulation of ATCT motifs in mice has alternatively led to AGT motifs in rats. More rat sequences would be useful, but allelic diversity in this genus also appears to include motifs that can be produced by templated, concerted mutagenesis.

The idea that concerted mutagenesis can also provide an explanation for allelic variation in humans can also be preliminarily suggested. An example drawn from the human DQβ gene is illustrated (Figure 8). This human gene sequence is closely related to the *Mus* alleles considered here. Again, the DNA sequences suggest that quasipalindromic mutations are responsible for amino acid variation in exactly the same portion of the sequence as seen for rats and mice. Notably, the concerted sequence changes produced at positions 63-66 were not represented among rodent sequences and indeed there is no nearby template in rodents. However, the human gene carries a silent polymorphism at 84 that can template the observed change exactly. An important role for selection in the accumulation of this allele could be suggested by the fact that both the rat and human alleles have tyrosine (TAT) codons at the same site that seem to be produced by different concerted mutational pathways.

Additional motifs between 40 and 90 in the B2 exon

Some lineages defined in this region were defined on the basis of a single nucleotide change. Specifically, five alleles have a C->T transition at position 70; (Figure 6, *mol1* and *dom9*) three other alleles have a C->T transition at position 65 (Figure 6, *dom5*). A single allele *k*, has both changes. In view of the relatively high frequency of C->T transitions among spontaneous mutants, it may not be necessary to assume that the combination of transitions in allele *k* would most likely have arisen by recombination, as suggested by She et al (1990). In particular, it can be noticed that at position 65, the initial C is part of a CpG doublet context. Thus, C 65 represents a site that is often methylated in mammalian DNA sequences. Methylated cytosines are routinely found to be hotspots for transitions in both bacteria and mammalian cells (Duncan and Miller, 1980). The mutation at position 65 produces no amino acid change, so it is unlikely to be directly selected. Because mutation at position 65 can be predicted to be extraordinarily high, the accumulation of this change may well occur independently in a variety of lineages.

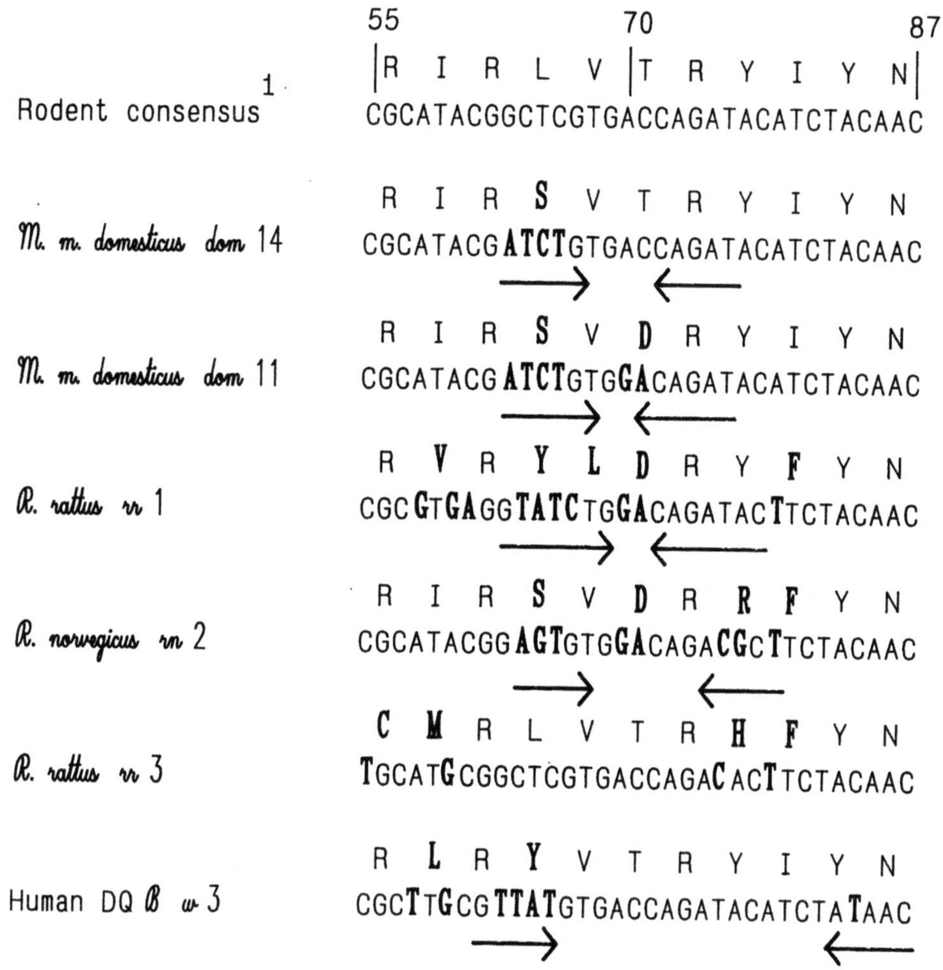

Figure 8. Concerted production of mutations in mouse, rat and human alleles.

[1]The illustrated consensus is from alleles examined by She et al. (1990).

In the analysis of She et al. (1990), the *dom9* allele was relegated to its own miscellaneous category of lineage between nucleotides 41 to 90 because of a unique GCG motif at positions 54-56. Although, it may be that the *dom9* allele accumulated these substitutions individually, all are transversions and are thus anticipated to be rare. The three base change can be explained by concerted mutagenesis. Both dislocation and quasirepeat mechanisms offer potential explanations. The quasirepeat, a GGCGA template (120-124), is in both ***dom9*** and the consensus sequence. Extended, but imperfect, additional complementarity is available to stabilize the misaligned pairing of the DNA intermediate. Another copy of this template is in the consensus sequence at position 20, but this copy is not present in ***dom9***.

A sample of additional motifs attributable to concerted mutagenesis in exon 2

The potential for concerted mutagenesis to account for the DNA sequence motifs in exon B2 are found at numerous sites within the exon and are not limited to the BS2 segment analyzed here. For example, an ACTC motif occurring in ***dom5*** and ***nod*** at position 154 appears to be templated by a palindrome. A GG motif at positions 97-98 is found in 14 alleles and can be templated by a nearby quasidirect repeat. Several conserved motifs in the first 30 bp of the exon appear in multiple alleles and can be explained by concerted mutagenesis involving nearby intron sequence.

Without suggesting that all consecutive substitutions are necessarily concerted, it is possible to strikingly reduce the number of mutational events required to account for many DNA sequence motifs seen uniquely or repeatedly in the rodent class II gene alleles examined. An evaluation of likely lineage relationships taking into account the possibility of contributions by concerted mutagenesis would seem to be appropriate. Descriptions of the new parsimonious trees might bring into clearer focus, the degree to which intraexonic recombination over very short distances would continue to offer an attractive explanation for the distribution of short sequence motifs among the allelic sequences of rodents.

Discussion

Our initial studies of templated, concerted mutagenesis were spurred by finding mutant sequences that were clearly unexplained by any theory of mutation. The theories subsequently developed to explain these concerted mutations have been conversely applied in the studies described here. That is, the specificity predictions of the mutational theories were used to see if templated, misalignment-mediated mutagenesis could offer novel explanations for the sequence changes which had previously been assumed to be due to untemplated, individual misincorporation of DNA bases.

Some spontaneous and induced single base substitutions in ***E. coli*** were found to correlate to the presence of nearby quasidirect repeats more often than do randomly generated substitutions or transition mutations produced by misincorporation of DNA bases. The result suggests that a subset of spontaneous single-base substitutions may arise by templated mechanisms. *E. coli* presents an experimentally accessible system in which this proposition might be directly tested.

Even if the contribution of templated processes to spontaneous mutagenesis has been underestimated to this point, it is clear that the rate of concerted, templated spontaneous mutagenesis is low when compared to total product of spontaneous mutagenesis. Nonetheless, the impact of this mutagenesis on the molecular evolution of some DNA sequences may be large for several reasons.

One important potential for impact becomes more clear when mutation rates are compared as a function of extent of DNA sequence change produced. For example, in bacteriophage T4, a base substitution at a particular site can be found in a typical laboratory population at a site-specific frequency in the range of 10^{-9} to 10^{-7}. The three substitutions produced by the mutation illustrated in Figure 2, might thus be expected to occur at a frequency of about 10^{-24}. Independent production of the 2 bp deletion would clearly add another significant factor, at least 10^{-9}. In comparison with an expected frequency of 10^{-33} for independent mutations, the frequency of concerted mutations, in the range of 10^{-10} or 10^{-11}, is enormous. Thus, in instances when multibase changes can be produced by concerted mutation, and where they are not subject to negative selection, their contribution to the accumulation of DNA sequence changes in populations may be much larger than generally appreciated.

The impact of concerted mutagenesis on the molecular evolution of DNA sequences may be especially strong under certain selection conditions. The superior ability of concerted mutations to change amino acid sequences provides a rationale for a large contribution by concerted mutations to DNA sequences subject to positive Darwinian selection. The same argument predicts a small contribution of concerted mutation to neutrally accumulated sequence changes in protein coding regions. Limit estimates of the relative frequencies of concerted and individual sequence changes might be obtainable from sequence comparisons in DNA regions not expected to be under selection pressures based on the local sequence.

If concerted mutations contribute to allelic sequence diversity, classically computed estimates of mutation rates and lineage distances have been overestimated. Even if the total overestimation turns out to be modest, a reinterpretation based on mixtures of single and concerted mutations might lead to modified conclusions. This is because the overestimated rates are not uniformly distributed through the gene, but are overestimated, by definition, primarily in the most variable regions. For example, standard calculations of mutation rates in variable regions of different MHC sequences present a considerable range of ratios for synonymous and non-synonymous substitutions within variable regions (Nei and

Hughes, 1991). Analyses that take into account the fact that concerted substitutions may simultaneously produce synonymous and selected substitutions may give a clearer quantitative picture of variability.

Many DNA motifs characteristic of allelic diversity in the proposed antigen binding region of the class II β genes catalogued by She et al. (1990) are shown here to be highly consistent with the hypothesis that they arise by concerted, templated mutations. The general relevance of the hypothesis is supported by finding similar, concerted explanations for very different sequences in various mouse, rat and human alleles in orthologous positions (Figure 8).

Of some potential importance to evolutionary considerations is the ability of concerted, templated mutagenesis to provide a mechanistic explanation for convergent evolution of sequence motifs. A motif produced by a concerted change is possible in all lineages that maintain the template. Thus, not all like motifs need be directly related by descent or by recombinational mixing of ancient mutant motifs. It must still be determined to what degree the standard approaches to predicting lineage relationships of MHC alleles may have been compromised by mutationally-mediated convergence. Although mutationally-mediated convergence might pose some problems for the determination of lineage relationships, it may also provide some powerful new tools for determining lineage relationships. Examples like the *dom7* and *rn2* alleles, in which complementary changes in the template and at the site of concerted mutation are found, may provide clarifying views of lineage.

An important general property of concerted mutations, as described in detail for quasipalindromes, is the ability of a second concerted mutation to wipe out all or most evidence of the first mutation. We have demonstrated this directly at the mutational level in bacteriophage T4 (Figure 3) and an MHC example consistent with this is the AGGTĊ motif of the *crv1* allele in Figure 7. This points out an important difference between concerted mutations and accumulated substitutions. Alternative motifs at one site could be routinely shuffled in a particular lineage as long as the templates are maintained, and selection is available to fix the change. This shuffling is not quantitatively expected for accumulated mutations. Concerted mutagenesis represents a potentially effective, certainly parsimonious approach to creating a mix of selectable motifs and may turn out to offer an attractive alternative mechanism to non-reciprocal allelic recombination to explain observed patterns of allelic diversity. One attraction may be the possibility of reducing the number of alleles required at speciation (Klein, 1987) to account for present day patterns of allelic diversity.

Clearer identification of the instances of concerted mutagenesis should provide improved distinctions between DNA motifs that are most likely to be trans-genically inherited from those most likely to be newly arising. The resulting improvement in the understanding of allelic lineages and the relative contribution of mutation and recombination to the generation of allelic diversity should provide important tools for dissecting the evolutionary mechanisms that are responsible for the characteristics of MHC loci. A full consideration of the impact of mutational specificity, especially concerted mutagenesis, on allelic divergence may provide new clues to the selective forces that drive the unusual diversity of MHC genes.

Acknowledgements

I would like to acknowledge the considerable education in evoluntionary issues related to the MHC loci provided by Drs. J. Klein and M. Nei. I am deeply greatful for the generosity of Dr. E. Wakeland and his colleagues in sharing their data, spurring my specific analysis of these sequences with a challenge, and enthusiastically awaiting the results. I thank M. Conkling, J. Drake, V. Kaiser and J. Wilusz for critically reading the manuscript. I apologize to any scientists whose contributions to prevelant theories of MHC evolution addressed here may not have been properly acknowledged. I encourage you to make me aware of my errors of omission. Computer support for the work reported was provided in part by the Academic Computing Center of the University of Medicine and Dentistry of New Jersey. The manuscript was prepared by Nettie Cleveland, Tracy Mann, and Diane Muhammadi. The body of research from my laboratory relevant to the mutational mechanisms described in this manuscript has been funded by the American Cancer Society and the NIH.

References

Albertini, A.M., Hofer, M., Calos, M.P., and Miller, J.H.: On the formation of spontaneous deletions: the importance of short homologies in the generation of large deletions. *Cell 29*:319-328, 1982

Bebenek, K., Abbots, J., Roberts, J.D., Wilson, S.H., and Kunkel, T.A.: Specificity and mechanism of error-prone replication by Human Immunodeficiency Virus-1 reverse transcriptase. *J Biol Chem 264*:16948-16956, 1989

Bowcock, A.M., Kidd, J.R., Mountain, J.L., Hebert, J.M., Carotenuto, L., Kidd, K.K., and Cavalli-Sforza, L.L.: Drift, admixture, and selection in human evolution: A study with DNA polymorphisms. *Proc Natl Acad Sci USA 88*:839-843, 1991

Brown, J.H., Jardetsky, T., Saper, M.A., Samraoui, B., Bjorkman, P.J., and Wiley, D.C.: A hypothetical model of the foreign antigen binding site of class II histocompatibility molecules. *Nature 332*:845-850, 1988

Burnouf, D., Koehl, P., and Fuchs, R.P.P.: Single adduct mutagenesis: Strong effect of the position of a single acetylaminofluorene adduct within a mutational hotspot. *Proc Natl Acad Sci USA 86*:4147-4151, 1989

Cox, E.C. and Yanofsky, C.: Altered base ratios in the DNA of an Eschericia coli mutator strain. *Proc Natl Acad Sci USA 58*:1895-1902, 1967

de Boer, J.G. and Ripley, L.S.: Demonstration of the production of frameshift and base-substitution mutations by quasipalindromic DNA sequences. *Proc Natl Acad Sci USA 81*:5528-5531, 1984 de Boer, J.G. and Ripley, L.S.: An *in vitro* assay for frameshift mutations: Hotspots for deletions of 1 bp by Klenow-fragment polymerase share a consensus DNA sequence. *Genetics 118*:181-191, 1988

Duncan, B.K. and Miller, J.H.: Mutagenic deamination of cytosine residues in DNA. *Nature 287*:560-561, 1980

Farabaugh, P.J., Schneissner, U., Hoffer, M., and Miller, J.H.: Genetic studies ont he *lac* repressor. VII. On the molecular nature of spontaneous hotspots in the *lac I* gene of *Escherichia coli. J Mol Biol 126*:847-863, 1978

Figueroa, F., Günther, E., and Klein, J.: MHC polymorphism pre-dating speciation. *Nature 335*:265-267, 1988

Fix, D.F. and Glickman, B.W.: Asymmetric cytosine deamination revealed by spontaneous mutational specificity in an Ung$^-$ strain of *Escherichia coli. Mol Gen Genet 209*:78-82, 1987

Fix, D.F., Burns, P.A., and Glickman, B.W.: DNA sequence analysis of spontaneous mutation in a PolA1 strain of *Escherichia coli* indicates sequence-specific effects. *Mol Gen Genet 207*:267-272, 1987

Freund, A.-M., Bichara, M., and Fuchs, R.P.P.: Z-DNA-forming sequences are spontaneous deletion hotspots. *Proc Natl Acad Sci USA 86*:7465-7469, 1989

Glickman, B.W. and Ripley, L.S.: Structural intermediates of deletion mutagenesis: A role for palindromic DNA. *Proc Natl Acad Sci USA 81*:512-516, 1984

Golding, G.B. and Glickman, B.W.: Sequence-directed mutagenesis: Evidence from a phylogenic history of human α-interferon genes. *Proc Natl Acad Sci USA 82*:8577-8581, 1985

Gordon, A.J.E., Burns, P.A., Fix, D.F., Yatagai, F., Allen, F.L., Horsfall, M.J., Halliday, J.A., Gray, J., Bernelot-Moens, C., and Glickman, B.W.: Missense mutation int eh lac I gene of *Escherichia coli.* Inferences on the structure of the repressor protein. *J Mol Biol 200*:239-251, 1988

Gorski, J and Mach, B.: Polymorphism of human Ia antigens: gene conversion between DR B loci result in a new HLA-D/DR specificity. *Nature 322*:67-70, 1986

Gyllensten, U.B., Sundvall, M., and Erlich, H.A.: Allelic diversity is generated by intraexonic sequence exchange at the DRB1 locus of primates. *Proc Natl Acad Sci USA 88*:3686-3690, 1991

Hampsey, D.M., Ernst, J.F., Stewart, J.W., and Sherman, F.: Multiple base-pair mutations in yeast. *J Mol Biol 201*:471-486, 1988

Horsfall, M.J. and Glickman, B.W.: Mutational specificities of environmental carcinogens in the lac I gene of *Escherichia coli.* I. The direct-acting analogue N-nitroso-N-methal-N-α-acetoxymethylamine. *Carcinogenesis 10*:817-822, 1989

Hughes, A.L. and Nei, M.: Pattern of nucleotide substitution at major histocompatibility complex class I loci reveals overdominant selection. *Nature 335*:167-170, 1988

Hughes, A.L. and Nei, M.: Nucleotide substitution at major histocompatibility complex class II loci: Evidence for overdominant selection. *Proc Natl Acad Sci USA 86*:958-962, 1989

Kimura, M.: Preponderance of synonymous changes as evidence for the neutral theory of molecular evolution. *Nature 267*:275-276, 1977

Klein, J.: Origin of major histocompatibility complex polymorphism: The trans-species hypothesis. *Hum Immunol 19*:155-162, 1987

Klein, J. and Schönbach, C.: Origins of Mhc diversity. *In* B.G. Solheim, E. Möller, and S. Ferrone (eds): *The HLA System Basic Concepts and Clinical Importance in Transplantation Immunology.* Springer-Verlag, Heidelberg, in press 1991

Kreitman, M.: Detecting selection at the level of DNA. *In* R.K. Selander, A.G. Clark, and T.S. Whittam (eds): *Evolution at the Molecular Level*, pp. 204-221, Sinauer Associates Inc., Sunderland, MA 1991

Kunkel, T.A. and Alexander, P.S.: The base substitution fidelity of eucaryotic DNA polymerases. Mispairing frequencies, site preferences, insertion preferences, and base substitutions by dislocation. *J Biol Chem 261*:160-166, 1986

Kunkel, T.A., Hamatake, R.K., Motto-Fox, J., Fitzgerald, M.P., and Sugino, A.: Fidelity of DNA polymerase I and DNA polymerase I-DNA primase complex from *Saccaromyces cerevisiae*. *Mol Cell Biol 9*:4447-4458, 1989

Lawlor, D.A., Ward, F.E., Ennis, P.D., Jackson, A.P., and Parham, P.: HLA-A and B polymorphisms predate the divergence of humans and chimpanzees. *Nature 335*:268-271, 1988

Li, W.-H., Wu, C.-I., and Luo, C.: A new method for estimating synonymous and nonsynonymous rates of nucleotide substitution considering the relative likelihood of nucleotide and codon changes. *Mol Biol Evol 2*:150-174, 1985

Masurekar, M., Kreuzer, K.N., and Ripley, L.S.: The specificity of topoisomerase-mediated DNA cleavage defines acridine-induced frameshift specificity within a hotspot in bacteriophage *T4*. *Genetics 127*:453-462, 1991

McConnell, T.J., Talbot, W.S., McIndoe, R.A., and Wakeland, E.K.: The origin of MHC class II gene polymorphism with the genus *Mus*. *Nature 332*:651-654, 1988

Nei, M.: *Molecular Evolutionary Genetics*. Columbia University Press, New York 1987

Nei, M. and Hughes, A.L.: Polymorphism and evolution of the major histocompatibility complex loci in mammals. *In* R.K. Selander, A.G. Clark, and T.S. Whittam (eds.): *Evolution at the Molecular Level*, pp. 222-247. Sinauer Associates Inc., Sunderland, MA 1991

Papanicolaou, C. and Ripley, L.S.: Polymerase-specific differences in the DNA intermediates of frameshift mutagenesis. *In vitro* synthesis errors of *Escherichia coli* DNA polymerase I and its large fragment derivative. *J Mol Biol 207*:335-353, 1989

Papanicolaou, C. and Ripley, L.S.: An *in vitro* approach to identifying specificity determinants of mutagenesis mediated by DNA misalignments. *J Mol Biol 221*, in press 1989

Pribnow, D., Sigurdson, D.C., Gold, L., Singer, B.S., Napoli, C., Brosius, J., Dull, T.J., and Noller, H.F.: *rII* cistrons of bacteriophage T4: DNA sequence around the intercistronic divide and positions of genetic landmarks. *J Mol Biol 149*:337-376, 1981

Ripley, L.S.: Model for the participation of quasipalindromic DNA sequences in frameshift mutation. *Proc Natl Acad Sci USA 79*:4128-4132, 1982

Ripley, L.S.: Frameshift mutation: Determination of specificity. *Annu Rev Genet 24*:189-213, 1990

Ripley, L.S., Clark, A., and de Boer, J.G.: Spectrum of spontaneous frameshift mutations. Sequences of bacteriophage T4 *rII* gene frameshifts. *J Mol Biol 191*:601-613, 1986

Ripley, L.S. and Glickman, B.W.: The unique self-complementarity of palindromic sequences provides DNA structural intermediates for mutation. *Col Spring Harbor Symp Quant Biol 47*:851-861, 1983

Schaaper, R.M. and Dunn, R.L.: Spectra of spontaneous mutations in *Escherichia coli* strains defective in mismatch correction: the nature of *in vivo* DNA replication errors. *Proc Natl Acad Sci USA 84*:6220-6224, 1987

Schaaper, R.M., Danforth, B.N., and Glickman, B.W.: Mechanisms of spontaneous mutagenesis: an analysis of the spectrum of spontaneous mutants in the *Escherichia coli lac I* gene. *J Mol Biol 189*:273-284, 1986

Sette, A., Buus, S., Appella, E., Smith, J.A., Chestnut, R., Miles, C., Colon, S.M., and Grey, H.M.: Prediction of major histocompatibility binding regions of protein antigens by sequence pattern analysis. *Proc Natl Acad Sci USA 86*:3296-3300, 1989

She, J.X., Boehme, S., Wang, T.W., Bonhomme, F., and Wakeland, E.K.: The generation of MHC class II gene polymorphism in the genus *Mus*. *Biol J Linnean Soc 41*:141-161, 1990

Shinedling, S., Singer, B.S., Gayle, M., Pribnow, D., Jarvis, E., Edgar, B., and Gold, L.: Sequence and studies of bacteriophage *T4rII* mutants. *J Mol Biol 195*:471-480, 1987

Singer, B.S. and Westlye, J.: Deletion formation in bacteriophage T4. *J Mol Biol 202*:233-243, 1988

Stewart, J.W. and Sherman, F.: Yeast frameshift mutations identified by sequence changes in iso-1-cytochrome c. *In* L. Prakash, F. Sherman, M.W. Miller, C.W. Lawrence, and H.W. Taber (eds.): *Molecular Environmental Aspects of Mutagenesis*, pp. 102-127. C.C. Thomas, Springfield, IL 1974

Streisinger, G., Okada, Y., Emrich, J., Newton, J., Tsugita, A., Terzahi, E., and Inouye, N.: Frameshift mutations and the genetic code. *Cold Spring Habor Symp Quant Biol 31*:77-84, 1966

Streisinger, GH. and Owen, J.E.: Mechanisms of spontaneous and induced frameshift mutation in bacteriophage T4. *Genetics 109*:633-659, 1985

Tautz, D., Trick, M., Dover, G.A.: Cryptic simplicity in DNA is a major source of genetic variation. *Nature 322*:652-656, 1986

Wu, S., Saunders, T.L., and Bach, F.H.: Polymorphism of human Ia antigens generated by reciprocal intergenic exchange between two *DRβ* loci. *Nature 324*:676-682, 1986

TWO MODELS OF EVOLUTION OF THE CLASS I MHC

Austin L. Hughes
Department of Biology and Institute of Evolutionary Genetics, The Pennsylvania State University, University Park, PA 16802, USA

Two models have been proposed for the evolution of the class I MHC in mammals: (1) a model of inter-locus homogenization by gene conversion; (2) a model of frequent duplication and deletion. Data from both primates and rodents now strongly favor the latter hypothesis. The MHC evolves much as do other multigene families. Features unique to the MHC are consequences of positive selection on antigen-binding regions.

In mammals there are a number of striking differences between the class I and class II major histocompatibility complex (*MHC*) gene families. In the case of the class II *MHC*, there are orthologous relationships between subfamilies of genes (*DR, DQ*, etc.), representatives of which have been found in all placental mammals studied so far. In the class I *MHC*, on the other hand, orthologous relationships are not usually found between loci of distantly related mammalian species, although closely related species may share orthologous loci (Hughes and Nei 1989a). For example, the chimpanzee shares the class I classical *A*, *B*, and *C* loci with humans (Lawlor et al. 1988). But loci orthologous to these are not found in mammals of other orders such as Rodentia and Artiodactyla (Hughes and Nei 1989a). Even New World primates lack orthologues to *A*, *B*, and *C* (Watkins et al. 1990). In addition to the highly expressed, generally polymorphic classical class I loci, the class I *MHC* of mammals so far studied includes several poorly expressed (nonclassical) and pseudogene loci. These apparently nonfunctional loci outnumber the functional loci in all known mammalian class I *MHC*s (Klein and Figueroa 1986). In general, a species' nonclassical and pseudogene loci are more similar to its classical loci than they are to the nonclassical or pseudogene loci of distantly related species (Hughes and Nei 1989a).

There are two possible explanations for these observations about the class I MHC: (1) A high rate of interlocus genetic exchange (via "gene conversion" or some other mechanism) causes homogenization of class I loci within a species over time (Rada et al. 1990). (2) Over evolutionary time, there are much higher rates of duplication of loci, loss of function by previously functional loci, and deletion of nonfunctional loci in the case of the class I MHC than in the case of the class II MHC (Hughes and Nei 1989a). Under the latter hypothesis, nonclassical loci are seen as duplicated classical loci which are on the way to becoming pseudogenes (Klein and Figueroa 1986). The reason for the reduced expression of such loci is presumably that they have accumulated mutations in regulatory elements

NATO ASI Series, Vol. H 59
Molecular Evolution of the
Major Histocompatibility Complex
Edited by J. Klein and D. Klein

(Hughes and Nei 1989a). The reason that such loci are not shared by mammals of different orders is that they are of relatively recent origin; over long evolutionary periods such nonfunctional loci are gradually deleted from the genome or become unrecognizable due to mutation. As a corollary, we might expect to find some cases where a nonclassical locus of one species is orthologous to a classical locus of another species. On the model of concerted evolution, on the other hand, one would expect to see that orthologous relationships between classical and nonclassical loci would be obliterated at about the same rate.

In fact, it is now known that the classical class I loci of the cottontop tamarin *Saguinus oedipus* are orthologous to the nonclassical *HLA-G* locus of humans (Watkins et al. 1990). To explain this observation only in terms of the unique reproductive strategy of the callitrichids (production of bone-marrow-chimaeric fraternal twins) is to miss the more general point. The tamarin, which diverged from the human some 40 million years ago, also has a nonclassical locus orthologous to the human nonclassical locus *HLA-F*; yet no loci orthologous to the human classical loci can be found in the tamarin (Watkins et al. 1990). These results are not consistent with the model of concerted evolution but rather with the model of repeated duplication, silencing, and deletion of class I loci.

Of course, these results did not rule out the possibility that the class I MHC of rodents might evolve in a fundamentally different way than that of primates. These was some evidence in support of this idea. For example, Rogers (1985) on the basis of hybridization studies claimed that there were no orthologous relationships between the class I genes of the mouse and those of the rat (diverged 10-25 million years ago). However, this technique is not very sensitive for establishing orthologous relationships between loci of different species. Comparison of exons 2-4 (encoding the alpha-1, alpha-2, and alpha-3 domains) of the mouse class I classical loci (*H-2K, -D*, and *-L*) revealed considerably fewer locus-specific nucleotides than did a similar analysis of the human classical loci *HLA-A, -B*, and *-C* (Parham et al. 1989; Rada et al. 1990). These results might suggest a higher rate of interlocus genetic exchange in the mouse than in the human. However, in other exons, the numbers of locus specific residues are not any lower in the mouse than in the human; and even in the human, locus-specific residues are few in exons 2 and 3. This is not surprising since these exons encode the antigen recognition site of the class I molecule, in which the rate of nonsynonymous nucleotide substitution is enhanced by positive (overdominant) natural selection (Hughes and Nei 1988).

In order to discriminate further between the two models of evolution of the mammalian class I MHC, I conducted a phylogenetic analysis of class I classical and nonclassical class I MHC genes from mouse and rat. Nonclassical class I genes tend to have a consistently lower G+C content at third positions, especially in exons 2 and 3, than do classical class I genes (Hughes and Nei 1989a; Hughes 1991). Such a difference in G+C content can markedly bias rates of evolution at synonymous nucleotide sites (Wolfe et al.

1989). Therefore, I based phylogenetic trees on the numbers of nucleotide substitutions per nonsynonymous site. (For further details see Hughes 1991).

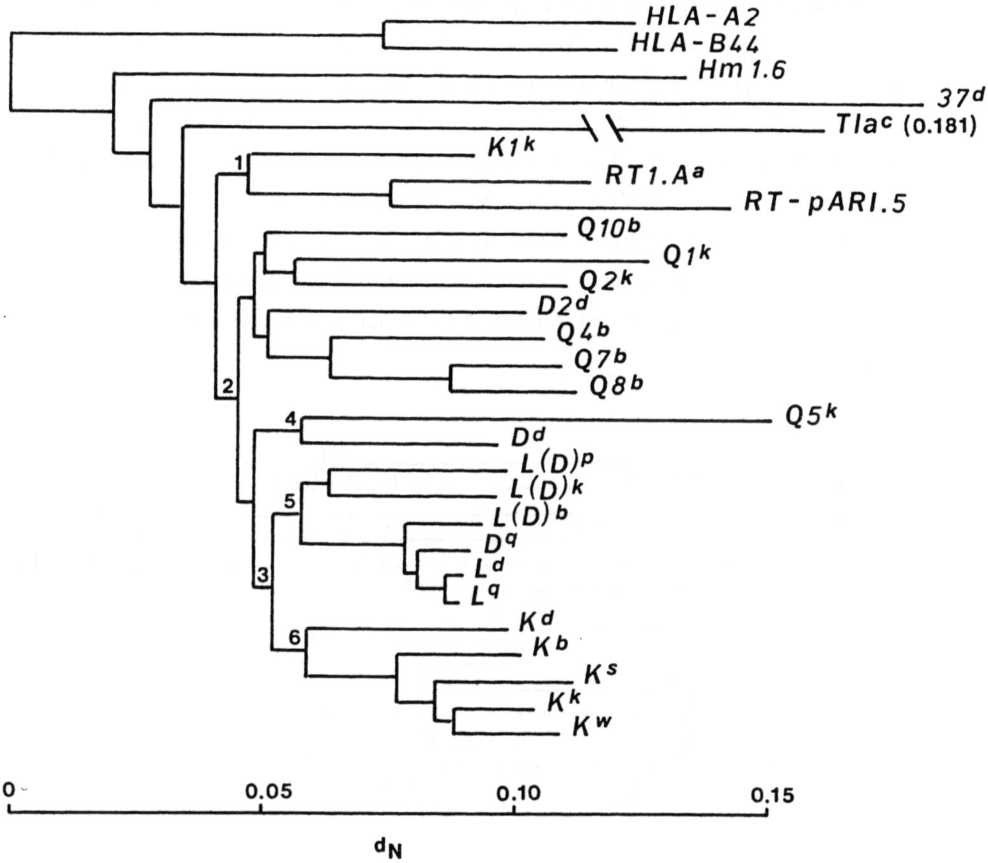

Fig. 1. Phylogenetic tree for rat and mouse class I MHC genes, constructed on the basis of number of nonsynonymous substitutions per site (d_N; Nei and Gojobori 1986) by the neighbor-joining method (Saitou and Nei 1987). Rat genes have the prefix *RT*. Human (*HLA*) and hamster (*Hm*) genes are used as outgroups to root the tree. The significance of internal branches was tested by Li's (1989) method. The branch (1-2) separating the cluster of $K1^k$ and rat genes from other mouse genes is significantly different from zero at the 1% level. The branch (3-4) between $D^d/Q5^k$ and other mouse classical genes is significantly different from zero at the 1% level. The length of the branch (5-6) separating the *K* alleles from the *L* alleles (plus D^q) is significantly different from zero at the 1% level.

The phylogenetic tree (Figure 1) shows that the rat locus *RT1.A*, which is the only known classical class I locus in the rat, is orthologous to the mouse *H-2K1* pseudogene. Both of these loci are located centromeric to the class II/class III genes in the MHC. In the mouse, the classical *H-2K* locus is also located centromeric to class II/class III. In the rat, there are several nonclassical or pseudogene loci located centromeric to class II/class III, including the gene designated *RT-pARI.5* (Radjcic et al. 1990). The existence of a group of class I genes centromeric to class II/class III is a characteristic of the rodents of the subfamily Murinae (Klein and Figueroa 1986). The results of the phylogenetic analysis suggest that at least two independent events of translocation of class I genes centromeric to class II/class III have occurred in the history of the Murinae. The first event, which occurred prior to the divergence of rat and mouse, involved translocation of the ancestor of *RT1.A* and of *H-2K1*. Subsequently, in the mouse *H-2K1* became a pseudogene, while in the rat *RT1.A* duplicated several times giving rise to a group of related genes (including *RT-pARI.5*). The second translocation event, which may have occurred after rat and mouse separated, involved the *H-2K* locus, which is clearly closely related to the other mouse classical loci, *H-2D* and *H-2L* (Figure 1).

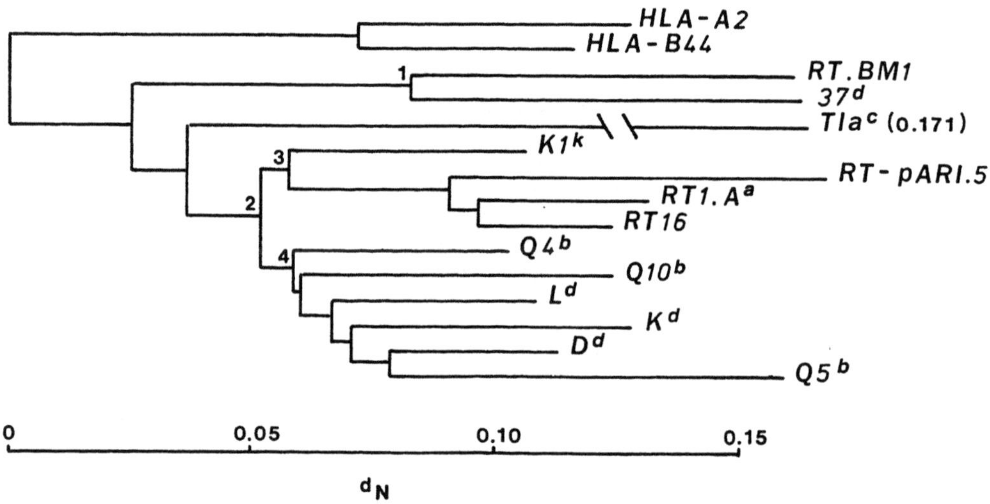

Fig. 2. Phylogenetic tree based on numbers of nonsynonymous substitutions per site (d_N), showing the relationships of rat partial class I MHC sequences *RT.BM1* and *RT16*. The branch (1-2) separating the RT.BM1/37d cluster from the other main group of rat and mouse class I genes in significantly different from zero at the 0.1% level. The branch (3-4) separating the cluster of *K1*^{k} and three rat genes from the mouse *Q* and classical genes is significantly different from zero at the 5% level.

In general, alleles at each of the three mouse classical class I loci form a separate cluster (Figure 1). The only exception is that D^q does not cluster with D^d. However, D^q is known to represent a recent duplication of the ancestor of L^q, and thus us not truly allelic to

D^d (Rubocki et al. 1990). Thus, in spite of the probability that the fourth exon has been exchanged among the three mouse classical class I loci (Hughes and Nei 1988), alleles at each locus have retained a locus-specific character. Phylogenetic analysis of rat class I MHC genes for which only a partial sequence was available revealed a second case of an orthologous relationship between a rat gene and a mouse gene. The rat nonclassical gene *RT.BM1* was found to be orthologous to the mouse nonclassical gene *37d* (Figure 2).

These results support the hypothesis that in the case of the rat and the mouse, the class I MHC has evolved by repeated duplications and deletions. The presence of orthologous relationships between class I genes of these two species is evidence against the hypothesis of concerted evolution. Cases of putative interlocus genetic exchange have been reported in the case of the mouse (Weiss et al. 1983; Hughes and Nei 1988). But these results suggest that such events do not occur at a rate high enough to cause within-species homogenization of loci.

In fact, there have also been reports of interlocus genetic exchange in the class I MHC of humans and other primates (Hughes and Nei 1989b; Parham et al. 1989). There is no real evidence that these events are more or less likely to occur in rodents than in primates. Rather, they seem to be characteristic of most multi-gene families. However, the fact that such events are clearly detectable by statistical analysis of DNA sequences, even millions of years after the event occurred, suggests that they are relatively rare (Hughes and Nei 1989b).

Indeed, in spite of the apparent wish of many workers to believe otherwise, there is nothing particularly remarkable or unusual about the way the types of interlocus interactions that occur in the MHC. Interlocus genetic exchange and other types of recombination occur in the MHC in much the same way as in other multi-gene families. What is unusual about the MHC is the extraordinary polymorphism of certain loci. Furthermore this polymorphism is maintained by natural selection, in contrast to the neutral polymorphisms which are found at most known polymorphic loci. Because they are maintained by selection, MHC polymorphisms can last much longer than neutral polymorphisms (Takahata and Nei 1990). These facts must be kept in mind whenever we attempt to unravel the history of MHC loci. When one allele at a polymorphic locus is duplicated, the new locus will be more closely related to that one allele at the parent locus than it is to other alleles at the same locus. The close relationship between *Dq* and *Ld* and *Lq* (Figure 1) provides an example. In the absence of a historical perspective, such a resemblance might well be attributed to gene conversion of *Dq* by *Ld* or *Lq*. Nonfunctional class I genes (nonclassical genes and pseudogenes), even though they are no longer subject to positive selection, show clear evidence of having been subject to such selection in the past (Hughes and Nei 1989a). Past selection may have caused nucleotide substitutions which parallel those at currently functional loci, leading to the sort of similarity often attributed to a "micro-conversion" event. Clearly researchers should be cautious in making such claims; instead no such claim should be accepted as established until other hypotheses relating to the unique features of the MHC are ruled out.

The features of the MHC that are unique among vertebrate multi-gene families (including the high polymorphism and "trans-species" polymorphisms) are only explainable as a result of overdominant selection relating to antigen binding (Hughes and Nei 1988). Aside from the effects of such selection, the MHC seems to evolve much as any other multi-gene family does. Often it seems that those who imagine mysterious and previously undescribed mechanisms of evolution in the MHC do so because have not considered in detail the consequences of overdominant selection at the DNA level.

Acknowledgment. This research was supported by a grant from the National Institutes of Health.

References

Hughes, A.L.: Independent gene duplications, not concerted evolution, explain relationships among class I MHC genes of murine rodents. *Immunogenetics*, in press 1991

Hughes, A.L. and Nei, M.: Pattern of nucleotide substitution at major histocompatibility complex loci reveals overdominant selection. *Nature 355*:167-170, 1988

Hughes, A.L. and Nei, M.: Evolution of the major histocompatibility complex: independent origin of nonclassical class I genes in different groups of mammals. *Mol Biol Evol 6*:559-579, 1989a

Hughes, A.L. and Nei, M.: Ancient interlocus exon exchange in the history of the HLA-A locus. *Genetics 122*:681-686, 1989b

Klein, J., and Figueroa, F.: Evolution of the major histocompatibility complex. *CRC Crit Rev Immunol 6*: 295-386, 1986

Lawlor, D.A., Ward, F.E., Ennis, P.D., Jackson, A.P., and Parham, P.: HLA-A and -B polymorphisms predate the divergence of humans and chimpanzees. *Nature 335*: 268-271, 1988

Li, W.-H.: A statistical test of phylogenies estimated from sequence data. *Mol Biol Evol 6*:424-435, 1989

Nei, M., and Gojobori, T.: Simple methods for estimating the numbers of synonymous and nonsynonymous nucleotide substitutions. *Mol Biol Evol 3*: 418-426, 1986

Parham, P., Lawlor, D.A., Lomen, C.E., and Ennis, P.D.: Diversity and diversification of HLA-A, B, C alleles. *J Immunol 142*:3937-3950, 1989

Rada, C., Lorenzi, R., Powis, S.J., van den Bogaerde, J., Parham, P., and Howard, J.C.: Concerted evolution of class I genes in the major histocompatibility complex of murine rodents. *Proc Natl Acad Sci USA 87*:2167-2171, 1990

Rogers, J.H.: Mouse histocompatibility-related genes are not conserved in other mammals. *EMBO J 4*:749-753, 1985

Radojcic, A., Kunz, H.W., and Gill, T.J., III.: Expression and analysis of the rat placental class I cDNA clone encoding the Pa antigen. *Immunogenetics 31*:326-332, 1990

Saitou, N., and Nei, M.: The neighbor-joining method: a new method for reconstructing phylogenetic trees. *Mol Biol Evol 4*: 406-425, 1987

Takahata, N., and Nei, M.: Allelic genealogy under overdominant and frequency-dependent selection and polymorphism of major histocompatibility complex loci. *Genetics 124*:967-978, 1990

Watkins, D.I., Chen, Z.W., Hughes, A.L., Evans, M.G., Tedder, T.F., and Letvin, N.L.: Evolution of the MHC class I genes of a New World primate from ancestral homologues of human non-classical genes. *Nature 346*:60-63, 1990

Weiss, E.H., Golden, L., Zakut, R., Mellor, A., Fahrner, K., Kvist, S., and Flavell, R.A.: The DNA sequence of the H2Kb gene: Evidence for gene conversion as a mechanism for the generation of polymorphism in histocompatibility antigens. *EMBO J 2*:453-462, 1983

Wolfe, K.H., Sharp, P.M., and Li, W.-H.: Mutation rates vary among regions of the mammalian genome. *Nature 337*:283-285, 1989

EVOLUTION OF MHC DOMAINS: STRATEGY FOR ISOLATION OF *MHC* GENES FROM PRIMITIVE ANIMALS

Keiichiro Hashimoto and Yoshikazu Kurosawa
Institute for Comprehensive Medical Science, Fujita Health University, Toyoake, Aichi 470-11, Japan

To understand the long-term evolution of MHC molecules, studies on MHC-like molecules of phylogenetically primitive animals are required. We have succeeded in isolating MHC-like genes from fish using polymerase chain reaction (PCR). The principle of the experimental strategy is presented in conjunction with a brief survey on the evolution of MHC domains.

1. Introduction

Recently, using polymerase chain reaction (PCR) we demonstrated the existence of both *MHC* class I and class II genes in teleost fish (carp), the most primitive class of vertebrates with the capacity for acute allograft rejection (Hashimoto et al. 1990). It took more than thirty years to isolate *MHC* genes from fish after the acute allograft rejection was reported in teleost fish (reviewed by Hildemann 1970). It has been difficult to identify *MHC*-like genes or molecules in fish with conventional approaches, including the cross-hybridization of fish DNA or RNA with probes of higher vertebrates, the cross-reaction of fish cell-surface molecules with antibodies against mammalian and avian MHC molecules, and the utilization of alloantisera (reviewed by Kaufman et al. 1990; Stet and Egberts 1991). In this short article, we would like to discuss the strategy for isolation of *MHC* genes from phylogenetically primitive vertebrates such as fish and even more primitive animals.

2. Strategy

The membrane-proximal domains of the MHC molecules clearly belong to C1-set of the immunoglobulin (Ig) superfamily (Williams and Barclay 1988), while the membrane-distal domains have a unique structure that cannot be classified into any of the established protein families (Bjorkman et al. 1987). Several amino acid residues surrounding two cysteines are well conserved not only in membrane-proximal domains of MHC molecules, but also in other Ig superfamily C1-set members including the constant domains of Ig and T-cell receptor (TcR). These characteristic amino acid residues should have existed in the primordial molecule more than several hundred million years ago when a C1-set member appeared in evolution. Even with a limited number of conserved residues, an application of PCR seemed to be a promising

NATO ASI Series, Vol. H 59
Molecular Evolution of the
Major Histocompatibility Complex
Edited by J. Klein and D. Klein

approach to isolate MHC genes from primitive vertebrates such as fish. The amino acid residues conserved in C1-set members can constitute a frame for designing PCR primers since they are located at the restricted regions. If primitive extant animals have actually retained the conserved amino acid residues in the MHC molecules, we should be able to isolate MHC genes from those animals using primers made of nucleotides for the conserved amino acid residues and some appropriate nucleotides for the variable amino acid positions. As hybrid formation between a primer and a template DNA at the priming site is important in PCR, choice of amino acid residues for the 3' end region of a primer is crucial. In the isolation of carp MHC genes, we chose two conserved regions surrounding two cysteines in the MHC membrane-proximal domains as primer sites. We designed two primers with the nucleotides for conserved proline and cysteine as 3' ends of the two primers, and with the nucleotides for several conserved and variable amino acid residues, which were selected based on the mammalian and avian MHC sequences. With these primers, we could obtain amplified DNA fragments of expected length in PCR products of primitive vertebrates including several kinds of fish (Hashimoto et al. 1990; unpublished results). As for locations of primers, other regions can be useful, for example, the regions conserved only in MHC molecules or in specific MHC domains. To further discuss the general applicability and limitation of these approaches in the following part, we take a brief look at the evolution of MHC molecules. As the membrane-proximal domains and the membrane-distal domains have different characteristics, they are discussed separately.

3. Evolution of MHC domains

Membrane-proximal MHC domains. Little is known about the long-term evolution of Ig superfamily members. According to Williams' classification (Williams and Barclay 1988), Ig superfamily members are divided into three main categories: V-set, C1-set, and C2-set. Some Ig superfamily members are found in invertebrates, for example, amalgam (Seeger et al. 1988) and neuroglian (Bieber et al. 1989) of *Drosopohila*, fasciclin II of grasshopper (Harrelson and Goodman 1988), hemolin of silk moth (Sun et al. 1990), and twitchin of *Caenorhabditis elegans* (Benian et al. 1989). These molecules possess domains which share characteristics with the C2-set or V-set members of the Ig superfamily. Although most Ig superfamily molecules are found at the cell-surface and are involved in recognition and/or adhesion in cell-cell interactions, twitchin, a muscle protein of *C.elegans,* and avian skeletal muscle C-protein (Einheber and Fischman 1990) are examples of intracellular members of the Ig superfamily. One Ig superfamily member namely, chaperone PapD, a product of uropathogenic *E.coli*, has been found in *Escherichia coli*. The three-dimensional structure of this protein has been shown to be similar to the Ig-fold (Holmgren and Bränden 1989). Since PapD has sequence similarity with human CD5 which shows fast rate of divergence among mammals, it was speculated that horizontal gene transfer may have taken place (Holmgren and Bränden 1989). *C.elegans* seems to be the most primitive animal in which the Ig superfamily domains have been identified to date. However, Ig superfamily members probably exist in more primitive animals.

Molecules which possess amino acid residues characteristic of C1-set have not been found in invertebrates as yet. Presently, elasmobranchs are the most primitive animals in which the genes of C1-set members, namely Ig genes (Hinds and Litman 1986) and MHC-like genes (Hashimoto et al. submitted), have been identified. Since the primitive MHC/TcR-equivalent system would play a role in self-nonself recognition in primitive animals, C1-set members, to which membrane-proximal MHC domains and TcR constant domains belong, should exist in more primitive animals than elasmobranchs.

Membrane-distal MHC domains. As for the origin of peptide-binding domains of MHC, two possibilities are discussed below. The first is that the membrane-distal domains might have been derived from a protein family other than the Ig superfamily. This idea appears to be supported by the X-ray crystallographic studies on the three-dimensional structure of an HLA molecule. Its membrane-distal domains turned out to be entirely different from the Ig-fold (Bjorkman et al. 1987).

Recently, the structural similarities between MHC peptide-binding domains and a putative peptide-binding region of the hsp70 family have been proposed (Flajnik et al. 1991a; Rippmann et al. 1991). These proposals are based on the peptide-binding properties of hsp70, the comparison of its amino acid sequence with those of MHC molecules, the estimation of the secondary structure of the putative peptide-binding region of hsp70 and its superimposition onto the HLA three-dimensional structure. The essential role of hsp70-like molecules is to assist protein folding and/or assembly even in bacteria. These molecules interact with gd TcR and moreover, genes for human hsp70 are located within the MHC region. These observations may further support the hypothesis that hsp70 is a donor of peptide-binding domains of an MHC molecule. However, there are several proteins which adopt a structure composed of a platform of ß-strands topped by α-helices similar to the HLA molecules, for example, bovine platelet factor 4 (Charles et al. 1989) and interleukin 8 (Baldwin et al. 1991), both of which belong to the same protein family, and bacteriophage coat protein MS2 (Valegård et al. 1990). Bovine platelet factor 4 and interleukin 8 form a dimer like class II molecules. Between these and the HLA molecule, however, there are differences in the angle between the ß-strands and the α-helical axis. The helices in HLA-A2 cross the ß-sheet at an angle of ~-45°, while in platelet factor 4, this angle is ~+52° (Charles et al. 1989). Probably many molecules will be found in the future to have a structure composed of a platform of ß-strands topped by α-helices. This structure could assume a stable conformation and many molecules controlled by independent genes coincidentally could form it. Determination of three-dimensional structures of many other molecules including peptide-binding region of hsp70 will be required to clarify the relation between these proteins and peptide-binding domains of MHC molecules.

There are several Ig superfamily molecules that consist of a combination of distinct domains: for example, neural cell adhesion molecules, N-CAM, and L1, contain both Ig superfamily C2-set domains and fibronectin type III domains (Cunningham et al. 1987; Barthels et al. 1987; Moos et al. 1988). If exon-shuffling created a primordial MHC gene, it must have occurred after the membrane-proximal domain had established its amino acid residues characteristic of C1-set.

The second possibility, which we prefer, is that the membrane-distal domains might have been derived from an Ig superfamily member. Membrane-distal domains of MHC molecules have several characteristics shared by Ig superfamily members. In Ig superfamily members, each domain is encoded by a single exon with few exceptions. In MHC genes, not only membrane-proximal domains but also membrane-distal domains are encoded by a single exon. Furthermore, in Ig superfamily genes, a splicing at the boundary of two domains occurs between the first and the second base of the boundary codon. This rule is also followed at the exon-intron boundaries of MHC genes.

Both the MHC class I α2 and class II ß1 domains contain a disulfide bond reminiscent of that found in Ig superfamily members. The two conserved cysteines have not been accidentally acquired in these domains by mammals. They must have existed in ancient animals because in the chicken (Bourlet et al. 1988; Guillemot et al. 1988), the amphibians *Xenopus* (Flajnik et al. 1991), and even in teleostean fish (Hashimoto et al. 1990), MHC molecules possess these cysteine residues. However, there is no cysteine at the equivalent positions in the class I α1 domain and class II α1 domain. Although two domains, class I α1 and α2 as well as class II α1 and ß1, which form the peptide-binding cleft, seem to have been derived from the same primordial domain, it is not clear yet whether one of the two domains acquired two cysteines or the other domain lost these residues. It is certain, however, that such a change occurred more than several hundred million years ago.

The differences between the Ig-fold and the structure of the membrane-distal domains of the MHC molecules are obvious. However, in the region of ß-strands, we found some similarities in amino acid residues between carp MHC class II ß1 domain and V-set members of the Ig superfamily (Hashimoto et al. 1990). Ig superfamily members for which three-dimensional structure data are available include Ig (reviewed by Amzel and Poljak 1979), MHC molecules (Bjorkman et al. 1987), two domains of the CD4 molecule (Wang et al. 1990; Ryu et al. 1990), and PapD (Holmgren and Bränden 1989). The number of ß-strands, their length and their topology varies among these molecules, but there is no gross change of the Ig-fold such as a change from a ß-strand to an α-helix. However, our knowledge on the evolution of protein-folding is very much limited. Although a small number of folding patterns may represent most of the known protein structures (Chothia and Finkelstein 1990), one could not have predicted the three-dimensional structure of the membrane-distal MHC domains before the crystallographic studies of the HLA molecules were carried out. Furthermore, in general it is difficult to predict the precise relations between already known, apparently different protein-folding patterns.

MHC class I heavy chains are reported to be stably expressed at the cell-surface only after their association with an appropriate peptide (resulting in a conformational change) and the ß2-microglobulin (Townsend et al. 1989; Townsend et al. 1990; Cerundolo et al. 1990). The class I heavy chain might contain non-helical structures instead of α-helices when it exists without the peptide. In fact, peptides that form a noncovalent α-helical dimer, including "leucine zipper" peptides, exist in an equilibrium between nonhelical monomers and α-helical dimers (O'Neil and DeGrado 1990; O'Shea et al. 1989a; O'Shea et al. 1989b). It is not clear whether the different conformation of MHC class I heavy chain without a peptide suggests the original characteristics of primordial peptide-binding domains or whether it reflects a transient state from a more stable primordial folding such as the Ig-fold.

In molecules capable of binding antigens, the simplest unit consists of two different kinds of domains, namely an antigen-binding variable V-set domain and a constant C1-set domain (e.g., Ig light chain and the TcR chain). Similar to such a basic structure, a half of an MHC class I molecule or an MHC class II chain has a polymorphic domain, which participates in the binding of a peptide, and a non-polymorphic domain. As described above, the origin of the MHC peptide-binding domain is not clear as yet. Irrespective of their history, these domains are polymorphic in nature, and the basic design of MHC molecules seems to be similar to that of other antigen-binding molecules in some respects.

4. Conclusion

Until recently, our knowledge of MHC genes had been restricted to the genes of higher vertebrates. Now we have a way of directly investigating MHC genes of lower vertebrates and possibly of some invertebrates. In a few years, a great deal of information on MHC genes of various primitive animals will become available and we will be able to gain some insight into the divergence of the two MHC classes and the divergence of the two peptide-binding domains by studying MHC-like genes in primitive animals. The results have already indicated that class I and class II sequences tend to share several amino acids and some characteristics in primitive vertebrates (Hashimoto et al. 1990; Hashimoto et al. submitted).

Using the knowledge of the conserved amino acid residues, it might become possible to find primitive extant animals in which MHC-like molecules still have retained the characteristics of the original primitive molecule that emerged several hundred million years ago when the primordial MHC molecule (at least its membrane-proximal domain) diverged from the TcR/Ig lineage. As already discussed, the origin of MHC peptide-binding domains is controversial. Characterization of MHC-like molecules in primitive animals as well as clarification of the three-dimensional structure of proteins other than MHC molecules might become essential for resolving this issue. With a strategy similar to that used in the isolation of carp MHC genes, it may be possible to isolate genes encoding the molecules of the TcR/Ig lineage in primitive animals. It will then become possible to investigate primitive self-nonself recognition systems, as well as to apply similar approaches to the study of the evolution of other Ig superfamily members.

There may be various strategies to isolate MHC-like genes or molecules from primitive animals. The approaches using PCR will be one of the most promising and effective ways of studying genes for MHC-like molecules and other immunologically important molecules in phylogenetically primitive animals.

References

Amzel, L.M. and Poljak, R.J.: Three-dimensional structure of immunoglobulins. *Annu Rev Biochem 48:* 961-997, 1979

Baldwin, E.T., Weber, I.T., Charles, R.St., Xuan, J.-C., Appella, E., Yamada, M., Matsushima, K., Edwards, B.F.P., Clore, G.M., Gronenborn, A.M., and Wlodawer, A.: Crystal structure of interleukin 8: Symbiosis of NMR and crystallography. *Proc Natl Acad Sci USA 88:* 502-506, 1991

Barthels, D., Santoni, M.-J., Wille, W., Ruppert, C., Chaix, J.-C., Hirsch, M.-R., Fontecilla-Camps, J.C., and Goridis, C.: Isolation and nucleotide sequence of mouse NCAM cDNA that codes for a M_r 79000 polypeptide without a membrane-spanning region. *EMBO J 6:* 907-914, 1987

Benian, G.M., Kiff, J.E., Neckelmann, N., Moerman, D.G., and Waterston, R.H.: Sequence of an unusually large protein implicated in regulation of myosin activity in *C. elegans*. *Nature 342:* 45-50, 1989

Bieber, A.J., Snow, P.M., Hortsch, M., Patel, N.H., Jacobs, J.R., Traquina, Z.R., Schilling, J., and Goodman, C.S.: Drosophila neuroglian: A member of the immunoglobulin superfamily with extensive homology to the vertebrate neural adhesion molecule L1. *Cell 59:* 447-460, 1989

Bjorkman, P.J., Saper, M.A., Samraoui, B., Bennett, W.S., Strominger, J.L., and Wiley, D.C.: Structure of the human class I histocompatibility antigen, HLA-A2. *Nature 329:* 506-512, 1987

Bourlet, Y., Béhar, G., Guillemot, F., Fréchin, N., Billault, A., Chausse, A.-M., Zoorob, R., and Auffray, C.: Isolation of chicken major histocompatibility complex class II (B-L) ß chain sequences: comparison with mammalian ß chains and expression in lymphoid organs. *EMBO J 7:* 1031-1039, 1988

Cerundolo, V., Alexander, J., Anderson, K., Lamb, C., Cresswell, P., McMichael, A., Gotch, F., and Townsend, A.: Presentation of viral antigen controlled by a gene in the major histocompatibility complex. *Nature 345:* 449-452, 1990

Charles, R.St., Walz, D.A. and Edwards, B.F.P.: The three-dimensional structure of bovine platelet factor 4 at 3.0-Å resolution. *J Biol Chem 264:* 2092-2099, 1989

Chothia, C. and Finkelstein, A.V.: The classification and origins of protein folding patterns. *Annu Rev Biochem 59:* 1007-1039, 1990

Cunningham, B.A., Hemperly, J.J., Murray, B.A., Prediger, E.A., Brackenbury, R., and Edelman, G.M.: Neural cell adhesion molecule: Structure, immunoglobulin-like domains, cell surface modulation, and alternative RNA splicing. *Science 236:* 799-806, 1987

Einheber, S. and Fischman, D.A.: Isolation and characterization of a cDNA clone encoding avian skeletal muscle C-protein: An intracellular member of the immunoglobulin superfamily. *Proc Natl Acad Sci USA 87:* 2157-2161, 1990

Flajnik, M.F., Canel, C., Kramer, J., and Kasahara, M.: Which came first, MHC class I or class II? *Immunogenetics 33:* 295-300, 1991a

Flajnik, M.F., Canel, C., Kramer, J., and Kasahara, M.: Evolution of the major histocompatibility complex: Molecular cloning of major histocompatibility complex class I from the amphibian *Xenopus. Proc Natl Acad Sci USA 88:* 537-541, 1991b

Guillemot, F., Billault, A., Pourquié, O., Béhar, G., Chaussé, A.-M., Zoorob, R., Kreibich, G., and Auffray, C.: A molecular map of the chicken major histocompatibility complex: the class II ß genes are closely linked to the class I genes and the nucleolar organizer. *EMBO J 7:* 2775-2785, 1988

Harrelson, A.L. and Goodman, C.S.: Growth cone guidance in insects: Fasciclin II is a member of the immunoglobulin superfamily. *Science 242:* 700-708, 1988

Hashimoto, K., Nakanishi, T., and Kurosawa, Y.: Isolation of carp genes encoding major histocompatibility complex antigens. *Proc Natl Acad Sci USA 87*: 6863-6867, 1990

Hildemann, W.H.: Transplantation immunity in fishes: Agnatha, Chondrichthyes and Osteichthyes. *Transpl Proc 2:* 253-259, 1970

Hinds, K.R. and Litman, G.W.: Major reorganization of immunoglobulin V_H segmental elements during vertebrate evolution. *Nature 320:* 546-549, 1986

Holmgren, A. and Brändén, C-I.: Crystal structure of chaperone protein PapD reveals an immunoglobulin fold. *Nature 342:* 248-251, 1989

Kaufman, J., Skjoedt, K., and Salomonsen, J.: The MHC molecule of nonmammalian vertebrates. *Immunol Rev 113:* 83-117, 1990

Moos, M., Tacke, R., Scherer, H., Teplow, D., Früh, K., and Schachner, M.: Neural adhesion molecule L1 as a member of the immunoglobulin superfamily with binding domains similar to fibronectin. *Nature 334:* 701-703, 1988

O'Neil, K.T. and DeGrado, W.F.: A thermodynamic scale for the helix-forming tendencies of the commonly occurring amino acids. *Science 250:* 646-651, 1990

O'Shea, E.K., Rutkowski, R., and Kim, P.S.: Evidence that the leucine zipper is a coiled coil. *Science 243:* 538-542, 1989a

O'Shea, E.K., Rutkowski, R., Stafford, W.F.,III, and Kim, P.S.: Preferential heterodimer formation by isolated leucine zippers from Fos and Jun. *Science 245:* 646-648, 1989b

Rippmann, F., Taylor, W.R., Rothbard, J.B., and Green, N.M.: A hypothetical model for the peptide binding domain of hsp70 based on the peptide binding domain of HLA. *EMBO J 10:* 1053-1059, 1991

Ryu, S.-E., Kwong, P.D., Truneh, A., Porter, T.G., Arthos, J., Rosenberg, M., Dai, X., Xuong, N.-H., Axel, R., Sweet, R.W., and Hendrickson, W.A.: Crystal structure of an HIV-binding recombinant fragment of human CD4. *Nature 348:* 419-426, 1990

Seeger, M.A., Haffley, L., and Kaufman, T.C.: Characterization of *amalgam*: A member of the immunoglobulin superfamily from Drosophila. *Cell 55:* 589-600, 1988

Stet, R.J.M. and Egberts, E.: The histocompatibility system in teleostean fishes: from multiple histocompatibility loci to a major histocompatibility complex. *Fish & Shellfish Immunol 1:* 1-16, 1991

Sun, S-C., Lindström, I., Boman, H.G., Faye, I. and Schmidt, O.: Hemolin: An insect-immune protein belonging to the immunoglobulin superfamily. *Science 250:* 1729-1732, 1990

Townsend, A., Öhlén, C., Bastin, J., Ljunggren, H.-G., Foster, L. and Kärre, K.: Association of class I major histocompatibility heavy and light chains induced by viral peptides. *Nature 340:* 443-448, 1989

Townsend, A., Elliott, T., Cerundolo, V., Foster, L., Barber, B. and Tse, A.: Assembly of MHC class I molecules analyzed in vitro. *Cell 62:* 285-295, 1990

Valegård, K., Liljas, L., Fridborg, K., and Unge, T.: The three-dimensional structure of the bacterial virus MS2. *Nature 345:* 36-41, 1990

Wang, J., Yan, Y., Garrett, T.P.J., Liu, J., Rodgers, D.W., Garlick, R.L., Tarr, G.E., Husain, Y., Reinherz, E.L., and Harrison, S.C.: Atomic structure of a fragment of human CD4 containing two immunoglobulin-like domains. *Nature 348:* 411-418, 1990

Williams, A.F. and Barclay, A.N.: The immunoglobulin superfamily - domains for cell surface recognition. *Annu Rev Immunol 6:* 381-405, 1988 [illegible], M.F., [illegible] C., [illegible] J., and [illegible] (19[illegible]) Which comes first, MHC class I or class [illegible] [illegible] 708 [illegible]

[illegible], M.T., [illegible] [illegible] ... [illegible] of [illegible] major histocompatibility complex ... [illegible] [illegible] class I ... [illegible]

[illegible]

GENERATION OF ALLELIC POLYMORPHISM AT THE DRB1 LOCUS OF PRIMATES BY EXCHANGE OF POLYMORPHIC DOMAINS: A PLAUSIBLE HYPOTHESIS?

Ulf B. Gyllensten*, Mats Sundvall*, and Henry A. Erlich+

*Department of Medical Genetics Biomedical Center, Box 589, University of Uppsala, S-751 23 Uppsala, Sweden

+Department of Human Genetics, Cetus Corporation, 100 Fifty-Third Street, Emeryville, CA 94608 USA.

The allelic polymorphism at the DRB1 locus of chimpanzee and gorilla was studied by nucleotide sequence analysis of amplified genomic fragments from the second exon. None of the combinations of polymorphic motifs found on chimpanzee and gorilla alleles have similar allelic counterparts in humans, indicating that many HLA-DRB1 alleles have been generated since the divergence of the three hominoid species. Phylogenetic tree analysis of different parts of the second exon indicate that allelic variation at the DRB1 locus could have been generated in part by inter-allelic and inter-locus sequence exchanges. A putative point for such exchange is located at the transition point between sequences encoding the ß-sheet and those encoding one of the α-helices. These results imply that allelic variation in the ß-sheet of the antigen is conserved between species, while the part encoding one of the α-helices has accumulated a number of species specific residues. The segments of genes encoding parts of the DR antigen involved in the peptide binding, and those encoding parts involved in the interaction with the T-cell receptor, may thus be under different selection pressures.

Introduction

The origin and maintenance of the allelic polymorphism at HLA class II loci has been the cause of much controversy and speculation (Trowsdale et al. 1985; Kappes and Strominger 1988; Horn et al. 1988; Todd et al. 1987; Hughes and Nei 1989; Bodmer 1972). Early proposals for a mechanism generating the allelic polymorphism at *DRB1* locus by sequence exchange were based exclusively on a few human sequences. Gorski and Mach (1986) suggested based on sequence similarities between *HLA-DRB1* and *HLA-DRB3* alleles that second exon sequences from *DRB3* on a *DRw6* haplotype were donated by gene conversion to the *DRB1* locus, thereby creating the *DR3* allele from the ancestral *DRw6* allele of *DRB1*. Similarly, Wu et al. (1986) have reported putative exchanges between sequence on *HLA-DRB1* and *HLA-DRB5* on the *DR2* haplotype. However, at the time of these reports the age of the polymorphism at both *DRB1*, *DRB3* and *DRB5* was not known and along with the

NATO ASI Series, Vol. H 59
Molecular Evolution of the
Major Histocompatibility Complex
Edited by J. Klein and D. Klein

limited knowledge of the allelic spectrum within humans, these data were merely indicative of the possibility of such exchanges.

In addressing the mechanism for generating the allelic polymorphism we choose to examine the chimpanzee and the gorilla; the two species most closely related to humans. The short divergence time of the hominoides makes it unlikely that mutational events are masked by multiple changes, thus allowing individual mutations to be identified. Also, comparisons of chimpanzee and gorilla alleles to those of humans, whose allelic spectrum is to a large extent known, may permit the mutations accumulating on individual allelic types to be determined. Thus, the relative selection pressures on different alleles can be studied. Finally, the pattern of allelic variation in non-human primates may shed light on the evolutionary age of the residues implicated in the susceptibility to autoimmune diseases in man.

Analyses of the allelic variation at *DQA1* (Gyllensten and Erlich 1989), *DQB1* (Gyllensten et al. 1990) and class I loci (Lawlor et al. 1988) of nonhuman hominoids have all provided support for the trans-species hypothesis, postulating that the majority of polymorphism is inherited from the ancestral species and only minor modifications have accumulated since (Klein 1987; Figueroa et al. 1988). Thus, there is no support for the notion that the polymorphism has been generated independently in each species since speciation, through mechanisms such as high mutation rates or a unique mechanism for generating polymorphism (Nathenson et al. 1986). For example, changes between hominoid *DQA1* alleles can be explained almost exclusively by the accumulation of single point mutations occuring on ancestral allelic types and high mutation rates or recombinational exchanges do not need to be invoked. Only for one allele at *DQA1* (the *DQA1*0201*) is there an indication, based on it sequence homology and exclusive presence in the human species, that it could have been generated after the divergence of the hominoids by a recombinational process (Gyllensten and Erlich 1989).

At the *DQB1* locus, evolutionary lineages or allelic types found in the human species can be traced into several other hominoids, although their distribution may differ between species (Gyllensten et al. 1990). A high degree of clustering of individual sequences by species at the terminal twigs of the phylogenetic tree rather than with sequences from other hominoids, indicate that *DQB1* allelic forms have accumulated additional changes subsequent to the separation of the species (Gyllensten et al. 1990). This post-speciation diversification among *DQB1* alleles suggest that continued modification of the basic allelic types are occuring at this locus. In general, the strong similarity of allelic types encoding the α- and ß-chain of the DQ antigen between hominoid species, both at the protein and silent nucleotide position level, is consistent with the allelic types (or allelic lineages) being due to common ancestry, rather than recent origin. Again, only for a limited number of the alleles does it seem necessary to invoke exchanges among other allelic types.

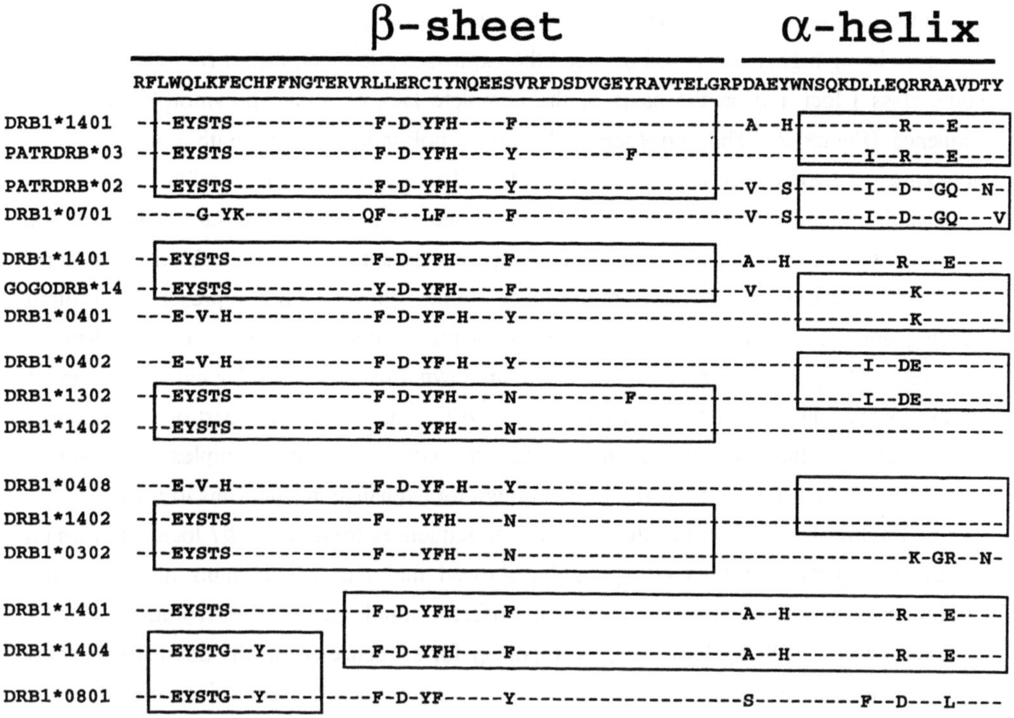

Fig. 1. Alignment of second exon sequences from the *DRB1* locus of primates. The boxes indicate the extent of sequence similarity between alleles and could represent putative sequences exchanged.

The availability of the large spectrum of HLA-DRB1 alleles, along with a partial spectrum for other hominoids have permitted the evolutionary antiquity of alleles to be studied in more detail. The general patchwork pattern of sequence polymorphisms characteristic for human alleles, is also present in chimpanzee and gorilla (Fan et al. 1989; Gyllensten et al. 1991a,b). However, sequence motifs encoded by the left part of the second exon have, with a few exceptions, identical or similar counterparts in several or all hominoid species. By contrast, the sequences in the right half of the same exon show much less similarity between species, and display a higher frequency of sequence motifs that are unique to each individual species (Gyllensten et al. 1991a). This pattern of variation support the notion that alleles at this locus could have been generated by combining different variant motifs encoded by the second exon (Figure 1; Gyllensten et al. 1991a). The left portion of the exon encode part of the ß-sheet and the right part one of the α-helices in the postulated 3-D model for the class II antigen (Bjorkman et al. 1987; Brown et al. 1988), and the exchanges of sequence motifs may thus involve combining variants of these two functional domains. A common point of sequence exchange could be located at the transition point between sequences encoding the two domains. We have postulated one such putative transistion point on the basis of the sequence being highly conserved between the expressed *DRB* loci, the *DPB1* locus and even some class I loci and on the basis of the complete lack of allelic polymorphism in this sequence (Figure 2). The sequence is however, polymorphic both within and between nonexpressed *DRB* loci (Gyllensten et al. 1991b). This sequence shows simlilarity to human minisatellite sequences as well as to the *E.coli* chi sequence. In addition, our putative site for exchange has been found to contain the sequence elements necessary to bind the general transcription activator, SP1, and preliminary studies shows that it represents a protein binding domain (Sundvall et al., manuscript in preparation). The molecular mechanism involved in the putative events are at present unknown. Possibly, a similar sequence element in the downstream intron, or in the case of the *DPB1*, in the righthand part of the α-helical encoding part of the exon, may act as second target site for a protein complex that will in a loop-like fashion bring the two sites close enough for exchange mechanisms to operate.

Phylogenetic analysis of the second exon sequences from the *DRB1* locus, performed on the two parts of the exon separately, showed that the two domains have different evolutionary histories, as reflected in the different topologies of the evolutionary trees (Gyllensten et al. 1991a). However, the topology for sequences from the downstream third exon was to a large extent similar to that for the segment of the second exon encoding the part of the ß-sheet, indicating that the sequence elements exchanged are limited to parts of the second exon and possibly also short segments of the downstream intron (Gyllensten et al. 1991a). Alternate phylogenetic trees were compared by considering the sequences in groups of three and comparing all possible topologies. The number of positions favoring each topology was then calculated for the two parts of the second exon and the third exon (Table 1). Given that the rate of evolution is not significantly different between the different genomic segments, the probability for each distribution can be used for statistical inferences (Table 1; Gyllensten et al. 1991a,b).

The phylogenetic tree analysis showed that the part encoding the ß-sheet was to a large extent identical between species, and conserved with very few species-specific changes. By contrast, the sequences encoding one of the α-helices or the antigen clusters to a large extent with sequences from the same species. This suggests that the sequences encoding one of the α-helices have to a higher degree accumulated substitutions after the divergence of the species. These contrasting patterns of evolution for different parts of the same exon indicate partly different functions for the polymorphism in the two parts of the molecule. The pattern of post-speciation divergence found for the sequences encoding one of the α-helices may indicate a need for local or unique mutational patterns in this part of the molecule for each species. Such mutations may be selected for as a response to the need for specific immune recognition patterns, or as a reflection of lack of constraints on this part of the molecule.

Table 1. Distribution of sites in the ß-sheet and α-helix of exon 2 and in exon 3 favoring each of three possible branching orders of groups of three *DRB1* sequences.

Sequences compared	Domain/ Exon	Topology (A+B) vs C	(A+C) vs B	(B+C) vs A	Probability (a)
A. DRB1*0404	2nd exon β-sheet	1	1	13	>0.05
B. DRB1*1302	2nd exon α-helix	0	6	0	>0.05
C. DRB1*1402	3rd exon, whole	0	0	9	>0.01
A. DRB1*0301	2nd exon β-sheet	18	0	1	>0.01
B. DRB1*1301	2nd exon α-helix	0	0	9	>0.01
C. DRB1*0103	3rd exon, whole	9	0	1	>0.05
A. DRB1*0103	2nd exon β-sheet	0	18	0	>0.01
B. DRB1*1301	2nd exon α-helix	6	0	0	>0.05
C. DRB1*0102	3rd exon, whole	0	10	0	>0.01

```
RFLWQLKFECHFFNGTERVRLLERCIYNQEESVRFDSDVGEYRAVTELGRPDAEYWNSQKDLLEQRRAAVDTY

                    R  A  V  T  E  L  G  R  P
HLA-DR1-6,8-10      CCGGGCGGTGACGGAGCTGGGGCGGCC
Pan,Gorilla         ---------------------------
Macaca, Papio       ----------T----------------

HDRB2               --A---A--T-----A-----------
Pan troglodytes     --A---------A--A-----------
Pan troglodytes     --A------------A-------A---
Pan paniscus        --A------------A-------A---
Pan paniscus        --A------A-----A-----------
Pan paniscus        --A---A-----A--A-----------
Gorilla             --A---A-----A--A-----------
Gorilla             --A------A-----A-----------
Gorilla             --A--------T---A----A------
Papio               --AA-----------A-----------

HDRB3*0101          -----------G---------------
Non-human hom.      ---------------------------

Myoglobin                         ----G----CA--
33.11                              -T-G----CA--
33.1                                A------CA--
```

Fig. 2. Alignment of nucleotide sequences from the second exon of the *Mhc* genes of primates, postulated to be involved in the interallelic and interlocus exchanges of sequences.

The frequency of interallelic exchanges occurring at the putative site is unknown, but as judged from the allelic sequences that can be explained by such exchanges, they represent a majority, albeit not all cases (McClure et al. 1991).

The two alternate hypotheses to sequence exchange for generating the *DRB1* allelic divergence, i.e., ancestral alleles and post-speciation divergence or convergent evolution, have little in favor of them. The lack of *DRB1* alleles in the nonhuman hominoids that share similarity to specific human alleles across the whole exon, and therefore can be traced over species boundaries indicates that while sequence motifs in certain parts of the allele are conserved, entire allelic sequences are to a much lesser extent. Thus, the ancestral allele hypothesis appears insufficient to explain the allelic polymorphism at *DRB1*, unless the unlikely assumption is made that the comparison is biased since by chance none of the alleles derived from common ancestral alleles have yet been found.

The similarity of amino acid sequences between species could, theoretically, be explained as the result of convergent evolution or parallell mutations. However, the prediction of convergent evolution, namely the appearance of similar amino acid sequences independent of the evolutionary origin of the nucleotide sequence, also imply that different codons are likely to be used in different phylogenetic groups. The codon usage in primate *DRB* alleles give little indication of but a single of degenerate codons used in several hominoids. Also, the pattern of very limited polymorphism at silent nucleotide positions among primate *DRB1* alleles do not indicate independent origins for the allelic types with similar amino acid motifs. This does not preclude the existence of convergent evolution operating on other MHC genes, or over other evolutionary timescales.

The conservation of the putative signal sequence between expressed *DRB* genes, as well as the fact that it is monomorphic within species, indicate that sequence exchanges may occur also between loci (Gyllensten et al. 1991b). The exception being the *DRB2* pseudogene which is quite polymorphic within and between species in the putative exchange signal (Table 1; Gyllensten et al. 1991b). In fact, nonhuman hominoids contain several sequences that could have been generated by such inter-locus exchanges (Gyllensten et al. 1991b). This does not imply that exchange events between loci necessarily are frequent. For example, alleles similar to those expected by recombination between *DRB1* and *DRB3*, discussed previously, have been found in both human and gorilla, strongly suggesting that these events predate speciation and may be relatively rare. A number of additional unique changes have accumulated in the gorilla sequence encoding the α-helix, in support of the putative *DRB1-DRB3* exchange predating speciation (Gyllensten et al. 1991b). Thus, in contrast to the allelic variation at the *DRB1* where combinations of sequence motifs could not be traced between species, some putative inter-locus exchanges are found in several species.

Acknowledgements. This work was supported by a Knut and Alice Wallenberg Foundation Fellowship (U.G.) and grants from the Swedish Natural Sciences Research Council (U.G.) and the NIH (H.E.).

References

Bjorkman, P.J., Saper, M.A., Samraoui, B., Bennett, W.S., Strominger, J.L., and Wiley, D.C.: *Nature 329*: 506-512. 1987

Bodmer, W.F.: Evolutionary significance of the HL-A system. *Nature 237*: 139- 1972.

Brown, J.H., Jardetzky, T., Saper, M.A., Samaraoui, B., Bjorkman, P.J., and Wiley, D.C.: A hypothetical model of the foreign antigen binding site of class II histocompatibility molecules. *Nature 332*: 845-850. 1988

Fan, W., Kasahara, M., Gutknecht, J., Klein, D., Mayer, W.E., Jonker, M., and Klein, J.: Shared class II Mhc polymorphism between human and chimpanzees. *Hum Immunol 26*: 107-121, 1989

Figueroa, F., Günther, E., and Klein, J.: Mhc polymorphism predating speciation. *Nature 335*: 265-268. 1988

Gorski, J. and Mach, B.: Polymorphism of human Ia antigens: gene conversion between two DRb loci results in a HLA-D/DR specificity. *Nature 322*: 67-70, 1986

Gyllensten, U. and Erlich, H.A.: Ancient roots for polymorphism at the DQα locus of primates. *Proc Natl Acad Sci USA 86*: 9986-9990. 1989

Gyllensten, U., Lashkari, D., and Erlich, H.A.: Allelic diversification at the class II DQB locus of the mammalian major histocompatibility complex. *Proc Natl Acad Sci USA 87*: 1835-1839, 1990

Gyllensten, U., Sundvall, M., and Erlich, H.: Allelic diversity is generated by intra-exon exchange at class II Mhc DRB1 locus of primates. *Proc Natl Acad Sci USA 88*: 3686-3690, 1991a

Gyllensten, U., Sundvall, M., Ezcurra, I., and Erlich, H.: Genetic diversity at class II DRB loci of the primate major histocompatibility complex. *J Immunol*, in press 1991b

Horn, G.T., Bugawan, T.L., Long, C., and Erlich, H.A.: Allelic variation of HLA-DQ loci: Relation to serology and to insulin-dependent diabetes susceptibility. *Proc Natl Acad Sci USA 85*: 6012-6016, 1988

Hughes, A. and Nei, M.: Nucleotide substitution at major histocompatibility complex class II loci: Evidence for overdominant selection. *Proc Natl Acad Sci USA 86*: 958-962, 1989

Kappes, D. and Strominger, J.L.: Human class II major histocompatibility complex genes and proteins. *Annu Rev Biochem 57*: 991-1028, 1988

Klein, J.: Origin of the major histocompatibility complex polymorphisms: The trans-species hypothesis. *Hum Immunol 19*: 155-162, 1987

Lawlor, D.A., Ward, F.E., Ennis, P.D., Jackson, A.P., and Parham, P.: HLA-A and B polymorphism predate the divergence of human and chimpanzees. *Nature 335*: 268-271, 1988

McClure, G.R., Ruberti, G., Fathman, C.G., Erlich, H.A., and Begovich, A.B.: DRB1*LY10 - A new DRB1 allele and its haplotypic association. *Immunogenetics*, in press 1991

Nathenson, S.G., Geliebter, J., Pfaffenbach, G.M., and Zeff, R.A.: *Annu Rev Immunol 4*: 471-502, 1986

Todd, J.A., Bell, J.I., and McDevitt, H.O.: HLA-DQß gene contributes to susceptibility and resistance to insulin-dependent diabetes mellitus. *Nature 329*: 599-604, 1987

Trowsdale, J., Young, J.A.T., Kelly, A.P., Austin, P.J., Carson, S., Meunier, H., So, A., Erlich, H.A., Spielman, R., Bodmer, J., and Bodmer, W.F.: Structure sequence and polymorphism in the HLA-D region. *Immunol Rev 85*: 5-43, 1985

Wu, S., Saunders, T., and Bach, F.: Polymorphism of human Ia antigens generated by reciprocal exchange between two DRß loci. *Nature 324*: 676-679, 1986

A PHYLOGENETIC INVESTIGATION OF MHC CLASS II *DRB* GENES REVEALS CONVERGENT EVOLUTION IN THE ANTIGEN BINDING SITE

K. Gustafsson, U. Brunsberg, S. Sigurdardottir, and L. Andersson
Department of Animal Breeding and Genetics, Swedish University of Agricultural Sciences, Biomedical Center, Box 596, S-751 24 Uppsala, Sweden

The evolution of Mhc *class II polymorphism has been studied by comparing the nucleotide and amino sequences of a number of pig, cattle, horse, dog, rodent, and human* DRB *alleles. Very similar sets of residues were found at several highly polymorphic amino acid positions in the putative antigen recognition site (*ARS*). As a consequence of this, in the* ARS, *certain alleles in several species have counterparts in other species that are more similar to each other than to any other intra-species allele. In contrast, the frequencies of silent nucleotide substitutions were found to be much higher in comparison between species than within species, implying that none of these alleles originated before the divergence of these distantly related species. An analysis of codon usage, as well as the use of silent nucleotides in individual codons, suggests that many of these similarities are due to convergent evolution.*

The major histocompatibility complex (*MHC*) class II genes encode cell surface molecules expressed on antigen presenting cells. Foreign antigens are 'processed' intracellularly to enable the binding of the resulting peptides to the MHC class II molecules. This complex is recognized by helper T cells forming the initiation of an antigen specific immune response (Buus et al. 1987). Likewise, 'self antigens' are bound and presented to naive T cells subsequently deleted in order to avoid autoimmune phenomena (Lorentz and Allen 1988). It is in this context the polymorphism of the class II molecule, residing mainly in the postulated antigen binding site (ABS; Brown et al. 1988), should be seen. The origin of and the forces promoting this polymorphism have been in the focus of much attention and debate. It is by now clear that a majority of this polymorphism originates by point mutations accumulating at a rate not markedly different from that at other loci in the genome (Hayashida and Miyata 1983). A number of examples indicate that a gene conversion-like mechanism occasionally contributes to the generation of polymorphism by shuffling parts of the class II genes between alleles (Parham et al. 1989; Wakeland et al. 1990). In addition, gene conversion-like mechanisms have been proposed as an explanation for non-allelic class II B gene similarities (Denaro et al. 1984; Gorski and Mach 1986). We have recently suggested that similarities between the regions encoding the α-helices of *DRB* and *DQB* homologues in several species are due to such mechanisms (Andersson et al. 1991).

NATO ASI Series, Vol. H 59
Molecular Evolution of the
Major Histocompatibility Complex
Edited by J. Klein and D. Klein

However, irrespective of what mechanism is generating *MHC* diversity it has been clearly shown that the extreme polymorphism is favoured by natural selection (Hughes and Nei 1989; Jonsson et al. 1989). In addition, as previously hypothesized (Klein 1980), it has been shown that *MHC* class II polymorphism is evolutionarily conserved such that individual alleles may be shared between closely related species (Figueroa et al 1988; Lawlor et al. 1988; Mayer et al. 1988; McConnel et al. 1988; Gyllensten et al 1990). We and others have recently noted that the polymorphic residues in individual positions of the class II DRB chain are frequently shared between even distantly related species (Sarmiento et al. 1990; Andersson et al. 1991). Moreover, in two isolated examples, individual alleles of these different species were shown to have constellations of almost identical ABS residues (Gustafsson et al 1990; Andersson et al. 1991). In order to study further the generality of and mechanisms underlying these intriguing similarities, as well as to address the question of how old the polymorphism and individual alleles can be, we decided to study the phylogenetic relationships of DRB alleles from distantly and closely related mammalian species.

Structure of *DRB* sequences in different mammalian species

The polymerase chain reaction (PCR) was used to amplify a number of first domain exon *DRB*-like sequences from several mammalian species. These were then cloned into M13 mp18/19 phage vectors to facilitate DNA sequencing of individual alleles. The primers used as well as the detailed protocols for amplification and sequencing have been described elsewhere (Sigurdardottir et al. 1991; Gustafsson et al., manuscript in preparation). The assignment of each amplified sequence as *DRB*-like was straightforward since at the nucleotide as well as the amino acid level the similarity to human DQB and DPB sequences was always markedly lower (data not shown). In the swine and horse amplifications *DRB*-like sequences containing pseudogene criteria (such as stop codons and frame shifts) appeared, complicating the analysis of the *DRB* genes in these species. These tentative pseudogenes were, with one exception (see below), omitted from the study. Figure 1 depicts an alignment of the deduced amino acid sequences from the obtained nucleotide sequences together with a number of published *DRB* sequences. Nine *DRB*-like sequences are from swine of which three (*DRB*1*, *DRB*2A* and *DRB*2B*) were amplified from wild boars and four from domestic Swedish Landrace swine. The remaining two (*DRB*c* and *DRB*d*) are from miniature swine cDNA clones previously described (Gustafsson et al. 1990). From the *Bovidae* family one sequence was amplified from an Indian Zebu cow, a bison, a yak, and a water-buffalo, respectively. The remaining nine bovine alleles have been described earlier (Andersson et al. 1991; Sigurdardottir et al. 1991). In addition, one Scandinavian elk and

```
            10        20        30        40        50        60        70        80
            +.+ +      .       + +      ++ .         .      +  +      +  ++  +   + . +   +
DRB1*0101  >WQLKFECHFFNGTERVRLLERCIYNQEESVRFDSDVGEYRAVTELGRPDAEYWNSQKDLLEQRRAAVDTYCRHNYGVGES
DRB1*0401  >E-V-H------------F-D-YF-H---Y---------------------------------K-----------------
DRB1*0701   --G-YK----------QF---LF-----F-------------------V--S------I--D--GQ---V----------
DRB1*0801   EYSTG--Y---------F-D-YF-----Y-------------------S---------F--D---L------------
DRB1*0901   K-D--------------Y-H-G------N-------------------V--S------F--R---E---V----------
DRB1*1001  >EEV------------------RVH----YA-Y-----------------------------R------------------
DRB1*1101   EYSTS------------F-D-YF-----Y---------F----------E--------F--D------------------
DRB1*1201   EYSTG--Y-------------HFH----LL--------F---------V--S------I--D--------------AV--
DRB1*1301  >EYSTS------------F-D-YFH----N---------F-------------------I--DE--------------V--
DRB1*1501   --P-R------------F-D-YF---------------F-------------------I---A--------------V--
DRB3*0101   ELR-S------------Y-D-YFH----FL------------------V--S----------K-GR--N-----------
DRB3*0201   EL--S------------F---HFH----YA------------R-------------------K-GQ--N--------V--
DRB4*0101   E-A-C----L------WN-I-Y------YA-YN--L---Q---------------------R---E------Y----V--
DRB5*0101   Q-D-Y------------F-H-D------DL----------------------------F--D------------------
            +.+ +      .       + +      ++ .         .      +  +      +  ++  +   + . +   +
PIG DRB*c  >H------------------Q-NC--G--Y---------------------K-R----------------------RILDT
PIG DRB*d   HL-------------------QY--G--FL--------------------KD-------------E---------RILDT
PIG DRB*1   YL-------------------QY--G--Y-------------------V-KD-------------E---------RTSDT
PIG DRB*2A  FLV-D----------------QY--G--H---------------------K-------I---K--E---V-----RILDT
PIG DRB*2B  FME----------------D-HY--G--IL-----E------------E-KD---R--I------E---------RISDT
PIG DRB*4   FLR-A--R-----------D-YF--GD-Y---------F-E---F-----K----L--F-----T------------SDG
PIG DRB*5   FLG-----------Q------QY--G--F---------------------KNY------------E---------RTSDT
PIG DRB*6A  FLG-A--------------D-YF--GD-Y---------F-E---F---S-K-------FM--K--E---V-----EIS-T
PIG DRB*6B >HLV-H--R--------L--D-YF--G--Y---------F-----------K-------I--DS--S-----I---RILDT
            +.+ +      .       + +      ++ .         .      +  +      +  ++  +   + . +   +
COW DRB3*1A EYS-S------------F-D-YYT-G--T------W--F--------Q----------F--EK--E--RV------GM--
COW DRB3*1B QYH-G--------------D-HF--G--F------WD-F---------A--Q------F---K--E--RV------GV--
COW DRB3*3 >EYC-S------------F---SF--G--F------W--------------K-----------K--N--------------
COW DRB3*4A>EYSTS------------F-D-YF--G--Y------W--FQ----------K-------I--RE-------------GV--
COW DRB3*5  EYH-S----------L-Y-D-YF--G--Y------W--------------K------EI--RKR-N-----------V--
COW DRB3*6  EYC-R--------------D--FH-G--F------W--F--------RV--HL----EI--RK--E---V----------
COW DRB3*8A EYATS------------F-D-YFH-G--L------W--F---------S-VHL-----F--DE--S-----------V--
COW DRB3*12 EYT-K------------F-N-YFH-G--F------W---------------------EI--RA-----------------
CO DRB3*13A EY--S------------F---YF--G--Y------W--------------K----------RK--N-----------V--
ZEB DRB     EYC-R------------F-D-YF--G--Y------W--F------------------EI--E---E--RV----------
BIS DRB    >EYC-S------------F-D--FH-G--Y------W---------------------EI--RA--------------V--
YAK DRB    >EYSTS------------Y-D-YFH-G--F------W-----------RV--QL-G---T--RE--Y-----------V--
WBU DRB    >QYA-S----S-------F--GHF--G--L-A----W-----A-----QI--Q------F--RK--N--------------
            +.+ +      .       + +      ++ .         .      +  +      +  ++  +   + . +   +
ELK DRB    >EYA-S------------F-D-Y---R--Y---------------------K-----E-I------E--------------
            +.+ +      .       + +      ++ .         .      +  +      +  ++  +   + . +   +
HOR DRB*1ps ELV-H----S---Q---F-D-YF--R--Y--------------K---T----#-G---V-DDA--------------TDT
HOR DRB*2  >EYSTS------------Y-D-YF--GK-Y------------L------------G-Q-I---K--K---------A-S--
HOR DRB*3  >EAV----R-S-------F---RFH-G--YA------------------------G-Q-I-DE---------------IDG
            +.+ +      .       + +      ++ .         .      +  +      +  ++  +   + . +   +
DOG DRB*Dw3>EVA-S--Y-T-------FV--Y-H-R--F---------F---------V--S--G--EI---E--T-----------I--
DOG DRB*D4  YQF-P----T-------FV--H-H-R--F----------------------S--G--E----E--T-----------I--
DOG DRB*D7  EVA-A----T-------FV--Y---R--F-------------------------P--E---RG-----------------
DOG DRB*D8  KMY-A----T-------YLM-D---R--F---------F---------------G--EI--RK-----------------
DOG DRB*Dw8>EMV------T---------V-D---R--H-P-----------------------G--E-------E---V-------I--
            +.+ +      .       + +      ++ .         .      +  +      +  ++  +   + . +   +
MOU Eb*b   >EYC-S----Y---Q-------YF--L--NL--------F------------N----PEF---K--E---V-----E-S-K
MOU Eb*d    EYVTS----Y---QH--F---FI--R--NL---------------------N----PEI--DA--S---------E-S-K
MOU Eb*s   >EYSTS----Y---Q-------YF--L--NL--------F------------N----PEF----------------E-L-K
MOU Eb*u    GYSTS----Y---Q---F-D-YF--R--W-------------------E--N----PEI---T------------E-S-K
            +.+ +      .       + +      ++ .         .      +  +      +  ++  +   + . +   +
RAT Db*u   >GY-------Y---Q-----A-L---R--YA------------------S---R-K--EFM-R-------------E-F-R
            + + +              + +      ++                  +  +      +  ++  +   +   +   +
No of diff. 8  12              8       10                   6  5      5  4   9   2   1   8
residues     16                12       3                                 7
```

Fig. 1. DRB first domain amino acid sequences from 11 mammalian species. The top 14 sequences are from human *DRB* genes. ZEB is an abbreviation for Zebu cattle, BIS for Bison, WBU for Water-Buffalo, HOR for Horse and MOU for Mouse. HOR *DRB*1ps* indicates that this is likely to be a pseudogene. + indicates positions in the postulated ABS. Below the sequences are the respective total numbers of polymorphic residues found in the ABS positions. Boxed sequences represent cross-species similarities in the ABS described in the text. Arrows to the left of sequences mark those included in the analyses in figures 2, 3, 5, and 6. References to published sequences are in the text.

three horse *DRB*-like sequences were amplified and sequenced. The remaining sequences have been published previously and are from man (Marsh and Bodmer 1989, 1990 and references therein), dog (Sarmiento et al. 1990), mouse (Mengle-Gaw and McDevitt 1983; Saito et al. 1983; Denaro et al. 1984; Ayane et al. 1986), and rat (Robertson and McMaster 1985).

It is evident from Figure 1 that a major part of the polymorphism resides in the residues belonging to the putative ABS. Position 11 appears to be the most polymorphic one with 16 different residues used followed by positions 13, 30, and 37 with 12, 12, and 10 different residues, respectively. Position 82 is the only monomorphic residue in the ABS. We feel it is likely, in view of the number of alleles and the wide representation of *DRB* sequences from different mammalian orders included, that the relative numbers of polymorphic residues in Figure 1 represent the situation at large among mammalian species. Figure 1 also suggests that previously described isolated examples of sharing of almost entire ABS constellations between alleles of different species (Gustafsson et al. 1990; Andersson et al. 1991) are general phenomena rather than exceptions. The most striking example of allelic similarity between species consists of the human *DRB1*1301*, cow *DRB3*4A* and horse *DRB*2* alleles where the human and the cattle alleles share 14 of 16 ABS residues and the human and horse 11 of 16 residues. This similarity has been noted by us earlier in a comparison of 14 cattle alleles with the human allelic series when another human allele (*DRB1*1102*) actually shared 15 of 16 ABS residues with the same cattle allele (Andersson et al. 1991). Other examples include: human *DRB1*0101* and pig *DRB*c* where the pig and human alleles share 10 of 16 ABS residues (Gustafsson et al. 1990), human *DRB1*1001* and horse *DRB*3* where 13 of 16 residues are shared and another horse allele (*DRB*1*ps) that shares 12 of 16 ABS residues with the human *DRB1*0401* allele.

Phylogenetic relationships of DRB alleles in distantly and closely related species

In order to determine the evolutionary relationships of the various *DRB* alleles studied we constructed phylogenetic trees using the nucleotide sequences of the first domain exons. Nineteen sequences were selected to represent major allelic groups in various species. The trees were rooted with the human *DOB* sequence. The frequencies of replacement and silent substitutions in all pairwise comparisons were determined using a computer program kindly provided by Dr. M. Nei, University of Texas Health Science Centre, USA (Nei and Gojobori 1986), as was the program for the construction of all phylogenetic trees according to the 'neighbour-joining' method (Saitou and Nei 1987).

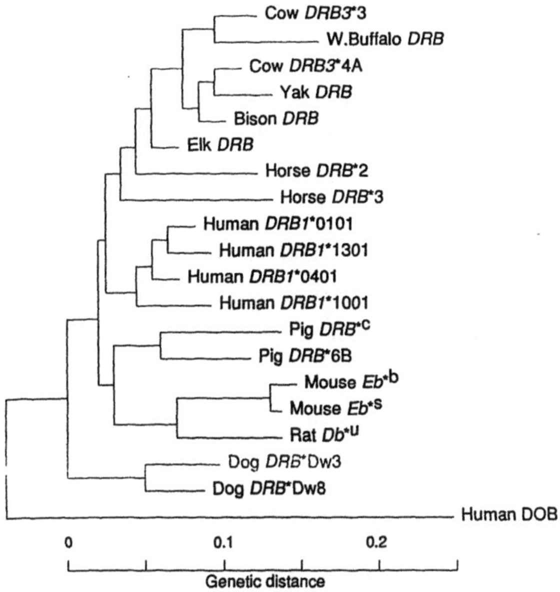

Fig. 2. Phylogenetic tree of *DRB* nucleotide sequences from the first domain exon of alleles indicated in Figure 1. Distances were based on data from all substitutions.

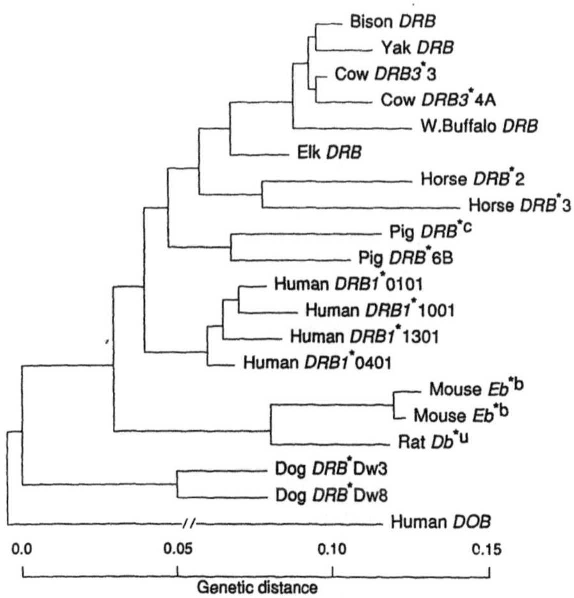

Fig. 3. Phylogenetic tree of *DRB* nucleotide sequences indicated in Figure 1. Distances were based on the first domain exon except ABS codons.

The tree in Figure 2 was constructed using all substitutions (silent as well as replacement) of the entire sequenced portion in the first domain exon. This type of tree connects the alleles of almost all species on one branch each. The fact, however, that the two pig alleles unexpectedly branched together with the rodents as well as the pattern of cross-species sharing of ABS residues described above prompted us to try alternative approaches. Figure 3 shows a tree constructed using data from all substitutions in the first domain exon excluding the 16 codons for the ABS (see Figure 1). Here the horse alleles are branching together as are all other intra-species alleles. In addition, the pig alleles are now branching with other *artiodactyls* as expected.

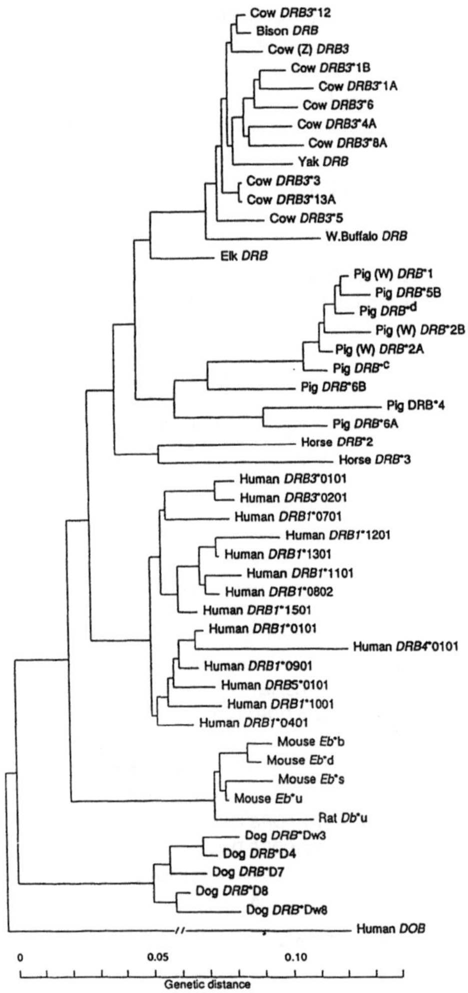

Fig. 4. Phylogenetic tree of 49 *DRB* nucleotide sequences from first domain exons (all substitutions) except ABS codons.

We therefore decided to use this type of data to construct a phylogenetic tree consisting of all *DRB* sequences from Figure 1 except the suspected horse *DRB* pseudogene and the Zebu cattle *DRB* sequence. Thus, Figure 4 constitutes a phylogenetic tree of 50 *DRB* sequences from 11 different species/subspecies. *DRB* alleles from several closely related species in the *Bovidae* family cluster together on the same main branch indicating that a 'trans-species' sharing of these alleles is the most reasonable explanation for their similarities (see Figure 1). This agrees well with published data from other closely related species, within the primate and rodent orders (Figueroa et al 1988; Lawlor et al. 1988; Mayer et al. 1988; McConnel et al. 1988; Gyllensten et al 1990). The tree clearly indicates that all alleles within respective distantly related species must have diverged well after the general mammalian radiation some 70-80 million years ago. It is a well known fact that negative selection against replacements is operating at a majority of non-ABS residues (Hughes and Nei 1989; Jonsson et al. 1989). If this type of selection differs in its requirements between species it is conceivable this could affect the tree in Figure 4. To test for this possibility we constructed a tree based on the 19 sequences used in Figures 2 and 3 using data only from silent sites.

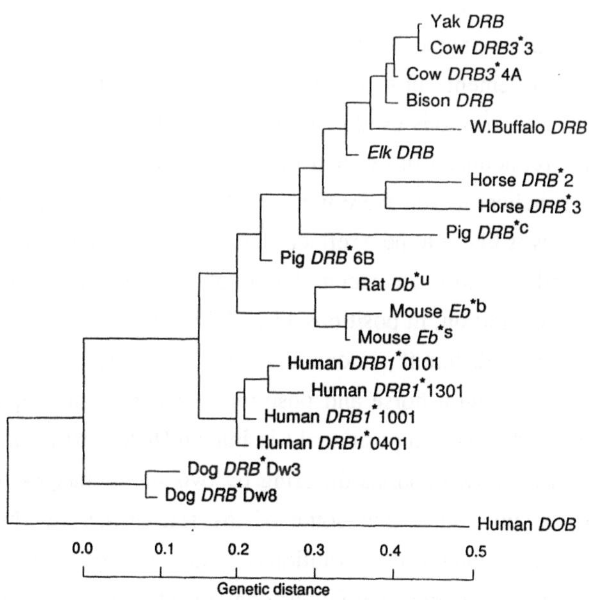

Fig. 5. Phylogenetic tree of synonymous sites except from ABS codons in the indicated *DRB* first domain exon sequences in Figure 1.

Figure 5 shows that such a phylogenetic tree has a very similar general structure, as compared to the tree in Figure 3. The only main difference worth noticing is the different branching pattern of rodent alleles visavi human alleles. The results from these trees agree well with the theoretical expectations of a 'maximum' age of an individual allele falling well short of the main mammalian radiation (Takahata and Nei 1990) and leaves us to explain by other means than 'trans-species transfer' why the striking cross-species allelic similarities persist after this long evolutionary time.

Convergent evolution in part of the ABS

We can envision two alternative explanations for this pattern of similarities. A conservation of individual residues of the ABS in combination with a gene conversion-like mechanism where parts of the class II *DRB* genes are shuffled between alleles could conceivably lead to a situation where short ancestral motifs actually antedate the main mammalian radiation. This would also involve a certain degree of homogenization of silent sites within a species. The alternative explanation is that the sequence similarities are due to convergent evolution.

To study the validity of these two possibilities systematically we aligned the nucleotide as well as amino acid sequences of the ABS residues and respective codons of the 20 *DRB* first domain exons used in Figure 2 representing striking similarities between species (Figure 6). In many silent site positions (i.e. commonly the third position in a codon) the same nucleotide is shared between alleles in one species despite amino acid differences caused by replacement substitutions in the same codon. In contrast, between species, even if the same amino acid is used in a position the silent site often has a different nucleotide. This is most pronounced in the part of the ABS lining the floor of the antigen binding groove. In the α-helical region there is a very low frequency of silent substitutions even between species. This is likely to be a reflection of the biased base composition of this region which is expected to cause an accumulation of parallel nucleotide substitutions (Andersson et al. 1991). In addition, in positions 11 and 13 a number of *Serine* (S) residues are encoded for by entirely different codons in different species. For example, when comparing position 11 in cow with horse and position 13 when comparing mouse/human with cow as well as cow with horse. In position 11 in human *DRB1*0101* and pig *DRB*^{C} the *Leucine* (L) is encoded for by two codons differing by two silent changes. Whilst we can presently not exclude the possibility that some of the similarities are due to ancestral sequence motifs conserved throughout the evolution of mammals, due to the differing codon usage and the silent substitution pattern we favour convergent evolution as the explanation for many of the cross-species allelic similarities.

	9	11	13	28	30	37	38	57	60	67	70	71	74	78	82	86
DRB1*	W	L	F	E	C	S	V	D	W	L	Q	R	A	Y	N	G
0101	TGG	CTT	TTT	GAA	TGC	TCC	GTG	GAT	TGG	CTC	CAG	AGG	GCG	TAC	AAC	GGT
DRB1*	E	V	H	D	Y	Y	V	D	W	L	Q	K	A	Y	N	G
0401	GAG	GTT	CAT	GAC	TAC	TAC	GTG	GAT	TGG	CTC	CAG	AAG	GCG	TAC	AAC	GGT
DRB1*	E	S	S	D	Y	N	V	D	W	I	D	E	A	Y	N	V
1301	GAG	TCT	TCT	GAC	TAC	AAC	GTG	GAT	TGG	ATC	GAC	GAG	GCG	TAC	AAC	GTG
DRB1*	E	V	F	E	R	Y	A	D	W	L	R	R	A	Y	N	G
1001	GAG	GTT	TTT	GAA	CGC	TAC	GCG	GAT	TGG	CTC	CGG	AGG	GCG	TAC	AAC	GGT
PIG	H	L	F	Q	N	Y	V	D	R	L	Q	R	A	Y	N	L
DRB*c	CAC	TTG	TTC	CAG	AAT	TAC	GTG	GAC	CGG	CTC	CAG	AGG	GCG	TAC	AAC	TTG
PIG	H	V	H	D	Y	Y	V	D	W	I	D	S	S	Y	N	L
DRB*6A	CAC	GTG	CAC	GAC	TAC	TAC	GTG	GAC	TGG	ATC	GAC	TCA	TCA	TAC	AAC	TTG
COW	E	C	S	E	S	F	V	D	W	L	Q	K	N	Y	N	G
DRB3*3	GAG	TGT	AGC	GAG	TCC	TTC	GTG	GAC	TGG	TTG	CAG	AAG	AAT	TAC	AAC	GGT
COW	E	S	S	D	Y	Y	V	D	W	I	R	E	A	Y	N	V
DRB3*4A	GAG	TCT	AGC	GAC	TAC	TAC	GTG	GAC	TGG	ATC	CGG	GAG	GCG	TAC	AAC	GTG
BIS	E	C	S	D	C	Y	V	D	W	I	R	A	A	Y	N	V
DRB	GAG	TGT	AGC	GAC	TGC	TAC	GTG	GAC	TGG	ATC	CGG	GCG	GCG	TAC	AAC	GTG
YAK	E	S	S	D	Y	F	V	V	L	T	R	E	Y	Y	N	V
DRB	GAG	TCT	AGC	GAC	TAC	TTC	GTG	GTC	TTG	ACC	CGG	GAG	TAT	TAC	AAC	GTT
WBU	Q	A	S	E	H	L	V	I	W	F	R	K	N	Y	N	G
DRB	CAG	GCT	AGC	GAG	CAC	TTG	GAG	ATC	TGG	TTT	CGG	AAG	AAT	TAC	AAC	GGT
ELK	E	A	S	D	Y	Y	V	D	W	I	Q	R	E	Y	N	G
DRB	GAG	GCT	AGC	GAC	TAC	TAC	GTG	GAC	TGG	ATC	CAG	AGG	GAG	TAC	AAC	GGC
HOR	E	S	S	D	Y	Y	V	D	W	I	Q	K	K	Y	N	S
DRB*2	GAG	AGT	TCC	GAC	TAC	TAC	GTG	GAC	TGG	ATC	CAG	AAG	AAG	TAC	AAC	AGC
HOR	E	V	F	D	R	Y	A	D	W	I	E	R	A	Y	N	I
DRB*3	GAG	GTG	TTC	GAG	CGC	TAC	GCG	GAC	TGG	ATC	GAA	CGG	GCG	TAC	AAC	ATT
DOG	E	A	S	E	Y	F	V	V	W	I	Q	E	T	Y	N	I
DRB*Dw3	GAG	GCA	TCC	GAA	TAC	TTC	GTG	GTC	TGG	ATC	CAG	GAG	ACG	TAC	AAC	ATT
DOG	E	V	F	V	D	H	V	D	W	L	Q	R	E	V	N	I
DOG*Dw8	GAG	GTA	TTC	GTG	GAC	TTC	GTG	GAC	TGG	CTC	CAG	AGG	GAG	GTG	AAC	ATT
MOU	E	C	S	E	Y	N	V	D	W	F	Q	K	E	V	N	S
Eb*b	GAA	TGT	TCT	GAA	TAC	AAC	CTG	GAC	TGG	TTC	CAA	AAG	GAG	GTG	AAC	TCG
MOU	E	S	S	E	Y	N	V	D	W	F	Q	R	A	Y	N	L
Eb*s	GAA	TCT	TCT	GAA	TAC	AAC	CTG	GAC	TGG	TTC	CAA	AGG	GCG	TAC	AAC	TTG
RAT	G	L	F	A	L	Y	A	S	R	F	R	R	A	Y	N	F
Db*u	GGA	CTT	TTT	GCC	TTA	TAC	GTG	TCA	AGG	TTC	CGG	AGG	GCG	TAC	AAC	TTT

Fig. 6. Alignment of nucleotide sequences and amino acid residues belonging to the putative ABS. Numbering above sequences refers to respective position in the *DRB* first domain. Boxes highlight examples of different codon usage or codons differing by two individual synonymous changes. Sequences used are the same as in Figures 2, 3, and 5. Abbreviations used are: BIS - Bison, WBU - Water-Buffalo, HOR - Horse and MOU - Mouse.

Evolution of the ABS

It is our firm belief that none of the different mechanisms previously invoked in the development and maintenance of *MHC* class II polymorphism (see the introductory paragraph) contradict each other. Nor does *convergent evolution* contradict any other suggested mechanism. However, some of these factors are clearly more important as seen in certain time perspectives. Single mutations fixed by positive selection for polymorphism, fine tuned for each of the 16 ABS residues, sets the scene. Gene conversion-like events can drastically change the structure of an individual class II allele. This could occasionally be beneficial and lead to new allelic variants. However, in a majority of cases it is likely to be a lost cause due to the action of selection. We as well as others can best explain the allelic similarities between closely related species with individual allelic origins predating speciation events. In parallel, striking sequence similarities are also observed between distantly related species. Here, we believe convergent evolution is an important contributing factor. This has implications, not only for the evolution of MHC class II genes *per se*, but also for the function of the class II molecule since it must be phenotypic selection that is chiefly responsible for *such* similarities. The observations suggest that the structural repertoires of antigenic peptides bound by the similar alleles in different species also are similar. If so, our data would imply that distantly related mammals coevolve with parasites in a parallel fashion.

Acknowledgments. We thank Dr. Masatoshi Nei for providing us with computer programs used in the phylogenetic analyses. This study was supported by the Swedish Cancer Society and the Swedish Research Council for Forestry and Agriculture.

References

Andersson, L., Sigurdardottir, S., Borsch, C., and Gustafsson, K.: Evolution of *MHC* polymorphism: extensive sharing of polymorphic sequence motifs between human and bovine *DRB* alleles. *Immunogenetics 33*:188-193, 1991

Ayane, M., Mengle-Gaw, L., McDevitt, H.O., Benoist, C., and Mathis, D.: $E\alpha^u$ and $E\beta^u$ chain association: where lies the anomaly? *J Immunol 137*:948-951, 1986

Brown, J.H., Jardetzky, T., Saper, M.A., Samraoui, B., Bjorkman, P.J., and Wiley, D.C.: A hypothetical model of the foreign antigen binding site of class II histocompatibility molecules. Nature 332:845-850, 1988

Buus, S., Sette, A., and Grey, H.: The interaction between protein-derived immunogenic peptides and Ia. *Immunol Rev 98*:115-141, 1987

Denaro, M., Hammerling, U., Rask, L., and Peterson, P.A.: The $E\beta^b$ gene may have acted as the donor gene in a gene conversion-like event generating the $E\beta^{bm12}$ mutant. *EMBO J 3*:2029-2032, 1984

Figueroa, F., Gunther, E., and Klein, J.: *MHC* polymorphism predating speciation. *Nature 335*:265-267, 1988

Gorski, J. and Mach, B. Polymorphism of human Ia antigens: gene conversion between two DR ß loci results in a new HLA D/DR specificity. *Nature 322*:67-70, 1986

Gustafsson, K., Germana, S., Hirsch, F., Pratt, K., LeGuern, C., and Sachs, D.H.: Structure of miniature swine class II *DRB* genes: conservation of hypervariable amino acid residues between distantly related mammalian species. *Proc Natl Acad Sci USA 87*:9798-9802, 1990

Gyllensten, U.B., Lashkari, D., and Ehrlich, H.A.: Allelic diversification at the class II *DQB* locus of the mammalian major histocompatibility complex. *Proc Natl Acad Sci USA 87*:1835-1839, 1990

Hayashida, H. and Miyata, T.: Unusual evolutionary conservation and frequent DNA segment exchanges in class I genes of the major histocompatibility complex. *Proc Natl Acad Sci USA 80*:2671-2675, 1983

Hughes, A.L. and Nei, M.: Nucleotide substitutions at major histocompatibility complex class II loci: evidence for overdominant selection. *Proc Natl Acad Sci USA 86*:958-962, 1989

Jonsson, A.-K., Andersson, L., and Rask, L.: Selection for polymorphism in the antigen recognition site of major histocompatibility complex class II molecules. *Scand J Immunol 30*:409-417, 1989

Klein, J.: Generation of diversity at *MHC* loci: implications for T-cell repertoires. *In* K. Fougereau and J. Dausset (eds) *Immunology 80*, pp. 239-253, Academic Press, London 1980

Lawlor, D.A., Ward, F.E., Ennis, P.D., Jackson, A.P., and Parham, P.: HLA-A and B polymorphisms predate the divergence of humans and chimpanzees. *Nature 335*:268-271, 1988

Lorentz, R.G. and Allen, P.M.: Processing and presentation of self proteins. *Immunol Rev 106*:115-127, 1988

Marsh, S.G.E. and Bodmer, J.G.: *HLA-DR* and *-DQ* epitopes and monoclonal antibody specificity. *Immunol Today 10*:305-312, 1989

Marsh, S.G.E. and Bodmer, J.G.: *HLA-DRB* nucleotide sequences. *Immunogenetics 31*:141-144, 1990

Mayer, W.E., Jonker, M., Klein, D., Ivanyi, P., van Seventer, G., and Klein, J.: Nucleotide sequences of chimpanzee MHC class I alleles: evidence for *trans*-species mode of evolution. *EMBO J 7*:2765-2774, 1988

McConnel, T.J., Talbot, W.S., McIndoe, R.A., and Wakeland, E.K.: The origin of *MHC* class II polymorphism within the genus *Mus*. *Nature 332*:651-654, 1988

Mengle-Gaw, L. and McDevitt, H.O.: Isolation and characterization of a cDNA clone for the murine I-Eß polypeptide chain. *Proc Natl Acad Sci USA 80*:7621-7625, 1983

Nei, M. and Gojobori, T.: Simple methods for estimating the numbers of synonymous and non-synonymous nucleotide substitutions. *Mol Biol Evol 3*:418-426, 1986

Parham, P., Lawlor, D.A., Lomen, C.E., and Ennis, P.D.: Diversity and diversification of *HLA-A,B,C* alleles. *J Immunol 142*:3937-3950, 1989

Robertson, K.A. and McMaster, W.R.: Complete structure of a rat RT1 Eß chain: Extensive conservation of *MHC* class II ß chains. *J Immunol 135*:4095-4099, 1985

Saito, H., Maki, R.A., Clayton, L.K., and Tonegawa, S.: Complete primary structures of the E ß-chain and gene of the mouse major histocompatibility complex. *Proc Natl Acad Sci USA 80*:5520-5524, 1983

Saitou, N. and Nei, M.: The neighbor-joining method: A new method for reconstructing phylogenetic trees. *Mol Biol Evol 4*:406-425, 1987

Sarmiento, U.M., Sarmiento, J.I., and Storb, R.: Allelic variation in the *DR* subregion of the canine major histocompatibility complex. *Immunogenetics 32*:13-19, 1990

Sigurdardottir, S., Borsch, C., Gustafsson, K., and Andersson, L.: Cloning and sequence analysis of 14 *DRB* alleles of the bovine *MHC* by using the polymerase chain reaction. *Anim Genet 22*, in press 1991

Takahata, N. and Nei, M.: Allelic genealogy under overdominant and frequency-dependent selection and polymorphism of major histocompatibility complex loci. Genetics 124:967-978, 1990

Wakeland, E.K., Boehme, S., and She, J.X.: The generation and maintenance of *MHC* class II gene polymorphism in rodents. *Immunol Rev 113*:207-226, 1990

DIVERSIFICATION OF CLASS II Aα WITHIN THE GENUS *Mus*

Ying Ye, Jin-Xiong She, and Edward K. Wakeland
Laboratory of Molecular Biology, Division of Basic Sciences, Department of Pathology and Laboratory Medicine, University of Florida, Gainesville, FL 32610, USA

The exon that encodes the antigen binding site (ABS) of Mhc class II molecules is extremely polymorphic. To understand the molecular and genetic mechanisms responsible for the generation of this diversity, we have assessed the ABS diversity of mouse Aa by determining and comparing nucleotide sequences of exon 2 for 30 alleles derived from 9 species or subspecies. The diversity is concentrated in 5 discrete segments that constitute the interior of the ABS, and the majority of the polymorphisms in these 5 segments arose prior to the divergence of many contemporary mouse species, which have diverged from 0.2 to 2.4 MYR. Comparisons of synonymous and nonsynonymous substitutions confirmed previous observations that strong diversifying selection for ABS diversity is involved in maintaining Mhc polymorphism. Our sequence analyses suggest that, unlike Ab, Aa does not experience extensive genetic exchanges through recombination or gene conversion. Thus, Aa diversity is generated predominantly via the accumulation of point mutations through long evolutionary periods, i.e. via the retention of ancestral polymorphism.

Introduction

Mhc class II molecules are cell membrane glycoproteins mainly expressed on macrophages and B lymphocytes. These molecules present processed antigenic fragments to regulatory T-lymphocytes during the initiation of immune responses. Consequently, the ability of class II molecules to bind and display antigenic peptides influences immune responsiveness of organisms to foreign antigens. The importance of these functions has strongly influenced Mhc evolution. Class II genes exhibit extensive nucleotide sequence diversity (Gyllensten and Erlich 1989; Gyllensten et al. 1991; Gyllensten et al. 1990; Lawlor et al. 1988; She et al. 1990; She et al. 1991). Previous studies indicated that Mhc *Ab* diversity resulted from the retention of ancestral polymorphism, which is further amplified by intra-exonic recombination (She et al. 1990; She et al. 1991). The primary goal of this study is to characterize the diversity of *Aa* exon 2 which encodes the alpha subunit of the molecule. Comparison of *Ab* and *Aa* diversity may give us some insight about the mechanisms generating Mhc diversity and Mhc evolution in general.

NATO ASI Series, Vol. H 59
Molecular Evolution of the
Major Histocompatibility Complex
Edited by J. Klein and D. Klein

Materials and methods

Mice. Genomic DNA was extracted from ethanol preserved tissues of wild-mouse-derived strains provided by F. Bonhomme (Montpellier, France). A total number of 22 alleles (including 9 inbred strains) were analyzed in this study. The geographical origin of these strains is listed in Table 1.

DNA sequencing. Polymerase chain reaction (PCR) methodology (Saiki et al. 1985) was used to amplify *Aa* exon 2. The two oligonucleotide primers used for PCR are: **ATCGGGATCCTAAATCCATCAGCCGACC** (5' primer) and **GCCGGAATTCGGAGAGCGGTGTGTAGCACG** (3' primer). The 5' primer contains a *Bam* HI site (underlined) and the 3' primer contains an *Eco* RI site. PCR reactions were performed as previously described (She et al. 1990; She et al. 1991). After 35 cycles of PCR amplification, the products were digested with both *Bam* HI and *Eco* RI and then cloned into M13mp19 digested with the same restriction enzymes. Standard dideoxynucleotide sequencing was performed to determine the sequences.

Computer programs. Sequence alignment and editing was done using Microgenie (Beckman). Allelic phylogenies were constructed using the UPGMA method (Sokal and Sneath 1963). Nucleotide divergence and diversity were calculated with Nei and Jin's programs (Nei and Jin 1989).

Results

Nucleotide sequence diversity of *Aa* exon 2

Twenty-two alleles of *Aa* exon 2 were amplified and sequenced. Eight sequences previously published (Bodmer et al. 1990 and reference there in) were also included in our analyses. These alleles are derived from 9 species or subspecies of the genus *Mus* (Table 1). The 5' primer used for PCR is complementary to an *Aa* intron 1 sequence adjacent to exon 2 and was originally designed to amplify the entire exon 2 (268 bp). Unfortunately, the last 7 nucleotides of the 5' primer (with high G-C contents) are identical to the sequence at position 16 - 22 of exon 2. Thus, the sequences from nt 1 - 22 of the exon was not obtained. This short sequence segment is very conserved among alleles, and therefore, our results will not be affected by the lack of sequences in this short segment.

Sequence comparisons of 30 alleles of *Aa* exon 2 revealed 26 different alleles and 4 pairs of identical alleles (Fig. 1). The nucleotide diversity between alleles was computed with Nei and Jin's program (data not show). Sequence diversity among most alleles is from

4 to 10%. Some alleles differ by 11.2% of their nucleotide sequences. We used the same panel of alleles (25 allelic sequences for both *Aa* and *Ab* from various species) to compare the degree of diversity in *Aa* and *Ab* exon 2. Our results suggest that *Aa* is slightly less divergent than *Ab* (the mean nucleotide diversity of *Aa* is 6.11 $\pm$ 0.7%, while the value for *Ab* is 8.5 $\pm$ 0.8%).

Table 1. List of alleles analyzed in this study, their taxonomy and geographical origin

Alleles	Species	Strains	Geographic Origin
d	*M. m. domesticus*	BALB/C	Lab inbred
b	*M .m. domesticus*	C57Bl/6	Lab inbred
k	*M. m. domesticus*	B10.BR	Lab inbred
u	*M. m. domesticus*	B10.PL	Lab inbred
q	*M. m. domesticus*	B10.6R	Lab inbred
f	*M. m. domesticus*	B10.M	Lab inbred
r	*M. m. domesticus*	B10.RIII	Lab inbred
p	*M. m. domesticus*	C3H.NB	Lab inbred
s	*M. m. domesticus*	A.SW	Lab inbred
$MudoAa_7$	*M .m. domesticus*	AZROU3	Morocco
$MudoAa_9$	*M. m. domesticus*	ERFOUD	Morocco
$MudoAa_1$	*M. m. domesticus*	24CI	Italy
$MudoAa_1$	*M. m. domesticus*	38CH	Italy
$MudoAa_1$	*M. m. domesticus*	METKOV1	Yugoslavia
$MumuAa_2$	*M. m. musculus*	MBS	Bulgaria
$MumuAa_4$	*M. m. musculus*	MBT	Bulgaria
$MumuAa_5$	*M. m. musculus*	MYL	Yugoslavia
$MumuAa_6$	*M. m. musculus*	BRNO4	Czechoslavakia
$MumuAa_7$	*M. m. musculus*	BEL1	Yugoslavia
$MucaAa_2$	*M. m. castaneus*	THONBURI1	Thailand
$MumoAa_1$	*M. m. molossinus*	MOL	Japan
$MuspAa_2$	*M. spretus*	SEI	Spain
$MuspAa_4$	*M. spretus*	SPE	Spain
$MusiAa_4$	*M. spicilegus*	ZRU	Bulgaria
$MusiAa_5$	*M. spicilegus*	ZYD	Yugoslavia
$MusiAa_6$	*M. spicilegus*	ZBN	Bulgaria
$MustAa_1$	*M. spretoides*	XBJ	Bulgaria
$MustAa_2$	*M. spretoides*	XBS	bulgaria
$MucoAa_1$	*M. cookii*	COK	Thailand
$MuceAa_1$	*M. cookii popaeus*	CRP	Thailand

Phylogenetic relationships of *Aa* exon 2 alleles

Phylogenetic relationships among *Aa* alleles have been studied by constructing phylogenetic trees. As shown in Fig. 1, most alleles are clustered into allelic lineages. As expected, several allelic lineages consist of alleles of the same species such as $MumuAa^4$ and

MumuAa^7. We can also commonly find allele lineages containing multiple species. Alleles in these lineages are very close to alleles of other species but very different from other alleles of the same species. For instance, allele "q" is very similar to *MuceAa*^{1} (1 base difference), but it is very different from other alleles of *Mus musculus domesticus* (for example, "q" is different from *MudoAa*^{13} and *MudoAa*^{15} in 8% of their nucleotide sequences).

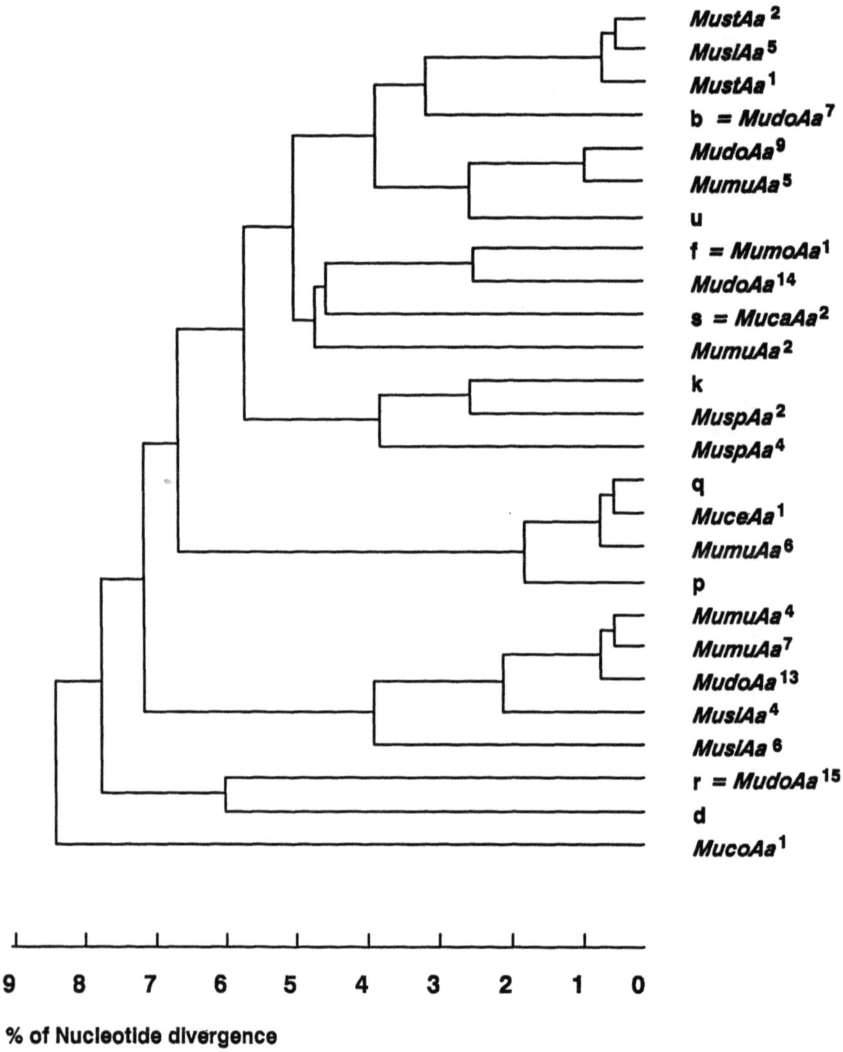

Fig. 1. Phylogenetic tree of Aa alleles constructed with the UPGMA method

These results indicate that the genealogy of *Mhc* alleles are not consistent with the phylogenetic relationships of species and that *Aa* polymorphisms are ancestral.

Aa diversity is concentrated in the segments encoding the antigen binding site

Sequence analyses revealed that the majority of the sequence polymorphisms are concentrated in 5 discrete segments. Each encodes a portion of the putative ABS of A molecule identified by the hypothetical model of Brown et al. (1988). Several highly divergent sequence motifs are found in each segment and these motifs are conserved in multiple species. For example, AS1 segment (*Aa* ß-strand 1) contains 9 different sequence motifs.

Aa diversity results from accumulation of point mutations

To understand if *Aa* has experienced intra-exonic recombination, we have performed sequence comparisons in different segments of exon 2. Our data indicate that the diversity in all five *Aa* segments is consistent, i.e. the allelic genealogies are similar in different segments. However, the allele "u", which is derived from an inbred strain, seems to be a recombinant allele combining an AS1 motif of alleles *MudoAa*^{9} and *MumuAa*^{6} and an α-helix motif of allele "k" allele (Fig. 2). Several alleles from various lineages share two identical nucleotides, but they can not be considered to be a result of gene conversion or recombination.

Sequence comparisons revealed that there are not species-specific substitutions at 54 polymorphic nucleotide positions and there is an extensive sharing of residues (and also nucleotides) found at polymorphic sites between species. The sharing of identical residues in various species may be explained by convergent evolution or sharing of short sequence motifs due to gene conversion. However, only one or at most two continuous polymorphic positions are involved in most cases, suggesting that convergent evolution is the most reasonable explanation.

```
                  30         40         50                 160        170        180
CONSENSUS  CCACGTAGGC TTCTATGGTA TAAGTGTATA   ...   GAGTTTGGCC AATTGACAAG CTTTGACCCC
MumuAa5    ---T------ -C-------- --GT------         ---------- ---------A ----------
MudoAa9    ---T------ -C-------- --GT------         ---------- ---------- ----------
u          ---T------ -C-------- --GT------         -------CT- --C---G--- ----------
k          ---------- -C-------- ---C------         -------CT- --C---G--- A-----G---
```

Fig. 2. Nucleotide sequences of four alleles in the ASI segment and the α-helix segment showing that allele "u" may be a recombinant allele between *MumuAa*^{5} or *MumdoAa*^{9} and K.

Positive selection operating in the diversification of *Aa* exon 2 alleles.

It has been shown that diversifying selection has been involved in Mhc evolution (Hughes and Nei 1988; Wakeland et al. 1990; Wakeland et al. 1989). To test this hypothesis in *Aa*, non-synonymous substitutions and synonymous substitutions have been computed using Jin and Nei's method (Nei and Gojobori 1986). As shown in Table 2, the non-synonymous mutations are higher than the synonymous mutations for the contact amino acids (d_N = 22.6%, d_S = 2.6%, d_N/d_S = 8.8), whereas the non antigen binding sites of exon 2 and exon 3 accumulate more synonymous substitutions. These findings confirm previous work (Hughes and Nei 1988) and indicate that diversifying selection operates only on the antigen binding site.

Table 2. Mean number of nucleotide substitutions per synonymous (d_S) and per nonsynonymous sites (d_N) expressed as percentages with their standard errors.

Aa Codons	Number of Codons	% of Substitutions d_s	d_n	d_n/d_s	P
Exon2	83	3.2±0.6	7.0±0.8	2.2	< 0.001
Contact amino acids	16	2.6±1.2	22.6±3.8	8.8	< 0.001
Non-Contact amino acids	58	2.7±0.7	3.6±0.7	1.3	n.s.
Exon 3	94	8.2±1.0	0.9±0.4	0.1	<0.001

Discussions and conclusions

Our sequence data of the murine Mhc class II *Aa* gene revealed extensive sequence diversity both within and between species. The mechanisms responsible for this diversity seem to be different from those generating diversity in the tightly-linked *Ab* gene which encodes the beta subunit of the A molecule. It has been shown that *Ab* diversity results from the retention of ancestral polymorphism, which is further amplified through shuffling of sequence motifs in various segments in exon 2 by intra exonic recombination. In contrast, *Aa* exon 2 does not seem to experience extensive genetic exchanges. Our results indicate that diversity at *Aa* may be generated by the accumulation of point mutations through long evolutionary periods, i.e. the polymorphisms are ancestral.

Aa and *Ab* encode the alpha and beta subunits of the A molecule. Previous data showed that these two genes must co-evolve because functional molecules cannot be assembled between inappropriate alpha and beta subunits (Germain et al. 1985). Why do these two genes generate their diversity through different mechanisms? Is it due to the intrinsic constraints of the genes? Accumulated data have suggested that all beta genes are involved in genetic exchanges between different genes or different alleles of the same gene (Gyllensten et al. 1991; She et al. 1990; She et al. 1991) and that a putative recombination signal sequence has been identified in the middle of exon of all beta genes (Wu et al. 1986; Gyllensten et al. 1991); She et al. 1991). This sequence may play an important role in promoting recombination in the beta genes. However, this sequence does not exist in the alpha genes, and the diversity is less extensive and recombination seems to be rare in the alpha genes. These observations seem to suggest that the mechanisms for generating diversity are not so crucial for Mhc genes. Any kind of mechanisms which can generate diversity could be potentially used, and subsequently, selection plays a very important role in maintaining the diversity by selecting functionally appropriate alleles or haplotypes.

References

Bodmer, J.G., Marsh, S.G.E, Parham, P., Erlich, H.A., Albert, E., Bodmer, W.F., Dupont, B., Mach, B., Mayr, W.R., Sasazuki, T., Schreuder, G.M.Th., Strominger, J.L., Svejgaard, A. and Terasaki, P.I.: Nomenclature for factors of the HLA system, 1989. *Hum Immunol 28*: 326-342, 1990

Brown, J.H., Jardetzky, T., Saper, M.A., Samraoui, B., Bjorkman, P.J. and Wiley, D.C.: A hypothetical model of the foreign antigen binding site of class II histocompatibility molecules. *Nature 332*: 845-850, 1988

Germain, R.N., Bentley, D.M. and Quill, H.: Influence of allelic polymorphism on the assembly and surface expression of class II MHC (Ia) molecules. *Cell 43*: 233-242, 1985

Gyllensten, U.B. and Erlich, H.A.: Ancient roots for polymorphism at the HLA DQα locus in primates. *Proc Natl Acad Sci USA 86*: 9986-9990, 1989

Gyllensten, U.B., Lashkari, D. and Erlich, H.A.: Allelic diversification at the class II DQB locus of the mammalian major histocompatibility complex. *Proc Natl Acad Sci USA 87*: 1835-1839, 1990

Gyllensten, U.B., Sundvall, M. and Erlich, H.A.: Allelic diversity is generated by intraexon sequence exchange at the DRB1 locus of primates. *Proc Natl Acad Sci USA 88*: 3686-3690, 1991

Hughes, A.L. and Nei, M.: Pattern of nucleotide substitution at major histocompatibility complex class I loci reveals overdominant selection. *Nature 335*: 167-170, 1988

Lawlor, D.A., Ward, F.E., Ennis, P.D., Jackson, A.P. and Parham, P.: HLA-A and B polymorphisms predate the divergence of humans and chimpanzees. Nature 335: 268-271, 1988

Nei, M. and Gojobori, T.: Simple methods for estimating the numbers of synonymous and nonsynonymous nucleotide substitutions. *Mol Biol Evol 3*: 418-426, 1986

Nei, M. and Jin, L.: Variances of the average numbers of nucleotide substitutions within and between populations. *Mol Biol Evol 6(3)*: 290-300, 1989

Saiki, R.K., Scharf, S., Faloona, F., Mullis, K.B., Horn, G.T., Erlich, H.A. and Arnheim, N.: Enzymatic amplification of ß-globin genomic sequences and restriction site analysis for diagnosis of sickle cell anemia. *Science 230*: 1350-1354, 1985

She, J.X., Boehme, S., Wang, T.W., Bonhomme, F. and Wakeland, E.K.: The generation of mhc class II gene polymorphism in the genus *Mus*. *Biol J Linn Soc 41*: 141-161, 1990

She, J.X., Boehme, S.A., Wang, T.W., Bonhomme, F. and Wakeland, E.K.: Amplification of major histocompatibility complex class II gene diversity by intra-exonic recombination. *Proc Natl Acad Sci USA 88*: 453-457, 1991

She, J.X. and Wakeland, E.K.: Molecular and genetic mwchanisms involved in the generation of Mhc diversity. *In* J. Klein and D. Klein (eds.): *Molecular Evolution of the Major Histocompatibility Complex*. Springer-Verlag, Heidelberg 1991

Sokal, R.R. and Sneath, P.H.A.: *Principles of Numerical Taxonomy*, Freeman, San Francisco, 1963

Wakeland, E.K., Boehme, S., She, J.X., Lu, C.C., McIndoe, R.A., Cheng, I, Ye, Y. and Potts, W.: Ancestral polymorphisms of MHC class II genes: Divergent allele advantage. *Immunol Res 9*: 123-131, 1989

Wakeland, E.K., Boehme, S. and She, J.X.: The generation and maintenance of MHC class II gene polymorphism in rodents. *Immunol Rev 113*: 207-226, 1990

Wu, S., Saunders, T.L. and Bach, F.H.: Polymorphism of human Ia antigens generated by reciprocal intergenic exchange between two DR loci. *Nature 324*: 676-679, 1986

MOLECULAR AND GENETIC MECHANISMS INVOLVED IN THE GENERATION OF *Mhc* DIVERSITY

Jin-Xiong She and Edward K. Wakeland
Laboratory of Molecular Genetics, Department of Pathology and Laboratory Medicine, University of Florida, Gainesville, FL 32611, USA

Class II genes in the major histocompatibility complex (Mhc) are extremely polymorphic, with many alleles differing in >10% of their nucleotide sequences. The mechanisms responsible for the generation of this diversity are controversial. We have analyzed the diversification pattern in the Mhc class II Ab gene by sequencing exon 2, exon 3 and a segment of intron 2 in alleles derived from 14 species of mice and rats. Our data indicate that the diversity of exon 2, which encodes the antigen binding site of the class II molecule, is localized to five highly polymorphic segments. This diversity was generated by the slow accumulation of point mutations over long evolutionary periods and has been amplified by repeated intra-exonic recombination. Sequence analysis of exon 3 and intron 2 indicates that these recombinational events occur specifically within exon 2, where a sequence highly homologous to a minisatellite common core sequence has been identified. This sequence may play a role in promoting recombination in Ab, DPb and DRb. The sequence data for Ab, together with other data, suggest that various molecular and genetic mechanisms are involved in the generation of Mhc diversity. The most important mechanism affecting the diversity of class II beta genes is the shuffling of highly conserved ancestral sequence segments.

Introduction

The polymorphism of *Mhc* genes has fascinated both immunogeneticists and evolutionary biologists. Two striking features of these genes are the existence of a large number of alleles at appreciable frequencies in natural populations and a high degree of nucleotide divergence between alleles of the same species. For example, 33 alleles have been sequenced for the human *DRB1* locus (Bodmer et al. 1990) and 75 alleles for the murine *Ab* locus (She et al. 1991; She and Wakeland, unpublished data), and more alleles will be described in the future. Some of these alleles differ in 25% or more of their nucleotide sequences in exon 2. The origin of this extensive diversity has been the subject of speculation for over a decade. Although a tremendous amount of data have been collected, a consensus explanation has not been reached. One reason may be that the mechanisms responsible for the generation of this diversity may differ between genes of the same

NATO ASI Series, Vol. H 59
Molecular Evolution of the
Major Histocompatibility Complex
Edited by J. Klein and D. Klein

organism and in homologous genes of different organisms. From a theoretical point of view, genetic diversity can be generated either through the accumulation of point mutations, or, through sequence segment rearrangements (among different genes or different alleles of the same gene). Therefore, one can easily imagine two categories of mechanisms involved in the generation of diversity at *Mhc* loci. In the first category, the diversity could result from the accumulation of point mutations by an increased mutation rate (hypermutation) or through long evolutionary time (ancestral polymorphism). In the second category, diversity could be generated by accumulating sequence rearrangements between different loci in the same gene family (gene conversion) or among different alleles of the same locus through crossing over or micro-recombination. In the following sections, we will try to summarize the data that we and others have accumulated in the past years and will speculate about the mechanisms involved in the generation of *Mhc* class II gene diversity.

There is no evidence supporting high point mutation rate in *Mhc* genes

A high rate of point mutations was the first mechanism proposed to explain *Mhc* polymorphism. However, accumulated evidence does not support such an explanation. First, *Mhc* genes with identical or very similar sequences are commonly found in species which diverged several million years (MYR) ago (Gyllensten and Erlich 1989; She et al. 1991; She et al. 1990a; Gyllensten et al. 1990). Second, many alleles with identical or extremely similar sequence segments in the highly polymorphic regions have been found in mice and rats (She et al. 1991; She et al. 1990a), which diverged 10 MYR ago (Jaeger et al. 1986) as well as in various primate species (Gyllensten et al. 1991). Third, the nucleotide divergence is discontinuous in many species (i.e. there are not intermediate alleles connecting very different alleles) and a continuous divergence pattern would be expected if the rate of mutation and fixation is high. These observations are all inconsistent with a high rate of point mutations.

Retention of ancestral polymorphisms in natural populations

If *Mhc* genes do not accumulate mutations faster than other nuclear genes, what mechanisms could be responsible for generating their extraordinary polymorphism? An alternative explanation is that mutations were accumulated through very long evolutionary periods (much longer than the life time of a species, i.e. the polymorphism is ancestral). This hypothesis, first proposed by J. Klein (Klein 1980; Klein 1987), was termed "trans-species evolution". Under the assumption of this hypothesis, ancestral species commonly pass a large component of their genetic variability to nascent species. Thus alleles can survive several or many speciation events and persist in populations for long evolutionary periods.

During their long persistence in evolution, point mutations are accumulated and the divergence between alleles become greater and greater. This hypothesis has received a lot attention although the term "trans-species evolution" is somewhat flawed (the term is not descriptive of the phenomenon and was previously used by other evolutionary biologists (Mayr 1970) to describe horizontal gene transfer from one species to another). Therefore, we suggest the term using "ancestral polymorphism" or "retention of ancestral polymorphism". The first genetic evidence supporting retention of multiple allelic lineages through speciation was obtained by Wakeland and co-workers (McConnell et al. 1988). Subsequently, many other investigators have obtained additional data supporting the "trans-species" hypothesis (Figueroa et al. 1988; Mayer et al. 1988; Lawlor et al. 1988; Gyllensten and Erlich, 1989; Gyllensten et al. 1990; Gyllensten et al. 1991; She et al. 1990a; She et al. 1991). McConnell et al. (1988) used the presence or absence of a retroposon in *Ab* intron 2 as a genetic marker to trace lineages of *Ab* alleles through several mouse species, suggesting that this particular polymorphism was created in an ancestral mouse population and had persisted for at least several MYR. Soon after this publication, Figueroa et al. (1988) used a two codon deletion polymorphism in the second exon of *Ab* to trace polymorphism back to the divergence of mouse and rat. Both of these pioneering studies utilized special genetic markers to analyze the age of the *Mhc* polymorphism. More recently, a large amount of sequence data has been accumulated (She et al. 1991; She et al. 1990a; Gyllensten and Erlich, 1989; Gyllensten et al. 1990; Gyllensten et al. 1991; Lawlor et al. 1990; Andersson et al. 1991; Mayer et al. 1988; Sarmiento et al. 1990) and the powerful techniques of phylogenetic analysis familiar to evolutionary biologists have been introduced to the study of *Mhc* evolution. The construction of phylogenetic trees for *Mhc* sequences generally revealed that alleles from different species can be more similar than alleles from the same species, indicating the ancestral nature of *Mhc* alleles. As a consequence, most immunogeneticists have now accepted the concept of ancestral polymorphism at *Mhc* loci. However, statistical tests were rarely provided in most of these studies and little attention was paid to the real meaning of the phylogenetic trees obtained. In some cases, more careful examination of the sequence data may lead to a different interpretation of the data.

How old are *Mhc* polymorphisms?

Two different methods can be used to estimate the age of a polymorphism. The first method is to analyze shared polymorphisms in different species with known divergence time, while the second is to estimate the divergence time based on nucleotide divergence. These two methods can lead to very different estimations of divergence times. For example, consider our results with *Mus* class II *Ab*. The retroposon polymorphism in intron 2 is found in both *Mus musculus domesticus* and *M. caroli* (Lu et al. 1991), suggesting that this polymorphism is at least as old as the divergence time for these two species, 3.5 MYR (She et al. 1990b). The oldest *Mhc* polymorphisms shared by rodent species are the two codon deletion polymorphism and some other short segments characterized in *Rattus* and *Mus* (divergence time approximately 10 MYR (Jaeger et al. 1986)). However, a very different result is obtained through the analysis of nucleotide divergence. Under the assumption of a constant rate of evolution, the divergence time between two alleles or genes equals the nucleotide divergence divided by the rate of evolution. For most nuclear genes, the rate of evolution is less than 0.5% per MYR. If *Mhc* genes evolve at a similar rate, then some *Mhc* alleles in *M. m. domesticus* with 25% divergence (She et al. 1990a, 1991) would have been maintained as polymorphisms for over 50 MYR. Among the 75 alleles analyzed in our laboratory, the nucleotide divergence among most alleles is from 5 to 15%, indicating that most *Mhc* polymorphisms are at least 10 to 30 MYR old.

There are two possible explanations for the distinctly different age estimates made by these methods: 1) the rate of evolution used for the calibration of divergence is too low, or, 2) the sample size analyzed to detect shared polymorphisms is too small and shared polymorphisms do exist between more distantly related species. The estimation of age based on nucleotide divergence is very difficult for *Mhc* genes, because the rate of evolution can not be exactly estimated. Furthermore, the estimation is very different when synonymous or nonsynonymous substitution rates are used for the calibration. For example, when synonymous substitution is used for the calibration, the estimated age is 8 MYR (4% average nucleotide divergence between *Mhc* alleles and an average rate of evolution of 0.5% per MYR for synonymous substitutions). However, when nonsynonymous substitution rate is used, the age is estimated to be more than 10 times older (10% average divergence and 0.1% divergence per MYR, that is 100 MYR). This value is too big to be true, thus indicating that *Mhc* accumulates nonsynonymous substitutions much faster than other nuclear genes. This is probably results from of a higher fixation rate due to diversifying selection. The age estimation is further complicated by intra-exonic recombination (She et al.; see below), which mixes up the phylogenetic information. In such cases, the estimations of divergence time are totally meaningless.

Estimations based on shared polymorphisms seem to be correct, although they are dependent on sample size. Data accumulated in rodents indicate that most polymorphisms are not older than the divergence of mouse and rat which share only a few sequence segment polymorphisms. However, data from primate species suggest that some *DQb* allele lineages are as old as 50 MYR (Bontrop et al. 1991). These studies clearly indicate that the age of *Mhc* polymorphisms can differ between various loci, allele lineages and organisms.

Intra-exonic recombination amplifies the diversity in *Ab*

The extensive diversity of *Ab* has been shown to result from intra-exonic recombination (She et al. 1991; She et al. 1990a; Wakeland et al. 1990). A 340 bp fragment containing *Ab* exon 2 was amplified from genomic DNA using PCR primers homologous to evolutionarily conserved intron sequences. A total of 65 *Ab* alleles were amplified, cloned into M13mp19, and sequenced using standard dideoxy sequencing methodology and combined with 10 previously published sequences from laboratory mouse and rat strains for analyses. The nucleotide divergence between most alleles is 5 to 15% (She et al. 1990a), although some alleles differed by as much as 25% in comparisons both within and between mouse species. The mean diversity of exon 2 within *Mus* species ($D = 7.57 \pm 0.7\%$) was comparable to the divergence observed between species ($D = 7.73 \pm 0.3\%$) and between all alleles in the genus ($D = 7.68 \pm 0.3\%$).

The majority of the polymorphisms in exon 2 occur in 5 specific regions, each of which encodes a specific element of the hypothetical ABS of the A molecule. The first 3 segments (termed BS1, BS2 and BS3) encode 3 ß-pleated sheet strands that form the floor of the antigen binding groove. The fourth encodes amino acids 56-67 of the α-helix and the fifth encodes the carboxyl end of the ABS whose tertiary structure could not be predicted (Brown et al. 1988). These findings indicate that *Ab* polymorphisms are focused predominantly in segments of exon 2 that encode the interior and sides of the antigen binding groove.

As illustrated by the data in Table 1, from 4 to 11 highly divergent sequence motifs were found in the five polymorphic segments. The nucleotide divergence between these motifs is very high, many differing by 20 - 30% of their nucleotides (She et al. 1990a). The majority of the motifs are shared by alleles from multiple mouse species which have diverged from 1 to 5 MYR, and some of the motifs are found in both *Rattus* and *Mus* species. These findings indicate that many of the motifs arose in ancestral rodent species. These motifs commonly persist for several MYR and in some cases persist for more than 10 MYR.

However, sequence comparisons for the whole exon indicate that similar sequences throughout the exon are only found in closely related species. Alleles from distantly related species are rarely identical throughout the exon although they may share identical sequence motifs for one or two segments. These observations can be explained by the shuffling of ancestral sequence motifs into different combinations by intra-exonic recombination. Figure 1 gives examples of such events. The *Ab* alleles were organized into 2 lineages according to the presence or absence of a double codon deletion and then divided into sub-lineages according to the sequence motifs in the carboxyl end. Four distinct sequence motifs of the carboxyl end are found in alleles with either deleted or undeleted lineages of the α-helix segment.

Table 1. Nucleotide sequence motifs in the BS1, BS2, BS3, 5' α-helix and the carboxyl end segments.

AB Region	Line-ages	Nucleotide Sequences	No. Alleles	Species
		13 23		
		TACCAGTTCAAGGGCGAG		
	1	---------C--CC-TTC	14	1,2,3,6,7,9,11
	2	GC------G---------	8	1,2,6
BS1	3	---------G-----CT-	5	12,13
	4	GT----------------	3	1,6
	5	------------------	23	1,2,3,4,5,6,8,10
		61 71 80		
		CGGCTCGTGACCAGATACATC		
	1	--ATCT----A----------	12	1,2,3,8
	2	--ATCT---GA----------	6	1,2,4,6
	3	--ATAT---------------	5	1,2,3,9
	4	A--TATC--GA-------T--	2	12
	5	---------------------	11	1,2,4,6
BS2	6	----------T----------	4	5,6,7
	7	-----T---------------	4	1,3
	8	----AT----G---T------	1	11
	9	---AGT---GA----CG-T--	1	13
	10	--TA-A---------------	1	10
		97 107		
		TACGTGCGCTTC		
	1	-GG---------	13	1,2,4,5,6,8,10
	2	---C--------	13	1,2,4,6,7,10
	3	---C------A-	4	1
BS3	4	----------A-	4	1,2,3
	5	------------	7	1,2,3,11,13
	6	-T-A--------	4	1,3,6
	7	----C-----A-	2	12
		155 165 175 185		
		CCAGACGCCGAGTACTGGAACAGCCAGCCGGAGATC		
	1	----------------AC--T-***---A-T-***-	29	1,2,3,4,5,6,7,8,9,10
	2	------------------------------------	14	1,2,4,10,11
	3	--CTCA--------------T-***---A-T-***-	4	1,2,13
α-helix	4	--CACA------------------------------	2	2
	5	-GG----------------------T-AA----T--	2	12
	6	--GTGG-----CG--------------AA-------	1	13
	7	--G----T-------------------A--------	1	12
		242 252 262 272		
		GGGCCGGAGACCCACACCTCCCTGCGGCGGCTT		
	1	---------------------------------	16	1,2,3,4,6,11
	2	-A-A-----GT--C-------------------	10	1,2,12,13
	3	---GT----------------------------	9	1,2,4,7
	4	AA-A---------C-------------------	6	1,2,3
	5	----T--------C--------------***--	3	1,6
3'segment	6	----T--------C-------------------	2	5,6
	7	---------G-T-T---T---------------	3	12,13
	8	---------T-AT----T---------------	1	8
	9	*****T----T--T--T-------A------C-	1	10
	10	---A--------CG---------A---------	1	10
	11	---GT--------G--------------***--	1	9

No. alleles indicates the number of alleles in each lineage among 53 Aß 1 exon alleles sequenced in mouse and rat (She et al. 1991, 1990a). These motifs are stable over a long evolutionary time and shared by multiple species as indicated by "species". 1, *M.m. domesticus*; 2, *M.m. musculus*; 3, *M. spretus*; 4, *M.m. castaneus*; 5, *M.m. molossinus*; 6, *M. spicilegus*; 7, *M. spretoides*; 8, *M. cervicolor*; 9, *M. cookii*; 10, *M. caroli*; 11, *M. platythrix*; 12, *Rattus rattus*; 13, *R. norvegicus*.

Although it is possible to argue that these blocks of polymorphic sequence could have been repeatedly generated due to convergent evolution (given the fact that diversifying selection is involved in *Mhc* evolution), this explanation requires many more mutational steps than would be required by intra-exonic recombination. Furthermore, similar events are also observed when any other pair of polymorphic segments are analyzed in this fashion.

	α-helix		Caboxyl end	Intron 2
Cons	GGAACAGCCAGCCGGAGATCCTGG	...	GGGCCGGAGACCCACACCTCCCTGCGGCGGC	2
MumuAb^{2}	------------------------		AA-A---------C-----------------	?
p	------------------------		-A-A-----GT--C-----------------	1
MudoAb^{2}	---C--------------------		-A-A-----GT--C-----------------	1
MumuAb^{1}	------------------------		-A-A-----GT--C-----------------	1
MudoAb^{1}	------------------------		---A-----GT--C-----------------	1
MudoAb^{3}	------------------------		-A-A-----GT--C-----------------	#
RT-1^{u}	-----------AA-----------		-A-A-----GT--C-T--***----------	#
RanoAb^{3}	-----------A------------		-A-A-----GT--C-----------------	#
q	------------------------		---GT--------------------------	1
MudoAb^{8}	------------------------		---GT--------------------------	1
MumuAb^{6}	------------------------		---GT--------------------------	#
MudoAb^{4}	------------------------		-------------------------------	1
MucaAb^{1}	-----------------C------		------------------------------T	1
MuplAb^{1}	------------A-----------		---T---------------------------	1
d	------------------------		------------AG-----------------	1
b	------------------------		-------------------------------	2
MumuAb^{5}	------------------------		-------------------------------	2
MudoAb3^{1}	------------------------		-------------------------------	2
MudoAb^{10}	----T-***---A-T-***-----		AA-A---------C-----------------	2
MudoAb^{11}	----T-***---A-T-***-----		AA-A---------C-----------------	2
MuspAb^{2}	----T-***---A-T-***-----		AA-A---------C----------------T	2
k	----T-***---A-T-***-----		AA-A---------C-----------------	3
MudoAb^{5}	----T-***---A-T-***-----		AA-A---------C-----------------	#
nod	AC--T-***---A-T-***-----		-A-A-----GT--C-----------------	1
MumuAb^{4}	TC--T-***---A-T-***-----		-A-A-----GT--C-----------------	2
u	AC--T-***---A-T-***-----		-A-A-----GT--C-----------------	3
MucaAb^{2}	AC--T-***---A-T-***-----		---GT--------------------------	1
MustAb^{1}	AC--T-***---A-T-***-----		---GT--------------------------	1
MucoAb^{1}	TC--T-***---A-T-***-----		---GT--------G--------------***	1
MudoAb^{6}	AC--T-***---A-T-***-----		---GT--------------------------	1
s	AC--T-***---A-T-***-----		---GT--------------------------	2
MumuAb^{3}	TC--T-***---A-T-***-----		---GT--------------------------	2
f	AC--T-***---A-T-***-----		--TGT--------C-----------------	2
r	AC--T-***---A-T-***-----		-------------------------------	1
MudoAb^{15}	AC--T-***---A-T-***-----		---T---------------------------	1
MusiAb^{6}	TC--T-***---A-T-***-----		-------------------------------	1
MudoAb^{12}	----T-***---A-T-***-----		-------------------------------	2
MusiAb^{4}	TC--T-***---A-T-***-----		-------------------------------	2
MuspAb^{1}	----T-***---A-T-***-----		-------------------------------	2
MuspAb^{3}	-C--T-***---A-T-***-----		-------------------------------	?

Fig. 1. Nucleotide sequences in the α-helix and the carboxyl end segments showing sequence shuffling between these two segments. The alleles were organized into two lineages in the α-helix segment and four lineages in the carboxyl end segment according to the sequence motifs. Four distinct sequence motifs of the carboxyl end are found in alleles with the two codon deletion (with "*") and without the two codon deletions in the α-helix segment. The alleles are assigned to three lineages based on intron 2 retroposon polymorphism (# indicates no data available).

How stable is the ABS in *Ab*?

These sequence motifs occur in many different combinations, possibly suggesting that a hyper-recombinational mechanism is involved. However, when sequences from sub-species such as *M. m. musculus* and *M. m. domesticus* are compared, identical or very similar alleles are often found, suggesting that intra-exonic recombination may not be a rapid process. Recombinational events appear to accumulate over time and ultimately play an important role in the generation of allelic diversity of this exon.

In order to obtain a more precise picture of the rate of recombinational events, we compared 23 alleles from *M. m. domesticus* and 22 alleles from *M. spretus*. These two mouse species diverged about 1.1 MYR ago based on DNA-DNA hybridization (She et al. 1990b). Sequence divergence analyses revealed that 8 out of 45 alleles shared 3 segments (divergence less than 5%), and 10 alleles share 4-5 segments. This analysis indicates that 22% of the alleles from the two species remain intact or are only affected once by recombination during the last MYR.

Sequence shuffling is exon 2 specific

Previous RFLP studies of the genomic stucture of *Ab* (McConnell et al. 1988; Lu et al. 1991) have demonstrated that alleles are clustered into three lineages according to the presence or absence of a retroposon insertion within *Ab* intron 2. Lineage 2 alleles are characterized by a 861 bp retroposon insertion in the middle of intron 2 and lineage 1 alleles do not have this insertion. Lineage 3 alleles have an additional insertion in comparison to the lineage 2 alleles. These polymorphisms are ancestral in the genus *Mus*. The three evolutionary lineages characterized by the retroposon are further confirmed by nucleotide sequence analysis using asymmetric PCR sequencing methodologies. Phylogenetic analysis of 15 intron 2 sequences indicates that three allelic lineages can also be defined by several lineage specific nucleotides (She and Wakeland, unpublished data). No clear-cut recombinational events could be found among theses sequences.

Nucleotide sequences of the entire *Ab* exon 3 (282 bp) derived from 13 mouse strains have also been determined. Ten published mouse and rat sequences were also included in our analyses. Among these 23 sequences, 19 different alleles were revealed. Three pairs of identical alleles were found in independent samples from the *Mus musculus* complex and one pair from *M. m. molossinus* and *M. spicilegus*. Evolutionary relationships among the exon 3 alleles were studied by constructing phylogenetic trees using both Fitch and Margoliash's least square distance and parsimony methods. The tree constructed by the least square distance method is presented (Fig. 2). According to these analyses, all mouse alleles can be divided into three distinct allele lineages and lineage 1 can be subsequently divided into two sublineages, 1A and 1B. Alleles within each lineage are very similar and commonly share several lineage-specific nucleotides across the whole exon. For example, lineage 2 differs from lineage 1 by 5 lineage-specific nucleotides, and lineage 3 differs from lineage 1 and 2 by 8 and 3 nucleotides respectively. Lineage 1A and 1B share 4 nucleotides which are different from lineage 2 and 3. These two sub-lineages are, however, different at 6 nucleotides. It should be mentioned that the two rat sequences are highly divergent from the mouse sequences (with 19 genus specific nucleotides) and cluster as a distinct lineage. Further sequence analyses indicate that all alleles can be assigned to a specific lineage without any doubt except one allele derived from *M. cervicolor*. The 5' end of this allele (nt 1 - 135) is identical to other lineage 1A alleles and the 3' end (nt 171 - 283) is very similar to lineage 2 alleles, suggesting that this allele is a recombinant product between lineage 1 and lineage 2 alleles.

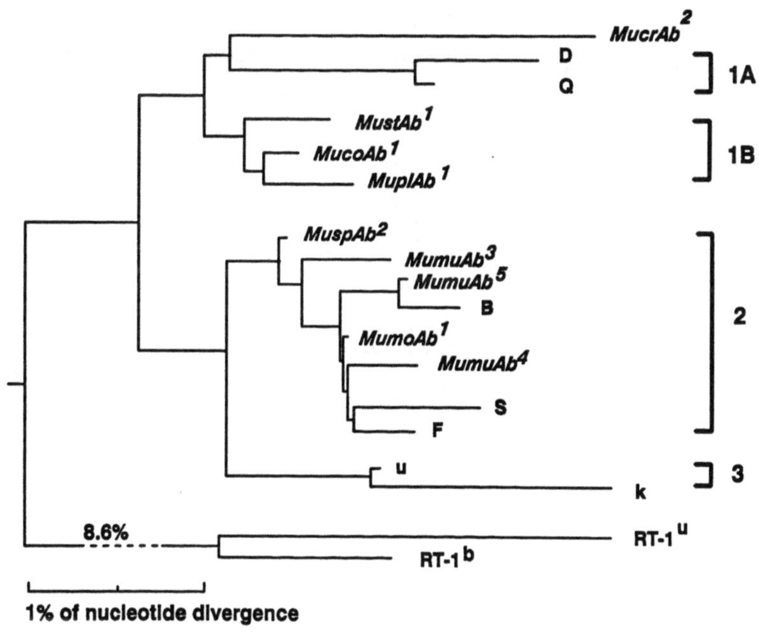

Fig. 2. Phylogenetic tree of exon 3 sequences constructed with Fitch and Margoliash' least square method.

We have further compared the diversity pattern of the genomic segment of *Ab* from intron 2 to exon 3, which is a fragment of 2.0 to 3.5 Kb in length (variable in different lineages). As shown in Fig. 3, the three segments analyzed exhibit consistent diversity patterns. Three identical allele lineages can be defined in all three segments without exception. This observation indicates that this entire region has evolved as a single entity and that recombination has not occurred within the region. These results together with the data in exon 2, further suggest that recombination is exon 2 specific.

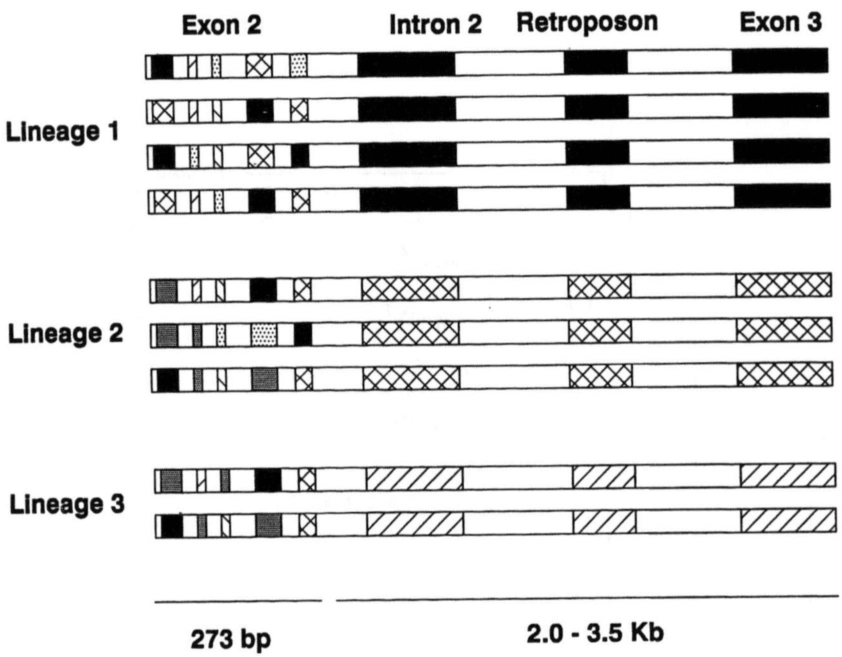

Fig. 3. Schematic diagram of *Ab* diversity patterns in exon 2, intron 2, and exon 3. Three allelic lineages in intron 2 and exon 3 are represented by different fill patterns. This region exhibits consistent diversity patterns suggesting that it has evolved as a single entity. However, the diversity pattern in this region is not consistent with the diversity pattern in exon 2, where recombination has occurred. This diagram illustrates that intra-exonic recombination in exon 2 specific.

Hitchhiking maintains neutral polymorphisms in regions adjacent to exon 2

It is generally believed that the importance of *Mhc* genes to antigen presentation has dramatically influenced their evolution and that it is advantageous for a species to maintain many different alleles due to a rapidly changing environment of pathogens. If this hypothesis is true, diversity in the domain encoding the ABS must be different from the rest of the gene because the ABS is the determinant domain for antigen presentation. The ratio of non-synonymous substitutions and synonymous substitutions can be used as an indicator of different types of selection (Hughes and Nei, 1988). As shown in Table 2, the non-synonymous mutations are higher than the synonymous mutations for the antigen binding sites (d_N = 13.6%, d_S = 3.7%, d_N/d_S = 3.7) and especially for the contact amino acids (d_N/d_S = 22.1), whereas, the non antigen binding sites of exon 2 and exon 3 accumulate more synonymous substitutions. These findings confirm previous work (Hughes and Nei, 1988) and indicate that diversifying selection operates only on the antigen binding site.

Table 2. Mean number of nucleotide substitutions per synonymous (d_S) and per nonsynonymous sites (d_N), expressed as percentages with their standard errors.

Segment	Codons	d_S	d_N	d_N/d_S	p
Whole Exon	91	3.9 ± 0.6	9.7 ± 0.7	2.5	< 0.001
Non-Antigen Binding Site	38	4.9 ± 1.3	4.1 ± 0.9	0.8	n.s.
Antigen Binding Site	42	3.7 ± 0.9	13.6 ± 1.4	3.7	< 0.001
Contact amino acids	14	1.6 ± 0.6	35.4 ± 3.9	22.1	< 0.001
Non-Contact amino acids	75	4.4 ± 0.7	6.3 ± 0.6	1.4	< 0.01
Exon 3	94	9.9 + 0.9%	2.1 + 0.4%	0.2	< 0.001

The high ratio of d_N/d_S has been interpreted as indicating that *Mhc* polymorphism is maintained by diversifying selection. Our RFLP data and sequence data show that polymorphisms of intron 2 and exon 3 are also ancestral and have persisted in mouse population for several MYR (Figs. 2, 3). Since selection does not seem to operate on these regions, how are the polymorphisms maintained for such a long evolutionary time? The most reasonable explanation is that the sequences in these regions have been physically held together and that recombinational events in exon 2 does not affect regions flanking the exon. Therefore, neutral polymorphisms such as retroposon insertion and silent mutations (in intron 2 and exon 3) can be maintained in natural populations for long evolutionary time through hitchhiking. This explanation implies that recombination in exon 2 only affects short segments.

What is the significance of the putative recombination signal sequence in exon 2?

Polymorphisms of many *Mhc* genes have been characterized in many organisms. In general, alpha genes are less polymorphic and do not exhibit sequence shuffling (Gyllensten et al. 1989; Ye et al. this issue). Retention of ancestral polymorphisms may be the major mechanism involved in the origin of alpha gene diversity. In contrast, several beta genes (*Ab, DPb* and *DRb*) have been shown to experience extensive sequence exchanges. Wu and co-workers (1986) have identified a sequence highly homologous to the minisatellite common core sequence within exon 2 of *DRb* gene and postulated that this sequence may play a role in promoting sequence exchanges in *Mhc* genes. As shown in Table 3, this sequence is also present and conserved in other class II beta genes, and a good correlation can generally be found between the degree of recombination and the presence of this sequence. For example, the sequence is very conserved in *Ab*, *DPb* and *DRb* genes where extensive sequence exchanges have occurred, and is polymorphic in *DQb* gene, where no clear cut recombinational events have been found among human alleles (some recombinational events have recently been observed among highly divergent primate species, Bontrop et al. 1991). All alpha genes lack the core sequence and are very polymorphic in the homologous segment of the exon (Table 3). It is tempting to suggest that the recombination signal sequence might play a role in recombination, although more experimental data are required to confirm this hypothesis.

Gyllensten et al. (1991) postulated that the recombination signal sequence defines a general break point of recombination in the *DRb* genes in primates. However, our data with mouse *Ab* clearly demonstrate that the break points in this exon occur throughout the entire exon, especially in the 3' portion of the exon. Some *DRb* allelic sequences suggest that *DRb* sequences may have experienced break points in other locations of the exon. These results suggest that the break point of recombination is not site specific, although the recombinational signal sequence may play an important role in promoting recombination. This may indicate that the enzymes or proteins involved in recombination that can bind to the minisatellite core sequence (Collick and Jeffreys, 1990) may facilitate recombination anywhere in close proximity of the signal sequence. Thus, the sequence itself may not represent an exact recombinational break point.

Table 3. Recombination signal sequence and degree of recombination

Genes	Rec. signal sequence*	Recombination
	GGAGCTGGGCAGG	
Myoglobin	----G--------	+++
33.3	----G--------	+++
33.5	----GY-------	+++
DRß	---------GC--	+++
DPß	---------GC--	++
Aß	C--------GC--	+++
Eß	A--------GC--	?
DQB1*0201	-CT------GCT-	-
*03	-CC------GCC-	-
*04	-CC------GC--	-
*05, 06	-CC--A---GC--	-
*0601	CCC--A---GC--	-
DOß	CA-------GCA-	?
Aß2	CAGC----AGCAC	-
DQα	T-CCTG-C-GT-- (polymorphic)	-
Aα	T-TCTG-ATGCTT (polymorphic)	-

* The sequence is located in nucleotide #140 - 152 in the second exon. "+++" indicates devotes sequence shuffling. "-" devotes no shuffling. "?" indicates that data are not enough to make any judgement.

Convergent evolution in *Mhc* genes

An important observation of *Mhc* diversity is that genes in very divergent species (i.e. human, bovine and dog) may have identical polymorphic residues at specific polymorphic sites (Andersson et al. 1991; Sarmiento et al. 1990). Some of these polymorphisms may be due to the common ancestry, i.e ancestral polymorphism, however, many of them can not be explained by ancestral polymorphisms or by gene conversions (It is very difficult to prove or disapprove gene conversion where only single amino acids are involved). Convergent evolution has been proposed to be responsible for the generation of these identical amino acids. This explanation is quite attractive and reasonable, given the fact that diversifying selection is involved in *Mhc* evolution. However, it is surprising that these genes not only share the same amino acids, but also in most cases, the same codons. If convergent evolution is the responsible mechanism, identical amino acids could be selected, however, the nucleotide sequences could be different due to the redundancy of the genetic code. To fully understand this question, more statistical analyses of codon usage and the distribution of substitutions should be performed.

Concluding remarks

The extensive polymorphism of *Mhc* genes has been a puzzle and the subject of debate among immunologists and evolutionary biologists for almost two decades. Several hypotheses have been proposed to explain the generation of these polymorphisms. Have we reached a consensus for the explanation even after extensive studies and exchanges of views among different groups? It seems that different opinions still exist in the field. However, it is clear that the issue is quite complicated and that different mechanisms may be involved in different genes and organisms.

1. Different genes exhibit different patterns of diversity: It is clear that most class I genes undergo extensive sequence exchanges between different loci and some exchanges between alleles at the same locus. Most class II genes encoding the beta subunit undergo genetic exchanges between different alleles through recombination and possibly some exchanges among different loci through gene conversion. The class II alpha genes do not seem to participate extensively in genetic exchanges.

2. Most polymorphisms at *Mhc* loci are ancestral. However, the unit of ancestral polymorphism is different in various genes. For example, the *Aa* gene is conserved for the entire exon 2 for several MYR in mouse (Ye et al., this issue) and the *DQa* genes are conserved in several primate species (Gyllensten et al. 1989). In contrast, only small segments of *Ab* are conserved between divergent mouse species or between *Rattus* and *Mus*. The intact exon is only shared in closely related species (mostly diverged for 1 - 2 MYR). Thus, the age of ancestral polymorphism may vary from one gene to another, and even from one allele lineage to another for the same gene.

3. The extensive diversity of the *Mhc* is a result of both mutational accumulation over evolutionary periods and recombination or gene conversion. This is a very efficient way to generate extensive diversity while conserving functionally important sequence segments.

4. Other mechanisms such as convergent evolution, gene duplication and deletion, and exon shuffling also have played a role in *Mhc* diversification for some genes and organisms.

5. Selection for ABS diversity is very important in maintaining *Mhc* polymorphism. *Mhc* is one of the few genes which is clearly affected by diversifying selection. Selection for the ABS diversity could also contribute to the maintenance of ancestral polymorphisms in other regions of the *Mhc* genes through hitchhiking.

References

Andersson, L., Sigurdardottir, S., Borsch, C. and Gustafsson, K.: Evolution of MHC polymorphism: extensive sharing of polymorphic sequence motifs between human and bovine DRB alleles. *Immunogenetics 33*:188-193, 1991

Bodmer, J.G., Marsh, S.G.E, Parham, P., Erlich, H.A., Albert, E., Bodmer, W.F., Dupont, B., Mach, B., Mayr, W.R., Sasazuki, T., Schreuder, G.M.Th., Strominger, J.L., Svejgaard, A. and Terasaki, P.I.: Nomenclature for factors of the HLA system, 1989. *Hum Immunol 28*: 326-342, 1990

Bontrop, R.E., Otting, N., Kenter, M., Van Weeren, P., and Jonker, M.: Mhc DQB repertoire variation in hominoid and Old World primate species. Abstract of Mhc evolution, Key Biscayne Conference, Key Biscayne, Florida 1991

Brown, J.H., Jardetzky, T., Saper, M.A., Samraoui, B., Bjorkman, P.J. and Wiley, D.C.: A hypothetical model of the foreign antigen binding site of Class II histocompatibility molecules. *Nature 332*:845-850, 1988

Collick, A. and Jeffreys, A.C.: Detection of a novel minisatellite-specific DNA-binding protein. *Nucleic Acids Res 18* (3):625-629, 1990

Figueroa, F., Gunther, E. and Klein, J.: MHC polymorphism pre-dating speciation. *Nature 335*: 265-267, 1988

Gyllensten, U.B. and Erlich, H.A.: Ancient roots for polymorphism at the HLA DQα locus in primates. *Proc Natl Acad Sci USA 86*:9986-9990, 1989

Gyllensten, U.B., Lashkari, D. and Erlich, H.A.: Allelic diversification at the class II DQB locus of the mammalian major histocompatibility complex. *Proc Natl Acad Sci USA 87*: 1835-1839, 1990

Gyllensten, U.B., Sundvall, M. and Erlich, H.A.: Allelic diversity is generated by intraexon sequence exchange at the DRB1 locus of primates. *Proc Natl Acad Sci USA 88*: 3686-3690, 1991

Hughes, A.L. and Nei, M.: Pattern of nucleotide substitution at major histocompatibility complex class I loci reveals overdominant selection. *Nature 335*: 167-170, 1988

Jaeger, J.-J., Tong, H. and Denys, C.: The age of *Mus-Rattus* divergence: paleontological data compared with the molecular clock. *CRAS 302* (2): 917-922, 1986

Klein, J. Generation of diversity at MHC loci: implications for T-cell receptor repertoires. In: Fougereau, M. and Dausset, J. (eds): *Immunology 80*, Academic Press, London, 1980

Klein, J.: Origin of major histocompatibility complex polymorphism: The trans-species hypothesis. *Hum Immunol 19*: 155-162, 1987

Lawlor, D.A., Ward, F.E., Ennis, P.D., Jackson, A.P. and Parham, P.: HLA-A and B polymorphisms predate the divergence of humans and chimpanzees. *Nature 335*: 268-271, 1988

Lawlor, D.A., Warren, E., Ward, F.E. and Parham, P.: Comparison of class I alleles in huaman and apes. *Immunol Rev 113*: 147-185, 1990

Lu, C.C., Ye, Y. She, J.X., and Wakeland, E.K.: Ancestral polymorphism of the Mhc Ab gene in the genus Mus. 1991 (Mol Biol Evol, submitted 1991)

Mayer, W.E., Jonker, M., Klein, D., Ivanyi, P., van Seventer, G. and Klein, J.: Nucleotide sequences of chimpanzee MHC class I alleles: evidence of trans-species mode of evolution. *EMBO J 7*:2765-2774, 1988

Mayr, E.: *Population, Species and Evolution*, Vol. 2. Harvard University, Boston 1970

McConnell, T.J., Talbot, W.S., McIndoe, R.A. and Wakeland, E.K.: The origin of MHC class II gene polymorphism within the genus *Mus*. *Nature 332*: 651-654, 1988

Sarmiento, U.M., Sarmiento, J.I. and Storb, R.: Allelic variation in the DR subregion of the canine major histocompatibility complex. *Immunogenetics 32*: 13-19, 1990

She, J.X., Boehme, S., Wang, T.W., Bonhomme, F. and Wakeland, E.K.: The generation of mhc class II gene polymorphism in the genus Mus. *Biol J Linn Soc 41*: 141-161, 1990a

She, J.X., Bonhomme, F., Boursot, P., Thaler, L. and Catzeflis, F.: Molecular phylogenies in the genus Mus: comparative analysis of electrophoretic, scn DNA hybridization, and mt DNA RFLP data. *Biol J Linn Soc 41*: 83-103, 1990b

She, J.X., Boehme, S.A., Wang, T.W., Bonhomme, F. and Wakeland, E.K.: Amplification of major histocompatibility complex class II gene diversity by intra-exonic recombination. *Proc Natl Acad Sci USA 88*: 453-457, 1991

Wakeland, E.K., Boehme, S. and She, J.X.: The generation and maintenance of MHC class II gene polymorphism in rodents. *Immunol Rev 113*: 207-226, 1990

Wu, S., Saunders, T.L. and Bach, F.H.: Polymorphism of human Ia antigens generated by reciprocal intergenic exchange between two DRß loci. *Nature 324*: 676-679, 1986

Ye, Y., She, J.X., and Wakeland, E.K.: Diversification of class II Aa within the genus Mus. *In* J. Klein and D. Klein (eds).: *Molecular Evolution of the Major Histocompatibility Complex*. Springer-Verlag, Heidelberg 1991

EVIDENCE FOR MULTIPLE MUTATIONAL MECHANISMS WHICH GENERATE POLYMORPHISM AT *H-2K*

John Martinko and Khaled Girgis
Department of Microbiology, Southem Illinois, University, Carbondale, IL 62901, USA

Several serologically similar H-2K alleles which are closely related to H-2K^d have been isolated. These variants were originlly isolated from Mus domesticus found in several geographically distant locations. We undertook the nucleotide sequence analysis of these naturally occurring variants to define their relationship to one another and to look for evidence of mutational mechanisms which may have given rise to these variants. Data were obtained by direct RNA sequencing and from PCR-generated cDNA clones. Data for K^{dvl}, K^{dv4} and the standard K^d are most informative and will be discussed for the antigen-binding αl and α2 domains. First, K^{dvl} has a total of six nucleotide (five amino acid) alterations as compared to K^d. K^{dv4} has identical nucleotide substitutions at all positions except one, where K^{dv4} has the same nucleotide as K^d. Thus, we believe that K^{dv4} is more related to K^d than is K^{dvl}. Second, three nucleotide substitutions in a span of eleven, coupled with identification of several D and Q region genes possessing the same substitutions, suggest that intergenic recombination events may account for some of the variations seen, a situation similar to the well-documented H-2 mutants. However, at least one variation is clearly the result of a point mutation and results in a unique substitution in K^d. Finally, these variations are concentrated in two groups on opposite sides of the antigen-binding pocket and may be useful for comparing binding to defined peptide antigens.

Several K^d-like molecules have been discovered in independent haplotypes derived from several "unrelated" mouse populations. These MHC molecules all display serological epitopes similar to K^d, but at least some have defined structural differences from K^d (Arden et al. 1982) while others are derived from either "wild" strains (w18) or from strains with no known genealogical relationship to the standard DBA/2 strain from which the standard K^d sequence was derived (Kvist et al. 1983).

A number of independently derived K^d-like genes have been chosen for structural analysis to, first, examine their structural relationship to one another. We expect several of these independently derived molecules to exhibit minor structural variations. Analysis of these variations will allow definition of the evolutionary origins of these molecules, relative to one another. Second, variants will provide a source of biologically relevant "mutant" molecules which were found as independent alleles. They represent the same type of system

NATO ASI Series, Vol. H 59
Molecular Evolution of the
Major Histocompatibility Complex
Edited by J. Klein and D. Klein

seen in HLA-A2 and are, essentially, subtypes. The K^d subtype clones will then be used, via transfection and peptide feeding experiments, to explore differences in antigen-binding specificity due to subtle alterations in class I structure.

Here we describe the cloning, sequencing and analysis of a number of H-2K^d-like mouse class I molecules to define and compare the structure and, in so far as possible, to define and compare the origins of diversity within these molecules. The haplotypes containing the K^d-like molecules are: *w5*(K^{dv1}), *w14*(K^{dv2}), and *w34*(K^{dv4}). The *K* alleles encode independly derived, serologically similar molecules, some of which are already known to possess structural alterations as compared to K^d.

Animals. Table 1 documents the independent origin of strains to be used in this study. All animals used in this study were provided by Dr. D. Shreffler, Washington University, St. Louis, MO.

Table 1. *H-2*K^d and subtypes

Mouse strain	Haplotype	*K*-locus	Origin	Reference
DBA/2	*d*	K^d	standard	Kvist et al. 1983
B10.KEA5	*w5*	K^{dv1}	wild	Duncan and Klein 1980
B10.STC77	*w14*	K^{dv2}	wild	Duncan and Klein 1980
STU	*w34*	K^{dv4}	European	Figueroa et al. 1982

Sequencing strategy. Sequence information from mRNA derived from the *K* locus of w5 (K^{dv1}), w14 (K^{dv2}), and w34 (K^{dv4}) has been obtained using direct sequencing methods for poly A^+ RNA template (Geliebter *et al.* 1986). Table 2 shows allele-specific oligonucleotides used to obtain this data.

The data for the K^{dv} variants are summarized in Table 3. The data suggest that the K^{dv} genes are closely related to the standard K^d: For 690 nucleotide positions compared in the coding region, there is at least 98.8% homology with the K^d standard allele (7 differences for K^{dv2} and K^{dv4}, 8 differences for K^{dv1})

Table 2. *H-2K*^{d}-allele-specific oligonucleotides used for direct sequence analysis. Primers were synthesized based on the K^d sequence reported by Kvist et al. (1983).

d1: 193-199	5'	GAC	ATC	AAC	TTG	AGA	TCT	GGG	3'
d4: 342-348	5'	TCA	CGC	TAG	AGA	ATG	AGG	GTC	3'
d6: 326-332	5'	AGA	CAG	ATC	AGA	GGT	CTG	GGA	3'
d7: 270-276	5'	AAG	CTT	CCA	TCT	CAG	GGT	GAG	3'
d8: 141-147	5'	CCA	CTT	GCG	TCT	GGT	GAT	CAG	3'
d9H: 131-119	5'	GTC	GTA	GGC	GAA	CTG	CTG	GTA	3'
d10H: 72-78	5'	CAG	GCT	CCC	TCG	GAA	ACT	CTG	3'
d13H: (+14)-(+20)	5'	ATG	TCA	CAG	GAG	ATG	TGT	GAA	3'

Clearly, this analysis shows at least two different structural patterns for the dv subtypes suggesting a minimum of three different variants - K^d, K^{dv1}-K^{dv2}, and K^{dv4}, all of which are different class I molecules.

Table 3. Polymorphisms in H-2K^{dv} variant alleles as compared to the standard H-2K^d allele.

Standard	Variant	Substitution	Location
K^d	*dv1, dv2, dv4*	CCC(Pro) -> CCG (Pro)	silent ,50
K^d	*dv1, dv2, dv4*	TGG(Trp) -> AGT (Ser)	α helix, 73
K^d	*dv1, dv2, dv4*	GTG (Val) -> GGG (Gly)	α helix, 76
K^d	*dv1, dv2, dv4*	TTC(Phe) -> ATC (Ile)	ß sheet, 95
K^d	*dv1, dv2, dv4*	CAG(Gln) -> GAG (Glu)	ß sheet, 114
K^d	*dv1, dv2*	AGG(Arg) -> AAG (Lys)	α helix, 157

This data is shown as it projects onto the HLA-A2 model of Bjorkman et al. (1987a, b) in Fig. 1. This data suggests that the alterations seen in each subtype are potentially involved in interactions with peptide antigens, presumably at two different interaction sites, one on each helix involved in the class I binding site. Sequence and analysis of a number of independently segregated *H-2K*^{d} alleles will define more polymorphic positions in the murine H-2 class I alleles and allow us to compare this information to studies of HLA class I, noteably studies of HLA-A2 and HLA-B27 subtypes (Parham et al.1988; 1989).

Analysis of the different substitutions shown in Table 3 leads to several interesting observations. First, each substitution seen in the variant molecules is found in at least one other mouse class I molecule (Kuhner and Goodenow 1989). The fact that all of the substitutions have been observed in other mouse class I molecules is consistent with a copy mechanism, analogous to gene conversion, where other molecules are used as a template for the substitutions. One set of substitutions at codons 73 and 76, where three nucleotides are altered in a span of eleven, is reminiscent of the complex single-event structural changes which characterize many of the well-defined spontaneous K^b mutants (Geliebter et al. 1986) as well as K^{km2} (Vogel et al. 1988) and K^{fm1} (Horton et al. 1990). All of the other substitutions are single nucleotide changes in widely separated structural locations which, of course, might be generated by a point mutation mechanism. However, there identity with other sequences at analogous positions in other mouse class I molecules is certainly consistent with a copy mechanism. Finally, the standard K^d codon at position 50, which is CCC, is unique when compared to all mouse class I molecules (Kuhner and Goodenow 1989). A random point mutation mechanism apparently generated this unique, allele-specific codon. Thus, it appears that at least two mechanisms are working to generate diversity in class I: point mutations and copy mechanisms or gene conversions. Unfortunately, it is not possible to assess which of these general mechanisms is most important in this limited model system, since most of the substitutions can be explained by invoking either hypothesis.

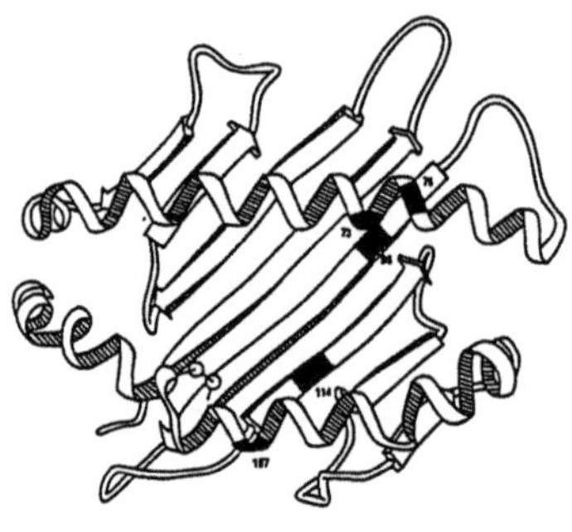

Fig. 1. Class I antigen binding site (after Bjorkman et al. 1987 a,b). The alterations in the K^d structural variants are shown in black. Altered amino acid residues are numbered.

The evolution of the variants within the context of the K^d standard molecule is fairly clear. Figure 2 depicts one possible relationship of the three unique molecules analyzed. Clearly, K^{dv4} is intermediate between K^{dv1} and K^d, and is much more closely related to K^{dv1} than to K^d. The direction of the mutations, i.e., which molecule - K^d or K^{dv1} - is more similar to an ancestral allele - is not clear from the limited data obtained.

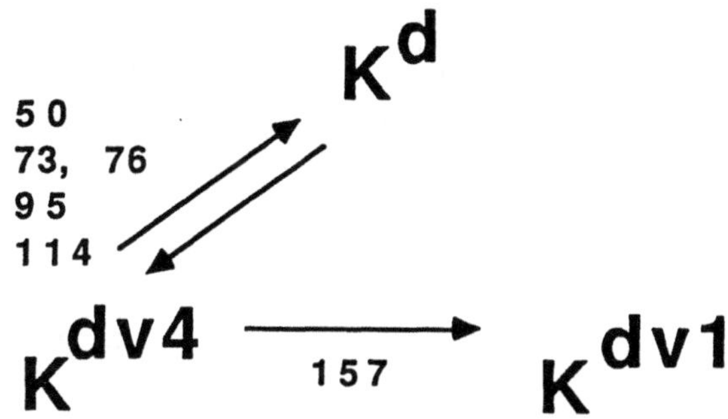

Fig. 2. Evolution of K^d-like antigens. The numbers indicate the amino acid position of each substitution relative to the nearest neighbor. Thus, K^{dv1} differs from K^d at the same positions as K^{dv4} and at an additional position, 157.

This work represents the first systematic attempt to investigate minor structural variations using a group of naturally occurring alleles which are found segregating in the mouse population: these molecules are functional and "work" in the real-world populations from which they were derived, as compared to more traditional investigations of structure-function relationships involving spontaneous, biologically functional, but evolutionarily untried, mutants, or the technology-derived site directed mutants (which have no obvious biological functions). A future goal is to expand investigations of these variants to include more closely related alleles.

Table 4. Known K^d-restricted peptides

Peptide source	Peptide	Reference
CW3	RYLKNGKETLQRA	Maryanski et al. 1990
NP	TYQRTRALVRTG	Rothbard et al. 1989
NP	TYQRTRALV	Rotzschke et al. 1990
HA	IYSTVASSLVL	Kuwano et al. 1988

Finally, perhaps the most interesting aspect of this work involves the potential structure-function questions that can be addressed using these variants. K^d is known to bind a number of specific peptides, as shown in Table 4. The positions of the K^{dv1} and K^{dv4} substitutions on the alpha helices and in the beta sheet structure thought to be involved in the antigen binding site, suggest that each substitution will effect class I peptide binding (Bjorkman et al. 1987a, b). Thus, these variants will be useful for investigating parameters of peptide - class I interactions, especially as they relate to specific antigen restriction patterns in K^d.

Acknowledgments. This work was supported in part by grant AI-21738 to J.M. from the National Institutes of Health and by a Disseratation Research Award to K.G. from Southern Illinois University.

References

Arden, B., Wakeland, E.K., and Klein, J.: Minor structural variants of H-2K-controlled molecules in wild mice. *Immunogenetics 16:* 491-493, 1982

Bjorkman, P.J., Saper, M.A., Samraoui, B., Bennett, W.S., Strominger, J.L., and Wiley, D.C.: Structure of the human class I histocompatibility antigen, HLA-A2. *Nature 329:* 506-512, 1987a

Bjorkman, P.J., Saper, M.A., Samraoui, B., Bennett, W.S., Strominger, J.L., and Wiley, D.C. The foreign antigen binding site and T cell recognition regions of class I histocompatibility antigens. *Nature 329:* 512-518, 1987b

Duncan, W.R. and Klein, J.: Histocompatibility-2-system in wild mice IX. Serological analysis of 13 new B10.W congenic lines. *Immunogenetics 10:* 45-65, 1980

Figueroa, F., Tewarson, S., Neufeld, E., and Klein, J.: H-2 haplotypes of strains DBR7, B10.NZW, NFS, BQZ, STU, T01, and T02. *Immunogenetics 15:* 431-436, 1982

Geliebter, J., Zeff, R.A., Melvold, R.W., and Nathenson, S.G.: Mitotic recombination in germ cells generated two major histocompatibility complex mutant genes shown to be identical by RNA sequence analysis: K^{bm9} and K^{bm6}. *Proc Natl Acad Sci USA 83:* 3371-3375, 1986

Horton, R.M., Hildebrand, W.H., Martinko, J.M., and Pease, L.R.: Structural analysis of H-$2K^f$ and H-$2K^{fm1}$ using H-2K locus-specific sequences. *J Immunol 145:* 1782-1787, 1990

Kuwano, K., Braciale, T.J., and Ennis, F.A.: Localization of a cross-reactive CTL epitope to the transmembrane region of the hemagglutinin of influenza H1 and H2. *FASEB J* 22221 (abstract), 1988

Kuhner, M.K. and Goodenow, R.S.: DNA sequences of mouse H-2 and Qa genes. *Immunogenetics 30:* 458-464, 1989

Kvist, S., Roberts, L., and Dobberstein, B.: Mouse histocompatibility genes: structure and organization of a K^d gene. *EMBO J 2:* 245-254, 1983

Maryanski, J.L., Verdini, A.S., Weber, P.C., Salemme, F.R., and Carradin, G.: Competition analogs for defined T cell antigens: peptides incorporating a putative binding motif and polyproline or polyglycine spacers. *Cell 60:* 63-72, 1990

Parham, P., Lawlor, D.A., Lomen, C.E., and Ennis, P.D.: Diversity and diversification of HLA-A,B,C alleles. *J. Immunol. 142:* 3937-3950, 1989

Parham, P., Lomen, C.P., Lawlor, D. A., Ways, J.P, Holmes, N., Coppin, H.L, Salter, R.D., Wan, A.M., and Ennis, P.D.: Nature of polymorphism in HLA-A,-B, and -C molecules: *Proc Natl Acad Sci USA 85:* 4005-4009, 1988

Rothbard, J.B., Pemberton, R.M., Bodmer, H.C., Askonas, B.A., and Taylor, W.R.: Identification of residues necessary for clonally specific recognition of a cytotoxic T cell determinant. *EMBO J 8:* 2321-2328, 1989

Rötschke, O., Falk, K., Deres, K., Schild, H., Norda, M., Metzger, J., Jung, G., and Rammensee, H.G.: Isolation and analysis of naturally processed viral peptides as recognized by T cells. *Nature 348:* 252-254, 1990

Vogel, J.M., Davis, A.C., McKinney, D.M, McMillan, M., Martin, J.W., and Goodenow, R.S.: Molecular characterization of the C3HfB/HcN H-2K^{km2} mutation. *J Exp Med 168*:1781-1800, 1988

CONTRIBUTIONS OF INTERLOCUS EXCHANGE TO THE STRUCTURAL DIVERSITY OF THE *H-2K*, *D*, AND *L* ALLELES

L.R. Pease, J.K. Pullen, Z. Cai, and R.M. Horton

Department of Immunology, Mayo Clinic, Rochester, MN 55905, USA

The nature of diversity that distinguishes alleles form each other and from sequences of other class I loci was analyzed using 24 H-2K, D *and* L *sequences, providing statistical evidence that alleles of the* K *and* D/L *loci are evolving independently throughout most of their coding regions. However, sequences determining the structure of the antigen binding domains of both allelic series appear to be co-evolving. Sequence motifs that distinguish alleles in these series are present at both the* K *and* D/L *loci, a pattern of diversity that is reminiscent of the class I mutants that have been described in the mouse.*

Among the many genes of the major histocompatibility complex are those encoding antigen presenting molecules. These genes are unique within the genomes of mammals in that they are highly polymorphic (Klein and Figueroa 1981) and their alleles are distinguished from each other by extensive amino acid diversity (Nathenson et al. 1981). The *H-2K*, *D*, and *L* class I antigen presenting molecules of the mouse are of particular genetic interest because they are comprised of very large sets of alleles that are present within populations at fairly low frequencies.

Early studies in the mouse suggested that alleles of the class I loci that encode antigen presenting molecules (*K* and *D*) were not distinctive and that given the structure of a specific molecule, assignment to one locus or the other could be difficult (Nathenson et al. 1981). Subsequently, the fact that alleles of the *K* and *D/L* loci can be distinguished on the basis of a greater overall similarity among alleles has been established on the basis of comparisons made using entire gene sequences (Watts et al. 1987). An extensive analysis of the structure of diversity among the mouse class I genes has not been undertaken because only eleven mouse sequences have been available. We have determined the structure of 13 additional *K* and *D/L* locus alleles and have used these new sequences, together with the previously reported *K* and *D* alleles, to examine the nature of diversity among the antigen presenting molecules of the mouse.

NATO ASI Series, Vol. H 59
Molecular Evolution of the
Major Histocompatibility Complex
Edited by J. Klein and D. Klein

Locus specific strategies for the isolation of *K* and *D* alleles

We have recently developed strategies for the direct sequence analysis of the alleles of the mouse *K*, *D* and *L* loci (Horton et al. 1990; Cai and Pease 1990). These three loci belong to a larger multigene family consisting of approximately 30 genes. Attempts have been made to identify locus specific probes for these genes (Kress et al. 1983); however, their utility in identifying the alleles from the numerous defined *H-2* genotypes has not resulted in a significant increase in the number of class I sequences that have been characterized in the mouse.

Our approach has been based on the locus specific amplification of cDNA by the polymerase chain reaction. We have identified locus associated sequences that are located in the 5' and 3' untranslated regions of the mRNA encoded by the targeted loci. A distinctive deletion present within a B2 SINE element that is characteristic of the *D* and *L* loci, and largely absent from all other class I loci, provided a convenient marker for these two related loci. We reasoned that the appearance of a polyadenylation signal in the SINE element would truncate the mRNA derived from these loci, making the sequences 3' of the SINE insertion more specific for the full length *K* mRNA, despite the fact that both *D*/*L* and *K* alleles share these most 3' sequences. Inspection of the 5' untranslated regions of *K* and *D*/*L* alleles revealed that there tended to be *K* and *D*/*L* associated motifs just upstream from the translation initiation codons.

PCR amplification of cDNA using these two sets of primers has allowed us to isolate single cDNA sequences representing the *K* and *D*/*L* loci in most haplotypes. PCR associated artifacts have been rare and limited to occasional nucleotide substitutions that are unique to individual clones isolated from amplified products. The major exception to the straight forward analysis of *K* and *D*/*L* sequences using this approach has been the analysis of haplotypes containing more than one *D*/*L* gene. We detected three genes of this class in $H\text{-}2^v$ (Pease et al. 1991). Thirteen class I sequences in the mouse have been determined using this approach, doubling the number of sequences available for the analysis of *K* and *D*/*L* diversity.

Patterns of diversity

We analyzed 9 *K* (*b, d, f, k, p, r, s, v, w28*) and 14 *D*/*L* (*b*, D^d, L^d, *dx, f, k, p*, D^q, L^q, *r, s, v1*, *v2, v3*) nucleotide sequences to determine the basis of conflicts between earlier published observations that indicated, on the one hand, *K* and *D/L* sequences were indistinguishable from each other on the basis of amino-terminal amino acid sequence (Nathenson et al. 1981), and on the other, that the sequences could be clearly subdivided according to locus designation using nucleotide sequence data (Watts et al. 1987). To assess locus specificity, the sequences were compared to each other in a matrix design to determine the average divergence of each sequence from its alleles in the panel, as well as, from the nonallelic sequences encoded by the sister class I loci. Comparisons were made between the coding segments that determine the entire antigen presenting molecule and for segments of the coding blocks that correspond to different regions of the encoded glycoproteins. The resulting matrices were analyzed nonparametrically, using a Wilcoxon rank sum test (Wilcoxon 1945). The results of these comparisons are shown in Table 1.

Table 1. Statistical evaluation of locus specificity

	Median % nucleotide differences					
	K vs K		K vs D/L	D/L vs D/L		D/L vs K
Leader	5.3	***	10.1	1.2	*****	5.2
Alpha 1	9.3	NS	9.8	6.6	NS	6.5
Alpha 2	8.8	**	10.2	6.5	NS	6.3
Alpha 3	2.7	*	5.0	1.9	**	3.1
TM	6.7	**	17.0	4.0	**	9.6
CYT	2.1	*	6.0	1.6	*****	4.0
Entire sequence	5.6	****	9.4	4.2	***	6.3

Median percent variation among alleles is indicated for comparisons of entire sequences, as well as for subregions of the compared sequences. Comparisons of distributions of averages were made using the Wilcoxon rank sum test (Wilcoxon 1945). Levels of significance are indicated as $P < .05$ (*), $P < .01$ (**), $P < .001$ (***), $P < .0001$ (****), $P < .00001$ (*****), and NS indicates not significantly different.

While nucleotide differences are evident throughout the sequences, the highest level of diversity was localized to the coding segments that determine the antigen binding domain of the antigen presenting molecules (alpha 1 and alpha 2 regions). As a test for locus specificity, the average distance of each *K* allele to other *K* alleles was compared to the average distance of each *K* allele to each *D*/*L* allele. A parallel analysis comparing *D*/*L* alleles among themselves and with *K* alleles also was performed. As can be seen in Table 1, significant differences between the distributions of averages were observed in comparisons of the entire sequences, as well as in the comparisons of subregions of the molecules, with the exception of the sequences encoding the antigen binding domain. In the case of the sequences determining the antigen binding domain (alpha 1 and alpha 2), three of the four comparisons did not reveal significant differences between the interlocus and intralocus distributions of averages. One comparison did reveal significant *K*-locus specificity in the alpha 2 region, although the median % differences distinguishing intralocus (8.8%) from interlocus (10.2%) comparisons were not dramatically different.

Assuming that sequence diversity evolves by the accumulation of point mutations, one would expect the regions of highest diversity to exhibit the greatest level of locus specificity. The fact that the regions of highest diversity in these molecules exhibit the least locus specificity can be taken as evidence that these genes are not evolving in a traditional manner.

A series of mutants have been described in the mouse that provide the basis for a hypothesis explaining the low level of locus specificity in the antigen binding domain sequences (Pease et al. 1983). These mutants are characterized by complex substitutions, representing sequence motifs that have been shown to be identical to sequences present in other class I loci within the mutant haplotypes. The hypothesis suggests that the mutants arose by a gene conversion-like process that is capable of transferring sequences between loci. Several of the mutants contain substituted motifs that suggest sequences have been exchanged between the *K* and *D* loci (Table 2).

The *K* and *D*/*L* sequences were examined to determine whether differences between alleles can be explained by the gene conversion hypothesis. There are several examples of sequence motifs that distinguish the *K* alleles and are also present among the *D*/*L* alleles and also sometimes among the related *Q* class I loci. Representative cases are shown in Table 3. These examples support the hypothesis that sequence motifs found in the coding sequences for the antigen binding domain and that distinguish alleles from each other are transferred between loci by a gene conversion-like mechanism.

Table 2. *H-2* gene conversion mutants

Mutant	Donor gene	Substitutions at codons				
K^{bm11}	D^{b}	77	80			
		AG-	-A-			
		Ser	Asn			
K^{km2}	D^{k}	95	96	98	99	
		A--	--G	T--	C--	
		Ile	*	Leu	Cys	
K^{fm1}	D^{f}	94	95	97		
		--G	A--	CG-		
		*	Ile	Arg		
D^{bm13}	K^{b}	114	116	118	119	
		A--	-AC	--A	--C	
		Ser	Tyr	*	Asp	
D^{bm24}	K^{b}	63	70	73	77	80
		--G	A-T	A-T	GA-	-C-
		*	Asn	Ser	Asp	Thr

Selected mutants exhibiting substitutions consistent with interlocus exchange between the *K* and *D* loci (Geliebter et al. 1988; Vogel et al. 1988; Hemmi et al. 1988; Horton et al. 1990).

For the most part, extensive sharing of motifs that mark allelic diversity by members of different loci is limited to sequences of the antigen binding domain. Some similar sequence sharing is evident in the transmembrane region and the 3' noncoding sequences of the genes. Two hypothesis for this skewed distribution of shared motifs have been advanced. These hypotheses are not mutually exclusive. The first is that gene conversion may be preferentially targeted to the exons that encode the antigen binding domain (Weiss et al. 1983). The second is that genetic exchanges may take place throughout the length of the genes, but the impact of exchanges on the structure of alleles in the population is determined by selection. While there is evidence that only diversity in the antigen binding domain is selectively advantageous (Hughes and Nei 1988), no information about the relative rates of gene conversion in different parts of the genes is available.

Table 3. Evidence for interlocus exchange among *K*, *D*, and *L* alleles

	SHARED MOTIFS			
	22	23	24	25
$K^f/K^p/L^q/L^d$	TAC	ATC	TCT	GTC
L^d/D^f	-T-	---	G--	---
$K^b/K^s/D^d$	---	--G	GAA	---
$K^k/K^{w28}/D^{v2}/D^k$	-T-	---	---	---
	80	81	82	83
$K^d/Q6^b/Q8^b$	ACC	GCA	CAG	AGA
$K^b/K^f/L^q/Q10^k$	---	CTG	-TC	G-G
$K^k/K^p/K^s/D^d$	---	--G	-T-	C-C
K^{w28}/D^k	---	CTG	-TC	G-C
	114	115	116	117
K^s/D^k	GAG	CAG	TTC	GCA
K^f/D^s	C--	---	CA-	---
K^k/D^f	---	---	-A-	---
	154	155	156	157
$K^b/K^f/D^{v2}$	GAG	AGA	CTC	AGG
$K^d/L^d/D^{v1}$	---	TAT	TA-	---
$K^k/K^p/D^k$	---	---	GA-	C--
	172	173	174	175
$K^b/K^{w28}/D^b/D^d/D^q$	CTG	AAG	AAC	GGG
$K^d/K^s/D^k/D^p/D^s$	---	G---	CT-	---
$K^k/D^f/Q7^b$	---	C--	CT-	---

Examples of sequence motifs that distinguish alleles and are shared by allelic series from different loci are indicated. The sources of *Q* region sequences are listed in Kuhner and Goodenow (1989).

Conclusions

The sequences of 13 additional *K* and *D*/*L* alleles were determined using a locus specific PCR-based cloning strategy. Together with the eleven previously described sequences, the new sequences were used to describe the nature of diversity that distinguish alleles from each other and from sequences of other class I loci. The comparisons provided statistical evidence that alleles of the *K* and *D*/*L* loci are evolving independently throughout most of the coding regions for the antigen presenting molecules. However, sequences determining the structure of the antigen binding domains of both allelic series appear to be co-evolving. Sequence motifs that distinguish alleles in these series are present at both the *K* and *D/L* loci. This sharing of motifs is reminiscent of the class I mutants that have been described in the mouse that provide clear evidence of genetic mechanisms capable of mediating interlocus exchanges of genetic information within the nucleotide sequences encoding the antigen binding domain. We take these observations to mean that similar genetic events have played a major role in shaping the structure of the mouse alleles that are present in inbred mice, and also presumably the alleles present in wild populations.

References

Cai, Z. and Pease, L.R.: Locus-specific cDNA cloning in the class I multigene family: Structure of H-2D^r and H2D^s. *Immunogenetics 32*: 456-459, 1990

Geliebter, J. and Nathenson, S.R.: Microrecombinations generate sequence diversity in the murine major histocompatibility complex: Analysis of the K^{bm3}, K^{bm4}, K^{bm10}, and K^{bm11}. *Molec and Cell Biol 8*: 4342-4352, 1988

Hemmi, S., Geliebter, J., Zeff, R.A., Melvold, W., and Nathenson, S.G.: Three spontaneous H-2D^b mutants are generated by genetic micro-recombination (gene conversion) events: Impact on the H-2-restricted immune responsiveness. *J Exp Med 168*: 2319-2335, 1988

Horton, R.M., Hildebrand, W.H., Martinko, J.M., and Pease, L.R.: Structural analysis of H-2K^f and H-2K^{fm1} by using H-2K locus-specific sequences. *J Immunol 145*: 1782-1787, 1990

Hughes, A.L. and Nei, M.: Patterns of nucleotide substitution at major histocompatibility complex class I loci reveals overdominant selection. *Nature 335*: 167-170, 1988

Klein, J. and Figueroa, F.: Polymorphism of the mouse H-2 loci. *Immunol Rev 60*: 24-57, 1981

Kress, M., Liu, W.-Y., Jay, E., Khoury, G., and Jay, G.: Comparison of class I (H-2) gene sequences. *J Biol Chem 258*: 13929-13936, 1983

Kuhner, M.K. and Goodenow, R.S.: DNA sequences of mouse H-2 and Qa genes. *Immunogenetics 30*: 458-464, 1989

Nathenson, S.G., Uehara, H., Ewenstein, B.M., Kindt, T.J., and Coligan, J.E.: Primary structural analysis of the transplantation antigens of the murine H-2 major histocompatibility complex. *Ann Rev Biochem 50*: 1025-1052, 1981

Pease, L.R., Horton, R.M., Pullen, J.K., and Cai, Z.: Structure and diversity of class I antigen presenting molecules in the mouse. *CRC Critical Reviews in Immunology*, 1991 (in press)

Pease, L.R., Schulze, D.H., Pfaffenbach, G.M., and Nathenson, S.G.: Spontaneous H-2 mutants provide evidence that a copy mechanism analogous to gene conversion generates polymorphism in the major histocompatibility complex. *Proc Natl Acad Sci USA 80*: 242-246, 1983

Vogel, J.M., Davis, A.C., McKinney, D.M., McMillan, M., Martin, W.J., and Goodenow, R.S.: Molecular characterization of the C3HfB/HeN H-$2K^{km2}$ mutation: Implications for the molecular basis of alloreactivity. *J Exp Med 168*: 1781-1800, 1988

Watts, S., Vogel, J.M., Harriman, W.D., Itoh, T., Hans, J.S., and Goodenow, R.S.: DNA sequence analysis of the C3H H-$2K^k$ and $2D^k$ loci evolutionary relationships to H-2 genes from four other mouse strains. *J Immunol 139*: 3878-3885, 1987

Weiss, E., Golden, L., Zakut, R., Mellor, A., Fahrner, K., Kvist, S., and Flavell, R.A.: The DNA sequence of the H-$2K^b$ gene: Evidence for gene conversion as a mechanism for the generation of polymorphism in histocompatibility antigens. *EMBO J 2*: 453-462, 1983

Wilcoxon, F.: Individual comparisons by ranking methods. *Biometrics Bulletin* I: 80-83, 1945

EVOLUTION OF GREAT APE MHC CLASS I GENES

Z.W. Chen*, A.L. Hughes°, N.L. Letvin*, and D.I. Watkins*

*Harvard Medical School, New England Regional Primate Research Center, One Pine Hill Drive, Southborough, MA 01772, USA

°Department of Biology, 208 Mueller Laboratory, Penn State Univ., University Park, PA 16802, USA

To investigate whether the classical HLA *class I loci have been preserved in the great apes and to determine when the A/E recombination event took place, we have sequenced and expressed eight MHC class I gene products expressed by lymphocytes from an orangutan (*Pongo pygmaeus*) and a gibbon (*Hylobates lar*). Three different* HLA-B*-related gene products were found in the orangutan, suggesting that the ancestral homologue of the* HLA-B *locus had undergone a duplication in this species. Since the* HLA-A *homologues found in the orangutan and the gibbon were recombinants between the ancestors of* HLA-A *and* HLA-E, *the recombination event must have occurred before gibbons and humans last had a common ancestor. Thus, the A/E recombination event occurred over 38 million years ago. Preservation of* HLA-A *and* -B *homologues in orangutans and gibbons suggests that the* HLA-A *and* -B *loci are extremely stable and that recombination between them is rare. Furthermore, the data presented here strongly argue against the role of concerted evolution in the generation of diversity in great ape MHC class I molecules.*

The products of the classical human MHC class I loci (*HLA -A*, *-B*, *-C*) are among the most polymorphic glycoproteins known (Parham et al. 1988). These diverse molecules play a crucial role in the immune system by binding peptides and presenting them to cytotoxic T lymphocytes (Townsend et al. 1986). The classical human MHC class I genes are part of a large multigene family on chromosome six that includes other MHC class I genes which code for molecules whose functions have not as yet been defined. These other MHC class I genes [*HLA-AR* (Zemmour et al. 1990), *-E* (Koller et al. 1988), *-F* (Geraghty et al. 1989) and *-G* (Geraghty et al. 1987)] are relatively non-polymorphic and do not encode restricting elements for the majority of T cells.

NATO ASI Series, Vol. H 59
Molecular Evolution of the
Major Histocompatibility Complex
Edited by J. Klein and D. Klein

The allelic products of each of the human *HLA-A*, *-B* and *-C* loci are easily differentiated one from one another. Locus-specific nucleotide substitutions are present throughout these genes. This suggests that these loci are phylogenetically old and may have been preserved during evolution. The *HLA-A* locus is unusual in that it appears to be the result of a recombination between the ancestral homologues of *HLA-A* and *HLA-E*. To determine the age of the *HLA-A*, *-B* and *-C* loci and to find out when the *HLA-A/E* recombination took place, we have been cloning and sequencing MHC class I genes from several species of Great Apes.

Limited diversity of *HLA-A* homologues in the chimpanzee

As part of this effort, PCR was used to amplify two MHC class I cDNAs from the EBV-transformed lymphocytes of Gay, a wild born male chimpanzee (*Pan troglodytes*). Sequence analysis indicated that the two chimpanzee MHC class I cDNAs were *HLA-A* homologues. Moreover, these two cDNAs differed very little from each other in their sequences. Comparison of these two cDNAs to the other four chimpanzee MHC class I cDNAs which have been cloned and sequenced (Mayer et al. 1988; Lawlor et al. 1988) demonstrated that they all are very similar. There are only twenty nucleotide differences between the two most divergent of these 6 chimpanzee MHC class I cDNAs.

HLA-A genes are comprised of five distinct allele families: the *A1/A3/A11*, *A9*, *A10*, *A19*, and *A2/28* families (Kato et al. 1989; Lawlor et al. 1990). The 6 MHC class I cDNAs of the chimpanzee which have been sequenced to date are all related to the *HLA-A1/A3/A11* family (Lawlor et al. 1988; Parham et al. 1989; Lawlor et al. 1990). This finding is consistent with earlier studies by Dorf and Metzgar (1970) in the United States and Balner et al. (1978) in Holland demonstrating a high frequency of cross-reactivity of human *HLA-A1/A3/A11*-specific alloantisera with chimpanzee leukocytes.

We are currently cloning, sequencing and expressing more chimpanzee as well as pygmy chimpanzee *A* locus cDNAs. Initial sequence data suggest that all of these cDNAs sequenced to date are related to the *HLA-A1/A3/A11* family. Precipitation and subsequent one-dimensional isoelectric focusing of chimpanzee and pygmy chimpanzee *A* locus glycoproteins with mAbs specific for *HLA-A* alleles demonstrates that there is polymorphism at the chimpanzee's *A* locus. Therefore, it is possible that the diversification of the *HLA-A1/A3/A11* family has taken place in chimpanzees subsequent to their split from the human lineage.

An unusual *HLA-A* homologue in the gorilla

Cloning, sequencing and expressing of two gorilla *HLA-A* homologues resulted in the isolation of a rather unusual gorilla *A* locus cDNA (Watkins et al. 1991). This cDNA (*Gogo-A3*) was the product of an MHC class I gene that resulted from a recombination between the ancestral homologues of *HLA-AR* and *HLA-A*. The recombination has affected the antigen (Ag)-recognition site in the alpha 1 domain, introducing into it four unique residues (not found in any other *HLA-A* molecule). Additionally, six other unique amino acids have been introduced into the framework of *Gogo-A3*. Unfortunately, no data are available as to whether *HLA-AR* is present as an inactivated locus in the gorilla. However, if it were, this would suggest that recombination can occur between inactivated and functional MHC class I genes. This would be a novel way of generating Ag-recognition site diversity.

HLA-A and *-B* homologues are found in the orangutan and gibbon

We have also cloned, sequenced and expressed 8 homologues of human *HLA-A* and *-B* genes from an orangutan and a gibbon. This demonstrates that the *HLA-A* and *-B* loci must be at least 30 million years old, the last time humans and gibbons had a common ancestor. Surprisingly, we could find no evidence for *HLA-C*-related MHC class I cDNAs into the species. Indeed, the orangutan appears to have duplicated its *HLA-B*-related locus. Finally, since the *HLA-A* genes of gibbons and orangutans are the product of a recombination between the ancestors of the *HLA-A* and *HLA-E*, this recombination event must have occurred before the ancestors of the gibbon and human diverged.

The human *HLA-A* and *-B* loci are more than 30 million years old. That they have been maintained as separate gene loci during evolutionary history suggests that recombinational mechanisms between loci are not a major driving force in the evolution of primate MHC class I molecules. Furthermore, the *HLA-G* and *HLA-F* gene loci have been well preserved in New World primates (Watkins et al. 1990). Thus. while inter-locus recombination events do occur at a low frequency in both humans and great apes, such segmental exchanges must be either very rare or selected against. This stands in apparent contrast to the situation in rodents where it has been suggested that a high frequency of recombination between MHC class I locus genes accounts for the homogenization of these genes (Dover 1982; Rada et al. 1990). However, if the rat MHC genes had evolved by duplication of an ancestral homologue of one of the mouse pseudogenes, this might explain why the classical mouse MHC class I genes are more similar to each other than they are to the rat genes (Hughes 1991).

Acknowledgments. This work was supported by Division of Research Resources grant RR00168 and DK36350 from the National Institutes of Health. Dr. Letvin is a recipient of an American Cancer Society Faculty Research Award. We thank S. Kotlikoff for preparation of this manuscript. We also thank Carol Allen at the Yerkes Primate Center for providing us with the chimpanzee, gorilla, orangutan and gibbon blood samples.

References

Balner, H., van Vreeswijk, W., Roger, J.H., and D'Amaro, J.: The major histocompatibility complex of chimpanzees: Identification of several new antigens controlled by the A and B loci of ChLA. *Tissue Antigens 12*: 1-18, 1978

Dorf, M.E. and Metzgar, R.S.: The distribution of human *HLA-A* antigens in chimpanzees and gorillas. Medical Primatology Proc. 2nd Conf. Exp. Med. Surg. Primates, pp. 12-23. New York 1969

Dover, G.A.: Molecular drive: A cohesive mode of species evolution. *Nature 299*: 111, 1982

Geraghty, D.E., Koller, B.H., and Orr, H.T.: A human major histocompatibility complex class I gene that encodes a protein with a shortened cytoplasmic segment. *Proc Natl Acad Sci USA 84*: 9145-9149, 1987

Geraghty, D.E., Wei, X., Orr, H.T., and Koller, B.J.: *HLA-F*: An expressed *HLA* gene composed of a class I coding sequence linked to a novel transcribed repetitive element. *J Exp Med 171*: 1-18, 1989

Kato, K., Trapani, J.A., Allopenna, J., Dupont, B., and Yang, S.Y.: Molecular analysis of the serologically defined *HLA-A*w19 antigens. *J Immunol 143*: 3371-3378, 1989

Koller, B.H., Geraghty, D.E., Shimizu, Y., DeMars, R., and Orr, H.T.: *HLA-E*: A novel *HLA* class I gene expressed in resting T lymphocytes. *J Immunol 141*: 897-904, 1988

Lawlor, D.A., Ward, F.E., Ennis, P.D., Jackson, A.P., and Parham, P.: *HLA-A* and B polymorphism predate the divergence of humans and chimpanzees. *Nature 335*: 268-271, 1988

Lawlor, D.A., Warren, E., Ward, F.E., and Parham, P.: Comparison of class I MHC alleles in humans and apes. *Immunol Rev 113*: 147-185, 1990

Mayer, W.E., Jonker, M., Klein, D., Ivanyi, P., van Seventer, G., and Klein, J.: Nucleotide sequences of chimpanzee MHC class I alleles: Evidence for trans-species mode of evolution. *EMBO J 7*: 2765-2774, 1988

Parham, P., Lomen, C.E., Lawlor, D.A., Ways, J.P., Holmes, N., Coppin, H.L., Salter,

R.D., Wan, A.M., and Ennis, P.D.: Nature of polymorphism in *HLA-A*, *-B*, and *-C* molecules. *Proc Natl Acad Sci USA 85*: 4005-4009, 1988

Parham, P., Lawlor, D.A., Salter, R.D., Lomen, C.E., and Ennis, P.D.: Diversity and diversification of *HLA-A*, *-B*, *-C* alleles. *J Immunol 142*: 3937-3950, 1989

Rada, C., Lorenzi, R., Powis, S.J., van den Bogaerde, J., Parham, P., and Howard, J.C.: Concerted evolution of class I genes in the major histocompatibility complex of murine rodents. *Proc Natl Acad Sci USA 87*: 2167-2171, 1990

Townsend, A.R.M., Rothbard, J., Gotch, F.M., Bahadur, G., Wraith, D., and McMichael, A.J.: The epitopes of influenza nucleoprotein recognized by cytotoxic T lymphocytes can be defined with short synthetic peptides. *Cell 44*: 959-968, 1986

Watkins, D.I., Chen, Z.W., Hughes, A.L., Evans, M.G., Tedder, T.F., and Letvin, N.L.: Evolution of the MHC class I genes of a New World primate from ancestral homologues of human non-classical genes. *Nature 346*: 60-63, 1990

Watkins, D.I., Chen, Z.W., Garber, T.L., Hughes, A.L., and Letvin, N.L.: Segmental exchange between MHC class I genes in a higher primate: recombination in the gorilla between the ancestor of a human non-functional gene and an A locus gene. *Immunogenetics* (submitted 1991)

Zemmour, J., Koller, B.H., Ennis, P.D., Geraghty, D.E., Lawlor, D.A., Orr, H.T., and Parham, P.: *HLA-AR*, an inactivated antigen-presenting locus related to *HLA-A*. *J Immunol 144*: 3619-3629, 1990

EVOLUTION OF NEW WORLD PRIMATE MHC CLASS I GENES

D.I. Watkins[*], T.L. Garber[°], Z.W. Chen[*], A.L. Hughes[+], and N.L. Letvin[*]

[*]Harvard Medical School, New England Regional Primate Research Center, One Pine Hill Drive, Southborough, MA 01772, USA

[°]Texas A&M Univ., College of Veterinary Medicine, College Station, TX 77843, USA

[+]Department of Biology, 208 Mueller Laboratory, Penn State Univ., University Park, PA 16802, USA

We cloned, sequenced and transfected the MHC class I genes from three other species of bone-chimeric primates (the saddle-backed tamarin, S.fuscicollis*; the common marmoset,* Callithrix jacchus*; and the golden-headed lion tamarin,* Leontopithecus rosalia*) and a single species of a non-bone marrow-chimeric New World primate (the owl monkey,* Aotus trivirqatus*). The MHC class I genes of the bone marrow-chimeric species all displayed low levels of MHC class I variation; the MHC class I genes of the non-bone marrow-chimeric* A.trivirqatus *were highly variable. This finding suggests that bone marrow-chimerism may indeed be related to a reduction in MHC class I variation in New World primates. To investigate the relatedness of these New World primate MHC class I genes, gene trees were constructed. All of the New World primate classical MHC class I genes, including those of* A.trivirqatus, *were related to the human non-clasical* HLA-G *gene. The* S.fuscicollis *MHC class I genes were indistinguishable from those of* S.oedipus. *Surprisingly, the MHC class I genes of* C.jacchus *were only distantly related to the MHC class I genes of* S.oedipus *and* S.fuscicollis. *Interestingly, the functional MHC class I genes of* C.jacchus *were closely related to two pseudogenes of* S.oedipus. *These data suggest that MHC class I evolution in New World primates may take place by duplication and subsequent inactivation or expansion of certain genes.*

The class I major histocompatibility complex (MHC) of vertebrates is a multi-gene family that plays an important role in the immune system (Klein 1986). Cytotoxic T cells (CTLs) lyse virus infected cells, preventing widespread dissemination of the virus. Classical MHC class I genes encode glycoproteins which are expressed on the surface of all nucleated cells and

NATO ASI Series, Vol. H 59
Molecular Evolution of the
Major Histocompatibility Complex
Edited by J. Klein and D. Klein

present viral peptides to these CTLs, facilitating the specific regulation events which lead to this cell lysis. The classical MHC class I loci (for example, *HLA-A*, *-B*, and *-C* in humans and *H-2K*, *-D*, and *-L* in mice) are among the most highly polymorphic loci known. There is evidence that this polymorphism is maintained by some form of balancing selection such as overdominant selection (Hughes and Nei 1988). In recent years, DNA sequencing of MHC class I genes from a number of species has revealed an unusually complex pattern of evolutionary relationships among these loci and has raised a number of unsolved questions regarding the evolutionary mechanisms operating on this very important multi-gene family.

In addition to the polymorphic classical MHC class I loci, nonclassical MHC class I genes and pseudogenes are found in the MHC region. These nonclassical loci are not polymorphic, and have no known function. Pseudogenes are not usually transcribed. If they are, the mRNA contains deletions that result in frame shift mutations and subsequent early termination. In mammals, there are only 1-3 classical MHC class I loci. However, there are a much greater number of nonclassical MHC class I loci and pseudogenes. For example, in the mouse, there are about 30 nonclassical MHC class I loci and pseudogenes, and in the human there are approximately 15. The origin and evolutionary relationships of these nonclassical and pseudogene loci to the classical MHC class I loci is not understood. Closely related species are known to share orthologous classical and nonclassical MHC class I loci. For example, the human and chimpanzee share the classical *HLA-A*, *B*, and *C* loci and at least some nonclassical loci (Mayer et al. 1988; Lawlor et al. 1988). However, when mammals of different orders are compared, there are apparently no orthologous relationships among either classical or nonclassical class I loci (Hughes and Nei 1989). Thus, the human classical MHC class I *HLA-A*, *-B* and *-C* loci are not orthologous to the mouse classical MHC class I *H-2-K*, *-D*, and *-L* loci. This unusual evolutionary relationship is not found in most other vertebrate multi-gene families. Even for the MHC class II genes (which have a distant evolutionary relationship to the MHC class I genes and have similar functions), mammals of different orders share orthologous loci (Klein and Figueroa 1987). To explain this unusual relationship among MHC class I genes of mammals, it has been suggested that there has been a high frequency of inter-locus, non-reciprocal genetic exchange ("gene conversion") leading to a homogenization of MHC class I loci within evolutionary lineages ("concerted evolution") (Dover 1982; Rada et al. 1990). Alternatively, the pattern of relationships among class I genes has been explained as the result of a process of evolution by which loci are duplicated and deleted at a relatively rapid rate over evolutionary time (Hughes and Nei 1989).

The MHC class I genes of New World primates provide an ideal system for studying the mechanisms of evolution of these genes. In most species, the number of MHC class I genes is often quite large; however in the cotton-top tamarin, there appear to be only between seven and eleven MHC class I loci, including classical, nonclassical, and pseudogene loci (Watkins et al. 1991a). The cotton-top tamarin is also particularly interesting in terms of MHC evolution because its classical MHC class I genes are orthologous to the genes of the nonclassical human locus *HLA-G*. This species also possesses a nonclassical gene which is orthologous to the human nonclassical *HLA-F* gene. However, the cotton-top tamarin does not appear to possess any genes orthologous to the human classical *HLA-A*, *-B*, and *-C* genes (Watkins et al. 1990b). Thus, it appears that in the evolution of the tamarin, the *G* locus duplicated to form classical loci, while the homologues of *A*, *B*, and *C* were deleted.

The cotton-top tamarin is also of interest in that it has relatively little diversity at its MHC class I loci and this relative lack of diversity may facilitate the establishment of bone marrow-chimerism. The cotton-top tamarins' MHC class I loci appear to be much more similar to each other than are those of the mouse and human. It is possible that these unusual features may be related to the tamarins' remarkable reproductive biology. More than 70% of tamarins are born as dizygotic twins; due to vascular anastomoses between the placental circulation, bone marrow elements are exchanged between the developing fetuses, resulting in the establishment of stable bone marrow chimeras. Individual members of a twin pair are, therefore, born with bone marrow elements derived from their co-twins (Benirshke et al. 1962). Certainly, a relatively low level of polymorphism and of inter-locus variation at the MHC class I might facilitate the establishment of tolerance in the tamarin. This lack of MHC class I diversity may also predispose the cotton-tamarin to fatal infection from pathogens. The cotton-top tamarin is acutely susceptible to lethal infection after inoculation with viruses that are not lethal in human. Additionally, cotton-top tamarins suffer from an extraordinary incidence of ulcerative colitis and adenocarcinoma of the colon. Thus, the cotton-top tamarins' unusual MHC class I molecules may render it susceptible to a variety of pathological processes.

Certain MHC class I molecules are expressed at high frequencies in the cotton-top tamarin

Our initial studies of the MHC class I glycoproteins of the cotton-top tamarin demonstrated that these primates have relatively non-polymorphic MHC class I molecules (Watkins et al. 1988a). Lymphocytes from all animals tested expressed one common MHC class I gene product. Another MHC class I gene product was expressed by 39 of 41 tested animals. Four other molecules were also expressed on the lymphocytes of these animals at a frequency of 50%. Indeed, only 11 different MHC class I molecules were found in over seventy tamarins tested. Restriction fragment length polymorphism analysis confirmed that there was a limited diversity of cotton-top tamarin MHC class I genomic sequences. Recent studies of 24 tamarins trapped in Columbia have confirmed that the limited diversity found in captive animals is representative of the situation in the wild.

MHC class I sharing in other New World primates

To explore whether the cotton-top tamarin was unusual among New World primates in its sharing of MHC class I molecules, we used 1-D IEF to examine MHC class I diversity in other species of New World primates (Watkins et al. 1991b). 1-D IEF analysis of MHC class I molecules from nine unrelated red-crested, bare-faced tamarins (*S.geoffroyi*; found in Panama and Western Columbia) demonstrated that these primates are also limited in their MHC class I polymorphism. Two MHC class I bands were present on the lymphocytes of every *S.geoffroyi* examined. Interestingly, cells from these *S.geoffroyi* expressed an MHC class I molecule with an isoelectric point identical to that expressed by all cotton-top tamarins. This result was not surprising since *S.geoffroyi* is closely related to *S.oedipus*. Indeed some authors have described *S.geoffroyi* as a subspecies of *S.oedipus* (Hershkovitz 1977). Thus, another tamarin species, closely related to *S.oedipus*, also exhibits limited MHC class I diversity and MHC class I sharing.

We then focused MHC class I molecules from other callitrichid species to explore the possibility that limited MHC class I polymorphism might be common place in bone marrow-chimeric New World primates. Peripheral blood lymphocytes were collected from a single white-footed tamarin (*S.leucopus*), 12 saddle-backed tamarins (*S.fuscicollis fuscicollis* and

S.fusciollis illigeri), 6 spix's moustached tamarins (*S.mystax*), and 4 common marmosets (*Callithrix jacchus*). The 1-D IEF patterns of the MHC class I molecules of the other New World primate species were found to be quite diverse. Extensive MHC class I sharing within a species was not seen in these other species of callitrichidae. Indeed, no single MHC class I molecule was shared by the lymphocytes of all individuals in any other of these bone marrow-chimeric species. Additionally PBL from the owl monkey (*Aotus trivirgatus*) and the squirrel monkey (*Saimiri sciureus*) express polymorphic MHC class I molecules (Watkins et al 1988a,b). Indeed, only the two closely related species of *S.oedipus* and *S.geoffroyi* appear to be exceptions to the rule of MHC class I polymorphism in New World primate species examined to date.

Cloning and sequencing of the cotton-top tamarins' MHC class I cDNAs

To investigate whether the structure of the cotton-top tamarins' MHC class I molecules was related to the species' extraordinary susceptibility to pathogens, we then cloned and sequenced three tamarin MHC class cDNAs (Watkins et al. 1990a). These tamarin MHC class I cDNAs coded for MHC class I molecules which contained amino acid substitutions not found in any of the 39 previously sequenced human MHC class I alleles (Parham et al. 1988). Moreover, the majority of these unique amino acid substitutions were located in the antigen (Ag)-recognition site at positions that have been shown to be critical in the presentation of viral peptides to T cells in mice and humans. These data suggest that selective pressures on MHC class I molecules preferentially act on the Ag-recognition site and that the peptide binding or presenting functions of these molecules may drive the generation of MHC class I polymorphism. Furthermore, the unique amino acid substitutions found in the Ag-recognition site of the tamarins' MHC class I molecules may render it unable to bind the same peptides that can be bound by human MHC class I molecules. Cotton-top tamarins may not be able, therefore, to mount effective CTL against viruses which were not present in the original habitat in Colombia.

Limited nucleotide sequence variation of the cotton-top tamarins' MHC class I cDNAs

The expressed cotton-top tamarin MHC class I cDNAs display little nucleotide sequence variation (Watkins et al. 1990b; Watkins et al. 1991a). Indeed only three amino acid sites in the transmembrane and cytoplasmic domains showed any variation (Fig. 1).

```
Transmembrane domain                                          Cytoplasmic domains
           275                      300           314         315            330          342
CONSENSUS  EPPSQPTIPIMGIVAILAILGAVVTGAVVAAVMWRKKSS         DKKGGSYSQAARSDSAQGSDVSLTACKV
SO-24      ---------------------------------------         ----------------------------
SO-21      ---------------------------------------         ----------------------------
SO-47      ---------------------------------------         ----------------------------
SO-16      ---------------------------------------         ----------------------------
SO-32      ---------------------------------------         ----------------------------
SO-6       -----------------------------T---------         ----------------------------
SO-17      -----------------------------T---------         ----------------------------
SO-8       ---------------------V-----------------         ----------------------------
SO-5       ---------------------V-----------------         ----------------------------
SO-4       ---------------------------------------         ----------------------------
SO-46      ---------------M-------------T---------         ----------------------------
```

Fig. 1. Limited variation of the cotton-top tamarins' MHC class I molecules. The consensus sequence is shown and identity with it is indicated with dashes.

We have cloned and sequenced all 11 of the different MHC class I cDNAs expressed by more than 80 tamarins thus far screened. Comparison of the nucleotide sequences of these tamarin cDNAs to their human counterparts indicates that they are derived from the ancestral homologues of the human non-classical, non-polymorphic HLA-G locus. A deletion of the *HLA-A*, *-B*, and *-C* locus homologues and recent duplication of *HLA-G* locus homologue accounts for the limited variation and polymorphism of the tamarin MHC class I loci. Utilization of *HLA-G*-related, non-polymorphic MHC class I molecules may be important in the establishment of tolerance in these bone marrow-chimeras.

Cloning and sequencing of MHC class I cDNAs from other New World primates

We then sought to explore the possibility that limited MHC class I sequence variation might facilitate bone marrow-chimerism. To this end, MHC class I cDNAs of three other bone marrow-chimeric and one species of non-bone marrow-chimeric New World primates were cloned, sequenced and expressed. The three MHC class I cDNAs cloned from the PBL of the non-bone marrow-chimeric owl monkey (*A.trivirgatus*) were highly variable. In contrast, there was little variation from one MHC class I cDNA to another among the bone marrow-chimeric

New World primate MHC class I cDNAs, suggesting that bone marrow-chimerism may by selecting for the expression of MHC class I genes with limited sequence variation.

Evolution of MHC class I genes in New World primates

The MHC class I cDNAs of the saddle-backed tamarin (*S.fuscicolis*) were indistinguishable from those of the cotton-top tamarin, suggesting that they were closely related. Surprisingly, the common marmosets' MHC class I genes were not closely related to the cotton-top tamarins' MHC class I genes. Since common marmosets and cotton-top tamarins probably shared a common ancestor 5-10 million years ago (Hershkovitz 1977), it was somewhat surprising that the MHC class I genes of these two species were not orthologous. Chimpanzees and humans (who also shared a common ancestor 5-10 million years ago) express MHC class I products derived from nearly identical *HLA-A*, *-B* and *-C* loci (Lawlor et al. 1988; Mayer et al. 1988). Concerted evolution (Dover 1982; Rada et al. 1990) could homogenize the MHC class I genes of both the cotton-top and the common marmoset and this might explain the dissimilarity between the cotton-top tamarins' and the common marmosets' MHC class I genes. However, sequence comparison of two cotton-top tamarin pseudogenes to the common marmoset functional genes revealed that these tamarin pseudogenes were closely related to the functional common marmoset genes. Several nucleotide substitutions unique to the cotton-top tamarin pseudogenes and the common marmoset genes were found. These substitutions, which were not found in any of the twelve expressed cotton-top tamarin genes, were in both exon 2 and 3, suggesting that they were not the result of recombinational mechanisms.

These data suggest that the limited variation and *HLA-G*-relatedness of the tamarin and marmoset genes may be related to the unusual bone marrow-chimeric nature of these two species. Moreover, and perhaps more importantly, the relationships among the human, marmoset and tamarin genes suggest that MHC class I evolution is taking place by a continual process of duplication and subsequent inactivation of certain MHC class I genes. The close relationship between *HLA-G* and the tamarins' expressed MHC class I genes and the similarity of the two tamarin pseudogenes (N1 and N3) to the common marmoset's expressed genes are good examples of duplication and subsequent inactivation of gene loci. These data also tend to argue against the notion that "concerted" evolution is important in primate MHC class I evolution, since certain genes are well preserved between the tamarin and human (i.e., *HLA-F* and *HLA-G* homologues). If concerted evolution were taking place, then these loci would be lost due to a high frequency of recombination.

Why are the MHC class I genes of the bone-marrow chimeric callitrichids so unusual?

The exons encoding the transmembrane and cytoplasmic domains of Callitrichid MHC class I molecule show very little variation. This is unique among MHC class I genes of all other species examined thus far. The nucleotide similarities of the 3' region of Callitrichid MHC class I genes and the apparent independent origins of the common marmoset and cotton-top tamarin MHC class I genes suggest that MHC class I evolution is taking place very rapidly in the callitrichids. One possible explanation for this unusual situation is that intensive selection pressure is operating to minimize the evolution of diverse MHC class I loci. Therefore, MHC class I loci are continually deleted or inactivated as they become more divergent. Concomitant with this may be a process of duplication of certain MHC class I loci. Thus, as MHC class I loci diverge they may be deleted and the remaining loci are duplicated. Positive selection on the Ag-recognition site then result in divergent 5' regions while the 3' regions of the gene are conserved. MHC evolution in callitrichids may, therefore, be a rapid version of the duplication and deletion or inactivation that occurs at a more leisurely pace in great apes (Hughes and Nei 1989).

Selective pressures that might be responsible for limited MHC class I diversity in the bone marrow-chimeric callitrichidae

Bone marrow-chimerism may select for MHC class I loci with limited diversity. Similarity between the MHC class I molecules of donor and host might facilitate the co-existence of genetically disparate cells in the these chimeras. Comparison of callitrichids with other chimeras suggests that callitrichids appear to be unique in the apparent completeness of their tolerance. Even in naturally occurring bovine bone marrow-chimeras, slow rejection of co-twin skin grafts were reported in 50% of animals at 2 mo to 2 y postengraftment (Stone et al. 1965). However, in reciprocal skin grafts exchanged between members of six tamarin twin pairs, no signs of rejection were evident as late as 600 days postengraftment (Porter and Gengozian 1969). Similarly, tamarin lymphocytes have never been shown to respond in MLC to co-twin cells (Picus et al. 1985). Additionally, callitrichids are the only order of mammals where bone marrow-chimerism is a way of life rather than an exception to the rule. Thus, it is likely that the requirement for complete tolerance in the callitrichids may be an important selective pressure.

The requirements for tolerance induction in bone marrow engraftment appear to be quite different from those for parenchymal organs and skin grafts (Martin et al. 1987). In humans, HLA identity between host and donor are critical to the survival of the grafted bone marrow. Indeed, as little as a single amino acid difference between the MHC class I molecules of the host and donor can result in failure of the engrafted bone marrow (Fleischhauer et al. 1990). Although, classical CTL are involved in the recognition of the allogeneic bone marrow grafts, other poorly understood mechanisms may also be operating in such grafts. The phenomenon of hybrid resistance in which irradiated MHC heterozygous mice reject bone marrow grafts from homozygous parents has baffled immunologists for the last 25 years (Bennet 1987). Indeed, it now appears that natural killer cells may play a large role in this phenomenon. Introduction of the D^d MHC class I gene into C57BL donor bone marrow cells results in the acceptance of the bone marrow graft by D^d expressing F_1 hybrid mice (Ohlen et al. 1990). Furthermore, bone marrow cells deficient in beta-2 microglobulin expression, which therefore do not express MHC class I molecules are rejected by MHC identical mice who express beta-2 microglobulin and MHC class I molecules. NK cells seem to be important in the rejection of this MHC class I deficient bone marrow graft (Bix et al. 1991). Therefore, expression of similar if not identical MHC class I molecules between donor and host maybe critical to the success of bone marrow engraftment. Thus, bone marrow-chimerism in callitrichids may be an important selective pressure in the generation of MHC class I loci with limited diversity.

Unique nature of the cotton-top tamarins' MHC class I molecules and its susceptibility to viruses and pathogenic processes

The unusual MHC class I loci of the cotton-top tamarin probably represents the extreme of a spectrum of MHC class I variation and polymorphism present in the callitrichids. Speciation of *S.oedipus* may be a fairly recent event (Fig. 2). It has been hypothesized (on the basis of a scant fossil record and morphological characteristics) that in the late Oligocene or early

Miocene the basic callitrichid stock gave rise to the ancestors of *Callithrix* and *Leontopithecus* (Hershkovitz 1977). The *Callithrix* stock then gave rise to *C.argentata* and *C.jacchus*. *C.jacchus* is now found on the eastern coast of Brazil. The forebearers of the *Saguinus* species arose from the ancestors of *C.argentata* that crossed the Rio Madeira. Indeed, today, no *Callithrix* species is found west of the Rio Madeira and likewise no *Saguinus* species is found East of this river. The evolution and radiation of *Saguinus* then occurred during the Pliocene. It is thought that populations isolated by natural barriers (rivers for example) resulted in the speciation of the *Saguinus*. Indeed, the unique vocal characteristics of *S.oedipus* and *S.geoffroyi* when compared to the long calls of other callitrichids would support the notion that these two species are quite distinct (Snowdon 1991a). Thus, it is possible that *S.oedipus* and *S.geoffryoi* represent the end of the *Saguinus* radiation.

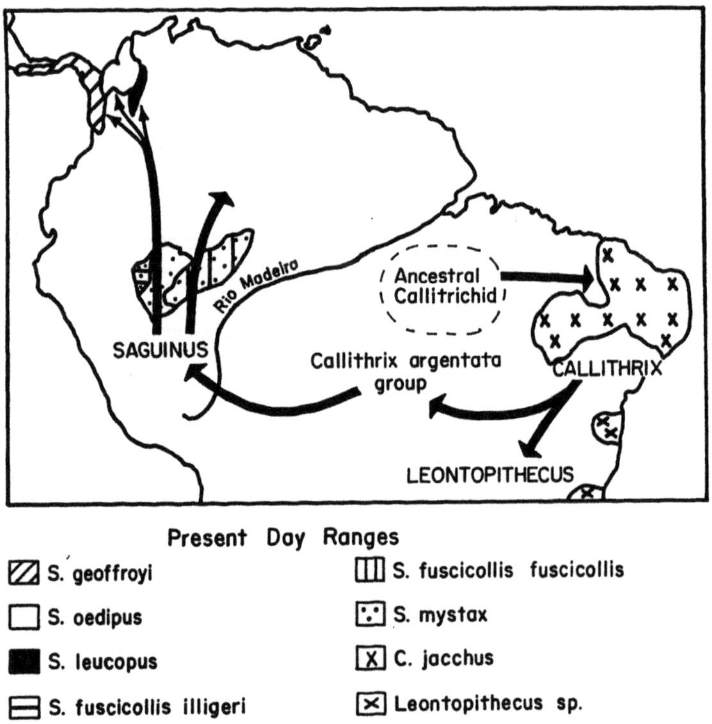

Fig. 2. Hypothetical scheme of callitrichid radiation.

Since all of the *Saguinus* species probably express MHC class I molecules with limited variation, it may be that the cotton-top tamarin and *S.geoffryoi* only inherited a limited number of MHC class I alleles. This might explain why there is such MHC class I sharing in these two species. This limited number of MHC class I alleles is superimposed on the limited variation among the MHC class I loci and thus the cotton-top tamarin may represent an extreme case of limited MHC class I diversity among the callitrichids. Somewhat contrary to the hypothesis that the present day cotton-top tamarin population arose from a limited number of ancestors is the observation that cotton-top tamarin PBLs express the products of polymorphic MHC class II loci (Watkins et al. 1988a). However, it must be noted that the extent of this polymorphism has not been determined.

There are several other mamalian species that express MHC class I molecules with low levels of polymorphism. The cheetah appears to have been through at least two genetic bottlenecks accounting for its reduced levels of polymorphism at all of its genetic loci (O'Brien et al. 1983,1985). The reasons for the limited polymorphism at the Syrian hamster's MHC class I loci are unclear since only fifteen animals have been captured from the wild (Darden and Streilein 1984; Watkins et al. 1990). There are, however, two species of mole-rat in which the genomic diversity of their MHC class I loci is low (Nizetic et al. 1988; Faulkes et al. 1990). The limited MHC class I polymorphism of the naked mole-rat (*Heterocephalus glaber*) is particularly interesting because there is also limited polymorphism at the minisatellite DNA loci of this species. This suggests that these animals are highly inbred and indeed the hierachial structure of their colonies might account for this. All reproduction is carried out by a single female (the queen) who inhibits ovulation by subordinate females (Jarvis 1981). Interestingly, dominant female cotton-top tamarins also suppression ovulation in subordinate females (Snowdon 1991b). Thus it may well be that groups of cotton-top tamarins in the wild may also exhibit limited genetic diversity at all genetic loci due to inbreeding.

The cotton-top tamarins' limited MHC class I diversity may predispose the cotton-top tamarin to various pathogenic processes (Deinhardt 1966; Miller et al. 1977; Wolfe and Deinhardt 1978). Humans express the products of three highly distinct loci; *HLA-A*, *-B*, *-C*. The pattern of amino acid substitution between alleles of the *HLA-A* locus is quite different from that of the alleles of the *HLA-B* locus, suggesting that they might bind quite different peptides (Parham et al. 1988). Although there is diversity in the Ag-recognition in the cotton-top tamarin, one MHC class I molecule is very similar to another. Since positive selection of

T cells occurs in the thymus (Zijlstra et al. 1990; Nikolic-Zugic and Bevan 1990), the limited number of self peptides bound in the thymus by the tamarins' MHC class I molecules may result in the selection of a restricted number of T cell receptors. This limited T cell receptor repertoire might severely compromise the tamarins' ability to defend itself against pathogens.

Finally, it is interesting to speculate as to why the cotton-top tamarin should suffer from an extraordinary incidence of ulcerative colitis (Chalifoux and Bronson 1981). The recent demonstration that rats transgenic for HLA-B27 suffer from inflammation of the GI tract (Hammer et al. 1990) may shed some light on this relationship. Over 90% of individuals with ankylosing spondylitis express HLA-B27 (Schlosstein et al. 1973). Of these individuals, more than 60% also experience an asymptomatic inflammation of the colon and ileum (Cuvelier et al. 1987). It is possible, therefore, that expression of certain MHC class I molecules might predispose an individual to the development of GI lesions. Two MHC class I molecules are expressed by the lymphocytes of nearly every cotton-top tamarin examined to date. It is conceivable that one of these molecules may be similar to HLA-B27 in its ability to predispose tamarins to ulcerative colitis.

Acknowledgments. This work was supported by Division of Research Resources grant RR00168 and DK36350 from the National Institutes of Health. Dr. Letvin is a recipient of an American Cancer Society Faculty Research Award. We thank S. Kotlikoff for help in preparation of this manuscript.

References

Benirschke, K., Anderson, J.M., and Brownhill, L.C.: Marrow chimerism in marmosets. *Science 138*: 513-515, 1962

Bennet, M.: Biology and Genetics of Hybrid Resistance. *Adv Immunol 41*: 333-446, 1987

Bix, M., Liao, N.-S., Zijlstra, M., Loring, J., Jaenisch, R., and Raulet, D.: Rejection of class I MHC-deficient haemopoietic cells by irradiated MHC-matched mice. *Nature 349*: 329-331, 1991

Chalifoux, L.V. and Bronson, R.T.: Colonic adenocarcinoma associated with chronic colitis in cotton top marmosets, Saquinus oedipus. *Gastroenterology 80*: 942-948, 1981

Cuveleir, C., Barbatis, C., Mielants, H., De Vos, M., Roels, H., and Veys, E.: Histopathology of intestinal inflammation related to reactive arthritis. *Gut 28*: 394-401, 1987

Darden, A.G. and Streilein, J.W.: Syrian hamsters express two monomorphic class I major histocompatibility complex molecules. *Immunogenetics 20*: 603-622, 1984

Deinhardt, F.: Neoplasms induced by Rous Sarcoma virus in New World monkeys. *Nature 10*: 443, 1966

Dover, G.A.: Molecular drive: A cohesive mode of species evolution. *Nature 299*: 111, 1982

Faulkes, C.G., Abbot, D.H., and Mellor, A.L.: Investigation of genetic diversity in wild colonies of naked mole-rats (Heterocephalus glaber) by DNA fingerprinting. *J Zool 221*: 87-91, 1990

Fleischhauer, K., Kernan, N.A., O'Reilly, R.J., Dupont, B., and Yang, S.Y.: Bone marrow-allograft rejection by T lymphocytes recognizing a single amino acid difference in HLA-B44. *New Eng J Med 323*: 1818-1822, 1990

Hammer, R.E., Maika, S.D., Richardson, J.A., Tang, J.P., and Taurog, J.D.: Spontaneous inflammatory disease in transgenic rats expresssion HLA-B27 and Human beta-2 microglobulin: An animal model of HLA-B27-associated human disorders. *Cell 63*: 1099-1112, 1990

Hershkovitz, P.: Living New World monkeys (Platyrrhini) with an introduction to Primates, Volume 1. University of Chicago Press, Chicago 1977

Hughes, A.L. and Nei, M.: Pattern of nucleotide substitution at major histocompatibility complex class I loci reveals overdominant selection. *Nature 335*: 167-170, 1988

Hughes, A.L. and Nei, M.: Evolution of the major histocompatibility complex: Independent origin of nonclassical class I genes in different groups of mammals. Mol. *Biol Evol 6*: 559-579, 1989

Jarvis, J.U.M.: Ecological studies on Heterocephalus glaber, the naked mole-rat in Kenya. *Natl Geogr Res Rep 20*: 429-437, 1985

Klein, J.: Natural History of the Major Histocompatibility Complex. Wiley, New York 1986

Klein, J. and Figueroa, F.: Evolution of the major histocompatibility complex. *CRC Crit Rev Immunol 6*: 295-384, 1987

Lawlor, D.A., Ward, F.E., Ennis, P.D., Jackson, A.P., and Parham, P.: HLA-A and B polymorphism predate the divergence of humans and chimpanzees. *Nature 335*: 268-271, 1988

Mayer, W.E., Jonker, M., Klein, D., Ivanyi, P., van Seventer, G., and Klein, J.: Nucleotide sequences of chimpanzee MHC class I alleles: Evidence for trans-species mode of evolution. *EMBO J 7*: 2765-2774, 1988

Martin, P.J., Hansen, J.A., Storb, R., and Thomas, E.D.:Human Marrow transplantation: An immunological perspective. *Adv Immunol 40*: 379- 438, 1987

Miller, G., Shope, T., Coope, D., Waters, L., Pagano, J., Bornkamm, G.W., and Henle, W.: Lymphoma in cotton-top marmosets after inoculation with Epstein-Barr virus: Tumor incidence, histologic spectrum antibody responses, demonstration of viral DNA, and characterization of viruses. *J Exp Med 145*: 948-967, 1977

Nikolic-Zugic, J. and Bevan, M.J.: Role of self-peptides in positively selecting the T-cell repertoire. *Nature 344*: 61-65, 1990

Nizetic, D., Stevanovic, M., Soldatovic, B., Savic, I., and Crkvenjakov, R.: Limited polymorphism of both classes of MHC genes in four different species of the Balkan mole rat. *Immunogenetics 28*: 91-98, 1988

O'Brien, S.J., Wildt, D.E., Goldman, D. Merril, C.R., and Bush, M.: The cheetah is depauperate in genetic variation. *Science 221*: 459-462, 1983

O'Brien, S.J., Roelke, M.E., Marker, L., Newman, A., Windler, C.A., Meltzer, D., Colly, L., Evermann, J.F., Bush, M., and Wildt, D.E.: Genetic basis for species vulnerability in the cheetah. *Science 227*: 1428-1434, 1985

Ohlen, C., Kling, G., Hoglund, P., Hansson, M., Scangos, G., Bieberich, C., Jay, G., and Karre, K.: Prevention of allogeneic bone marrow graft rejection by H-2 transgene in donor mice. *Science 246*: 666-667, 1989

Parham, P., Lomen, C.E., Lawlor, D.A., Ways, J.P., Holmes, N., Coppin, H.L., Salter, R.D., Wan, A.M., and Ennis, P.D.: Nature of polymorphism in HLA-A, -B, and -C molecules. *Proc Natl Acad Sci USA 85*: 4005-4009, 1988

Picus, J., Aldrich, W.R., and Letvin, N.L.: A naturally occurring bone-marrow-chimeric primate. *Transplantation 39*: 297-302, 1985

Porter, R.P. and Gengozian, N.: Immunological tolerance and rejection of skin allografts in the marmoset. *Transplantation 8*: 653-658, 1969

Rada, C., Lorenzi, R., Powis, S.J., van den Bogaerde, J., Parham, P., and Howard, J.C.: Concerted evolution of class I genes in the major histocompatibility complex of murine rodents. *Proc Natl Acad Sci USA 87*: 2167-2171, 1990

Schlosstein, L., Terasaki, P.I., Bluestone, R., and Pearson, C.M.: High association of an HL-A antigen, W27 with ankylosing sponkylitis, *New Engl J Med 288*: 704-706, 1973

Snowdon, C.T.: A vocal taxonomy of the callitrichids. *In* M. Thiago de Mello (ed.): Selected Papers from the XIth Congress of the International Primatological Society. Brasilia 1991a

Snowdon, C.T.: Mechanisms maintaining monogamy in monkeys. *In* D.A. Dewsbury (ed.): Contemporary issues in comparative psychology. Sinauer Associates, Sunderland, MA 1991b

Stone, W.H., Cragle, R.G., Swanson, R.E., and Brown, D.G.: Skin grafts: Delayed rejection between pairs of cattle twins showing erythrocyte chimerism. *Science 148*: 1335-1339, 1965

Watkins, D.I., Hodi, F.S., and Letvin, N.L.: A primate species with limited major histocompatibility complex class I polymorphism. *Proc Natl Acad Sci USA 85*: 7714-7718, 1988a

Watkins, D.I., Kannagi, M., Stone, M.E., and Letvin, N.L.: Major histocompatibility complex class I molecules of nonhuman primates. *Eur J Immunol 18*: 1425-1432, 1988b

Watkins, D.I., Letvin, N.L., Hughes, A.L., and Tedder, T.F.: Molecular cloning of cDNAs that encode MHC class I molecules from a New World primate (Saguinus oedipus): Natural selection acts at positions that may affect peptide presentation to T cells. *J Immunol 144*: 1136-1143, 1990a

Watkins, D.I., Chen, Z.W., Hughes, A.L., Evans, M.G., Tedder, T.F., and Letvin, N.L.: Evolution of the MHC class I genes of a New World primate from ancestral homologues of human non-classical genes. *Nature 346*: 60-63, 1990b

Watkins, D.I., Chen, Z.W., Hughes, A.L., Lagos, A., Lewis, A.M., Shadduck, J.A., and Letvin, N.L.: Syrian hamsters express diverse MHC class I gene products. *J. Immunol 145*: 3483-3490, 1990c

Watkins, D.I., Garber, T.L., Chen, Z.W., Toukatly, G., Hughes, A.L., and Letvin, N.L.: Unusually limited nucleotide sequence variation of the expressed major histocompatibility complex class I genes of New World primate species (Saguinus oedipus). *Immunogenetics 33*: 79-89, 1991a

Watkins, D.I. and Letvin, N.L.: Limited MHC class I polymorphism is not essential for bone marrow chimerism in New World primates. *Immunogenetics 33*: 194-197, 1991b

Wolfe, L.G. and Deinhardt, F.: Overview of viral oncology studies in Saguinus and Callithrix species. *Primate Med 10*: 96-118, 1973

Zijlstra, M., Bix, M., Simister, N.E., Loring, J.M., Raulet, D.H., and Jaenisch, R.: Beta 2-microglobulin deficient mice lack CD4-8+ cytolytic T cells. *Nature 344*: 742-746, 1990

POLYMORPHISMS OF THE MAJOR HISTOCOMPATIBILITY COMPLEX IN OLD AND NEW WORLD PRIMATES

J. Sullivan, C. Kennedy, J. Trejaut, and H. Dunckley
Red Cross Blood Transfusion Service, 153 Clarence St, Sydney NSW 2000, Australia

A fundamental problem confronting all species of vertebrates is how to exist and survive in the face of persistent exposure to pathogenic organisms. To combat this problem a wide range of sophisticated immunological strategies have evolved. Amongst these strategies, the wide spread species distribution of a Major Histocompatibility Complex (MHC)-like structure clearly illustrates its importance and early appearance during evolution. While MHC-like genes are found in a number of marine invertebrates (Reinisch and Litman 1989), it is in vertebrates that this genetic system has evolved to play a central role in cell-mediated immunity and hence determine the ability of a species to mount an immunological defence against a wide range of pathogens.

In humans the MHC (also called the Human Leucocyte Antigen (HLA) system) is located on the short arm of chromosome 6 at 6p21.3 (Spring et al. 1985) and consists of three major gene clusters (Altman and Trowsdale 1989). The class I region consists of approximately 20 genes and it is presently thought that up to half of these genes may be expressed (including *HLA-A, B, C, E, F*, and *G* (Lawlor et al. 1990). The class II genes are centromerically located to the class I genes and consist of four subregions, including *DP, DZ/DO, DQ*, and *DR* (Altman and Trowsdale 1989). Expressed class I and II molecules consist of noncovalently linked bimolecular complexes of proteins with the latter consisting of α and ß chains both encoded from within the MHC while class I molecules consist of a polymorphic MHC derived α-chain associating with a non-MHC encoded nonpolymorphic ß2 microglobulin. The extensive polymorphism found within the human MHC class I and class II region is intimately involved in immune recognition of both foreign and self antigens. The class III region contains genes for certain complement proteins and steroid 21-hydroxylase as well as the more recently described nonclassical class III genes (Altman and Trowsdale 1989). These include the genes for TNFα and -ß, certain heat shock proteins, and a number of novel gene transcripts which have yet to be ascribed a specific function (Altman and Trowsdale 1989). The presence of some of these class III genes, such as those encoding C2 or C4 complement components or TNF, can be explained in evolutionary terms by postulating that these genes in some way enhance specific immune responses. At some stage during evolution the association of class III genes with MHC class I and II genes might have resulted in an improved immune response and subsequent selective pressures might have

NATO ASI Series, Vol. H 59
Molecular Evolution of the
Major Histocompatibility Complex
Edited by J. Klein and D. Klein

been sufficient to maintain these gene associations. The class I and II genes of the MHC region are members of the immunoglobulin supergene family (Williams 1987). Estimates of the time of divergence between members of this family suggest that their common ancestor arose in chordates some 6-8 x 10^8 years ago. The MHC component of the immune system most likely emerged some 3 x 10^8 years ago as present day amphibians possess a structurally and functionally similar MHC region to that found in mammals (Du Pasquier et al. 1989).

Because of the evolutionary proximity of primates to humans, a large body of data exists on the MHC in a number of primate species. In chimpanzees both RFLP (Bontrop et al. 1989) and direct sequencing studies (Fan et al. 1989; Kasahara et al. 1989; Lawlor et al. 1990) of class I and II genes have shown that extensive polymorphisms at these loci clearly predated speciation (Fan et al. 1989; Lawlor et al. 1990). Both gene conversion (Lawlor et al. 1990) and point mutation followed by selection (Fan et al. 1989) have been proposed to explain the subsequent diversification found among related sets of alleles in these two species.

At the class I level chimpanzee genes display a similar overall organization and homologues of the human *HLA-A* and *-B* genes have been described (Lawlor et al. 1990). These studies have failed to identify any species specific substitutions at either loci and it appears that the individual *HLA-A* and *-B* alleles are more closely related to individual chimpanzee alleles than to other *HLA-A* or *-B* alleles. The class I products of other primate species have not been extensively studied but the recent development of biochemical techniques for MHC class I typing of primates (Watkins et al. 1988) provides a simple method for the characterization of class I products in most primate species. Limited class I polymorphism in the new world primate, the cotton-top tamarin, has been proposed to explain the high susceptibility of this endangered species to a number of common pathogens (Watkins et al. 1991). Serological and cellular techniques have also been applied in MHC class II typing of a limited number of primate species (Jonker and Balner 1980). The use of molecular biological techniques has greatly extended our ability to study the class II region in primates and both sequencing (Fan et al. 1989) and RFLP (Bontrop et al. 1989; Schepart et al. 1990; Watkins et al. 1989) approaches have been used. In chimpanzees the relatively nonpolymorphic *DRA* (Fan et al. 1989) and *DOB* (Kashara et al. 1989) genes are highly conserved (approximately 99% homology with the equivalent human genes) while the polymorphic chimpanzee *DRB* genes are more closely related to specific human *DRB* alleles than unrelated *DRB* alleles are to each other (Fan et al. 1989). Recently the polymerase chain reaction (PCR) coupled with direct sequencing has been applied to the polymorphic second exon of *DQA* and *DQB* in a number of primate species (Gyllensten and Erlich 1989; Gyllensten et al. 1990). In general, a given human *DQA* allelic type is more closely related to its equivalent gorilla or chimpanzee allelic homologue than it is to any other human allelic type. This suggests that most of the *DQA* allelic diversity in the human population was present in an ancestral species prior to the divergence of the human, chimpanzee, and gorilla lineages (Gyllensten and Erlich 1989). A similar conclusion has been drawn for the *DQB* locus but here some additional subtypic diversification has occurred after speciation (Gyllensten et al. 1990).

At the RFLP level, human derived HLA class II probes have also been used and this data basically supports the sequencing data. In addition, these probes have been used in a much wider range of new and old monkeys clearly extending the notion of extensive MHC homology beyond higher primates (which only diverged some 5-8 x 10^6 years ago) to primate species which diverged more than 30 x 10^6 years ago. To date, extensive MHC class II polymorphism has been demonstrated using RFLP based methodologies in chimpanzees, baboons, cynomolgus, rhesus, and squirrel monkeys (Devilee et al. 1984; Buntrop et al. 1989; Schepart et al. 1990; Watkins et al. 1989).

In this study we have identified MHC class I, II and III gene polymorphisms in hamadryas baboons (*Mhc-Paha*) and squirrel monkeys (*Mhc-Sasc*) and compared these with RFLP patterns found in humans and chimpanzees (*Mhc-Patr*).

Materials and methods

DNA was obtained from 5 chimpanzees, 7 squirrel monkeys, and 9 hamadryas baboons. The preparation of genomic DNA, *Taq* I restriction endonuclease digestions, electrophoresis and Southern blotting were as per standard methods (Maniatis et al. 1982). The following human-derived MHC probes were used: *DRB, DRA, DQB, DQA, C4*, and *TNF A/B*. All were 10th Histocompatibility Workshop probes with the exception of *DRB* (Bidwell et al. 1988), *C4* (Belt et al. 1984), and *TNF A/B* (Nedospasov et al. 1986). Specific inserts were recovered after digestion, electrophoresis and electrolution and individual inserts were labeled by random priming. Hybridizations were carried out overnight at 65°C in the absence of formamide and the final stringency wash was at 0.1% SSPE at 65°C for 15 min.

Results

Class I MHC genes

The human-derived MHC class I probe *DPOO1* corresponds to the full length cDNA of *HLA-B7* and was used to identify related sequences in squirrel monkeys and baboons. Hybridization with this probe identified extensive polymorphism in both species but as with other species (Devilee et al. 1984) nonpolymorphic bands predominated (results not shown). Surprisingly this probe hybridized with many more fragments in the baboon and squirrel monkey samples than with the chimpanzees but these fragments were predominantly nonpolymorphic and it is interesting to speculate that this reflects the predominance of nonclassical class I genes in the former two species.

Class II MHC genes.

The restriction profile obtained with the *HLA-DRA* probe was relatively simple with a maximum of four fragments identified in any particular species. Thus in chimpanzees we observed only two, nonpolymorphic fragments at 2.7 and 1.9 kb. In squirrel monkeys three polymorphic fragments were observed at 2.6, 2.4, and 1.8 kb while in baboons four fragments were noted at 2.2, 1.9, 1.7, and 1.6 kb, with only the 1.7 kb fragment displaying any polymorphism.

The short *DRB* probe pRTVI (Bidwell et al. 1988) identified extensive polymorphism in all of the species studied (Table 1). A total of eight fragments were identified in chimpanzees (three polymorphic), 15 fragments in squirrel monkeys (10 polymorphic) and 16 fragments in baboons (14 polymorphic; see Table 1).

SIZE (kb)	CHIMPANZEES 1	2	3	4	5
10.7	+	+	+	+	+
6.5	+	+	+	+	+
6.0	+	+	+	+	+
5.6	-	-	+	-	+
5.2	+	+	+	+	+
4.2	+	+	+	+	+
4.0	-	-	+	-	+
2.2	-	+	-	-	-

SIZE (kb)	SQUIRREL MONKEYS 1	2	3	4	5	6	7
10.7	+	+	-	-	+	-	+
8.9	+	+	+	+	+	+	+
6.3	-	-	-	-	+	-	-
5.1	-	+	-	-	-	-	-
4.7	+	+	+	+	+	-	+
4.0	-	-	-	-	+	-	-
3.8	+	+	+	+	+	+	+
3.5	+	+	+	+	+	+	+
3.3	-	-	(+)	+	-	-	-
2.6	+	+	+	+	+	+	+
2.5	+	-	+	+	+	(+)	-
2.4	-	-	-	-	+	-	-
2.0	+	+	+	-	+	-	+
1.8	-	+	+	+	-	-	-
1.6	+	+	+	-	+	+	+

SIZE (kb)	BABOONS 1	2	3	4	5	6	7	8	9
9.5	+	-	+	+	-	+	-	-	-
7.8	+	+	-	+	+	-	+	-	-
7.2	-	-	-	-	+	+	-	+	-
6.5	-	-	-	-	-	+	-	+	-
5.8	-	-	-	-	-	+	-	+	-
5.6	+	-	+	+	-	+	-	-	-
4.4	+	+	+	+	+	+	+	+	+
4.2	-	-	+	-	-	-	+	+	+
4.1	-	-	-	-	+	-	-	-	-
3.6	+	-	+	+	-	+	+	+	+
2.9	+	-	+	+	-	+	-	+	-
2.5	-	-	-	-	+	-	-	-	-
2.4	-	-	+	-	-	-	+	+	+
2.3	-	-	-	-	+	-	-	-	-
1.8	-	-	-	-	-	-	-	-	+
1.4	+	+	+	+	+	+	+	+	+

( ) = PRESENCE OF BAND UNCERTAIN

Table 1. *Taq* 1 *DRB* hybridizing fragments in chimpanzees, squirrel monkeys and baboons.

The *DQA* probe hybridized with three fragments of 5.5, 5.0, and 2.1 kb in chimpanzees (two polymorphic), in squirrel monkeys fragments were observed at 5.1, 3.1, and 2.3 (two polymorphic) while in baboons five fragments at 4.8, 3.8, 2.9, and 2.1 kb were observed (all polymorphic). Likewise the *DQB* probe also identified specific polymorphisms in all of the species studied. In chimpanzees three fragments at 4.6, 4.4, and 4.0 kb were observed (only the 4.0 kb fragment being polymorphic), in squirrel monkeys nonpolymorphic fragments were observed at 10.2 and 4.7 kb and polymorphic fragments at 7.8, 7.2, and 3.7 kb. Baboons were characterized by five polymorphic fragments at 5.5, 3.1, 2.3, 1.7, and 1.6 kb.

Class III MHC genes

The human-derived C4 probe identified the two nonpolymorphic 21-hydroxylase fragments (3.5 and 2.8 kb) and three C4-related fragments at 6.2, 5.2, and 4.2 kb (only the latter fragment was polymorphic) in chimpanzees. This probe failed to hybridize to any squirrel monkey DNA while in baboons three C4-related fragments were identified at 5.0, 3.9 and 3.4 kb and two nonpolymorphic fragments (21-hydroxylase ?) were also observed at 3.2 and 2.6 kb.

Tumor Necrosis Factor (TNF)-related hybridization profiles revealed a common nonpolymorphic fragment amongst all the primate samples (at 1.1 kb) and also indicated that each species was characterized by nonpolymorphic fragment(s) which were common to all individuals in that species. Thus both humans and chimpanzees were characterized by common fragments at 5.0, 1.9, and 0.9 kb, squirrel monkeys with common fragments at 2.1, 1.2 kb and baboons with a common fragment of 3.1 kb. Both squirrel monkeys and baboons also shared a common nonpolymorphic faint band at 1.4 kb. Despite the strong conservation of TNF-related *Taq* I fragments between humans and chimpanzees, the latter did not possess the human TNF-ß *Nco* I 10.5 or 5.5 kb fragments (Messler et al. 1991) but instead possessed two other nonpolymorphic fragments (results not shown).

Discussion

The results of this investigation confirm and extend earlier reports on MHC polymorphism in primates. Thus in this study we have documented extensive polymorphism in MHC class I, II, and to a lesser extent class III genes in representative species of old and new world primates. Using *Taq* I digested primate DNA we were able to demonstrate MHC identity in two groups of two chimpanzees while MHC identity was not observed between members of the baboon or squirrel monkey colonies. It is beyond the scope of this article to consider the possible influence of social and sexual hierarchies, captivity, and other undoubted pressures which might affect breeding preferences but this limited study has clearly demonstrated the value of human-derived MHC DNA probes in detecting the extent of inbreeding within primate colonies.

Comments on the extent of MHC polymorphism amongst these groups of primates must be tempered by the knowledge that the relatedness of founding members of each colony is not known but despite this limitation, MHC fingerprinting should prove useful in the breeding management of captive primates. As such populations are severely constrained as to the level of inbreeding that can be tolerated, an accurate objective approach to monitoring the truly successful breeding male (ie. in terms of offspring sired rather than frequency of mating) is essential so that breeding programmes can be organized to minimize extensive inbreeding.

At the specific gene level, MHC class I gene homologues are extensively maintained in the MHC regions of the squirrel monkey and baboon. Indeed, the predominance of strongly hybridizing nonpolymorphic fragments in both these species is suggestive of a large number of class I-related pseudogenes and/or equivalent homologues to the recently described *HLA-E, F*, and *G* genes (Lawlor et al. 1990). Sequencing of representative class I genes will be required to resolve this issue.

Using human-derived *HLA-DRA, DRB, DQA*, and *DQB* probes, polymorphic homologues for each of these gene loci were detected in both baboons and squirrel monkeys. Comparisons between fragment assignments for class II genes in this study and in other published reports (Bontrop et al. 1989; Watkins et al. 1989) is difficult, especially for complex loci such as *DRB*, but at the chimpanzee *DQA* level the fragments we have assigned at 5.5, 5.0, and 2.1 kb probably correspond to the 6.99, 5.53, and 2.09 kb in the report by Bontrop et al. 1989.

The *DQA2* equivalent of the 2.1 kb nonpolymorphic fragment in chimpanzees was not detected in either the squirrel monkey or the baboon. Crosshybridization between probes within a species was generally not observed, an exception being the common 4.7 kb nonpolymorphic fragment detected in all squirrel monkey samples with both the *DRB* and *DQB* probes.

At the class III gene level both the *C4* and *TNF* probes were generally informative with the exception of the C4 probe/squirrel monkey combination which failed to hybridize to either 21-hydroxylase or C4-related fragments. This contrasts to the report by Takata et al. 1989, where both genes were identified in a number of new world primates. Differences in both the probe complexity and the final stringency wash conditions may partly explain this discrepancy. The TNF probe, in addition to identifying a common primate and human nonpolymorphic fragment, also hybridized with at least one species unique fragment. Very limited polymorphism was demonstrated for the chimpanzee and squirrel monkey samples.

This study has demonstrated the usefulness of MHC probes in documenting extensive MHC gene polymorphisms in both old and new world primates. While the gross organizational details of the MHC are similar in the three groups studied here the sequences of representative genes from each loci will be required to provide more specific information on the evolutionary relatedness of individual MHC loci in primates.

Ackknowledgments. We would like to thank A.Phippard, A.Geczy, and the Veterinary Staff at Taronga Zoo for assistance. The donation of DNA material by individual animals is also greatfully acknowledged. Thanks also to E. Reader for typing this manuscript.

References

Altman, D.M. and Trowsdale. J.: Major histocompatibility complex structure and function. *Current Opin Immunol 2*: 93-98, 1989

Belt, K.T., Carroll, M.C., and Porter, R.R.: The structual basis of multiple forms of human complement component C4. *Cell 36*: 907-914, 1984

Bidwell, J.L., Bidwell, E.A., Savage, D.A., Middleton, D., Klouda, P.T., and Bradley, B.A.: A DNA-RFLP system that positively identifies serologically well-defined and ill-defined HLA-DR and DQ alleles, including DRw-10. *Transplantation 45*: 640-646, 1988

Bontrop, R.E., Otting, N., Broos, L.A.M., Noort, M.C., Kenter, M., and Jonker, M.: RFLP analysis of the HLA-, ChLA-, and RhLA-DQ alpha chain gene regions: conservation of restriction sites during evolution. *Immunogenetics 30*: 432-439, 1989

Devilee, P., Warnaar, J.N., and Giphart, M.J.: MHC homology between various mammalian species at the DNA level: Its relevance to MHC evolution. *Exp Clin Immunogenet 1*: 90-98, 1984

Du Pasquier, L., Schwager, J., and Flajnik, M.F.: The immune system of *Xenopus*. *Annu Rev Immunol 7*: 251-275, 1989

Fan, W., Kasahara, M., Gutknecht, J., Klein, D., Mayer, W.E., Jonker, M., and Klein, J.: Shared Class II MHC polymorphisms between humans and chimpanzees. *Hum Immunol 26*: 107-121, 1989

Gyllensten, U.B. and Erlich, H.A.: Ancient roots for polymorphisms at the *HLA-DQα* locus in primates. *Proc Natl Acad Sci 86*: 9986-9990, 1989

Gyllensten, U.B., Lashkari, D., and Erlich, H.A.: Allelic diversification at the class II DQB locus of the mammalian major histocompatibility complex. *Proc Natl Acad Sci USA 87*: 1835-1839, 1990

Jonker, M. and Balner, H.: Current knowledge of the D/DR region of the major histocomp atibility complex of Rhesus monkeys and chimpanzees. *Hum Immunol 4*: 305-316, 1980

Kasahara, M., Klein, D., and Klein, J.: Nucleotide sequence of a chimpanzee DOB cDNA clone. *Immunogenetics 30*: 66-68, 1989

Lawlor, D.A., Zemmour, J., Ennis, P.D., and Parham, P.: Evolution of Class-I MHC genes and proteins: From natural selection to thymic selection. *Annu Rev Immunol 8*: 23-63, 1990

Messer, G., Spengler, U., Jung, M.C., Honold, G.A., Blomer, K., Pape, G.R., Reithmuller, G., and Weiss, E.H.: Polymorphic structure of the Tumor Necrosis Factor (TNF) and polymorphism in the first intron of the human TNF-ß gene correlates with a variant amino acid in position 26 and a reduced level of TNF-ß production. *J Exp Med 173*: 209-219, 1991

Maniatis, T., Fritsch, E.F., and Sambrook, J.: *Molecular Cloning: A Laboratory Manual.* Cold Spring Harbor Press, New York 1982

Nedospasov, S.A., Shakhou, A.N., Turetskaya, R.L., Mett, V.A., Azizou, M.M., Georgiev, G.P., Kubobko, V.G., Dobrynin, V.N., Filippov, S.A., Bystrov, N.S., Boloreva,

E.F., Chuvpilo, S.A., Chumakov, A.M., Shingarova, L.N., and Ouchinnikov, Y.A.: Tandem arrangement of genes coding for Tumor Necrosis Factor (TNF-α) and lymphotoxin (TNF-ß) in the human genome. *Cold Spring Harbor Symp Quant Biol 51*: 611-624, 1986

Reinisch, C.L. and Litman, G.W.: Evolutionary immunobiology. *Immunol Today 10*: 278-281, 1989

Schepart, B.S., Buck, D.A., Manning, C.H., Keever, C.A., McMahon, M.R., and Heise, E.R.: Restriction fragment polymorphism of the Cynomolgus monkey major histocompatibility complex. *Exp Clin Immunogen 7:* 109-122, 1990

Spring, B., Fonatsch, C., Muller, C., Pawelec, G., Kompf, J., Wernet, P., and Ziegler, A.: Refinement of HLA gene mapping with induced B-cell mutants *Immunogenetics 21*: 277-291, 1985

Takata, H., Haranaka, M., Takenaka, O., Miwa, N., Murayama, Y., and Iri, H.: Evolutionary analysis of MHC genes in non human primates. In B. Dupont (ed) *Histocompatiblity Testing 1987*, p. 595, Springer-Verlag, New York 1989

Watkins, D.I., Garber, T.L., Chen, Z.W., Toukatly, G., Hughes, A.L., and Letvin, N.L.: Unusally limited nucleotide sequence variation of the expressed major histocompatibility complex class I genes of a New World Primate species (*Sanguinus vedipus*) *Immunogenetics 33*: 79-89, 1991

Watkins, D.I., Kannagi, M., Stone, M.E., and Letvin, N.L.: Major histocompatibility complex class I molecules of nonhuman primates. *Eur J Immunol 18*: 1425-1432, 1988

Watkins, D.I., Shadduck, J.A., Hodi, F.S., Stone, M.E., and Letvin, N.L.: Use of cDNA probes specific for the human MHC class II B loci for tissue-typing nonhuman primates at their class II B loci. *Transplantation 48*: 170-174.

Williams, A.F.: A year in the life of the immunoglobulin superfamily *Immunol Today 8*: 298-303, 1987

Mhc CLASS II GENES OF NEW WORLD MONKEYS AND THEIR RELATIONSHIP TO HUMAN GENES

Blazenka Grahovac, Colm O'hUigin, Herbert Tichy, Werner E. Mayer, and Jan Klein*
Max-Planck-Institut für Biologie, Abteilung für Immungenetik, Correnssstraße 42, 7400 Tübingen, Germany
*Department of Microbiology and Immunology, University of Miami School of Medicine, Miami, FL 33101, USA

At least two DRB *loci are transcribed in the B lymphocytes of the cotton-top tamarin,* Saguinus oedipus, *a representative of the Platyrrhini (New World monkeys). One of the loci,* Saoe-DRB1, *is functional, the other (*Saoe-DRB2*) seems to be occupied by pseudogenes. The polymorphism of the genes at the functional* DRB1 *locus may be low. Neither of the two loci is directly orthologous to any of the known* DRB *genes thus far found in the Catarrhini. We suggest that the Platyrrhini were founded by a small number of animals and that the switch from one set of functional genes to another occurred as a result of genetic drift in the founding population.*

The Platyrrhini

Paleobiologists agree that all monkey species living in the wild in South and Central America (the New World monkeys) are derived from a single ancestor which was not the ancestor of monkeys living in other parts of the world (Herskovitz 1977; Ciochon and Chiarelli 1980). Attesting to their common origin are a number of characteristics shared by all New World monkeys, one of them being the disposition of nostrils. In all New World monkeys, the nostrils are separated by a wide partition and they face forward; in virtually all other monkeys the nostrils are separated by a relatively narrow partition and face outward (Herskovitz 1977; Martin 1990). Because of this particular characteristic, the New World monkeys are called Platyrrhini (broad-nosed) and the Old World monkeys Catarrhini (narrow-nosed). The first fossil primates in South America date back to the earliest Oligocene, some 37 million years (my) before present (reviewed by Rose and Fleagle 1981). At that time, South America was an island continent separated by more than 1000 km of water from both North America and

NATO ASI Series, Vol. H 59
Molecular Evolution of the
Major Histocompatibility Complex
Edited by J. Klein and D. Klein

Africa; the land connection with North America did not form until 3 my ago (Tarling 1980). It is commonly assumed that the ancestors of Platyrrhini reached South America at some point before 37 my ago and, finding a continent without primates they would have had to compete with, they evolved rapidly into a variety of species adapted to different environmental niches (= adaptive radiation; Fig. 1).

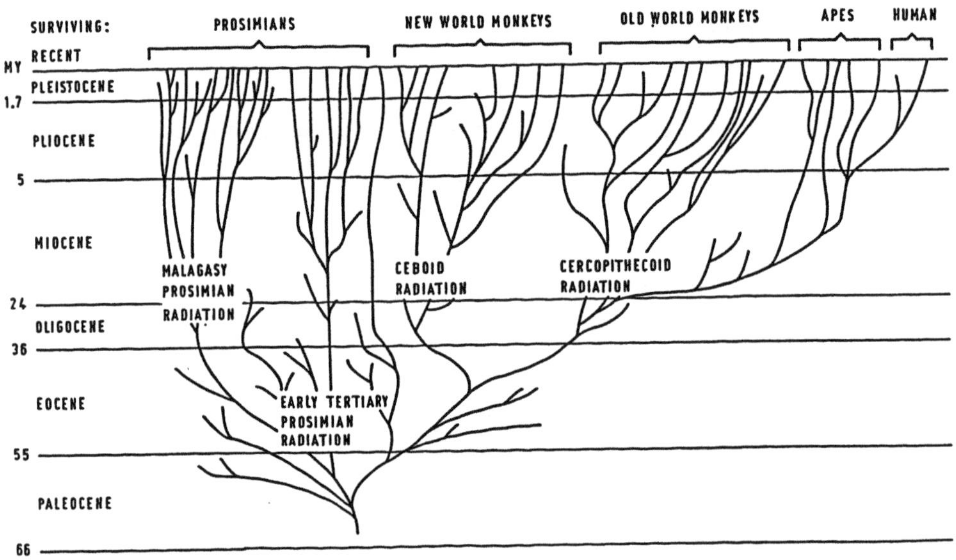

Fig. 1. The evolutionary origin of primates depicted as a series of adaptive radiations. Only some of the known branches are shown; many more probably existed so that the depicted bushes were much more dense. Geological epochs and their approximate time of duration in million of years (my) are shown on the left. Adapted and modified from Simpson (1949).

At the time the Platyrrhini ancestors arrived in South America, there were primates living in North America, Asia, and Africa, and it is not known from which of these three continents the ancestors arrived and how. Paleobiologists are equally divided between those who place the origin of Platyrrhini in Africa and those who consider North America to be the original homeland of the broad-nosed primate stock (Fleagle 1988; Martin 1990). The colonization probably occured via a chain of islands that may have existed at that time between Africa and South America or between North America and South America. The small, arboreal ancestors may have spread from one island to another by clinging to trees that drifted in the water after major storms. A similar route was also taken at approximately the same time by a group of rodents, but besides primates and rodents, no other mammals reached South America during its period of isolation. All other South American mammals evolved either from stocks that were already on this continent before it separated from Africa and North America more than 85 my ago or from immigrants that entered the continent after the emergence of the Panamian isthmus and the formation of the landbridge to North America some 3 my ago.

The implication from the island-hopping mode of colonization is that the South American primates were founded by a small population in which genetic drift was an important element of evolution. Dramatic departures from the genetic composition of the parental populations may have occurred in the founding population and the initial genetic upheavals may have left their mark on the entire platyrrhine stock. These marks may still be recognizable in comparative studies of Old World and New World monkeys, particularly in genetic systems that may have been relatively unstable at that time. One such system may have been the major histocompatibility complex (*Mhc*), especially the class II loci occupying the *DRB* region (Klein 1986). A characteristic feature of the *DRB* region, at least in apes, is that it displays not only allelic but also haplotype polymorphism and that both polymorphisms have evolved in a trans-specific manner (Klein et al. 1991).

The tamarins

The living New World monkeys can be divided into two groups, the marmosets (tamarins) and the cebids (Martin 1990). The former are small-bodied monkeys with claws on their fingers. They are further characterized by the absence of the third molar from both the upper and the lower jaws, production of dizygotic twins, and by infant care provided by group members other than the mother. In contrast to the marmosets and tamarins, the cebid monkeys are larger-bodied, they have nails on all their fingers, they possess the third molar in both jaws, and they produce only one infant at a time which is taken care of by the mother. Which of the two groups is more advanced and which is closer to the ancestral Platyrrhini stock is still being debated by primatologists. Martin (1990) argues rather convincingly that the small body size of the marmosets and tamarins is the result of secondary dwarfing which was accompanied by adaptations aimed at increasing reproductive potential by twinning; by the transfer of the burden of caring for two relatively large infants from the mother to all adults and subadults in the social group; by the loss of the third molar as an adjustment to the downscaling of the jaws; and by the reversal from nails to claws as an adaptation to locomotion on relatively broad tree trunks and branches.

Fig. 2. Cotton-top tamarin (*Saguinus oedipus*).

We have selected the cotton-top tamarin, *Saguinus oedipus* (Fig. 2) for our initial studies from which a B-cell line (SC2) has been derived and made available to us (Watkins et al. 1988). The cotton-top tamarin is easily identifiable by a nest of long, white hair growing in a narrow peak on the top of its head between areas of naked skin (German speakers refer to it as "Liszt monkey"). The species is now restricted to a few isolated areas of lowland tropical evergreen forest in North and Central Colombia, more than three-fourths of its original habitat having been deforested and converted into cattle pasture. It is a diurnal, arboreal monkey, living in groups of 2 to 13 and feeding on small fruits, insects, buds, and leaves (Eisenberg 1989; Emmons 1990).

The tamarin *DRB* region

In the cDNA library prepared from the mRNA of the SC2 cell line, we found only three different *DRB* sequences (Grahovac et al. 1991). Two of these were apparently derived from allelic genes because they differed by two substitutions only in the entire coding region: one synonymous and the other a nonsynonymous conservative substitution outside the segment coding for the putative peptide-binding region. The third sequence was derived from a pseudogene in which the stretch of nucleotides 176-270 has apparently been duplicated and transposed 5' of exon 2. This sequence also contained a stop codon at a position normally coding for the amino acid residue at position 94, so that even if the message were translated, the amino acid sequence would have been garbled beyond this point. More recently, we identified a fourth gene by sequencing part of a cosmid clone isolated from a SC2 genomic library. The sequence covers most of exon 3 (Fig. 3) of what is apparently also a pseudogene: it contains a single-nucleotide deletion in codon 130, destroying the correct reading frame, and two stop codons, one at position 147 and the other at position 180. No transcript of this gene could be found in the SC2 cDNA library. The two pseudogene sequences seem to be related to each other and may in fact represent alleles at a single locus. If so, there should be at least two loci in the *DRB* region of the cotton-top tamarin, one (*Saoe-DRB1*) occupied by functional genes and the other (*Saoe-DRB2*) occupied by pseudogenes. There seem to be at least two alleles at each of these two loci, *Saoe-DRB1*01, -DRB1*02, -DRB2*01*, and *DRB2*02*. The *DRB1* might be the only functional *DRB* locus in the cotton-top tamarin. Southern blot analysis of SC2 genomic DNA suggests that there might still be one or two additional loci in the *DRB* region, but these would presumably be pseudogenes like *DRB2*.

Polymorphism of the *Saoe-DRB* loci

As mentioned earlier, the coding sequences of the two *DRB1* alleles differ by two nucleotide substitutions only and the encoded proteins by one amino acid substitution only. For a functional locus this is a surprisingly small difference. Alleles at the human *DRB1* loci can differ at many nucleotide substitutions and the corresponding proteins at many amino acid substitutions (Marsh and Bodmer 1991). Although minor allelic variants differing by very few substitutions do occur in the human population, they are rare. The probability that two such minor variants would come together in a single heterozygous individual is very low. The finding of two very closely related *Saoe-DRB1* alleles in a single cell line might therefore be indicative of low *DRB* polymorphism in this species. Of course, even in humans there are some *DRB* loci which, although not pseudogenes, are monomorphic (i.e., the *HLA-DRB4*

	1	2	3	4	5	6	7	8	9	10	11	12	13	14	15	16	17	18	19	20	21	22	23	24	25	26	27	28	29	30
*Saoe-DRB1*01*	GGG	GAC	ACC	CGA	CCA	CGT	TTC	TTG	GAG	CAG	GCT	AAG	GCT	GAG	TGT	CAT	**CA**C	TTC	AAC	GGG	ACG	GAG	CGG	GTG	CGC	**C**TC	CTG	GAA	AGA	TAC
*Saoe-DRB1*02*	---	---	---	---	---	---	---	---	---	---	---	---	---	---	---	---	---	---	---	---	---	---	---	---	--G	---	---	---	---	---
*Saoe-DRB2*01*	-T-	C-G	CGG	T--	G--	---	---	---	---	---	TT-	---	TA-	---	---	-G-	---	C--	---	---	---	---	---	---	--G	---	---	-TC	---	---

	31	32	33	34	35	36	37	38	39	40	41	42	43	44	45	46	47	48	49	50	51	52	53	54	55	56	57	58	59	60
*Saoe-DRB1*01*	TTC	TA**C**	AAC	CAG	GAG	GAG	TTG	GTG	CGC	TTC	GAC	AGC	GAC	GTG	GGG	GAG	TAC	CGG	GCG	GTG	ACG	GAG	CTG	GGG	CGG	CCT	GAC	GAG	GAG	TAC
*Saoe-DRB1*02*	---	---	---	---	---	---	--C	---	---	---	---	---	---	---	---	---	---	---	---	---	---	---	---	---	---	---	---	---	---	---
*Saoe-DRB2*01*	A--	---	---	---	---	---	-AT	---	---	---	---	---	---	---	---	---	-T-	---	---	---	---	---	-G-	---	---	---	-TT	-CA	---	---

	61	62	63	64	65	66	67	68	69	70	71	72	73	74	75	76	77	78	79	80	81	82	83	84	85	86	87	88	89	90
*Saoe-DRB1*01*	TGG	AAC	AGC	CGG	AAG	GAC	ATC	CTG	GAG	CAG	ACA	CGG	GCC	GCG	GTG	GAC	ACC	TTC	TGC	AGA	CAC	AAC	TAC	G**A**G	**A**TT	**T**TA	GAG	AG**G**	TTC	**C**T**A**
*Saoe-DRB1*02*	---	---	---	---	---	---	---	---	---	---	---	---	---	---	---	---	---	---	---	---	---	---	---	---	---	---	---	---	---	---
*Saoe-DRB2*01*	-TC	---	---	-T-	---	---	TA-	G--	---	---	GAG	---	---	---	---	---	---	G-G	---	---	---	---	---	---	---	--T	---	---	---	---

	91	92	93	94	95	96	97	98	99	100	101	102	103	104	105	106	107	108	109	110	111	112	113	114	115	116	117	118	119	120
*Saoe-DRB1*01*	GTG	CCG	CGG	AGA	GTC	CAA	CCT	AAG	GTG	ACT	GTG	TAT	CCT	GCA	AAG	ACC	CAG	CCC	CTG	CAG	CAC	CAC	AAC	CTC	CTG	GTC	TGC	TCT	GTA	AGT
*Saoe-DRB1*02*	---	---	---	---	---	---	---	---	---	---	---	---	---	---	---	---	---	---	---	---	---	---	---	---	---	---	---	---	---	---
*Saoe-DRB2*01*	---	-A-	---	C--	---	---	---	---	---	---	---	---	---	---	---	---	---	---	---	---	---	---	---	---	---	---	---	---	--G	---
*Saoe-DRB2*02*	...	...	...	...	...	...	...	...	...	...	...	...	...	...	...	...	...	...	...	...	---	---	---	---	---	---	---	---	--G	---

	121	122	123	124	125	126	127	128	129	130	131	132	133	134	135	136	137	138	139	140	141	142	143	144	145	146	147	148	149	150
*Saoe-DRB1*01*	GGT	TTC	TAC	CCA	GGC	AGC	ATT	GAA	GTC	AAG	TGG	TTC	CGG	AAC	GAT	CAG	GAG	GAG	AAG	GCT	GGG	GTG	GTG	TCC	ACA	GGC	CTG	ATC	CAG	AAT
*Saoe-DRB1*02*	---	---	---	---	---	---	---	---	---	---	---	---	---	---	---	---	---	---	---	---	---	---	---	---	---	---	---	---	---	---
*Saoe-DRB2*01*	---	---	---	---	---	---	---	--G	---	---	---	---	---	---	--C	---	---	---	---	---	---	---	---	---	---	---	---	---	--C	---
*Saoe-DRB2*02*	---	---	--T	---	---	--T	---	---	---	*--	---	---	---	--T	-GC	---	---	--A	---	---	---	---	---	---	---	--T	---	---	--C	---

	151	152	153	154	155	156	157	158	159	160	161	162	163	164	165	166	167	168	169	170	171	172	173	174	175	176	177	178	179	180
*Saoe-DRB1*01*	GGA	GAC	TGG	ACC	TTC	CAG	ACA	CTG	GTG	ATG	CTG	GAA	ACT	GTT	CCT	CAG	AGT	GGA	GAG	GTT	TAC	ATC	TGC	CAA	GTG	GAG	CAC	GCA	AGC	ATG
*Saoe-DRB1*02*	---	---	---	---	---	---	---	---	---	---	---	---	---	---	---	---	---	---	---	---	---	---	---	---	---	---	---	---	---	---
*Saoe-DRB2*01*	---	---	---	--A	---	---	--C	---	---	---	---	---	---	---	---	---	---	---	---	---	---	-C-	---	---	---	---	---	C--	---	G--
*Saoe-DRB2*02*	---	---	---	---	---	---	--C	---	---	---	---	---	---	---	---	--A	---	---	---	---	-G-	-C-	--T	---	---	---	---	C--	---	G--

	181	182	183	184	185	186	187	188	189	190	191	192	193	194	195	196	197	198	199	200	201	202	203	204	205	206	207	208	209	210
*Saoe-DRB1*01*	ACG	AGC	CCT	CTC	ACA	GTG	CAA	TGG	AGA	GCA	CGA	TCC	GAA	TCT	GCA	CAG	AGT	AAG	ATG	CTG	AGT	GGA	GTC	GGG	GGC	TTT	GTG	CTG	GGT	CTG
*Saoe-DRB1*02*	---	---	---	---	---	---	---	---	---	---	---	---	---	---	---	---	---	---	---	---	---	---	---	---	---	---	---	---	---	---
*Saoe-DRB2*01*	--C	---	---	---	---	---	---	---	---	--G	-AG	--T	---	---	---	---	--C	---	---	---	---	---	---	---	---	---	---	---	--C	---
*Saoe-DRB2*02*	--C	---	---	-A-	---	---	---	---	...	...	...	...	...	...	...	...	...	...	...	...	...	...	...	...	...	...	...	...	...	...

	211	212	213	214	215	216	217	218	219	220	221	222	223	224	225	226	227	228	229	230	231	232	233	234	235	236	237	238
*Saoe-DRB1*01*	CTC	TTC	CTT	GGA	GCA	GGG	CTG	TTC	ATC	TAC	TTC	AGG	AAT	CAG	AAA	GGA	CAC	TCT	GGA	CTT	CAG	CCA	ACA	GGA	CTC	CTG	AGC	TGA
*Saoe-DRB1*02*	---	---	---	---	---	---	---	---	---	---	---	---	---	---	---	---	---	---	---	---	---	---	---	---	---	---	---	---
*Saoe-DRB2*01*	---	---	---	--G	---	--T	---	---	---	---	---	---	---	---	---	---	---	---	-C-	---	---	---	T--	---	---	---	---	---

Fig. 3. The nucleotide sequence of the four known *DRB* genes of the cotton-top tamarin, *Saguinus oedipus*. Dashes (-) indicate identity with the top sequence, dots (.) unavailability of information, and asterisks (*) postulated gaps (deletions) in the sequence. Codons are numbered starting with the codon specifying the first amino acid residue of the mature protein. The species-specific nucleotides are highlighted (virtually all other known primate *DRB* sequences have a differnt nucleotide at these positions).

locus at which, thus far, only one allele has been identified), but these exist in addition to a highly polymorphic locus (*HLA-DRB1*) on the same chromosome. In the tamarin, the *Saoe-DRB1* might be the only functional *DRB* locus and if so, its low polymorphism would be surprising.

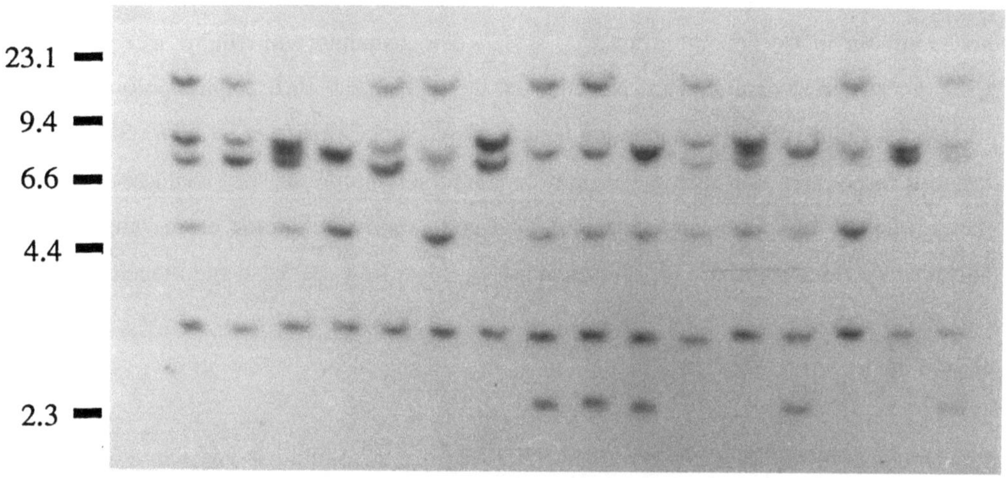

Fig. 4. Southern blot hybridization of a filter containing DNA from 16 cotton-top tamarins (1-16). The DNA was digested with the *Hin* dIII restriction endonuclease and the blot was hybridized with a probe encompassing the cotton-top tamarin 3' untranslated region. Marker sizes are given in kilobases (kb) on the left. The tissue samples came from animals maintained at the University of Bielefeld, Bielefeld, Germany.

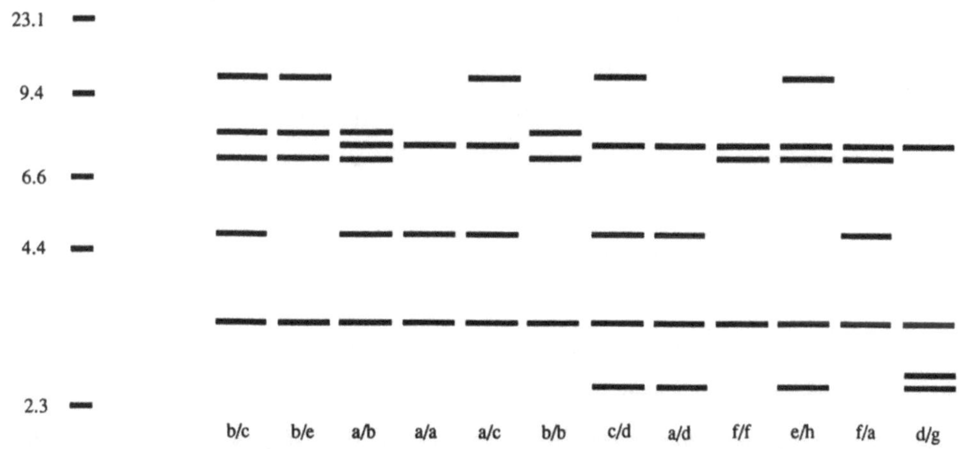

Fig. 5. Schematic representation of hybridization patterns of a total of 24 cotton-top tamarins. Hybridization was carried out as shown in Figure 4. The postulated haplotypes are indicated at the bottom of the diagram.

Southern blot analysis of genomic DNA from 24 cotton-top tamarins has revealed maximum of eight different haplotypes after digestion with the *Hin* dIII restriction endonuclease and hybridization with a probe specific for the 3' untranslated region (3'UT) of the *Saoe-DRB1* gene (Figs. 4, 5). Even fewer haplotypes could be resolved in DNA digested with the *Eco* RI endonuclease. None of these observations, however, demonstrates definitively low *DRB* polymorphism in the cotton-top tamarin; such a demonstration will require sequencing of *DRB* genes from several animals. Nonetheless, the hint of low *DRB* polymorphism squares well with a similar finding at cotton-top tamarin class I loci (Watkins et al. 1988). Should the tamarins indeed turn out to have generally low *Mhc* polymorphism, they could be added to the growing list of species that apparently manage to survive with little or no variability at this otherwise highly variable chromosomal region (for review see Klein and Figueroa 1986).

Origin of the *Saoe-DRB* genes

The human *DRB* region is highly variable. Not only are there multiple genes in a haplotype and multiple alleles at some of the loci, but the number of genes per haplotype is also not the same in different individuals (Klein et al. 1991). Similar variation is also found in the great apes (chimpanzee, gorilla, orang-utan) and in some Old World monkeys. Phylogenetic analysis suggests that the different Catarrhini *DRB* genes are all derived from a single ancestral gene. (Catarrhini are hominoids and Old World monkeys together, hominoids are humans and apes together.) Was this gene also the ancestor of the Platyrrhini *DRB* genes as represented by the genes of the cotton-top tamarin?

Most likely not. Figure 6 depicts the presumed phylogenetic relationship of *Saoe-DRB* genes to representative Catarrhini and dog *DRB* genes, as determined by the maximum parsimony method of dendrogram construction (Felsenstein 1988). The nature of the method is such that it often produces not one but several most parsimonious dendrograms. The tree in Figure 6 shows a strict consensus of two trees which differ only in the topology of the *DRB2* sequences. The dog (*Cafa*) gene is an outgroup (it stands outside the remaining collection of genes), which means that all the primate genes are derived from a common ancestral gene after this gene separated from the ancestral dog *DRB* gene. Furthermore, the tamarin genes constitute an outgroup to all the Catarrhini *DRB* genes, which indicates that the former are derived from one ancestor and the latter from another.

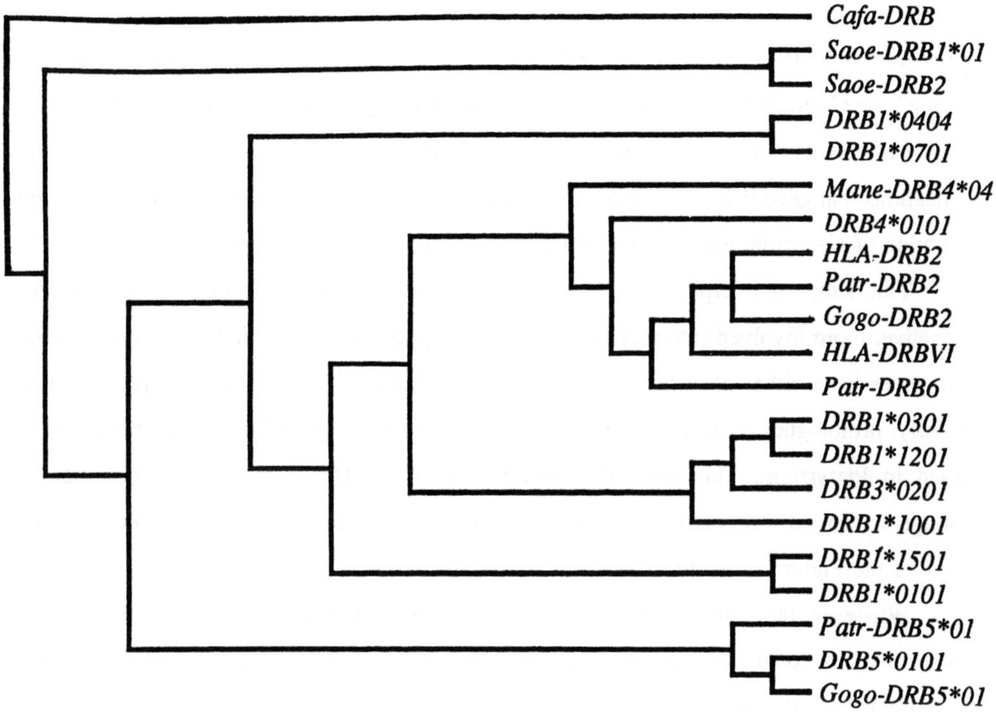

Fig. 6. Dendrogram depicting the relationship of the cotton-top tamarin *DRB* genes to those of other primates. The tree was constructed by the maximum parsimony method. For references of the individual sequences see Grahovac et al. (1991). *Cafa*, dog, *Canis familiaris; Gogo*, gorilla, *Gorilla gorilla; HLA*, human, *Homo sapiens; Mane*, pig-tail macaque, *Macaca nemestrina; Patr*, chimpanzee, *Pan troglodytes; Saoe*, cotton-top tamarin, *Saguinus oedipus.*

This conclusion is supported by the existence of several species-specific substitutions that are shared by all four tamarin genes but either do not occur at all in any of the other known *DRB* genes or occur only sporadically (highlighted in Figure 3). The species-specific residues set the tamarin *DRB* genes apart from the Catarrhini genes. The collection of Catarrhini sequences includes both orthologous and paralogous genes. For example, *HLA-DRBVI* and *Patr-DRB6* are genes at the same locus but are found in different species (human and chimpanzee, respectively): they are orthologous. By contrast, *HLA-DRB1*03* and *HLA-DRB3*01* are genes at two different loci in the same species (human), but are apparently derived from the same ancestral gene by duplication: they are paralogous. Hence, if the dendrogram reflects the true phylogeny of this collection, its simplest interpretation is that in which all the Catarrhini *DRB* genes, orthologous and paralogous, are derived from an ancestor which was not the ancestor of *Saoe-DRB*. Such interpretation would mean that all the duplications that gave rise to the different Catarrhini paralogous *DRB* genes occurred after the separation of Catarrhini and Platyrrhini lineages from their common ancestor.

This simplest interpretation, however, need not necessarily be the correct one. An alternative, more involved interpretation would presuppose that the orthologues of the *Saoe-DRB* genes have either been lost or have not been identified in Catarrhini as yet and that, similarly, orthologues of at least some of the Catarrhini *DRB* genes have either been lost or missed in Platyrrhini. This alternative interpretation is difficult to rule out completely, although it can be progressively rendered less likely as more and more data on the *DRB* region of the Platyrrhini and Catarrhini primates accumulate.

Whatever the correct interpretation, it is clear that the tamarin expressed and presumably functional *DRB* genes are not directly orthologous to the functional *DRB* genes of the Catarrhini: the two sets of functional genes have evolved from different ancestors. A similar nonorthologous relationship exists also between the tamarin and human class I genes. Watkins and his colleagues (1990) observed that the apparently functional cotton-top tamarin class I genes are more closely related to the nonclassical (and presumably nonfunctional) *HLA-G* (most) and *HLA-F* (few) than to the classical functional *HLA-A* and *HLA-B* class I genes. The difference between these two situations is that whereas the nonfunctional human orthologue of the functional tamarin gene has been identified for the class I loci, as far as the *DRB* loci are concerned there is thus far no sign of it, suggesting that it may, in fact, not exist.

Thus the tamarin relies on a different set of functional class I and class II (*DRB*) loci than the Catarrhini. While no information is available on other representatives of Platyrrhini, it would be very unusual if this change in deployment of functional genes were limited to the one species studied thus far. It can be expected that all the tamarins and marmosets at least have this change in common, and it would not be surprising if all the Platyrrhini did as well.

What significance does the change in usage of functional *Mhc* genes have? One could spin various elaborate tales of how it may be related to the unusual (for primates) reproductive behavior of tamarins, but to us the most likely explanation appears to be that the change is linked to the origin of the Platyrrhini. If the founding population of these primates were indeed as small as we assumed it to have been, genetic drift would have played a major part in the early history of the New World monkeys. Mutations that would have had little chance of spreading in large populations could have become fixed rapidly in the small founding populations. Among these mutations could have been those that de-emphasize the expression of one set of *Mhc* loci and bring others into focus for natural selection to act on. Since this change in emphasis could almost certainly have transpired only as a result of multiple mutations, a phase of radical upheavals in population structure could have been the only time they could all have been fixed in a relatively short interval. In this regard, there may not have been anything special about the tamarins or Platyrrhini and their innovative utilization of the resources available in the *Mhc*. Similar events may have transpired during the founding phases of other mammalian taxa. Abandoning old functional genes and calling up new genes from the reserve or from the pool of recently duplicated genes might the way the *Mhc* has been evolving from the times of its inception in the early vertebrates.

Acknowledgments. We thank Ms. Lynne Yakes for editorial assistance and Dr. H.J. Wolters, University of Bielefeld, Bielefeld, Germany, for tamarin tissue samples. The experimental work described in this publication was supported, in part, by a grant from the Fonds der Chemischen Industrie, Frankfurt, Germany (163 258).

References

Ciochon, R.L. and Chiarelli, A.B.: Paleobiographic perspectives on the origin of the Platyrrhini. In R.L. Ciochon and A.B. Chiarelli (eds.): *Evolutionary Biology of the New World Monkeys and Continental Drift*, pp. 459-493, Plenum Press, New York 1980

Eisenberg, J.F.: *Mammals of the Neotropics. The Northern Neotropis*. Vol.1. The University of Chicago Press, Chicago 1989

Emmons, L.H.: *Neotropical Rainforest Mammals. A Field Guide*. The University of Chicago Press, Chicago 1990

Felsenstein, J.: Phylogenies from molecular sequences. Inference and reliability. *Annu Rev Genet 122*:521-565, 1988

Fleagle, J.G.: *Primate Adaptation & Evolution.* Academic Press, San Diego 1988

Grahovac, B., Mayer, W., Vincek, V., Figueroa, F., O'hUigin, C., Tichy, H., and Klein, J.: Major histocompatibility complex *DRB* genes of a New World monkey, the cotton-top tamarin (*Saguinus oedipus*). *Mol Biol Evol*, in press 1991

Herskovitz, P.: *Living New World Primates (Platyrrhini), with an Introduction to Primates.* Vol.1. University of Chicago Press, Chicago 1977

Klein, J.: *Natural History of the Major Histocompatibility Complex.* John Wiley, New York 1986

Klein, J. and Figueroa, F.: Evolution of the major histocompatibility complex. *CRC Crit Rev Immunol 6*:295-386, 1986

Klein, J., O'hUigin, C., Kasahara, M., Vincek, V., Klein, D., and Figueroa, F.: Frozen haplotypes in *Mhc* evolution. *In* Klein, J. and Klein, D. (eds.): *Molecular Evolution of the Major Histocompatibility Complex.* Springer-Verlag, Heidelberg, in press 1991

Marsh, S.G.E. and Bodmer, J.G.: HLA-DRB nucleotide sequences 1990. *Immunogenetics 31*:141-144, 1991

Martin, R.D.: *Primate Origins and Evolution: A Phylogenetic Reconstruction.* Chapman and Hall, London 1990

Rose, K.D. and Fleagle, J.G.: The fossil history of non human primates in the Americas. In A.F. Coimbra-Filho and R.A. Mittermeier (eds.): *Ecology and Behavior of Neotropical Primates*, pp. 111-167. Academia Brasileira de Ciencias, Rio de Janeiro 1981

Simpson, G.G.: *The Meaning of Evolution. A Study of the History of Life and of its Significance for Man.* Yale University Press, New Haven 1949

Tarling, D.H.: The geologic evolution of South America with special reference to the last 200 million years. In R.D. Ciochon and A.B. Chiarelli (eds.): *Evolutionary Biology of the New World Monkey and Continental Drift*, pp. 1-41, Plenus Press, New York 1980

Watkins, D.I., Hodi, F.S., and Letvin, N.: A primate species with limited major histocompatibility complex class I polymorphism (*Saguinus oedipus*). *Proc Natl Acad Sci USA 85*:7714-7718, 1988

Watkins, D.I., Chen, Z.W., Hughes, A.L., Evans, M.G., Tedder, T.F., and Letvin, N.L.: Evolution of the MHC class I genes of a New World primate from ancestral homologues of human non-classical genes. *Nature 346*: 60-63, 1990

SELECTIVE INACTIVATION OF THE PRIMATE *Mhc-DQA2* LOCUS

Marcel Kenter*[1], Jaqueline Anholts*, Gabriela Ruff°, Nel Otting+, and Ronald Bontrop+

*Dept. of Immunohaematology and Blood Bank, Universital Hospital Leiden, P.O. Box 9600, 2333 AA Leiden, The Netherlands

°Division of Immunogenetics, Institute for Animal Breeding, University of Bern, Switzerland

+ITRI/TNO, Rijswijk, The Netherlands

Based on sequence analyses, phylogenetic studies and Southern blot data, a model for the evolution of the primate Mhc-DQA1 *and* DQA2 *loci is presented. The data suggest the existence of two primate* Mhc-DQA1 *allele families (i.e.* DQA1-I *and* II*) before the separation of the ancestors of the old world monkeys and hominoids (approximately 22-28 million years ago). The primate* Mhc-DQA2 *locus probably originates from a gene duplication that occurred before this separation. After the separation of the two primate lineages, the* DQA2 *locus was deleted along the old world monkey evolutionary line and was inactivated in the hominoid lineages. These findings suggest the presence of selective forces against a functional primate* Mhc-DQA2 *locus.*

The polymorphic *Mhc* class II molecules are expressed on B-cells, macrophages and monocytes and play a crucial role in the immune response as they present processed antigens to CD4+ T-cells. These molecules consist of two polypeptide chains, alpha and beta, encoded by *A* and *B* genes respectively (reviewed by Kappes and Strominger 1988). The different class II *A* and *B* genes are localized in separate regions on the short arm of chromosome 6, called *HLA-DP*, *DQ* and *DR*. The structural relation between several *Mhc* class II *A* and *B* genes of different species suggests they may have derived from a common ancestral *A* and *B* gene (Figueroa and Klein 1988). A comparison of alleles of orthologous class II loci of closely related species reveals that some alleles are more similar to certain alleles of different species than to other alleles within the species (Figueroa et al. 1988; Fan et al. 1989; Gyllensten and Erlich 1989). This feature is explained by a trans-species evolution of polymorphism (Klein 1980).

---

[1]To whom correspondence should be addressed

NATO ASI Series, Vol. H 59
Molecular Evolution of the
Major Histocompatibility Complex
Edited by J. Klein and D. Klein

The *HLA-DQ* subregion contains two gene pairs, designated *DQA1-DQB1* and *DQA2-DQB2*, which were probably generated by a duplication of a primordial *DQA-DQB* gene pair (Okada et al. 1980). The *DQA1-DQB1* gene pair is functional whereas no expression has been detected for the *DQA2-DQB2* genes which are considered to be pseudogenes (Kappes and Strominger 1988).

The *HLA-DQA1* locus is highly polymorphic. Most of its variation is confined to the second exon encoding the first extra-cellular domain, which is though to be involved in antigen binding and presentation processes (Bell et al. 1989). The *HLA-DQA2* locus is basically non-polymorphic. These data are in agreement with the current knowledge of *Mhc* genes where functional genes accumulate variation due to positive selection whereas pseudogenes lack positive selection and therefore do not gain allelic variation (Hughes and Nei 1988; Nei 1989; Klein and Takahata 1990).

In this paper the results of our studies on primate *Mhc-DQA1* and *DQA2* loci are combined with findings of others. Based on the data a model for the evolution of the primate *Mhc-DQA1* and *DQA2* locus is presented.

Results and Discussion

Variation of the primate Mhc-DQA1 and DQA2 loci

The second exons of *Mhc-DQA1* and *DQA2* loci of 5 hominoid species (human, chimpanzee, gorilla, orang-utan and gibbon) and 5 old world monkey species (rhesus monkey, crab-eating monkey, bear macaque, african green monkey and baboon) were amplified by PCR (Saiki et al. 1985) and sequenced after cloning. Based on a deletion of codon 56 and on sequence motives from position 45- 56, four different primate *Mhc-DQA1* allele families were defined (here provisionally called: *DQA1-I*, *II*, *III* and *IV*; figure 1). The *DQA1-I* and *II* allele families are present in hominoid and old world monkey species, whereas the *DQA1-III* and *IV* allele families were found in old world monkeys only (Kenter et al. manuscript in prep.; Fig. 1).

Phylogenetic analyses indicated that family *DQA1-I* and *II* originate from a primordial *DQA1-I* and *II* allele respectively, that existed before the separation of the ancestors of hominoids and old world monkeys (22-28 million years ago; Koop et al. 1986). The deletion of codon 56 (i.e. the *DQA1-I* allele family) may be specific for primates since it has not been found in other mammalian species like mouse, rat, rabbit, pig and goat (Figueroa et al. 1990 and references herein; M. Kenter and G. Ruff manuscript in prep.).

```
                      50
                      .
HLA-DQA1*0301   VWQLPLFRRFRR
         0101   A.RW.E.SK.GG ]
         0102   A.RW.E.SK.GG ]  DQA1-II
         0103   A.RW.E.SK.GG ]
         0401   ..C..VL.Q.*. ]
         0501   ..C..VL.Q.*. ]  DQA1-I
         0601   ..C..VL.Q.*. ]

Mamu-DQA1*02    .......SK..S ]
          03    .......SK..S ]  DQA1-IV
          04    .......SK..S ]
          05    ..R....SK.GG ]
          06    ..R....SK.GG ]  DQA1-III
          07    ..R....SK.GG ]
          08    A.RW.ELSK.GV ]
          09    A.RW.ELSK.GG ]  DQA1-II
          10    A.RW.ELSK.GG ]
          11    A.RW.ELSK.GG ]
          12    .....VLSQ.*S ]  DQA1-I
```

Fig. 1. Four different primate *Mhc-DQA1* allele families (*DQA1-I, II, III* and *IV*) from two representative primate species: human (hominoid) and rhesus monkey (old world monkey) are presented. The characteristic region (position 45-56) is shown. The deduced amino acid sequence of the *HLA-DQA1*0301* allele is used as a reference to illustrate the characteristic residues for each *DQA1* allele family. A "*" is introduced for optimal sequence alignment and represents the deletion at codon 56. Human sequences are obtained from (Gyllensten and Erlich 1988), rhesus monkey sequences are from (Kenter et al. manuscript in prep.). Nomenclature of the rhesus monkey sequences is according to (Klein et al. 1990).

Primate *Mhc-DQA2* second exon sequences were obtained from all hominoid species only and contain replacement substitutions at positions that are conserved in all primate *DQA1* alleles (n=44). The *DQA2* sequences appeared to be essentially non-variable within and very similar (>95%) across hominoid species. These data suggests that the *DQA2* gene may be a "dormant" *Mhc* class II gene in all hominoids. However more experimental data is required to enable us to make more solid conclusions regarding the status of the *DQA2* gene.

Presence or absence of the Mhc-DQA2 locus

PCR, sequence and Southern blotting data (Bontrop et al. 1989 and Kenter unpublished results) reveals the presence of the primate *Mhc-DQA2* locus on hominoid haplotypes but its absence on old world monkey haplotypes. This observation can be explained in two different ways: either the *DQA2* generating duplication occurred before the separation of the ancestors of hominoids and old world monkeys and was subsequently deleted in the old world monkey lineages, or the duplication occurred after this separation in the ancestors of the hominoids only. If the *DQA2* locus was generated by duplication after the split of the two primate lineages it must have been transferred independently to the different hominoid haplotypes. Alternatively, if the duplication occurred before the separation of the two primate lineages it must have been deleted from the pool of haplotypes present in the ancestors of old world monkeys.

The presence of a polymorphic *DQA1* and a non-polymorphic *DQA2* locus in the domestic goat (Kenter and Ruff manuscript in prep.) supports the second hypothesis, although independent *DQA* duplications along the primate and artiodactyl lineage can not be ruled out.

Alu-insertion in primate Mhc-DQA1 locus

Recently, an Alu-insertion has been described in the promotor region of the *HLA-DQA1*0301* allele and it was concluded that the *DQA1* and *DQA2* locus arose before this Alu-insertion (Del Pozzo and Guardiola 1990). Our results confirmed that conclusion since all hominoid *DQA1* loci tested possess the Alu-repeat in the corresponding integration site, whereas all hominoid *DQA2* loci lack the Alu-repeat. The Alu-repeat is also present in the corresponding target site in old world monkey species (Kenter et al. manuscript in prep.). If we accept that Alu-insertions are rare events and Alu-repeats are not cleared from the integration site without leaving a genetic footmark, one has to conclude that this Alu-insertion took place after the *DQA* duplication but before the present *DQA1* allele families were formed.

*Estimated divergence time of the DQA1*0301 and DQA2 sequences*

Using the two-parameter method (Kimura 1980) we estimated the divergence time of the primate *DQA1*0301* and *DQA2* sequences by calculating the genetic distance (K_s) between these sequences. The time of divergence T is given by the formula: $T=K_s/2V_s$, where V_s is the evolutionary rate. Based on the sequence data of globin pseudogenes, Li and coworkers computed V_s for the human-rhesus monkey equal to 1.5×10^{-9} subsitutions per site per year and for the human-chimpanzee equal to 1.2×10^{-9} substitutions per site per year. (Li et al.

1987). Using these formulas and 2282 nucleotides of non-coding sequences of the *HLA-DQA1*0301* allele and the *HLA-DQA2* locus (Auffray et al. 1987) we find a K_s of 0.16933 (S.E. 0.01) and a T of 56 - 71 million years. These numbers do not exclude a *DQA* duplication predating the separation of the two primate lineages (22 - 28 million years ago).

Whether this *DQA* duplication predates the arctiodactyl-primate separation cannot be concluded from these results, since the estimated time of separation of the ancestors of the arctiodactyls and primates is 65-80 million years (Li et al. 1987; Li et al. 1990 and references herein).

Evolution of the Mhc-DQA1 and DQA2 loci

Figure 2 summarizes all the above data. Briefly, a primordial *DQA* locus was duplicated which resulted in the *DQA1* and *DQA2* locus. Subsequently an Alu-repeat inserted in the promotor region of the *DQA1* locus. Next a deletion of codon 56 created the *DQA1-I* allele family after which the two primate lineages separated. Subsequently it became disadvantageous for the primate species to possess a functional *DQA2* gene. During the old world monkey lineage the *DQA2* gene was deleted, while in the hominoid lineages the *DQA2* gene was inactivated. Consequently, the primate *DQA2* locus did not develop into a functional polymorphic *Mhc*-class II locus like the primate *DQA1* locus.

Selection against a functional Mhc-DQA2 locus?

If the proposed evolutionary pathway is correct then the *DQA2* gene may have been functional. This is in agreement with the structure and sequence of the gene (Auffray et al. 1987). Later, changes in the action of selective forces acting on this gene may have resulted in a selection *against* a functional *DQA2* locus and led to the deletion (in the old world monkey lineage) or to silencing (in the hominoid lineage) of the *DQA2* locus.

This selective inactivation of a *Mhc* gene may not be restricted to the *DQA2* locus. In fact, the *HLA* class II region contains other loci with functional and non-functional related genes. For example the functional *HLA-DPA1-DPB1* and the non-functional *DPA2-DPB2* gene pairs closely resemble the situation of the *HLA-DQ* region. The E^0 alleles may represent an example of the same phenomenon in the mouse (Begovich et al. 1990; Figueroa et al. 1990), while the *HLA-AR* locus might be an example for the *HLA* class I region (Zemmour et al. 1990).

In concordance with the model proposed by Zemmour and coworkers for inactivation of the *HLA-AR* locus (Zemmour et al. 1990), it can be postulated that the *DQα2β2* restricted T-cell subpopulation that originally was deleted in the thymus later became essential for the recognition of "new" pathogens. The inactivation of the *DQA2* locus may represent a selective process to rescue these T-cells.

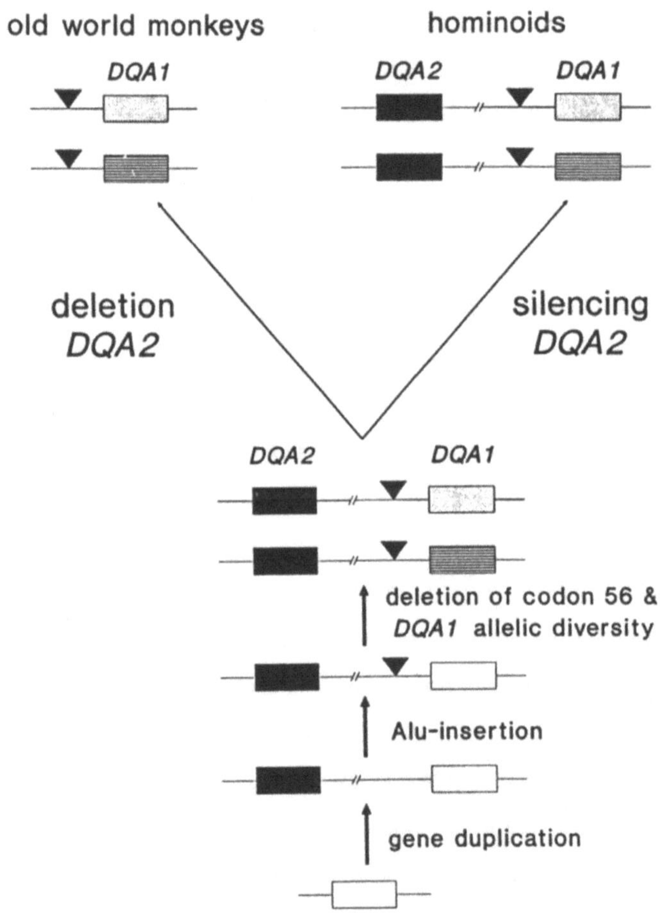

Fig. 2. Proposed evolution of the primate *Mhc-DQA1* and *DQA2* locus. Boxes present the *DQA1* and *DQA2* loci (closed box). The *DQA1-I* and *II* allele family are represented by a dotted and a hatched box. Alu-insertion is indicated by a triangle. Separation of the two evolutionary lines is approximately 22-28 million years ago (Koop et al. 1986).

In conclusion these data suggest the existence of selective forces against excessive functional *Mhc* genes.

Acknowledgments. We would like to thank prof. Van Rood, J. Howard and our colleagues for their encouraging discussions. This work was supported in part by the PWK-foundation.

References

Auffray, C., Lillie, J.W., Korman, A.J., Boss, J.M., Fréchin, N., Guillemot, F., Cooper, J., Mulligan, R.C., and Strominger, J.L.: Structure and expression of *HLA-DQα* and *-DXα* genes: interallelic alternate splicing of the *HLA-DQα* gene and functional splicing of the *HLA-DXα* gene using a retroviral vector. *Immunogenetics 26:* 63-73, 1987

Begovich, A.B., Vu, T.H., and Jones, P.: Characterization of the molecular defects in the mouse $Eß^f$ and $Eß^q$ genes. Implications for the origin of the MHC polymorphism. *J Immunol 144:* 1957-1964, 1990

Bell, J.I., Todd, J.A., and McDevitt, H.O.: Molecular structure of human class II antigens. *In* B. Dupont (ed.): Immunobiology of HLA, vol.II pp. 40-49, Springer-Verlag, New York 1989

Bontrop, R.E., Otting, N., Broos, L.A.M., Noort, M.V., Kenter, M., and Jonker, M.: RFLP analysis of the *HLA-*, *ChLA-*, and *RhLA-DQ* alpha chain gene regions: conservation of restriction sites during evolution. *Immunogenetics 30:* 432-439, 1989

Del Pozzo, G. and Guardiola, J.: A SINE insertion provides information on the divergence of the *HLA-DQA1* and *HLA-DQA2* genes. *Immunogenetics 31:* 229-232, 1990

Figueroa, F. and Klein, J.: The evolution of MHC class II genes. *Immunol Today 7:* 78-81, 1986

Figueroa, F., Günther, E., and Klein, J.: MHC polymorphism pre-dating speciation. *Nature 335:* 265-267, 1988

Figueroa, F., Gutknecht, J., Tichy, H., and Klein, J.: Class II *Mhc* genes in rodent evolution. *Imm Rev 113:* 27-46, 1990

Gyllensten, U.B. and Erlich, H.A.: Generation of single-stranded DNA by the polymerase chain reaction and its application to direct sequencing of the *HLA-DQA* locus. *Proc Natl Acad Sci USA 85:* 7652-7656, 1988

Gyllensten, U.B. and Erlich, H.A.: Ancient roots for polymorphism at the HLA-DQα locus in primates. *Proc Natl Acad Sci USA 86:* 9986-9990, 1989

Hughes, A.L. and Nei, M.: Pattern of nucleotide substitution at major histocompatibility complex class I loci reveals overdominant selection. *Nature 335:* 167-170, 1988

Hughes, A.L. and Nei, M.: Nucleotide substitution at major histocompatibility complex class II loci: Evidence for overdominant selection. *Proc Natl Acad Sci USA 86*: 958-962, 1989

Kappes, D. and Strominger, J.: Human class II major histocompatibility complex genes and proteins. *Ann Rev Biochem 57:* 991-1028, 1988

Kimura, M.: A simple method for estimating evolutionary rates of base substitutions through comparative studies of nucleotide sequences. *J Mol Evol 16:* 111-120, 1980

Klein, J.: Origin of major histocompatibility complex polymorphism: the trans species hypothesis. *Hum Immunol 19:* 155-162, 1987

Klein, J., Bontrop, R.E, Dawkins, R.L., Erlich, H.A., Gyllensten, U.B., Heise, E.R., Jones, P.P., Parham, P., Wakeland, E.K., and Watkins, D.I.: Nomenclature for the major histocompatibility complexes of different species: a proposal. *Immunogenetics 31:* 217-219, 1990

Klein, J. and Takahata, N.: The major histocompatibily complex and the quest for origins. *Immunol Rev 113:* 5-25, 1990

Koop, B.F., Goodman, M., Xu, P., Chan, K., and Slightom, J.K.: Primate η-globin DNA sequences and man's place among the great apes. *Nature 319:* 234-238, 1986

Li, W.-L., Tanimura, M., and Sharp, P.M.: An evaluation of the molecular clock hypothesis using mammalian DNA sequences. *J Mol Evol 25:* 330-342, 1987

Li, W.-L., Gouy, M., Sharp, P.M., O'hUihgin, C., and Yang, Y.-W.: Molecular phylogeny of Rodentia, Lagomorpha, Primates, Artiodactyla, and Carnivora and molecular clock. *Proc Natl Acad Sci USA 87:* 6703-6707, 1990

Okada, K., Boss, J.M., Prentice, H., Spies, T., Mengler, R., Auffray, C., Lillie, J., Grossberger, D., and Strominger, J.L.: Gene organization of DC and DX subregions of the human major histocompatibility complex. *Proc Natl Acad Sci USA 82:* 3410-3414, 1985

Saiki, K.R., Scharf, S., Faloona, F., Mullis, K.B., Horn, G.T., Erlich, H.A., and Arnheim, N.: Enzymatic amplification of ß-globin genomic sequence and restriction site analysis for diagnosis of sickle cell anaemia. *Science 230:* 1350-1354, 1985

Zemmour, J., Koller, B.H., Ennis, P.D., Geraghty, D.E., Lawlor, D.A., Orr, H.T., and Parham, P.: HLA-AR, an inactivated antigen-presenting locus related to HLA-A. *J Immunol 144:* 3619-3629, 1990

IS DQB2 FUNCTIONAL AMONG NONHUMAN PRIMATES?

Lakshmi K. Gaur*1, Eugene R. Heise*, Phillip S. Thurtle+ and Gerald T. Nepom+
*Department of Microbiology and Immunology, Bowman Gray School of Medicine, Winston-Salem, NC 27103, USA
+Immunology and Diabetes Programs, Virginia Mason Research Center, Seattle, WA 98101, USA

The non-human primate class II genes are highly homologous to human class II genes. Within the HLA-DQ region, the human genome contains an apparent gene duplication, in which the expressed DQA1 and DQB1 genes are found approximately 40 kb telomeric of the DQA2 and DQB2 genes. In spite of apparently normal gene organization and RNA splice sites, the DQA2 and DQB2 genes (also known as "DX"), are usually referred to as "pseudogenes", since their transcription has not been reported. This DQ-DX gene duplication has previously been thought to be a fairly recent evolutionary event, because DQB2 was not found in Old World monkeys (Gyllensten et al. 1990). However, using the polymerase chain reaction, the second exon of the DQB2 locus from four primate species i.e., a chimpanzee, (Pan troglodytes), a gorilla (Gorilla gorilla), a gibbon (Hylabates lar), and an owl monkey (Aotus trivirgatus) were amplified. Our observation that the DQB2 is present in an owl monkey suggests that this diversification could have occurred prior to diversification of owl monkey and Old World monkey lineages. Sequence analysis of the DQB2 amplified genes revealed an extremely high degree of conservation through primate evolution. The nucleotide patterns observed have several features with possible functional implications: [1] To date four distinct chimpanzee DQB2 alleles have been identified reminiscent of patterns of polymorphism found in human DQB1 alleles; [2] the ratio of productive to silent substitutions, particularly in the owl monkey and the chimpanzee, resembles that of a functional locus, and overall is 1:1.35 among the five species examined; and [3] the owl monkey DQB2 gene may be actively transcribed, since a DQB2 sequence was recovered using an RNA-PCR technique.

The evolution of the MHC gene family is under close scrutiny in recent years largely because of the remarkable degree of polymorphism both between and within species. This MHC class II polymorphism in humans can be seen as a three tier system in which the complex is divided into three main regions (*DR, DQ*, and *DP*), each region into multiple loci, and each

[1]Address correnspondence and reprint requests to Dr. Lakshmi K. Gaur: Puget Sound Blood Center, 921 Terry Ave, WA 98101, USA

NATO ASI Series, Vol. H 59
Molecular Evolution of the
Major Histocompatibility Complex
Edited by J. Klein and D. Klein

locus into multiple alleles. Based on the structural similarities between class I and class II molecules, a common ancestral gene is predicted to closely resemble the contemporary class II gene (Lawlor et al., 1990a). The tandem duplication of this gene followed by subsequent diversification presumably led to the contemporary complex of closely related genes. This glorious scheme of duplication-diversification cycles of MHC also resulted in subregions within regions; for example, there are two alpha and two beta genes in both the *DP* and *DQ* regions, and one alpha and up to five beta genes in the *DR* region. It also appears, however, that some of these genes were inactivated either at the time of duplication, or subsequently. Several such pseudogenes show high homology to their functional MHC counterparts but contain nucleotide changes such as frameshifts and nonsense mutations that prevent expression (Gojobori et al. 1982). In the case of the human *DQ* region, a primary distinction between the functional *DQB1* gene and the nonexpressed *DQB2* gene however, now appears to be due to alterations in transcriptional regulatory elements in the *DQ* promoter region (Shewey et al. 1991, submitted).

The *HLA-DQ* region contains two alpha and two beta genes; *HLA-DQA1* and *HLA-DQB1* are highly homologous (~85%) to their counterparts designated *HLA-DQA2* and *HLA-DQB2* and are presumed to have arisen from ancestral gene duplication events (Auffray et al. 1984; Okada et al. 1985). This duplication event was estimated to have occurred within the last 10 million years (MYr; Bodmer 1986; Gyllensten et al. 1990), since *HLA-DQB2* homologs have been reported in chimpanzees and gorillas (Gyllensten et al. 1990); they have not been found in the rodent MHC. We now report the observation of *DQB2* homologs in additional nonhuman primates, including a New World monkey, indicative of the presence of the *DQ* gene duplication for at least 40 MYr. In contrast to other class II beta genes, the *HLA-DQB2* gene is strikingly non-polymorphic in humans; since *DQB2* transcription has not been found in humans, this has been taken as evidence for an inactive gene not subject to the same selective pressures which maintain polymorphism at other class II loci. However, our sequence analysis of the amplified *DQB2* genes from additional primates suggests that evolutionary selective pressure may constrain the evolution of the *DQB2* gene as well, presumably maintaining an as yet unknown function.

Materials and Methods

Detailed methods have been reported elsewhere (Gaur et al. 1991, submitted). Briefly, genomic DNA was extracted from primate cell lines and used as a template for *DQB* gene amplification and sequencing. Cell lines used included: Tank (Lawlor et al. 1990b), an EBV transformed line from a chimpanzee, kindly provided by Dr. Francis Ward, Duke University Medical Center; Machi (Lawlor et al. 1990b), a gorilla B-LCL, and 531H (Strnad et al. 1979), an owl monkey lymphoblastoid cell line, obtained from Mr. Bob Ruitz, Duke

University Medical Center; 86107 (Regional Primate Research Center, Seattle) cynomolgus macaque cell line obtained from Chuck Estin (Oncogen, Seattle); a gibbon (Hylobates lar) B-LCL, B8080 (Yerkes Primate Center, Atlanta), an EBV transformed cell line, 1036 (Bowman Gray School of Medicine, Winston-Salem), a rhesus macaque (*Macaca mulatta*) cell line (Herpes virus papio transformed B-LCL) established by the author (LKG).

DNA samples were amplified using *DQB* specific primers, GH28 and GH29 (Scharff et al. 1988). XIth HLA workshop primers were used to amplify *DQA* homologous sequences from owl monkey. Amplifications used the PCR technique with *Taq* polymerase with 3 minute extensions at 72°C, 1 minute denaturation at 94°C, and primer annealing at 55°C for 30 cycles. The amplified DNA was visualized on a 4% agarose gel to identify bands of the predicted size. 30 µl of this PCR product was digested with 50 units of restriction enzyme *Pst* I for 2 hours at 37°C, and with 60 units of restriction enzyme *Bam* HI for an additional 2 hours at 37°C. This digestion mixture was then ligated into similarly digested M13 mp18 and M13 mp19 vectors used for preparation of single stranded DNA template, and nucleotide sequenced using the Sanger dideoxy sequencing technique (Sanger et al. 1977).

RNA PCR was performed using 1 µg total RNA (treated with RNase free DNase) from each of four nonhuman primate species, i.e., chimpanzee (Tank), rhesus macaque (1036), cynomolgus macaque (86107), and an owl monkey (531H). The first strand of cDNA was generated by reverse transcription of mRNA in the presence of reverse transcriptase RAV2 (Amersham, 17 u/µl), which then was used as a template for polymerase chain reaction, as described above. The amplified product was then ligated into a TA cloning vector (Invitrogene). The TA Cloning™ system was used since it allows direct cloning of PCR products into the multifunctional pCRTM 2000 vector without the use of restriction or modifying enzymes. Briefly, one µl of the PCR product was ligated into TA cloning vector, in the presence of T4 DNA ligase, at 12°C overnight. The ligated DNA was transformed using the vendor provided INVaF' competent cells, and after an overnight incubation at 37°C, white plaques were picked, and grown, double stranded DNA was prepared, and sequenced by the Sanger dideoxy sequencing method.

Results and Discussion

DQB2 nucleotide sequences were obtained from genomic DNA of the chimpanzee, gorilla, gibbon, and owl monkey, and an additional *DQB2* nucleotide sequence from owl monkey cDNA. Two sequences each from the chimpanzee and the owl monkey cell line were found, presumably representing alleles at the *DQB2* locus. These six sequences were each unique at the nucleotide level, although there was a high degree of homology, particularly at the amino acid level. Each of the primate *DQB2* sequences is also highly homologous to the published human sequence, with 94% to 99% identity at the nucleic acid level. Figure 1 lists the predicted amino acid sequences for each of these six *DQB2* genes, homologous to the first domain of class II polypeptides. These are aligned for comparison with selected previously reported human and primate *DQB2* predicted amino acid sequences.

```
CODON           21       30        40        50        60        70
                |        |         |         |         |         |
HLA-DQB2        TERVRGVARYIYNREEYGRFDSDVGEFQAVTELGRSIEDWNNYKDFLEQERAAVDK
Patr-DXB10      --------------------------------------------------------
Patr-DXB5       ------------------G-------------------------E----------E
Papa-PPDX       ------------------G------------------------------------G
Gogo-GDX        ------------------G----GW-------------------------------
Patr-DQB2*01    -----L-T---------V----------------------------------P---
Patr-DQB2*02    --------------------------------------------------------
Gogo-DQB2*01    --------------------------------------------------------
Hyla-DQB2*01    --------------------------------------------------------
Aotr-DQB2*01    -------G----------A--------R--------T------------Q----YT
Aotr-DQB2*02    ------------------A--------R--------T------------Q--P-YT
```

Fig. 1. Comparison of *DQB2* predicted amino acid sequences from various nonhuman primates with human *DQB2* sequence. Predicted amino acid sequences from the present study: *Gogo-DQB2*01* = *Gorilla gorilla* (cell line, Machi); *Patr-DQB2*01* = *Pan troglodytes,* and *Patr-DQB2*02* = *Pan troglodytes* (chimpanzee cell line, Tank); *Hyla-DQB2*01* = *Hylobates Lar* (gibbon cell line, B8080); *Aotr-DQB2*01* and *Aotr-DQB2*02*= *Aotus trivirgatus* (owl monkey cell line, 531H); from Okada et al. 1985: *HLA-DQB2*; from Gyllensten et al., 1990: DXB10, *DXB5* (chimpanzee *DQB2*), PPDX (Pygmy chimpanzee *DQB2*), and *GDX* (gorilla *DQB2*).

As expected, closely related species such as the chimpanzee and the gorilla have highly homologous *DQB2* genes, confirming previous reports (Gyllensten et al. 1990). In addition, *DQB2* genes from a more distantly related gibbon (divergence roughly 19 MYA), and surprisingly from an owl monkey (New World monkey, 40 MYA [Sarich and Cronin 1976]) were also identified. In each case, a remarkable degree of sequence conservation was found. Table 1 presents a listing of the percent similarities of nucleotides for each of the new sequences. These *DQB2* sequences are all less than 85% homologous with the nucleotide sequences for *DQB1* alleles in primates and humans, and all contain a characteristic sequence "fingerprint" of the *DQB2* gene, namely a deletion of three nucleotides in *DQB2* relative to the *DQB1* gene at codon 56 which creates an in-frame deletion of one codon triplet in the *DQB2* gene. This deletion is immediately followed by two *DQB2*-specific codons, AGC ATC, in *DQB2*.

Table 1. Comparison of percent similarities among various primates.

	HLA-DQB2	*Patr-DQB2*01*	*Patr-DQB2*02*	*Gogo-DQB2*01*	*Hyla-DQB2*01*	*Aotr-DQB2*01*	*Aotr-DQB2*02*
HLA-DQB2	-						
*Patr-DQB2*01*	95.7	-					
*Patr-DQB2*02*	99.0	95.7	-				
*Gogo-DQB2*01*	99.0	95.2	99.5	-			
*Hyla-DQB2*01*	99.0	95.2	99.5	99.0	-		
*Aotr-DQB2*01*	95.7	93.3	96.2	95.7	95.7	-	
*Aotr-DQB2*02*	94.3	92.3	94.7	94.3	95.2	99.0	-

In contrast to the remarkable conservation of most of the *DQB2* sequence, there is some notable nucleotide and amino acid variation. As summarized in Table 2, most of the sequences carry both silent and productive changes, relative to the human *DQB2* sequence. Four chimpanzee sequences are now known for the *DQB2* gene, two derived by us (Gaur et al. 1991 submitted) and two previously described (Gyllensten et al. 1990). All four differ slightly, and together they carry six silent and seven productive substitutions. Even more striking are the owl monkey sequences, *Aotr DQB2*01* and *-*02*, which show three to four silent substitutions and seven replacement substitutions each with respect to the human sequence. One of the two chimpanzee alleles, *DXB5* previously described by Gyllensten et al. (1990) and the pygmy chimpanzee *DQB2* allele, *PPDX* from this study are identical with one silent and three productive substitutions compared to human.

Table 2. Summary of nucleotide substitutions in ten primate *DQB2* alleles compared to a human *DQB2* allele.

Primate DQB2	Silent substitutions	Productive substitutions
Patr-DXB10	1	1
Patr-DXB5	1	3
Patr-DQB2*01	3	3
Patr-DQB2*02	1	0
Papa-PPDX	1	3
Gogo-GDX	2	3
Gogo-DQB2*01	2	0
Hyla-DQB2*01	2	0
Aotr-DQB2*01	3	7
Aotr-DQB2*02	3	7
Total	**20**	**27**

The composite variation seen in the nonhuman primate *DQB2* sequences is present as 20 silent substitutions and 27 replacement substitutions, giving a replacement/silent ratio a little over 1:1, implying that the protein is constrained from rapid evolution.

The remarkable conservation of the *DQB2* sequences at the amino acid level over 40 MYr, the presence of four different polymorphic sequences among four chimpanzee sequences known, and the low replacement to silent substitution ratio for nucleotide variation all suggest that the *DQB2* gene, rather than being a pseudogene, may be in fact subject to the same kinds of selective pressures that characterize functional class II genes. To further explore the possibility of expression of any of these nonhuman primate *DQB2* alleles, we have carried out a RNA-PCR (Fig. 2) from chimpanzee and the owl monkey, and sequenced the amplified products. To our surprise we have obtained a *DQB2* sequence , denoted *Aotr-DQB2*02,* from the owl monkey directly demonstrating the presence of a *DQB2* transcript in this species.

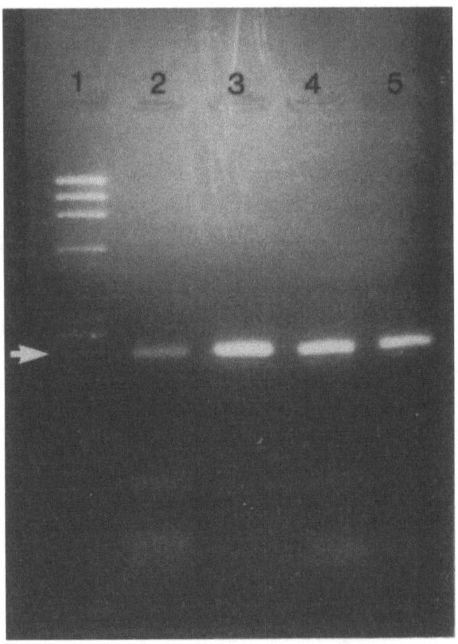

Fig. 2. *DQB* exon 2 amplification products from cDNA templates of four nonhuman primate species. The amplification reaction was carried out as described in the text; 10 μl of the reaction mixture were electrophoresed in a 3% agarose (NuSeive) gel, stained with Ethidium bromide to visualize the appropriate bands. **Lane 1** contains the molecular weight markers, and ~ 234 bp product (indicated by the arrow) can be seen for all the four lanes, owl monkey, 531H (**lane 2**), cynomolgus macaque, 86107 (**lane 3**) rhesus macaque, 1036 (**lane 4**) and chimpazee, Tank (**lane 5**).

In view of the unanticipated detection of an apparently expressed *DQB2* gene in the owl monkey, we also amplified and sequenced the DQA genes in this species. As shown in Figure 3, the *Aotr-DQA*01* sequence obtained is very similar to previously described *DQA1* sequences, both in humans and other primates; there is little resemblance to *HLA-DQA2*. This owl-monkey derived sequence very closely resembles each of the previously described New World monkey *DQA1* alleles, most closely that of *Ceca-1* (Gyllensten and Erlich 1989) with only two predicted amino acid differences.

```
Codon            20         30         40         50         60         70
                 |          |          |          |          |          |
HLA-DQA1*0101    FYGPSGQYTG EFDGDEEFYV DLERKETAWR WPEFSKFGGF DPQGALRNMA VAKHNLNIMI
HLA-DQA1*0102    ---------H ------Q--- ---------- ---------- ---------- ----------
HLA-DQA1*0103    -------F-H ------Q--- ---K------ ---------- ---------- ----------
HLA-DQA1*0201    S------F-H ---------- -------V-K L-L-HRL*R- ---F--T-I- -L------L-
HLA-DQA1*0301    S-------SH ---------- -------V-Q L-L-RR-RR- ---F--T-I- -L------V-
HLA-DQA1*0401    S--------H ------Q--- --G----V-C L-VLRQ-*R- ---F--T-I- -T------L-
HLA-DQA1*0501    S--------H ------Q--- --G----V-C L-VLRQ-*R- ---F--T-I- -L-----SL-
HLA-DQA1*0601    S------F-H ------Q--- --G----V-C L-VLRQ-*R- ---F--T-I- -T------L-
HLA-DQA2         SH-------H ---------- ---T---V-Q L-M----IS- ---T--R-M- -G--TLEFMM
HLA-DQA2         SH-------H ---------- ---T---V-Q L-M----IS- ---S--R-M- -G--TLEFMM
Patr-PDQA3       S--------- ---------- -------V-Q L-L-RS-R-- ------T-I- IT------LI
Patr-PDQA1       ---------- ------Q--- ---K------ ---------- ---------- -T--------
Patr-PDQA4.X     S------F-- ------Q--- --W----V-C L-VLRQ-*R- ---F--T-I- -L-----SL-
Patr-PDQA4.Y     S------F-- ------Q--- --G----V-C L-VLRQ-*R- ---F--T-I- -L-----SL-
Patr-PDXA        SH-------- ---------- ---T---V-Q L-M----IS- ---S------ -G--T-EF--
Gogo-GDQA3       S-------S- ---------- -------V-Q F-M--RLRR- ------T-I- -L------V-
Gogo-GDQA1       -------F-- ------Q--- ---------- ---------- ---------- ----------
Gogo-GDQA1.2     ---------- ------Q--- ---------- ---------- ---------- ----------
G0g0-GDQA4       S------F-- ------Q--- -------V-C L-VLRQ-*R- ---F--T-I- -L-----SL-
G0g0-GDXA        SH-------- -S-------- ---T---V-Q L-M----IS- ---S------ -G--T-EF-M
Pale-1           S--L----S- ------Q--- -------V-Q LLF--Q-*S- -A-F--T-I- -G-Q---YLM
Pale-2           S-------S- ---------- -------V-Q L-L------- --------L- -G------LM
Mamu-1           T--L------ G-----Q--- -------V-L -L-------- --------L- -S--------
Mamu-3           ---L----S- ------Q--- --G----V-- L-IPYRY*-- ---F----I- -LT----VL-
Pren-1           S--L------ ------Q--- --G----V-W L-VLSQ-*S- ---F--T-L- ---D---CL-
Pren-3           S--------- ---------- -------V-Q L-L----SS- --------L- -G-QT---L-
Casp-1           S------F-- ---------- --G----V-- L-V-N--AS- ------T-I- -T----D-L-
Casp-2           S------F-- ---------- ---K---V-- L-V--T-TS- ------T-I- -T----D-L-
Ceca-1           S--------- ---------- --GK---V-- L-M--T-TS- ------T-I- -T----D-L-
Ceca-3           S--------- ---------- --GK---V-- L-M--T-TS- ------T-I- LM--G-DLN-
AotrDQA*01       S------F-H ---------- --GK---V-- L-V--T-TS- ------T-I- -T----D-L-
```

Fig. 3. Primate *DQA* second exon amino acid sequence alignment. Predicted amino acid sequences from the present study: *Aotr-DQA1*01* = *Aotus trivirgatus* (cell line 531H); other nonhuman primate *DQA* sequences from Gyllensten and Erlich (1989) for primate species, *Patr*, chimpanzee (*Pan troglodytes*); *Gogo*, gorilla (*Gorilla gorilla*); *Pale*, baboon (*Papio leucophaeus*); *Mamu*, rhesus macaque (*Macaca mulatta*); *Pren*, langur (*Presbytis entellus*); *Casp*, marmoset (*Callithrix* species not specified); *Ceca*, capuchin (*Cebus capucinus*). *HLA-DQA* sequences were obtained from GenBank.

In summary, the present study provides evidence for [1] allelic variability of *DQB2* in nonhuman primates (two alleles from owl monkey and four from chimpanzee), with predicted amino acid substitution patterns similar to other functional class II genes; [2] remarkable conservation of the *DQB2* allele through at least 40 Myr of evolution; [3] a replacement-to-substitution ratio among old and New World monkeys which implies selective conservation of the *DQB2* gene; and [4] apparent transcription of *DQB2* in at least one New World monkey species, potentially encoding a class II product, possibly associated with *DQA1* to form a *DQA1-DQB2* dimer.

Based on the above results, we assume that the *DQ* region probably duplicated prior to diversification of owl monkey-human lineages; whereas the human lineage has retained a transcriptionally silent *DQB2* gene, the owl monkey may have conversely inactivated its *DQB1* gene. Future studies of the gene organization of class II genes in New World monkeys will clarify this possibility. In previously reported studies of Old World monkeys, where *DQB2* was not found (Gyllensten et al. 1990; Bontrop et al. 1989) it is similarly possible that those species had inactivated or deleted their DQB2 locus. However, Bontrop and colleagues (in work presented at the NATO MHC conference) were successful in finding one *DQB2* sequence from a rhesus macaque (*M. mulatta*), thus making it more likely that previous reports of not finding *DQB2* in Old World monkeys might be due to technical difficulties. If different species differentially use either their *DQB1* or *DQB2* locus, a question for human immunogenetics is raised: Is the *DQB2* gene in humans functional, or could it be in transition to become a pseudogene? According to the trans-species hypothesis (Klein 1987), sequence variation found in *MHC* genes predates the divergence of human and other primate lineages. This has been reported for class I *HLA-A* and *-B* loci (Lawlor et al. 1988; Mayer et al. 1988), as well as for class II loci (Gyllensten et al. 1990; Gyllensten and Erlich, 1989; Bontrop et al. 1990; Fan et al. 1989). The persistence of these ancestral alleles in present day humans and monkeys suggests the presence of selective mechanisms favoring the maintenance of established polymorphisms. The *DQB2* gene fits this same evolutionary paradigm, even as it is not known to be transcribed or translated in modern man. Present day sequences in different primate lineages appear to have derived from a common ancestor, with very limited or selected polymorphism since interspecies divergence. Although previously thought to represent about 10 MYA of evolutionary divergence (Bodmer 1986; Gyllensten et al.1990), the divergence of New World Monkey-human lineages is closer to 50 MYr (Sarich and Cronin, 1976), and would seem to indicate strong presumptive evidence for a functional maintenance of DQB2 in humans.

Acknowledgements. This work was supported in part by National Institutes of Health (NIH) grants R24RR02159-06 (E.R.H), and DK42732 (G.T.N).

References

Auffray, C., Lillie, J.W., Arnot, D., Grossberger, D., Kappes, D., and Strominger, J.L.: Isotypic and allotypic variation of human class II histocompatibility antigen α-chain genes. *Nature 308*:327-333, 1984

Bodmer, W.F.: HLA Today. *Hum Immunol 17*: 490-503, 1986

Bontrop, R.E., Otting, N., Broos, L.A.M., Noort, M.C., Kenter, M., and Jonker, M.: RFLP analysis of the *HLA-*, *ChLA-*, and *RhLA-DQ* alpha chain gene regions: conservation of restriction sites during evolution. *Immunogenetics 30*: 432-439, 1989

Fan, W., Kasahara, M., Gutknecht, J., Klein, D., Mayer, W.E., Jonker, M., and Klein, J.: Shared Class II MHC polymorphisms between humans and chimpanzees. *Hum Immunol 26*: 107-121, 1989

Gojobori, T., Li, W.-H., and Graur, D.: Patterns of nucleotide substition in pseudogenes and functional genes. *J Mol Evol 18*:360-369, 1982

Gyllensten, U.B. and Erlich, H.A.: Ancient roots for polymorphism at the HLA-DQα locus in primates. *Proc Natl Acad Sci USA 86*: 9986-9990, 1989

Gyllensten, U.B., Lashkari, D., and Erlich, H.A.: Allelic diversification at the class II DQB locus of the mammalian major histocompatibility complex. *Proc Natl Acad Sci USA 87*: 1835-1839, 1990

Klein, J.: Origin of major histocompatibility complex polymorphism: the *trans-species* hypothesis. *Hum Immunol 19*:155-162. 1987

Lawlor, D.A., Ward, F.E., Ennis, P.D., Jackson, A.P., and Parham, P.: HLA-A and B polymorphisms predate the divergence of humans and chimpanzees. *Nature 335*: 268-271, 1988

Lawlor, D.A., Zemmour, J., Ennis, P.D., and Parham, P.: Evolution of class-I MHC genes and proteins: from natural selection to thymic selection. *Annu Rev Immunol 8*:23-63, 1990a

Lawlor, D.A., Warren, E., Ward, F.E., and Parham, P.: Comparison of class I MHC Alleles in human and apes. *Immunol Rev 113*: 147-185, 1990b

Mayer, W.E., Jonker. M., Klein, D., Ivanyi, P., van Seventer, G., and Klein J.: Nucleotide sequences of chimpazee MHC class I alleles: evidence for trans-species mode of evolution. *EMBO J 7*: 2765-2774, 1988

Okada, K., Boss, J.M., Prentice, H., Spies, T., Mengler, R., Auffray, C., Lillie, J., Grossberger, D., and Strominger, J.L.: Gene organization of DC and DX subregions of the human major histocompatibility complexes. *Proc Natl Acad Sci USA 82*: 3410-3414, 1985

Sanger, F., Niklen, S., and Coulson, A.R.: DNA sequencing with chain terminating inhibitors. *Proc Natl Acad Sci USA 74*: 5463-5467, 1977

Sarich, V.M. and Cronin, J.E. In M. Goodman and R.E Tashian (eds.): *Molecular Anthropology. Genes and Proteins in the Evolutionary Ascent of Primates*, pp. 141-170, Plenum, New York 1976

Scharf, S.J., Friedman, A., Brautbar, C., Szafer, F., Steinman, L., Horn, G., Gyllensten, U., and Erlich. E.A.: HLA class II allelic variation and susceptibility to pemphigus vulgaris. *Proc Natl Acad Sci USA 85*:3504-3508, 1988

Strnad, B.C., Neubauer, R.H., and Rabin, H.: Biochemical Identification of Primate lymphoid cell-surface glycoproteins. *Int J Cancer 23*: 76-82, 1979

ALU REPEATS AND EVOLUTION OF THE *HLA-DQA1* LOCUS

Giovanna Del Pozzo, Maria Neve Ombra, Carmela Perfetto, Andrea De Lerma Barbaro, Monica Autiero, Antonella Maffei and John Guardiola
The International Institute of Genetics and Biophysics, via Guglielmo Marconi 10, I-80125 Naples, Italy

The HLA-DQA1 gene is highly polymorphic in the coding as well as in the non-coding regions. Analysis of polymorphisms at the DNA level by use of restriction enzymes shows that multiple restriction sites are simultaneously affected, thus suggesting that insertion/deletion mechanisms might be in part responsible for the generation of genetic variability at this locus. To study in detail this phenomenon we have cloned and sequenced DNA fragments derived from the HLA-DQA1 locus isolated from different haplotypes. This work has provided clues concerning DNA polymorphisms located in the 5' flanking region of the gene. Generation of polymorphism seems to be related to the presence of a cluster of repetitive elements of the Alu family; the possible influence of polymorphisms of the regulatory region on the control of gene activity is discussed. Furthermore, we report indications that Alu repeats can be used as tags to study the divergence of HLA-DQA1 and of its alleles from the strictly related HLA-DQA2 gene.

The ability of helper T-cells to respond to antigenic stimuli is controlled by a family of heterodimeric surface glycoproteins coded for by the MHC class II A and B genes (Giles and Capra, 1985). In man three different subsets of class II antigens are observed, namely HLA-DR, -DP and - DQ. The most abundantly represented of these antigens on the cell surface is HLA-DR. It appears that redundancy occurs in this system and that multiple genes having the same or very similar functions have been evolved as a safeguard against failure of one of the genes to be appropriately expressed. Therefore HLA-DR, -DP and -DQ may share the same functions in antigen presentation. It must be noted, however, that some of these antigens may specialize for particular tasks in the immune response; HLA-DQ, for example has been shown to be epistatic to HLA-DR in controlling the immune response to schistosomal antigens (Hirayama et al. 1987) and has been proposed to play a major role in the suppressor T-cell response to streptococcal cell wall antigen (Sasazuki 1989). The possibility that *HLA-DR* and *HLA-DQ* may have different functions in the mounting of the immune response is also strengthened by the finding that the *HLA-DQA* and *-B* genes may be regulated independently of the remaining class II genes (Ono et al. 1991). Functional differences may be reflected in the evolutionary history of genes of different subsets.

NATO ASI Series, Vol. H 59
Molecular Evolution of the
Major Histocompatibility Complex
Edited by J. Klein and D. Klein

The genes that code for MHC class II antigens constitute a polymorphic system. Notably the *A* gene coding for the heavy chain of the HLA-DQ antigen (namely *HLA-DQA1*) and the *B* genes exhibit allelic variability; conversely the *HLA-DRA* gene is monomorphic (Giles and Capra, 1985). Sequence diversity among allelic forms is found mainly in the second exon encoding for the aminoterminal domain of the subunits. At least eight alleles have been so far described for *HLA-DQA1*; these alleles have been classified in four major subfamilies designed A1-A4 (Gyllensten and Erlich 1989). The *HLA-DQA1* alleles are in strong linkage disequilibrium with the *DR* locus (Moriuchi et al. 1985). In fact comparison of nucleotide and protein sequences among DQA1 alleles reveals that the *HLA-DQA1* subunits derived from *DR4, -7* and *-9* haplotypes are closely related whereas a *DQA1* subunit derived from a DRw6 cell line was found to be rather divergent. A hypothetical evolutionary tree has been constructed for *HLA-DQA1* alleles (Moriuchi et al. 1985); subsequently, an exhaustive sequence comparison revealed that, in hominoids, allelic divergence at the *HLA-DQA1* locus preceded speciation and that most of the polymorphism at the *HLA-DQA1* locus was already accumulated at least 5 Myr ago in the common progenitor of chimpazee, gorilla and man (Gyllensten and Erlich 1989).

The *HLA-DQ* region contains besides the *HLA-DQA1* and the *HLA-DQB1* genes also other two genes, *HLA-DQA2* and *HLA-DQB2*. Sequence comparison of *HLA-DQA1* and *HLA-DQA2* shows a significant degree of similarity of the two genes; similarity is maximal in regions such as the promoter proximal region and in some exons including the 5' UT/SP exon, the alpha-2 exon, the intron 3 and the CP/TM/CY/3'UT exon (90.5 to 98.6 %) and sharply decreases in the 5' and 3' flanking region sequences (27.8 and 68.7 %, respectively; Auffray et al. 1987). Although no obvious defect was found in the structure of the gene or in its regulatory signals, expression of *HLA-DQA2* gene (and of the *HLA-DQB1* gene) remains to be documented (Auffray et al. 1987). *HLA-DQA2* appears to be monomorphic (Horn et al. 1988). Parsimony trees for *HLA-DQA1* and *HLA-DQA2* are compatible both with the hypothesis that *HLA-DQA2* diversification predates the diversification of the *HLA-DQA1* alleles (Gyllensten and Erlich 1989) or that it originated later.

DNA polymorphisms in the *HLA-DQA1* locus

Restriction fragment length polymorphisms of the *HLA-DQA1* gene have been detected by genomic blotting hybridization; correlations between restriction fragments and serologically defined epitopes were reported (Auffray et al. 1983; Trowsdale et al. 1983; Spielman et al. 1984; Le Gall et al. 1985); however it is very unlikely that the correlated restriction polymorphisms can directly affect epitope-encoding sequences (Cohen et al. 1984). Clustering of variant restriction sites for several enzymes is a relevant feature of the *HLA-DQA1* locus polymorphisms (Guardiola et al. 1988; Cohen et al. 1984). We and others

(Guardiola et al. 1988; Cohen et al. 1984) have previously suggested that such a clustering may be due to rearrangements of the locus rather than to linkage disequilibrium of independently arisen point mutation inactivating pre-existing restriction sites or creating new ones. Various mechanisms can be invoked to explain how a number of restriction sites may be simultaneously affected; multiple mutations may be generated by the correction of heteroduplex favoured by the insertion of short repetitive sequences (Ollo and Rougeon 1983); alternatively changes in the pattern of restriction enzymes may be consequent to intergenic transfer of short DNA sequences between non-allelic genes or to exchange of genetic material by recombination (Widera and Flavell 1984). Insertion or deletion events also cause pleiotropic variation in the pattern of restriction digestion. Insertion of repetitive sequences, which have the effect of introducing new restriction sites or of introducing longer distances between pre-existing ones, may also provide the same output (Rogers 1985). In addition repetitive elements may provide hot-spots for recombinational rearrangements by unequal crossing-over. Structural variability of genetic loci due to insertion of repetitive elements has been observed in many cases; for example polymorphism near the rat prolactin gene are caused by insertion of an Alu-like element (Schuler et al. 1983); in other instances, Alu sequences were related to recombinational events leading to deletion of the low density lipoprotein receptor gene (Lerhman et al. 1986) or to deleterious rearrangements of the human C1 inhibitor locus (Stoppa-Lyonnet et al. 1990). The presence of Alu elements has also been correlated to polymorphisms of the globin cluster (Ottolenghi and Giglioni 1982) and of the *mlvi-2* locus (Economu-Pachnis and Tsichlis 1985). Repetitive elements of the LINE family participate into the chromosomal rearrangements involving the *c-myc* locus in tumor cells (Morse et al. 1988).

The structure of the *HLA-DQA1* locus

It is very likely that most of the rearrangements from which restriction fragment length polymorphisms originate are localized in the intronic or flanking regions of the gene. Conversely, comparative analysis of gene sequences is mostly confined to the coding exons whose evolution is shaped by selective pressures stemming from the functional role of the HLA-DQ antigen (Moriuchi et al. 1985; Gyllensten and Erlich 1988; Horn et al. 1988). Therefore hints to the understanding of the mechanisms which generate insertion/deletion variability may derive from the comparative analysis of the structural features accross the whole locus. To this end, we have subcloned and sequenced a 8 kb DNA fragment encompassing the 5' UT/SP exon of the gene obtained from a DR4 haplotype. These data have shown the presence in the 5' flanking region of several repeated sequences (see Fig. 1).

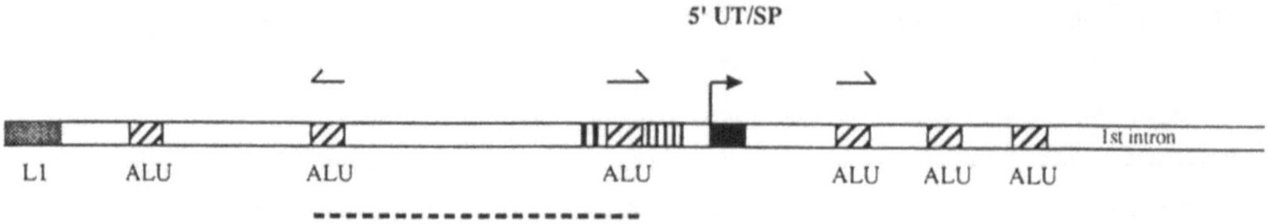

Fig. 1. Structure of the 5' end of the *HLA-DQA1* gene. The figure represents 10 kb of the DNA region surrounding the 5'UT/SP exon of *HLA-DQA1* (black box; the arrowhead indicates the direction of transcription of the gene). Arrows above Alu repeats indicate the head-to-tail orientation of the elements; the dashed bar shows the polymorphic region bracketed by two Alu repeats oriented in the opposite direction. The 5' most of the latter two corresponds to alu-q2 in Fig. 2, while that adjacent to the transcriptional start corresponds to alu-q; vertical bars preceding and following alu-q represent *HLA-DQA1* specific regulatory boxes (Ombra et al. 1990). The presence of three Alus inside the 1st intron was extrapolated from data by Pääbo 1985 and Del Pozzo and Guardiola 1989.

Notably one of the Alus is found at a location immediately adjacent to the promoter proximal sequences which control the expression of *HLA-DQA1* (Del Pozzo and Guardiola 1990). Other two Alus, one of which has been sequenced, are found further upstream to the promoter region (G. Del Pozzo and J. Guardiola, unpublished data). In addition three more Alus, one of which has been sequenced, are located in the 5' end of intron 1 (Pääbo 1985; Del Pozzo and Guardiola 1989). In summary, 6 Alu elements can be found in ca. 10 kb of DNA around the 5' end of *HLA-DQA1*. The frequency of Alu elements in this portion of the gene (1 in less than 2 kb) is somewhat higher than the average frequency of Alu elements in the genome (one element every 4-6 Kb). Hybridization experiments, moreover, indicate that other Alu repeats are present throughout the gene. Instances where clustering of Alu repeats in other loci leads to insertion/deletion polymorphisms have been reported (Stoppa-Lyonnet et al. 1990). Repetitive elements belonging to other families may also be present; for example we have found that a member of the L1 family is located 6 kb upstream to the transcriptional start of the gene (G. Del Pozzo and J. Guardiola, unpublished data).

Alu elements and polymorphisms of the *HLA-DQA1* gene

We have studied in particular a region of the *HLA-DQA1* gene located immediately upstream to the start of transcription of the gene. This region (about 3 kb in length, see Fig.

1) was found to be bracketed by two oppositely oriented Alu elements; each of these elements is in turn delimited by two short terminal repeats and their present location in the genome can be thus considered as the consequence of a translocation event (Rogers 1985). Genomic hybridization using single copy probes derived from the 5' flanking of the gene allowed to establish that the intervening region between the two Alus is polymorphic in several cell lines (G. Del Pozzo and J. Guardiola, unpublished data). PCR analysis, conversely, indicated that there was no sequence variation in the regions immediately outside the two Alus. We therefore conclude that the alterations of the pattern of restriction enzyme digestion observed are related, at least in part, to the presence of Alu elements in this region. Furthermore *HLA-DQA1* clones purified from a placental DNA bank were obtained; comparison of the restriction maps of these two clones and of clones described previously in the literature showed that the restriction pattern is substantially different. These data suggest that restriction fragment lenght polymorphisms extend over the entire genetic locus; it is also possible that the structure of the gene directly isolated from fresh tissues may be more polymorphic than that observed in cultered cell lines.

Polymorphisms of the 5' flanking region of *HLA-DQA1* and regulation

The upstream region of *HLA-DQA1* encompasses several signals that are crucial for gene regulation: these signals include a promoter proximal site containing the X, Y, and Z boxes (Accolla and Guardiola 1989; Benoist and Mathis 1990); an upstream region which acts as a weak enhancer onto heterologous promoters (Ombra et al. 1990) and as a negative control element when combined to the *HLA-DQA1* minimal promoter (reported in Accolla et al. 1991). An upstream controlling region not yet fully characterized seems to lie further upstream (Ombra et al. unpublished observation). The structure and the distribution of these signals (Ombra et al. 1990; Dorn et al. 1988) suggests that loop formation or other mechanisms of long distance cross-talk among different regulatory elements are possible. DNA polymorphisms of the regulatory region, although may not affect the regulatory sequences themselves, may alter the distances of these signals relative to each other, thus influencing the regulation of gene activity. Another mechanism whereby Alu sequences may alter the expression of adjacent genes is by direct interference in the transcriptional machinery (Wu et al. 1990); Alu repeats, moreover, may carry sequences that may interact at the regulatory level with the signals that control gene activity (Saffer and Thurston 1989). These types of "regulatory polymorphisms" seem to us plausible means to increase the variability of MHC class II mediated modulation of immune response.

```
        1                                                       57
CONS    GGCCGGGCGCGGTGG-------CTCACGCCTGTAATCCCAGCACTTTGGGAGGCCGA
ALUb    ***************-------***********************************
AFP     ***************-------***********************************
MLVI    T**************-------******T****************************
ALUc    ***************-------***********************************
ALUQ    ****A***A******-------****T***********A***************T**
ALUQ2   ***ATCTATT**GCCAGTCGTG**************T******************A*
        58                                                     114
CONS    GGCGGGCGGATCACCTGAGGTCAGGAGTTCGAGACCAGCCTGGCCAACATGGTGAAA
ALUb    **************--***CC******A*********T******T****C*******
AFP     **************--***********A*********T******T****C*******
MLVI    **************--***********A*********T**C***T**A*C*******
ALUc    **************--*******A***A*********T*******************
ALUQ    *******A*****T--**************A********G**********AG*****
ALUQ2   ***A***A*****T--**************A********G**********A***A**
        115                                                    171
CONS    CCCCGTCT-CTACTAAAAA-TACAAAAA----TTAGCCGGGCGTGGTGGCGCGCGCC
ALUb    ********-**********-********----*******************G*****
AFP     ********-**********A**-C****AAAA*************A***T*G*****
MLVI    **-*****-*******-**A**-C****AAAA***********A*A***T*-***-*
ALUc    ********-**********-********----*******************G*****
ALUQ    *T*T****-**********-****G**-----**********A*******CA***-*
ALUQ2   ***T****-**********-********A---******************AG*****
        172                                                    228
CONS    TGTAATC-CCAGCTACTCGGGAGGCTGAGGCAGGAGAATCGCTTGAACCCGGGAGGC
ALUb    ****G**-*******************************G**G**************
AFP     ****G**-*******************************G**G*******T******
MLVI    ****G**-*******************************G**G*******T******
ALUc    ********-**********-********----*************************
ALUQ    **C*G**A*****************************************T*******
ALUQ2   ****G**-*********T---------------------------------------
        229                                                    285
CONS    GGAGGTTGCAGTGAGCC-GAGATCGCGCCACTGCAC---TCCAGCCTGGGCGACAGA
ALUb    ****C************-******************---******************
AFP     ****C************-******************---******************
MLVI    *--**************-******************---******************
ALUc    *****************-******************---******************
ALUQ    AA***************-****C**T****T*****C--*-*********T******
ALUQ2   -------**********-*****TA**********C---******************
        286
CONS    GCGAGACTCCGTCTCA
ALUb    ****************
AFP     ****************
MLVI    ****************
ALUc    ****************
ALUQ    *T******A*******
ALUQ2   **A*****--******
```

Fig. 2. Comparison of the sequence of various Alu elements. Asterisks represent nucleotides identical to those of the consensus sequence (*CONS*, see Britten et al. 1988 and Jurka and Smith 1988); Alub and Aluc are the consensus for type b and type c repeats of the AluS subfamily, respectively (Jurka and Smith 1988). *AFP* and *MLVI* are individual type b repeats. Alu-q and alu-q2, whose location in the *HLA-DQA1* locus is illustrated in Fig. 1, have been identified as type a AluS family members. Dashes represent gaps; note that the enumeration of nucleotide positions is different than in Jurka and Smith 1988; when analyzing the sequences for diagnostic purposes, the convention of the latter authors was used.

Past history of the *HLA-DQA1* gene

Alu elements or structures generated by rearrangements triggered by the Alu recombinogenic activity may serve as tags for marking different alleles of the *HLA-DQA1* gene, thus allowing to trace the history of their divergence. The presence of an Alu element in one allelic form and of an empty integration site in another may be helpful to determine the time of divergence of the two forms (Del Pozzo and Guardiola 1990); similarly the genetic drift of the an Alu sequence located in the same site in two divergent alleles may be informative (McConnell et al. 1988). Studies of Alu family repeats have in the last few years allowed to identify different subfamilies and have also permitted to attribute an evolutionary pattern of separation among different members (Britten et al. 1988; Deininger and Slagel 1988; Jurka and Smith 1988). Sequence comparison lead to the development of a simple classification method based onto the analysis of diagnostic nucleotide positions that are invariant in members of the same branch of the Alu family (Britten et al. 1988; Jurka and Smith 1988). For example, analysis of the two Alu repeats which bracket the polymorphic region previously discussed, namely alu-q and alu-q2 (Fig. 1 and Fig. 2), indicates that both of them most probably belong to the type a branch of the AluS subfamily, thus placing their integration time in the 5' flanking of the gene at ca. 30 Myr ago.

As reported in Fig. 2, both Alu elements share nucleotide substitutions in positions that are invariant in other Alus (positions 65, 71, 88, 97, 108, and 118 in Fig. 2), suggesting therefore that they may belong to a particular subset of type a AluS repeats. Moreover, instead of carrying a 3' poly dA stretch as more than 83 % of Alu repeats do, they have a patterned tail whose sequence is (GA)n as in 6 % of Alu elements so far sequenced (Economou et al. 1990); a similar tail is observed in yet another Alu sequence, that immediately downstream to the 5' UT/SP exon. It is thus possible that the *HLA-DQA1* region has attracted Alu members of the same or strictly related branch. Knowledge of the sequence of other Alus of the region is obviously required to settle the argument. The analysis of Alu sequences and of their integration sites has been applied to the study of the divergence of the *HLA-DQA1* and *HLA-DQA2* gene (Del Pozzo and Guardiola 1990). As a matter of fact these two share sequence homology throughout their entire spans (Auffray et al. 1987); this homology terminates abruptly some 300 bp upstream to the transcriptional start; in this position we found that an Alu sequence, alu-q, lies in *HLA-DQA1*. As mentioned alu-q is bracketed by short terminal repeats derived from the duplication of the integration site of the SINE element (Del Pozzo and Guardiola 1990); an empty integration site was instead found in the same position in *HLA-DQA2* (Fig. 3).

```
                         -654
DQA1                      ATATAACTGATGACCTCATCTATCCATGAGGCTGCTCACG
                          **** * ********* * ******* * ********* *
DQA2      ACCCTGAAGGATGCCATATTATTGATGACCTTACCTATCCAAGAGGCTGCTCA-G
         -337

DQA1 aaatgcccTCccctGGCCAGGCACGGTGGCTCATGCCTGTAATCCAAGCACTTTGGGAGG

DQA2 ------------------------------------------------------------

DQA1 GTGAGGCAGGCAGATCATGAGGTCAGGAGTTCAAGACCAGCGTGGCCAACATAGTGAAAC

DQA2 ------------------------------------------------------------

DQA1 TCTGTCTCTACTAAAAATACAGAATTAGCCGGGCATGGTGGCAGCGCTGCAGTCACCAGC

DQA2 ------------------------------------------------------------

DQA1 ACTCGGGAGGGTGAGGCAGGAGAATCGCTTGAACCTGGGAGGCAAAGGTTGCAGTGAGCC

DQA2 ------------------------------------------------------------

DQA1 GAGACCATGCCATTGTACCTCAGCCTGGGTGACAGAGTGAGACTACGTCTCGAAAAGAAA

DQA2 ------------------------------------------------------------

                                                            -254
DQA1 AGAAAAAGAAAAGAAAAAGAAAAAAAAAAGAAaaatgcccATccctCTTGCGAATGGCAG
                                     **** ***  ******  *   * ****
DQA2 --------------------------------aaatCcccGCccctCTGACAC-TAGCAG
                                                            -254
```

Fig. 3. Comparison of the upstream sequences of *HLA-DQA1* and *HLA-DQA2*. Asterisks indicate nucleotide identity, dashes represent gaps. Bold characters show the short terminal repeats bracketing alu-q in *HLA-DQA1* and the postulated integration site in *HLA-DQA2*.

We therefore conclude that *HLA-DQA1* and *HLA-DQA2* diverged before the integration of the alu-q sequence; the latter in turn seems to have taken place before separation of Old World monkeys. An evolutionary tree based on these data and on those from Moriuchi et al. 1985 is presented in Fig. 4. It should be interesting to determine whether other *HLA-DQA1* alleles carry Alu sequences or vestigial sequences due to Alu-mediated rearrangements at the same site, because this would solve the problem as to whether the two genes diverged before accumulation of *HLA-DQA1* alleles.

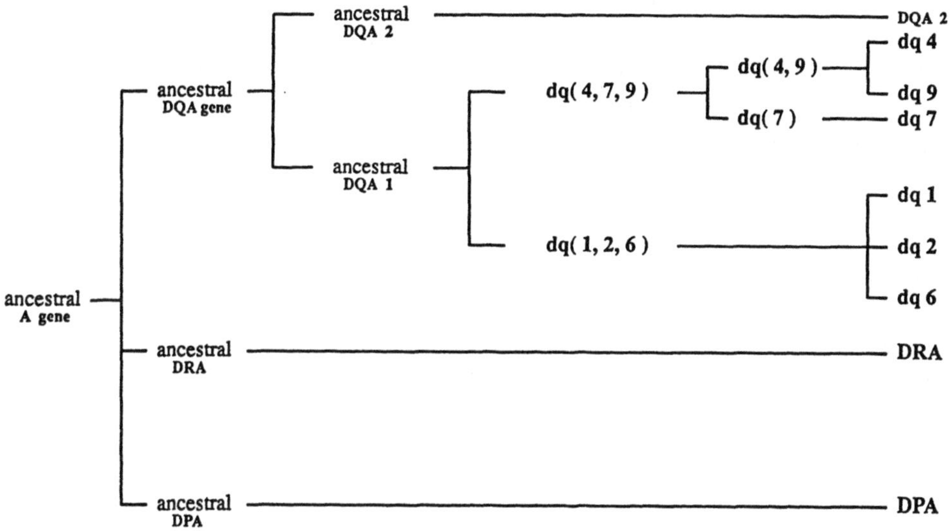

Fig. 4. Hypothetical evolutionary pathway of the *HLA-D A* genes. The pattern of divergence has been inferred from Moriuchi et al. 1985 and Del Pozzo and Guardiola 1990. Dq-1, -2, -4, -6, -7 and -9 indicate the *HLA-DQA1* alleles as designated by Moriuchi et al. 1985. Dq(4,7,9), dq(1,2,6), dq(4,9) and dq(7) are the evolutionary intermediates postulated by these authors.

Conclusions

The *HLA-DQA1* gene is highly polymorphic; generation of polymorphisms is in part related to the presence of clustered Alu sequences; integration of Alu family members at given genetic location have in fact been previously shown to cause DNA rearrangements of the adjacent loci (Lerhman et al. 1986; Stoppa-Lyonnet et al. 1990); the *HLA-DQA2* gene which lacks at least one of the Alu inserts is by far less polymorphic at the DNA level than *HLA-DQA1*. The two genes show a high level of homology and are placed almost at the same position on the chromosome; it is thus unclear why *HLA-DQA1* should be so prone to polymorphisms. Perhaps the biological properties of repetitive sequences play a role in this phenomenon; it is in fact possible that integration of an Alu family member in a given position may in a way catalize the nearby integration of repeats thus creating conditions for an exhuberant polymorphism as that exhibited by *HLA-DQA1*.

We believe that, by influencing gene expression, polymorphic traits outside the coding region of the *HLA-DQA1* gene may contribute to the heterogeneity of the immune response. Studies of the structure of these regions may provide information on the past history of the gene in addition to those obtained through the analysis of coding sequences, thus shedding light on the evolution of the MHC class II locus.

Acknowledgments. Research in this laboratory is supported by P.F. Biotechnology and Bioinstrumentation, CNR, by the Ministry of Health/Istituto Superiore di Sanit and by the Associazione Italiana per la Ricerca sul Cancro (AIRC). M.A. gratefully acknowledges a fellowship from AIRC. We also thank C. Sole, F. Cavaliere, P. Villano, V. Romano and N. Indaco for their technical support.

References

Accolla, R.S., Auffray, C., Singer, D., and Guardiola, J.: The molecular biology of MHC. *Immunol Today 12*: 97-99 1991

Accolla, R.S. and Guardiola, J.: Regulation of human MHC class II genes. *J Immunol Res 1*: 191-198, 1989

Auffray, C., Ben-Nun, A., Roux-Dosseto, M., Germain, R.N., Seidman, J.G., and Strominger, J.L.: Polymorphism and complexity of the human DC and murine I-A alpha chain genes. *EMBO J 2*: 121-124, 1983

Auffray, C., Lillie, J.W., Korman, A.J., Boss, J.M., Frèchin, N., Guillemot, F. Cooper, J., Mulligan, R.C. and Strominger, J.L.: Structure and expression of the HLA-DQA and -DXA genes: interallelic alternate splicing of the HLA-DXA gene using a retroviral vector. *Immunogenetics 26*: 63-73, 1987

Bell, G.I., Selby, M.J., and Rutter, W.J.: The highly polymorphic region near the human insulin gene is composed of simple tandemly repeating sequences. *Nature 295*: 31-35, 1982

Benoist, C. and Mathis, D.: Regulation of major histocompatibility complex class II genes: X and Y and other letters of the alphabet. *Annu Rev Immunol 8*: 681-715, 1990

Britten, R.J., Baron, W.F., Stout, D.B., and Davidson, E.H.: Sources and evolution of human Alu repeated sequences. *Proc Natl Acad Sci USA 85*: 4770-4774, 1988

Cohen, D., Le Gall, I., Marcadet, A., Font, M.-P., J.-M. Lalouel, and Dausset, J.: Clusters of HLA class IIß restriction fragments describe allelic series. *Proc Natl Acad Sci USA 81*:7870-7884, 1984

Deininger, P.L. and Slagel, V.K.: Recently amplified Alu family members share a common parental Alu sequence. *Mol Cell Biol 8*: 4566-4589, 1988

Del Pozzo, G. and Guardiola, J.: Mummy DNA fragment identified. *Nature 339*: 431-432, 1989

Del Pozzo, G. and Guardiola, J.: A SINE insertion provides information on the divergence of the *HLA-DQA1* and *HLA-DQA2* genes. *Immunogenetics 31*: 229-232, 1990

Dorn, A., Fehling, J.H., Koch, N., Lemeur, M., Gerlinger, P., Benoist, C. and Mathis, D.: B-cell control region at the 5' end of a MHC class II gene: sequences and factors. *Mol Cell Biol 8*: 3975-3987, 1988

Economou, E.P., Bergen, A.W., Warren, A.C., and Antonarakis, S.E.: The polydeoxyadenylate tract of Alu repetitive elements is polymorphic in the human genome. *Proc Natl Acad Sci USA 87*: 2951-2954, 1990

Economu-Pachnis, A. and Tsichlis, P.N.: Insertion of an Alu SINE in the human homologue of the Mlvi-2 locus. *Nucleic Acids Res 13*: 8379-8387, 1985

Giles, R.C. and Capra, J.D.: Structure, function and genetics of human MHC class II molecules. *Adv Immunol 37*: 1-71, 1985

Guardiola, J., Maffei, A., Carrel, S., and Accolla, R.S.: Molecular genotyping of the HLA-*DQA* gene region. *Immunogenetics 27*: 12-18, 1988

Gyllensten, U.B. and Erlich, H.A.: Ancient roots for polymorphism at the HLA-DQA locus in primates. *Proc Natl Acad Sci USA 86*: 9986-9990, 1989

Hirayama, K., Matsushita, S., Kikuchi, I., Iuchu, M., Ohta, N., and Sasazuki, T.: HLA-DQ is epistatic to HLA-DR in controlling the immune response to schistosomal antigen in humans. *Nature 327*: 426-430, 1987

Horn, G., Bugawan, T.L., Long, C.M., and Erlich, H.A.: Allelic sequence variation of the HLA-DQ loci: relationship to serology and to insulin-dependent diabetes susceptibility. *Proc Natl Acad Sci USA 85*: 6012-6016, 1988

Jurka, J. and Smith, T. A.: fundamental division in the Alu family of repeated sequences. *Proc Natl Acad Sci USA 85*: 4775-4778, 1988

Le Gall, I., Marcadet, A., Font, M.-P., Auffray, C., and Strominger, J.L., Lalouel, J.M., Dausset, J. and Cohen, D.: Exuberant restriction fragment length polymorphism associated with the DQA chain gene and the DXA chain gene. *Proc Natl Acad Sci USA 82*: 5433-5436, 1985

Lehrman, M.A., Russell, D.W., Goldstein, J.L., and Brown, M.S.: Exon-Alu recombination deletes 5 kilobases from low density lipoprotein receptor gene, producing a null phenotype in familial hypercholesterolemia. *Proc Natl Acad Sci USA 83*: 3679-3683, 1986

McConnell, T.J., Talbot, W.S., McIndoe, R.A., and Wakeland, E.K.: The origin of MHC class II gene polymorphism within the genus *Mus*. *Nature 332*: 651-654, 1988

Moriuchi, J., Moriuchi, T. and Silver, J.: Nucleotide sequence of an HLA-DQA clone derived from a DRw9 cell line: genetic and evolutionary implications. *Proc Natl Acad Sci USA 82*: 3420-3424, 1985

Morse, B., Rotherg, P.G., South, V.J., Spandorfer, J.M., and Astrin, S.M.: Insertional mutagenesis of the myc locus by a LINE sequence in a human breast carcinoma. *Nature 333*: 87-90. 1988

Ollo, R. and Rougeon, F. Gene conversion and polymorphism: generation of mouse immunoglobulin 2b chain alleles. *Cell 32*: 515-523, 1983

Ombra, M.N., Del Pozzo, G., Perfetto, C., Maffei, A., and Guardiola, J.: Effect of the AIR-1 locus on the activation of an enhancerless HLA-DQA1 promoter. *Immunogenetics 31*: 368-376, 1990

Ono, S.J., Bazil, V., Sugarawa, M., and Strominger J.L.: An isotype-specific trans-acting factor defective in a mutant B cell line that expresses HLA-DQ, but not -DR or -DP. *J Exp Med 173*: 629-637, 1991

Ottolenghi, S. and Giglioni, B.: The deletion in a type of thalassaemia begins in an inverted AluI repeat. *Nature 300*: 770-771 1983

Pääbo, S.: Molecular cloning of ancient egyptian mummy DNA. *Nature 314*: 644-645, 1985

Rogers, J.H.: Origin and evolution of retroposons. Part 2: The structure and evolution of retroposons. *Intern Rev Cytol 93*: 231-279, 1985

Safer, J.D. and Thurston, S.J.: A negative element with properties similar to those of enhancer is contained within an Alu sequence. *Mol Cell Biol 9*:355-364, 1990

Sasazuchi, T.: Immunological recognition. *Cold Spring Harbor Symp Quant Biol. 54*: 69-74, 1989

Schuler, L.A., Weber, J.L., and Gorski, J.: Polymorphism near the rat prolactin gene caused by insertion of an Alu-like element. *Nature 305*: 159-160, 1983

Spielman, R.S., Lee, J., Bodmer, W.F., and Trowsdale, J. Six HLA-D region alpha chain genes on human chromosome 6: polymorphisms and association of DC related sequences with DR types. *Proc Natl Acad Sci USA 81*: 3461-3465, 1984

Stoppa-Lyonnet, D., Carte, P.E., Meo, T., and Tosi, M.: Cluster of intragenic Alu repeats predispose the human C1 inhibitor locus to deleterious rearrangements. *Proc Natl Acad Sci USA 87*: 1551- 1555 1990

Trowsdale, J., Lee J., Carey, J., Grosveld, F., Bodmer, J.: and Bodmer, W. Sequences related to HLA-DRA chain on chromosome 6: restriction enzyme polymorphisms detected with DCA chain probes. *Proc Natl Acad Sci USA 80*: 1972-1976, 1983

Widera, G. and Flavell, R.A.: The nucleotide sequence of I-EßЬ immune response genes: evidence for gene conversion events in class II genes of the major histocompatibility complex. *EMBO J 3*:1221-1225, 1984

Wu, J., Grindlay, G.J., Bushel, P., Mendelsohn, L., and Allan, M.: Negative regulation of the human epsilon-globin gene by transcriptional interference: role of an Alu repetitive element. *Mol Cell Biol 10*: 1209-1216, 1990

THE *ALU* REPEATS OF THE PRIMATE *DRB* GENES

Christian Schönbach* and Jan Klein*°
*Abteilung Immungenetik, Max-Planck-Institut für Biologie, 7400 Tübingen, Germany
°Department of Microbiology and Immunology, University of Miami School of Medicine, P.O. Box 016960 (R-138), Miami, FL 33101, USA

*Eleven Alu elements have thus far been found in the primate DRB genes: HLA-DRB1*04-Alu1, HLA-DRBVII-Alu2, HLA-DRBVII-Alu3, HLA-DRBVII-Alu4, HLA-DRBVII-Alu5, HLA-DRBVIII-Alu6, HLA-DRBVIII-Alu7, HLA-DRBVIII-Alu8, Gogo-DRB8-Alu7, Gogo-DRB8-Alu8, and Patr-DRB6-Alu9. Comparison with consensus sequences indicates that the DRB-Alu elements belong to five different subfamilies -- Sp, Sx, Sq, Sc, and Sb (in the classification scheme of Jurka and Milosavljevic). The Alu-Sp elements are the oldest of the five subfamilies, most of them having been inserted at their current locations more than 38 million years (myr) ago. The Alu-Sb elements are the youngest, having spread through the genome in the last 18 myr. The presence of Alu7 and Alu8 at corresponding positions in orthologous genes provides evidence that they were inserted in the DRB8 gene more than 6 myr ago. The presence of Alu9 in the Patr-DRB6 but not in the orthologous HLA-DRBVI gene indicates that this element is less than 6 myr old. The presence of old Alu elements in several DRB pseudogenes supports the notion of the pseudogenes being very old. No evidence of sequence homogenization could be found for the DRB-Alu elements: they have all conserved their diagnostic nucleotides scattered over the entire length of each element. Hence, no gene conversion apparently occurred in the elements, even in those that are more than 30 myr old. Whether the Alu elements might be responsible for the truncation of some of the DRB pseudogenes cannot be decided at the present time.*

One of the most abundant short interspersed repeated DNA elements (SINEs) in the primate genome is the *Alu* family, so termed because many (though not all) sequences belonging to it contain the recognition site of the *Alu* I restriction endonuclease (Houck et al. 1979; Jelinek et al. 1980). The human haploid genome contains more than 500,000 copies of *Alu* elements scattered over all 46 chromosomes and a similarly large number of these repeats probably occurs in most other primates (Rinehart et al. 1980; Hwu et al. 1986). A typical *Alu* element is approximately 300 nucleotides long and consists of two monomers, one slightly longer than the other. The monomers are separated by an A-rich region and one of the monomers is flanked by a 3' oligo(dA) tail of variable length. The dimer is flanked by a short direct repeat sequence formed during the integration of the *Alu* element into its chromosomal locus (Rinehart et al. 1980; Daniels and Deininger 1983). Many *Alu* elements

NATO ASI Series, Vol. H 59
Molecular Evolution of the
Major Histocompatibility Complex
Edited by J. Klein and D. Klein

are, however, incomplete or truncated, containing only some of the above-described features.

Table 1. Division of *Alu* family into subfamilies according to different authors.

Authors	1	2	3	4	5	6
	J	J	I	D	Diverged	J
	Sx	Se	II	F	Major	Sa
	Sq	Sd	II	E	Major	Sa
	Sp	Sd	II	E	Major	Sa
	Sc	Sc	III	C	Major	Sc
	Sb	Sb	IV	A	Conserved (Mliv-2-Alu)	Sb

Equivalent designations are listed in each line of the Table.
1, Jurka and Milosavljevic 1991; 2, Jurka and Smith 1988, 3, Britten et al. 1988; 4, Quentin 1988; 5, Willard et al. 1987; 6, Labuda and Striker 1989.

Several *Alu* elements identified in various primates have been sequenced. When they are aligned and compared, it becomes apparent that they fall into several subfamilies, which were given different designations by different authors (Table 1). Here we will use the nomenclature of Jurka and Milosavljevic (1991). Each subfamily is characterized by certain diagnostic nucleotides at specified positions. The average sequence similarity between elements of the same subfamily is approximately 80%; that between members of different subfamilies is about 74%.

In about one-half of the cloned *Alu* elements, one of the monomers of the *Alu* dimer contains a promoter region which is used by the RNA polymerase III to transcribe the element into a short RNA that matches the length of the repeat (Duncan et al. 1979). If an *Alu* element is integrated in a protein-encoding gene, it can be transcribed co-linearly with the gene and later removed during mRNA maturation. Rarely (as for example in the case of the LDL receptor gene) does the *Alu* sequence remain in the mature mRNA (Yamamoto et al. 1984; Limborska et al. 1987).

All the *Alu* elements show striking sequence similarity to the highly conserved 7SL RNA gene which codes for a 294 nucleotide cytoplasmic RNA present in virtually all cells (Ullu et al. 1982; Ullu and Tschudi 1984). The RNA is part of the signal recognition ribonucleoprotein particle which aids in the secretion of newly formed polypeptides through the membranes of the endoplasmic reticulum (Gundelfinger et al. 1984). The evolution of the *Alu* repeats from the 7SL gene is envisioned in the following way (Britten et al. 1988, 1989): more than 75 million years (myr) ago, before the main mammalian radiation (Carroll 1988), the 7SL gene duplicated and the duplicated copy lost approximately 150 nucleotides through deletion in the middle of the gene . This duplicated-deleted element flanked, by a poly-A tail, became the basic *Alu* monomer from which Alu-like sequences of certain mammalian orders derive (in rodents, for example, the repeats of the *B1* family have the basic structure of an *Alu* monomer; see Britten et al. 1988, 1989; Deininger and Daniels 1986). Early in the evolution of the primate lineage, before the separation of prosimians and simians, the basic *Alu* monomer duplicated and shortly afterwards (or simultaneously with the duplication), the first monomer lost approximately 25 nucleotides through deletion. In this way, the basic dimer, which we can refer to as source ("master") gene I, was generated. Copies of source gene I were then, from time to time, integrated randomly by retroposition into various locations in the primate genome, preferentially into A-rich regions (Daniels and Deininger 1985). The integration occurred with a frequency of more than one retrotransposition every 100 years or so (Zuckerkandel et al. 1989). The rounds of retroposition from the source gene I generated the oldest (J) subfamily of *Alu* elements. What happened subsequently is not clear. According to one hypothesis (Jurka and Milosavljevic 1991; Deininger and Daniels 1986), after some time (after the separation of simians from prosimians more than 65 myr ago; see Martin 1990), a new source gene was generated from source gene I, presumably by retroposition. The source gene II began to disseminate a new batch of *Alu* elements into the primate genome, all reflecting the sequence divergence between source genes I and II and hence belonging to a new subfamily, Sx. According to another hypothesis (Willard et al. 1987), the new subfamily could have been generated from the original source gene after it was altered by accumulating mutations. Be that as it may, the process was repeated several times during primate evolution, with each new source gene giving rise to a new subfamily of *Alu* elements. Comparisons indicate that sequence similarity between the *Alu* subfamilies decreases in the order Sb-Sc-Sp-Sq-Sx. This order is believed to reflect the different age of the subfamilies, which has been estimated as follows: J, >55 myr; Sx, Sq, Sp, ~ 38 myr; Sc, ~ 24 myr; Sb ~ 18 myr. It appears that the rate of amplification has slowed down gradually during primate evolution so that it is now approximately ten to 100 times slower than it was 65 myr ago (Zuckerkandel et al. 1989; Labuda and Striker 1989).

These characteristics of the *Alu* family make the repeats convenient markers for the evolution of genes and gene families. Recent studies from our and other laboratories have revealed the presence of *Alu* repeats in several of the class II major histocompatibility complex (*Mhc*) genes belonging to the *DRB* family (Andersson et al. 1987; Larhammar et al.

1985; Klein et al. 1991; Figueroa et al. 1991). In this communication we analyze the *Alu* repeats of the primate *DRB* genes as part of an ongoing effort to decipher the complex evolutionary history of the *Mhc* region.

Alu sequences of *DRB* genes were retrieved from the GenBank Release 66.0 (12/90) and prealigned by the Pythia on-line identification program (Jurka and Smith 1988). They were then aligned with the general *Alu* consensus sequence of Jurka and Milosavljevic (1991), and nucleotides at the diagnostic positions were selected, and used to classify the sequences considering their weight. Genetic distance (Saitou and Nei 1987; Kimura 1980) and maximum parsimony (Felsenstein 1988) trees were constructed from the aligned sequences. Jukes and Cantor (1969) pairwise distances and Kimura's 2-parameter distances were calculated using the SANDN computer program (Nei and Gojobori 1986). The JC distances and Kimura's 2-paramter distances were used for the construction of trees by the neighbor-joining method of Saitou and Nei (1987); these were calculated and drawn by the NJTREE program. The unweighted pair group method with arithmetic means (Sokal and Sneath 1963) was executed using the UPGMA program. Maximum parsimony trees were constructed using the DNAPARSO program (Felsenstein 1988). Alternative trees were constructed either by including or excluding the CpG hotspot positions.

Table 2. List of *Alu* elements found in primate *Mhc-DRB* genes.

Alu element	GenBank accession number	*Alu* subfamily	Reference
*HLA-DRB1*0401-Alu 1*	M20555, M19556	Sx	Andersson et al. 1987
HLA-DRBVII-Alu 2	KO2772	?	Larhammar et al. 1985
HLA-DRBVII-Alu 3	KO2772	Sq	Larhammar et al. 1985
HLA-DRBVII-Alu 4	KO2772	Sb	Larhammar et al. 1985
HLA-DRBVII-Alu 5	KO2772	Sc	Larhammar et al. 1985
HLA-DRBVIII-Alu 6	M20556, M19556	Sb	Andersson et al. 1987
HLA-DRBVIII-Alu 7	M20556, M19556	Sc	Andersson et al. 1987
Gogo-DRB8-Alu 7		Sc	Klein et al. 1991
HLA-DRBVIII-Alu 8		Sp	Andersson et al. 1987
Gogo-DRB8-Alu 8		Sp	Klein et al. 1991
Patr-DRB6-Alu 9		Sb	Figueroa et al. 1991

Eleven *Alu* elements have thus far been identified in the *DRB* genes of humans and great apes, eight of them in the *HLA-DR4* haplotype (Fig. 1; Table 2). We designate the elements by the names of the genes in which they are found and a serial *Alu* number (1 through 9; see Table 2). The alignment of *Alu* sequences is given in Figure 2, the subfamily designations, deduced from the diagnostic positions, are listed in Table 2. The *Alu 2* element has only been sequenced partially and so it has been excluded from further analysis. Two *Alu* elements (*Alu7* and *Alu8*) are found at corresponding positions in orthologous genes (*DRB8*) of humans and gorillas. They are therefore designated by the same serial numbers and differentiated by the gene designations (*HLA* and *Gogo* for the human and gorilla genes, respectively).

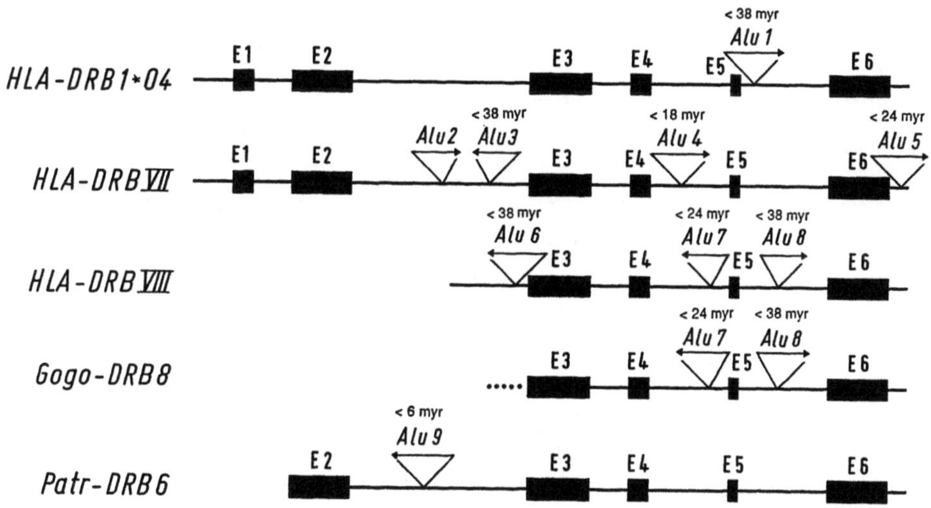

Fig. 1. Schematic representation of the *HLA-DRB1*04, HLA-DRBVII, HLA-DRBVIII, Gogo-DRB8,* and *Patr-DRB6* loci. Exons are shown as filled boxes. *Alu* elements and their orientation are depicted as arrows. The maximum estimated insertion time of the *DRB-Alu* elements is given in million years.

When were these nine *Alu* elements inserted into the *DRB* genes? There are several ways of answering this question, none of them, unfortunately, incontestable. The most reliable way of estimating the *minimum* (or alternatively, *maximum*) insertion time is by comparing *Alu* elements in orthologous genes. In the *DRB* genes, these are two pairs of *Alu7* and *Alu8* elements of the *HLA-DRBVIII and Gogo-DRB8 loci;* one *Alu* element (*Alu9*) is present in one (*Patr-DRB6*) gene but not in its orthologue (*HLA-DRBVI*). The fact that *Alu7* and *Alu8* are present, at corresponding positions, in both humans and gorillas signifies that

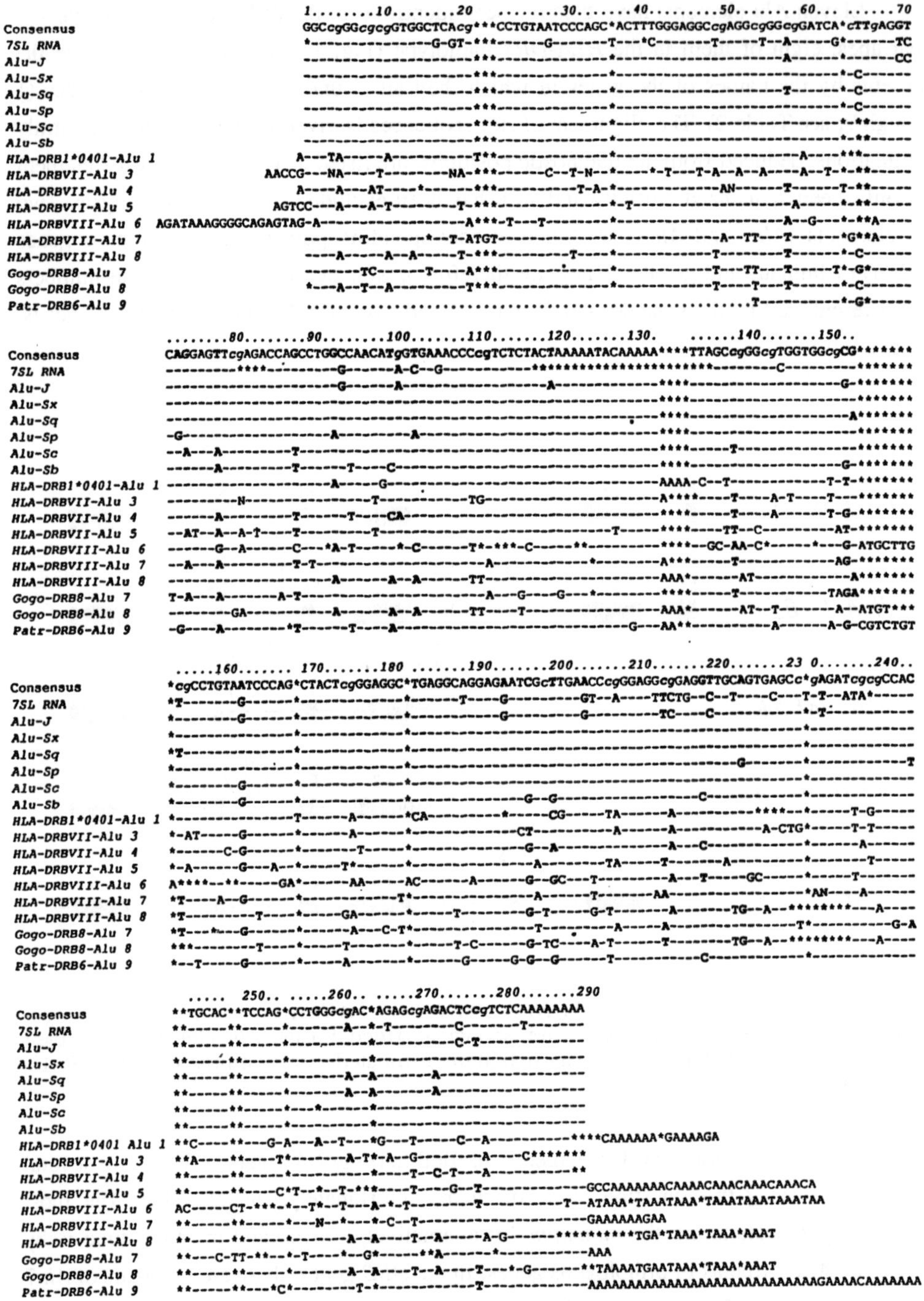

Fig. 2. Dash alignment of *DRB-Alu* elements to the general *Alu* consensus sequence, the *Alu-J, Alu-Sx, Alu-Sq, Alu-Sp, Alu-Sc, Alu-Sb* subfamily consensus sequences, and homologous regions of human *7SL RNA* (Jurka and Milosavljevic 1991). Diagnostic positions for the subfamilies are in bold. Hypermutable CpG hotspots are italicized. The numbering refers to the general consensus sequence. Gaps are not counted, and are indicatd by asterisks. Incomplete sequences are indicated by dots.

they were inserted into the *DRB8* gene more than 5-8 myr ago, the estimated time of human-gorilla divergence (Martin 1990). The presence of *Alu9* in *Patr-DRB6* and its absence in *HLA-DRBVI* indicates that the element was inserted *after* the separation of chimpanzee and human lineages; the element must therefore have been at this position for less than 8 myr.

Table 3. Genetic distance of *DRB-Alu* repeats from subfamilies consensus sequences and estimated insertion time.

Alu repeat	**Subfamily/** **age** (myr) *		**Distance** ** ± S.E.	**Insertion time** (myr)
*HLA-DRB1*0401-Alu 1*	*Alu-Sx*	**38**	.0637 ± .0175	**21** ± 6
HLA-DRBVII-Alu 3	*Alu-Sx*	**38**	.1041 ± .0228	**35** ± 8
HLA-DRBVIII-Alu 6	*Alu-Sx*	**38**	.1301 ± .0263	**43** ± 9
HLA-DRBVIII-Alu 8	*Alu-Sp*	**38**	.0778 ± .0191	**26** ± 6
Gogo-DRB8-Alu 8	*Alu-Sp*	**38**	.0774 ± .0191	**26** ± 6
HLA-DRBVII-Alu 5	*Alu-Sc*	**24**	.0669 ± .0175	**22** ± 6
HLA-DRBVIII-Alu 7	*Alu-Sc*	**24**	.0488 ± .0149	**16** ± 5
Gogo-DRB8-Alu 7	*Alu-Sc*	**24**	.0727 ± .0184	**24** ± 6
HLA-DRBVII-Alu 4	*Alu-Sb*	**18**	.0440 ± .0141	**15** ± 5
Patr-DRB6-Alu 9	*Alu-Sb*	**18**	.0423 ± .0151	**14** ± 5

* Based on (Labuda and Striker, 1989).
** The hypermutable *CpG* sites were omitted from the calculations.
S.E., standard error; myr, million years.

Another way of timing the *Alu* insertion events is by calculating the genetic distance of specific elements from the subclass consensus sequence and then converting the distance into time by referring to the evolutionary rate of the repeats (0.15% per million years; see Britten 1986). The two assumptions underlying this method are, first, that the consensus is reasonably close to the sequence of the subfamily source gene; and second, that the divergence rate following insertion is constant. These assumptions, however, may not always hold true. Estimates based on this procedure are given in Table 3 for the individual *DRB-Alu* elements. According to these calculations, the youngest element of the *DRB* genes is *Alu4* of the *HLA-DRBVII* gene (estimated insertion time 15 myr), while the oldest is *Alu3* in the same gene (estimated insertion time 35 myr). The *Alu5* element is estimated to have been inserted into the *DRB7* gene 22 myr ago (no information is available on the insertion time of the *Alu2* element). According to these calculations, the *DRB7* has been targeted by

Alu elements at least every 11 myr or so. Genetic distance comparisons of *DRB7* to other *DRB* genes do indeed point to an old age of this pseudogene (Klein et al. 1991). It may have arisen shortly after the separation of New World and Old World monkey lineages some 33 myr ago (Martin 1990). If so, the element could be younger than its estimated time of insertion suggests but the difference between the two estimates is within the error range of both the *Alu* element insertion times and the separation times of the monkey lineages. How large the error can be is highlighted by the *Alu9* element, whose insertion time is estimated at 14 myr ago, while in reality the element could not have been inserted more than 6 - 8 myr ago. All other *DRB* genes in which old *Alu* elements are present also seem old from genetic distance comparisons (Klein et al. 1991, Vincek et al. 1991; Figueroa et al. 1991). It may be significant that most of them are pseudogenes.

Yet another way of estimating the age of the *Alu* inserts is based on comparisons of the direct repeats and the A-rich regions. The flanking direct repeats arise from a single target sequence at the time of insertion and hence begin to diverge from this moment. The degree of divergence from the original sequence should therefore be proportional to the time elapsed since the insertion event. Similarly, the A-rich regions (central and 3' flanking) once consisted of poly (A) exclusively so that the amount of change in these regions should also reflect the age of the element. Since, however, both the direct repeats and the A-rich regions are quite short, it is difficult to evaluate statistically the changes they have incurred. The direct repeats of the *DRB-Alu* elements are aligned in Table 4; the sequences of the A-rich regions are given in Table 5. Since the standard error of the distances based on these sequences is greater than the distances themselves, their use for age estimates is precluded. However, the significant divergence of several of the sequences supports the notion that most of the *Alu* elements have been diversifying for quite some time and hence that their host genes separated from one another a long time ago.

The relationships betwen the *DRB-Alu* sequences were evaluated by the distance and maximum parsimony methods but only the results obtained by the latter method are shown here. In the dendrogram in Figure 3, the sequences group according to the subfamily classification and with the subfamily consensus sequences. Moreover, the order in which the branches separate in the dendrogram is that at which the individual subfamilies are believed to have emerged.

The fact that the *DRB-Alu* elements separate very clearly into subfamilies, just like *Alu* elements in other genes do, indicates that they have not been subjected to homogenization by gene conversion or by other forms of recombination (Dover 1982). This observation, too, is in agreement with similar observations made on other *Alu* repeats. Gene conversion is believed by some investigators to be of major importance in generating polymorphism at the major histocompatibility complex (*Mhc*) loci. The apparent absence of concerted evolution in the *DRB-Alu* repeats, however, must mean either that gene

Table 4. Alignment of the direct repeats flanking the *DRB -Alu* elements

Alu element	Direct Repeat	*Alu* subfamily and orientation of insert
*HLA-DRB1*0401-Alu 1*	AAACAGA*TTTGG	<-*Sx*-
	----G--A---CC	
HLA-DRBVII-Alu 3	AAAATACATAAAACAAT	<-*Sx*-
	A---C-A-A------G-	
HLA-DRBVIII-Alu 6	AATGTGGAT	-*Sx*->
	----A-C--	
HLA-DRBVII-Alu 4	AAAGAGTCCTT	-*Sb*->

Patr-DRB6-Alu 9	n.s.	<-*Sb*-
	AAAAAAACAAAAGA	
HLA-DRBVII-Alu 5	AAAAATCATCT	-*Sc*->

HLA-DRBVIII-Alu 7	AAAAGATATGGAAGCA	<-*Sc*-

Gogo-DRB8-Alu 7	AAAAGATATGGAAGCA	<-*Sc*-

HLA-DRBVIII-Alu 8	AAAAGGCCTGTGCTT	-*Sp*->
	----C---A-----C	
Gogo-DRB8-Alu 8	AAAAGGCCTGTGCTC	-*Sp*->
	----*----*-*--T	

n.s., not sequenced.

Table 5. The sequence of the A-rich and the Poly (A) regions of the *DRB-Alu* elements

	A-rich region between left and right monomer
Consensus	TACTAAAAATACAAAAA****
Alu-J	---A-------------****
Alu-Sx	-----------------****
Alu-Sq	-----------------****
Alu-Sp	-----------------****
Alu-Sc	-----------------****
Alu-Sb	-----------------****
*HLA-DRB1*0401-Alu 1*	-----------------AAAA
HLA-DRBVII-Alu 3	-----------------A***
HLA-DRBVIII-Alu 6	C------**--------****
HLA-DRBVII-Alu 4	-----------------****
Patr-DRB6-Alu 9	-----------------AA**
HLA-DRBVII-Alu 5	-----------------****
HLA-DRBVII-Alu 7	--------*--------A***
Gogo-DRB8-Alu 7	----G---*--------****
HLA-DRBVII-Alu 8	-----------------AAA*
Gogo-DRB8-Alu 8	-----------------AAA*
	Poly(A)
*HLA-DRB1*0401-Alu 1*	AAAAAA*CAAAAAA*GAAAAGA
HLA-DRBVII-Alu 3	C
HLA-DRBVIII-Alu 6	AAAAA**TAAATAAATAAATAAA*TAAATAAATAAATAA
HLA-DRBVII-Alu 4	AAAAAA
Patr-DRB6-Alu 9	AAAAAAAAAAAAAAAAAAAAAAAAAAAAGAAAACAAAAAAA
HLA-DRBVII-Alu 5	AAAAAAAAGCCAAAA*AAACAAAACAAACAAACAAACA
HLA-DRBVII-Alu 7	AAAAAAAAGAAAAAAGAA
Gogo-DRB8-Alu 7	AAAAAAAAAAA
HLA-DRBVII-Alu 8	AAAAAA**GA*TAAATAAA*AAA*T
Gogo-DRB8-Alu 8	ATAAAA*TGAATAAATAAA*AAA*T

conversion is far less frequent than these investigators would like us to believe, or that there are some mechanisms which restrict it to certain genes only and perhaps even to certain regions within a gene. It is nevertheless striking that within the last 30 myr or so, apparently not one single gene conversion event affected the 11 *DRB-Alu* elements.

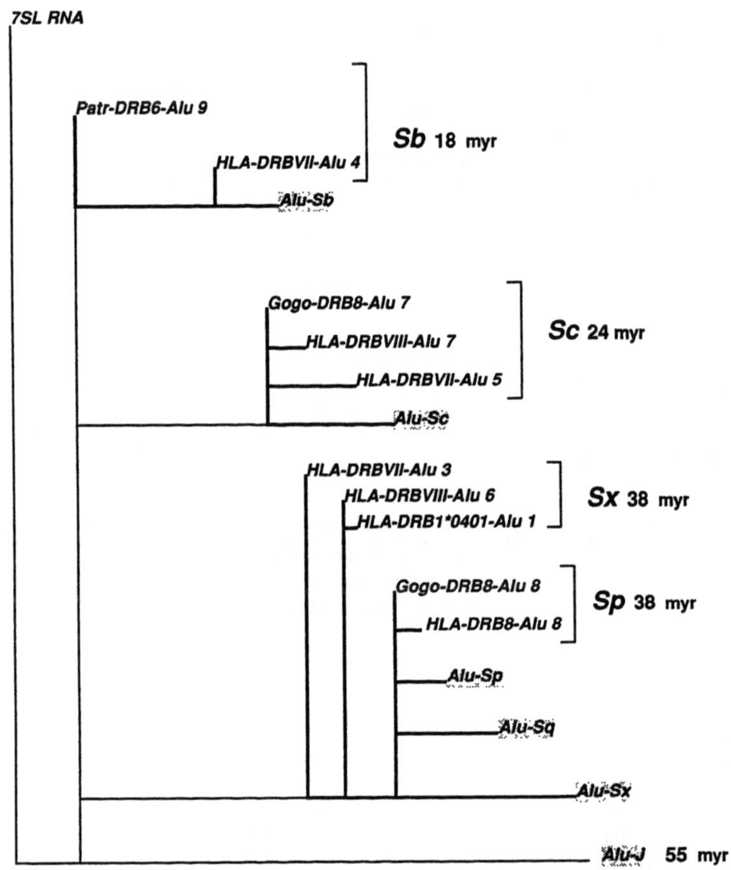

Fig. 3. Maximum parsimony tree of *DRB-Alu* elements and the *Alu* subfamily consensus sequences. For tree construction the hypervariable CpG hotspots were omitted.

The presence of *Alu* repeats is believed to have a destabilizing effect on the target gene. Numerous instances have been reported in which (unequal) crossing-over has occurred at or near the site of *Alu*-repeat insertion and several authors proposed that the *Alu* elements themselves are the cause of genetic exchanges in the region of their integration (e.g., Lehrman et al. 1985; Ottolenghi and Giglioni 1982). Crossing-over, however, occurs also in regions where no *Alu* repeats can be found so that these elements may at best be one of several factors increasing the probability of recombination at a particular site.

In the *Mhc*, the presence of *Alu* elements has thus far been documented predominantly for pseudogenes, of which several occur in a truncated form, presumably generated by some form of unequal crossing-over. It is therefore tempting to regard the *Alu* elements as the cause of the apparent instability of these genes. There are, however, two facts to consider before placing the blame on the *Alu* repeats. One is that more *DRB* pseudogenes than functional genes have been sequenced in the intron regions, in which *Alu* repeats are likely to be found. The second is that the *DRB* pseudogenes are apparently older than most functional *DRB* genes (Klein et al. 1991; Vincek et al. 1991; Figueroa et al. 1991), so that the former had more time to accumulate *Alu* repeats than the latter, and moreover, because of the slowdown in integration frequency, the old genes were more likely to become the target of *Alu* insertion than the younger ones. It is, therefore, difficult to decide at this time whether the *Alu* repeats are responsible for the functional inactivation of the *DRB* genes or the pseudogenes are replete with *Alu* elements simply because they have been around for a long time. Complete sequencing of additional functional *DRB* genes may help to choose between these two alternatives.

Acknowledgments. We thank Ms. Lynne Yakes for editorial assistance. The work was supported in part by grant AI 23667 from the National Institutes of Health, Bethesda, MD, USA and grant 163 258 from the Fonds der Chemischen Industrie, Frankfurt, Germany.

References

Andersson, G., Larhammar, D., Widmark, E., Servenius, B., Peterson, P.A., and Rask, L.: Class II genes of the human major histocompatibility complex: Organization and evolutionary relationship of the *DRβ* genes. J Biol Chem 262:8748-8758, 1987

Britten, R.J.: Rates of DNA sequence evolution differ between taxonomic groups. Science 231:1393-1398, 1986

Britten, R.J., Will, F.B., Stout, D.B., and Davidson, E.H.: Sources and evolution of human Alu repeated sequences. Proc Natl Acad Sci USA 85:4770-4774, 1988

Britten, R.J., Stout, D.B., and Davidson, E.H.: The current source of human Alu retrotransposons is a conserved gene shared with Old World monkey. Proc Natl Acad Sci USA 86:3718-3722, 1989

Carroll, R.L.: *Vertebrate Paleontology and Evolution.* W.H. Freeman, New York 1988

Daniels, G.R. and Deininger, P.L.: A second major class family of Alu family repeated DNA sequences in a primate genome. Nucleic Acids Res 11:7595-7610, 1983

Daniels, G.R. and Deininger, P.L.: Integration site preferences of the Alu family and similar repetitive DNA sequences. Nucleic Acids Res 13:8939-8954, 1985

Deininger, P.L. and Daniels, G.R.: The recent evolution of mammalian repetitive DNA elements. TIGS 2:76-80, 1986

Dover, G.: Molecular drive: a cohesive mode of species evolution. Nature 299:111-117, 1982

Duncan, C., Biro, P.A., Choudary, P.V., Elder, J.T., Wang, R.R.C., Forget, B.G., DeRiel, J.K., and Weissman, S.M.: RNA polymerase III transcriptional units are

interspersed among human non-a-globin genes. Proc Natl Acad Sci USA 10:5095-5099, 1979
Felsenstein, J.: Phylogenies from molecular sequences: Inference and reliability. Annu Rev Genet 22:521-565, 1988
Figueroa, F., O'hUigin, C., Inoko, H., and Klein, J.: Primate DRB6 pseudogene -- a clue to the evolutionary origin of the HLA-DR2 haplotype. Submitted, 1991
Gundelfinger, E.D., di Carlo, M., Zopf, D., and Melli, M.: Structure and evolution of the 7SL RNA component of the signal recognition particle. EMBO J 3:2325-2332, 1984
Houck, M., Rinehart, F.P., and Schmid, C.W.: Ubiquitous family of repeated DNA sequences in the human genome. J Mol Biol 132:289-306, 1979
Hwu, H.R., Roberts, J.W., Dawidson, E.H., and Britton, R.J.: Insertion and/or deletion of many repeated DNA sequences in human and higher ape evolution. Proc Natl Acad Sci USA 83:3875-3879, 1986
Jelinek, W.R:, Toomey, T.P., Leinwald, L., Duncan, C.H., Biro, P.A., Choundary, P.V., Weissman, S.M., Rubin, C.M., Houck, C.M., Deininger, P.O., and Schmid, C.W.: Ubiquitous interspersed repeated sequences in mammaliana genomes. Proc Natl Acad Sci USA 77: 1398-1402, 1980
Jukes, T.H. and Cantor, C.R.: Evolution of protein molecules. In *Mammalian Protein Metabolism III*. H.N. Munro (ed.), pp. 21-132, Acad. Press, New York, 1969
Jurka, J. and Milosavljevic, A.: Reconstruction and analysis of human Alu genes. J Mol Evol 32:105-121, 1991
Jurka, J. and Smith, T.: A fundamental division in the Alu family of repeated sequences. *Proc Natl Acad Sci USA 85*:4775-4778, 1988
Kimura, M.: A simple method of estimating evolutionary rates of base substitution through comparative studies of nucleotide sequences. J Mol Evol 16:111-120, 1980
Klein, D., Vincek, V., Kasahara, M., Schönbach, C., O'hUigin, C., and Klein, J.: Gorilla-*Mhc-DRB* pseudogene: implications for the evolution of the *HLA-DR4* haplotype. Submitted, 1991
Labuda, D. and Striker, G.: Sequence conservation in Alu evolution. Nucleic Acids Res 17:2477-2491, 1989
Larhammar, D., Servenius, B., Rask, L., and Peterson, P.A.: Characterization of an HLA-DRß pseudogene. Proc Natl Acad Sci USA 82:1475-1479, 1985
Lehrman, M.A., Schneider, W.J., Südhof, T.C., Brown, M.S., Goldstein, J.L., and Russell, D.W.: Mutation in LDL receptor: Alu-Alu recombination deletes exons encoding transmembrane and cytoplasmic domains. Science 227:140-146, 1985
Limborska, S.A., Korneev, S.A., Maleeva, N.E., Slominsky, P.A., Jincharadze, A.G., Ivanov, P.L., and Ryskov, A.P.: Cloning of Alu-containing cDNAs from human fibroblasts and identification of small Alu+ poly(A)+ RNAs in a variety of human normal and tumor cells. FEBS Lett 212 (2):208-212, 1987
Martin, R.D.: *Primate Origins and Evolution: A Phylogenetic Reconstruction.* Chapman and Hall, London 1990
Nei, M. and Gojobori, T.: Simple methods for estimating the numbers of synonymous and nonsynonymous nucleotide substitutions. Mol Biol Evol 3:418-426, 1986
Ottolenghi, S. and Giglioni, B.: The deletion in a type d^{o}-$ß^{o}$-thalassaemia begins in an inverted *Alu* I repeat. Nature 300:770-771, 1982.
Rinehart, F.P., Ritch, T.G., Deininger, P.L., and Schmid, C.W.: Biochemistry 20:3003-3010, 1980
Saitou, N. and Nei, M.: The neighbor-joining method. A new method for reconstructing phylogenetic trees. Mol Biol Evol 4:406-425, 1987
Sokal. R.R. and Sneath, P.H.A.: *Principles of Numerical Taxonomy.* W.H. Freeman, New York 1973
Ullu, E. and Tschudi, C.: Alu sequences are processed 7SL RNA genes. Nature 312:171-172, 1984
Ullu, E., Murphy, S., and Melli, M.: Human 7S RNA consists of a 140 nucleotide middle repetitive sequence inserted in an Alu sequence. Cell 29:195-202, 1982

Vincek, V., Klein, D., Hauptfeld, V., Kasahara, M., O'hUigin, C., Mach, B., and Klein, J.: The evolutionary origin of the *HLA-DR3* haplotype. Submitted, 1991

Willard, C., Nguyen, H.T., and Schmid, C.W.: Existence of at least three different Alu subfamilies. J Mol Evol 26:180-186, 1987

Yamamoto, T., Davis, C.G., Brown, M.S., Schneider, W.J., Casey, M.L., Goldstein, J.L., and Russel, D.W.: The human LDL receptor: a cysteine-rich protein with multiple Alu sequences in its mRNA. Cell 39:27-38, 1984

Zuckerkandel, E., Latter, G., and Jurka, J.: Maintenance of function without selection: Alu sequences as "cheap genes". J Mol Evol 29:504-512, 1989

INTERPRETING MHC DISEQUILIBRIUM

William Klitz and Glenys Thomson
Department of Integrative Biology, 345 Mulford Hall, University of California, Berkeley, CA 94720, USA

The record of non-random association of alleles between tightly linked loci in a population contains a wealth of information on the recent evolutionary past. Methods for the analysis of disequilibrium are presented and exemplified with application to HLA data.

Our purpose is to present a variety of methods suitable for unravelling the information on the evolutionary past contained in gametic or linkage disequilibrium -- the non random association of alleles -- suitable for application to the MHC of any species. The tight linkage and high polymorphism of MHC loci make them ideal candidates for such a venture.

The presence of linkage disequilibrium can be due to several causes including selection on one or more loci, admixture of populations having different allele frequencies, non-random mating, recent occurrence of a mutation, and random drift. Disequilibrium in tightly linked loci lasts many generations, declining steadily as a function of recombination. In contrast, genotype ratios return to Hardy Weinberg proportions each generation under random mating regardless of the intensity of selection or other forces altering allele frequencies. We have compiled information on five complementary measures of disequilibrium in Table 1, including a brief definition, properties of the measures, approximate sample size requirements, and uses. An example of the application of each method is given following its description below.

The disequilibrium measure, D, forms the basis for X^2 testing of deviations from linkage equilibrium on a single haplotype using a 2x2 table of the presence and absence of an allele at each of two loci. In addition D forms the basis for several other measures. It is defined simply as (the observed - the expected) frequency of a haplotype. The normalized disequilibrium measure D' adjusts for the influence of varying allele frequencies on D. D' is useful in comparing the disequilibrium values of different haplotypes. Hedrick (1987) and Lewontin (1988) have commented on some of the technical and philosophical issues underlying disequilibrium measurement efforts to construct an allele frequency independent measure. Begovich et al. (1991) used D' and X^2 testing of D to present the significant associations among the *HLA* loci *DRB1* and *DPB1* in a sample of Caucasians.

If interest is in the amount of disequilibrium between two multiallelic loci, a method must be employed that combines deviations present in each of the (k x l) haplotypes, where k and l are the number of alleles at each of the loci. The measure W uses absolute frequencies

NATO ASI Series, Vol. H 59
Molecular Evolution of the
Major Histocompatibility Complex
Edited by J. Klein and D. Klein

Table 1. Methods for analyzing MHC disequilibrium

Method	Definition	Properties	Sample size (log N)	Use
1. Single haplotype disequilibrium.				
D	$x_{ij}-p_iq_j$ (observed-expected)	$\Sigma D_i=0$, $\Sigma=D_j=0$ (tested with 2X2 Tables)	2	2X2 tests of single haplotypes in D' methods 3 and 4
D'	D/Dmax Dmax=maximum D givenallele frequencies	normalizes effect of differing allele frequencies (range -1 to +1)	2	Identify haplotypes in high positive disequilibrium
2. Total two-locus disequilibrium, W	$\sqrt{\Sigma\Sigma \frac{(x_{ij}-p_iq_j)^2}{p_iq_j}}$	Sample size free $NxW^2 = X^2$ statistic		Measures overall disequilibrium between two loci; determine extent and strength of disequilibrium with changing recombination
3. Disequilibrium pattern analysis (DPA).	pattern of all two-locus haplotypes sharing a single allele	The constant pressure of recombination moving all haplotypes towards D=0 is dynamically resisted by selection	3	Identify haplotypes subject to selection in the last few hundred generations
4. Constrained disequilibrium	Delta=diff. between (two-locus) normalized disequilibrium and normalized constrained (three-locus) disequilibrium	The hallmark of selection is a high positive Delta value for the two unselected loci when the third locus is selected	3	Identify the particular allele in a haplotype under selection
5. Primal haplotype	when the recombination is ≤ the mutation rate, then original linkage relationships are maintained	recognized by complete association of two-locus haplotypes	2-3	reconstruct order of appearance of alleles on haplotypes

Sources: 1, Lewontin 1964; 2, Cohen 1988; 3, Thomson and Klitz 1987; 4, Robinson et al. 1991; 5, Klitz et al. 1991, Begovich et al. 1991.

to sum the squared deviations of each haplotype. Squaring W and introducing the overall sample size N as a product results in the X^2 statistic, convenient for significance testing. With several tightly linked multiallelic loci, as is common in the MHC, W allows the comparison of disequilibrium to map distance. Apropos to the unit of selection debate, it is possible to observe, by plotting the W values of locus pairs having different recombination rates, how far disequilibrium extends along a chromosome. Klitz et al. (1991) and Begovich et al. (1991) each use W to analyze the relationship of disequilibrium to recombination map distance in the HLA region.

If the deviations (i.e., D) are calculated for a row by column table of haplotype frequencies, it can be noted that any row or column must (algebraically) sum to zero. This fact is the basis for the method Disequilibrium Pattern Analysis (DPA), which detects past evolutionary events by examining the pattern of disequilibrium among haplotypes in the population. If selection acts to favor lone haplotype the disequilibrium of that haplotype increases and all haplotypes sharing just one of the alleles of the favored haplotype i.e. those in the same row or column) all decrease. The force of recombination acting as constant pressure to pull the whole array towards gametic equilibrium (D=O) is dynamically resisted by selection. The pattern of a 'selected" haplotype, then, consists of one haplotype with strong positive disequilibrium with all related haplotypes having close to identical D' values in the negative space. This method has been used for example, to demonstrate the selective origin of the Caucasian haplotype A1-B8 (Klitz and Thomson 1987; Klitz et al. 1991).

Strong allelic associations between loci often extend beyond two loci to create what have been called "extended haplotypes." Constrained disequilibrium analysis examines the relationship of two locus disequilibrium values in a three locus system (Robinson et al. 1991), which can determine if one of the three loci may have been under selection relative to the other two. Examination of the northern European *HLA* haplotype *A1-B8-DR3* in this fashion, suggests that, of the three alleles as currently defined, *B8* may have been favored.

Very tightly linked loci create a special opportunity to reconstruct the order of appearance of new alleles in a population. When the recombination rate is so low as to be of the same order of Magnitude or less than the mutation rate, the generation of new alleles at one locus can be followed on the haplotype of a second locus having an invariant allele. For example in the HLA class II haplotypes involving *DR4* and *DQA3*, the many subtypes of DR4 are exclusively associated with the apparently ancient *DQA3* allele. This suggests that DR4 diversification occurred on the *DQA3* haplotype (Begovich et al. 1991).

Acknowledgments. This work was supported by NIH grant GM35326.

References

Begovich, A.B., McClure, G.R., Suraj, V., Helmuth, R.C., Feldes, N., Bagawan, T.L., Erlich, H.A., and Klitz, W.: Polymorphism, recombination and linkage disequilibrium within the HLA-class II region. *J Immunol*, submitted 1991

Cohen J.: *Statistical Power Analysis for the Behavioral Sciences*. 2nd. edn., pp. 215-271. L Erlbaum Associates, Hillsdale 1991

Hedrick, P.W.: Disequilibrium measures: proceed with caution. *Genetics 117*:331-341, 1987

Klitz, W. and Thomson, G.: Disequilibrium pattern analysis II. Application to Danish HLA A and B locus data. *Genetics 116*:633-643 1987

Klitz, W., Thomson, G., Borot, N., and Cambon-Thomsen, A.: Evolutionary and population perspectives of the human *HLA* complex. *Evol Biol*, in press 1991

Lewontin, R.C.: The interaction of selection and linkage I. General considerations: heterotic models. *Genetics 49*:49-67 1964

Lewontin, R.C.: On measures of gametic disequilibrium. *Genetics 120*:849-852, 1988

Robinson, W.P., Cambon-Thomsen, A., Borot, N., Klitz, W., and Thomson, G.: Selection, hitchhiking and disequilibrium analysis at three linked loci: application to the Provinces Francaises HLA data. *Genetics*, in press 1991

Thomson, G. and Klitz, W.: Disequilibrium pattern analysis I Theory. *Genetics 116*:623-632 1981

FROZEN HAPLOTYPES IN *Mhc* EVOLUTION

Jan Klein*+, Colm O'hUigin*, Masanori Kasahara+, Vladimir Vincek+ Dagmar Klein+, and Felipe Figueroa*

*Max-Planck-Institut für Biologie, Abteilung Immungenetik, 7400 Tübingen, Germany

oDepartment of Microbiology and Immunology, University of Miami School of Medicine, Miami, FL 33101, USA

An odd feature of the HLA-DRB *region is its remarkable haplotype polymorphism: At least five groups of haplotypes differing both in length and in number of genes exist in the human population at appreciable frequencies. We have demonstrated that similar* DRB *haplotype polymorphism exists also in the apes -- the chimpanzee and the gorilla. Moreover, certain haplotype segments have apparently remained intact since the human-chimpanzee (gorilla) divergence more than five million years ago. Hence, not only allelic but also haplotype polymorphisms evolve trans-specifically at the* DRB *region. The evolutionary origins of the individual haplotype groups and the possible reasons for the extreme linkage disequilibrium between certain* DRB *genes (the frozen haplotypes) are discussed.*

What is a haplotype?

The term *haplotype* was introduced by Cepellini and his associates in 1967 to describe "the combination of individual antigenic determinants that are positively controlled by an allele" (Cepellini et al. 1967, p. 165) and thus to replace the older term *phenogroup* (Stormont 1955). There are, however, only a few people today who still recognize haplotype by this definition and even fewer who realize that the meaning of the word changed. Since scientists generally do not read papers older than five years and historians of science are not interested in anything younger than 50 years, everybody now happily agrees that a haplotype is "a combination of alleles from closely linked loci (usually having functional affinity, such as *HLA* loci) found in a single chromosome" (Bodmer and Cavalli-Sforza 1976, p. 759).

The simplest haplotype consists of two adjacent loci (*1* and *2*) each having two alleles (1^a, 1^b, 2^a, and 2^b). In the double heterozygote $1^a2^a/1^b2^b$, the combination 1^a2^a is the *A* haplotype and the combination 1^b2^b is the *B* haplotype. At the other extreme, the most complex haplotype would be one encompassing all the loci of one chromosome. In the latter case, however, the requirement for "functional affinity" would have to be waived, as has in fact been done by those authors coining the term *extended haplotype* in which they sometimes group together closely linked but functionally unrelated genes (see, for example, Awdeh et al. 1983).

NATO ASI Series, Vol. H 59
Molecular Evolution of the
Major Histocompatibility Complex
Edited by J. Klein and D. Klein

What is a frozen haplotype?

Let us assume that the two adjacent loci in our hypothetical simplest haplotype are ten kilobase pairs (kb) apart and that when A ($1^{a}2^{a}$) and B ($1^{b}2^{b}$) arose, they were the only two haplotypes in existence for this particular pair of loci. If, as geneticists ever since Morgan believe, every site along a chromosome has an equal probability of being broken and reunited by the process of crossing-over at meiosis, then the two loci should eventually recombine and produce two new haplotypes C ($1^{a}2^{b}$) and D ($1^{b}2^{a}$). In humans, the frequency of recombination is approximately one percent per one million base pairs (bp), and so the A and B haplotypes should recombine with a frequency of one per 10,000 progeny. Of course, the recombinants need not survive and to determine the chances of finding them in the population, we need to take gene frequencies into account. Let us assume, for simplicity's sake, that the four alleles occur with equal frequencies of 0.5 ($P_{1a} = P_{1b} = P_{2a} = P_{2b} = 0.5$). According to theory, following the creation of the haplotypes, their frequencies should tend toward an equilibrium at which $P_{1a2a} = P_{1a}P_{2a}$, $P_{1a}P_{2b} = P_{1a}P_{2b}$, $P_{1b2a} = P_{1b}P_{2a}$, and $P_{1b2b} = P_{1b}P_{2b}$. In the transition phase, in which the frequencies have not yet reached equilibrium, the following equations should be valid: $P_{1a2a} = P_{1a}P_{2a} + D$, $P_{1a2b} = P_{1a}P_{2b} - D$, $P_{1b}P_{2a} = P_{1b}P_{2a} - D$, and $P_{1b2b} = P_{1b}\,P_{2b} + D$, where D is the measure of linkage disequilibrium, the departure from equilibrium (Hartl and Clark 1989). The D decreases by the factor of $1\text{-}r$ in each generation, where r is the recombination frequency, or by the factor of 10 in n generations $[(1\text{-}r)^{n} = 1/10]$.

Under the conditions we have set for our hypothetical haplotypes, the initial value of D is 0.25 which is the maximum it can reach. Since the two loci are 10 kb apart and recombination between them is expected to occur with a frequency of 0.01%, it should take 2.3×10^{4} generations to reduce D from 0.25 to 0.025. If we take the human generation time as some 25 years (Deevey 1960), we could expect that in 575,000 years -- that is within the tenure of the species *Homo sapiens* -- a substantial progress toward equilibrium of the haplotypes A, B, C, and D will occur. This progress could be counteracted by selection preferring certain gene combinations, by migration, or by other factors, which could either slow it down or even arrest it at some intermediate frequency; after 575,000 years, however, recombinant haplotypes should be found in the population at appreciable frequencies. If recombinants are *not* found, the original haplotypes can be regarded as evolutionarily "frozen" in their ancestral configuration. Here we examine whether such frozen haplotypes occur and if they do, why they remain frozen.

The *HLA-DR* haplotypes

The class II (*D*) region of the human major histocompatibility complex (*Mhc*) -- the *HLA* complex -- is at least 1100 kb long and consists of three main subregions, *DP, DQ,* and *DR*, arranged in this order in the direction from the centromere to the telomere (Klein 1986; Dausset and Pla 1989). Each subregion contains two kinds of loci, *A* and *B*, coding for the α and β polypeptides of the class II heterodimers, respectively. The *DQ* subregion is approximately 300 kb long and as far as is known, of the same length and organization in all individuals (Blanck and Strominger 1988; Dunham et al. 1989). It consists of two *DQA* and two *DQB* loci arranged in the order *DQB2...DQA2...DQB1...DQA1*, whereby the last locus is 50 kb from the first locus (*DRB1*) of the *DR* subregion. The *DQB2, DQA2* loci do not appear to be functional; the *DQB1, DQA1* loci, on the other hand, are both polymorphic (17 and 13 alleles have thus far been recognized at the two loci, respectively; WHO Committee 1991) and presumably capable of presenting peptides to helper T lymphocytes. The *DR* subregion varies in length from 60 kb to about 260 kb, and in the number of *DRB* genes from one to four, depending on the haplotype (Dunham et al. 1987; Tokunaga et al. 1988, 1989; Niven et al. 1990; Kendall et al. 1991); all *DR* haplotypes contain one nonpolymorphic *DRA* gene. In those *DR* haplotypes in which there are more than two *DRB* genes, only two *DRB* genes are expressed on the cell surface; the other one or two are pseudogenes. To date, 47 alleles have been identified at the *DRB1* locus and nine at the second expressed locus (*DRB3, DRB4*, or *DRB5*; WHO Committee 1991). Here we shall first consider briefly the chromosomal segment (haplotype) extending from *DQB1* to the most telomeric of the *DRB* genes, and then focus exclusively on the segment occupied by the *DRB* loci.

At equilibrium, one would expect the human population to contain 17 x 13 x 47 x 9 = 93,483 *DQ-DR* haplotypes and 47 x 9 = 423 *DR* haplotypes. The number actually found in a population is far below this expectation (Table 1). In Caucasians, the population studied most extensively, there are only 53 *DQ-DR* haplotypes and 48 *DR* haplotypes. Although all of the existing haplotypes may not have been identified, the number of unidentified haplotypes is probably not very large. Clearly there is a tremendous deficit of observed haplotypes in comparison to the numbers expected if all the *DQ* and *DR* genes were in equilibrium. This linkage disequilibrium encompasses both subregions, *DQ* and *DR*, so that there are preferred associations between certain *DQB1, DQA1*, and *DRB* alleles.

Turning our attention to *DR*, we shall divide the 48 known haplotypes into five groups -- *DR2, DR1, DRw52, DRw8*, and *DRw53* -- each characterized by a particular organization of its *DRB* genes (Table 1; Fig. 1). Each group contains one or more major *DR* haplotypes which can be subclassified further into sets of minor haplotypes. Generally, major haplotypes within a group are more related to one another than they are to haplotypes of another group; and minor haplotypes of a given major haplotype are more related to one another than they are to any of the minor ones of another major haplotype.

Table 1. The known *DQ-DR* haplotypes.

Group	Major haplo-types	Minor haplotypes				
DR1	*DRw10*	*DQw5(w1)*	*DQB1*0501*	*DQA1*0301*	*DRB1*1001*	
	DR1	*DQw5(w1)*	*DQB1*0501*	*DQA1*0101*	*DRB1*0101*	
DR2	*DR2*	*DQw6(w1)*	*DQB1*0602*	*DQA1*0102*	*DRB1*1501*	*DRB5*0101*
	DR2	*DQw6(w1)*	*DQB1*0601*	*DQA1*0103*	*DRB1*1502*	*DRB5*0102*
		DQw5(w1)	*DQB1*0502*	*DQA1*0102*	*DRB1*1601*	*DRB5*0201*
		DQw5(w1)	*DQB1*0502*	*DQA1*0102*	*DRB1*1601*	*DRB5*0202*
		DQw7(w3)	*DQB1*0301*	*DQA1*0501*	*DRB1*1602*	*DRB5*0202*
DRw52	*DR3*	*DQw2*	*DQB1*0201*	*DQA1*0501*	*DRB1*0301*	*DRB3*0101*
		DQw2	*DQB1*0201*	*DQA1*0501*	*DRB1*0301*	*DRB3*0201*
		DQw4	*DQB1*0401*	*DQA1*0401*	*DRB1*0302*	*DRB3*0101*
	DR5	*DQw7(w3)*	*DQB1*0301*	*DQA1*0501*	*DRB1*1101*	*DRB3*0201*
		DQw7(w3)	*DQB1*0301*	*DQA1*0501*	*DRB1*1102*	*DRB3*0201*
		DQw7(w3)	*DQB1*0301*	*DQA1*0501*	*DRB1*1103*	*DRB3*0201*
		DQw5(w1)	*DQB1*0502*	*DQA1*0102*	*DRB1*1104*	*DRB3*0201*
		DQw7(w3)	*DQB1*0301*	*DQA1*0501*	*DRB1*1201*	*DRB3*0201*
	DRw6	*DQw6(w1)*	*DQB1*0603*	*DQA1*0103*	*DRB1*1301*	*DRB3*0101*
		DQw6(w1)	*DQB1*0603*	*DQA1*0103*	*DRB1*1301*	*DRB3*0201*
		DQw6(w1)	*DQB1*0604*	*DQA1*0102*	*DRB1*1302*	*DRB3*0301*
		DQw5(w1)	*DQB1*0503*	*DQA1*0104*	*DRB1*1401*	*DRB3*0201*
		DQw7(w3)	*DQB1*0301*	*DQA1*0501*	*DRB1*1402*	*DRB3*0101*
DRw8	*DRw8*	*DQw4*	*DQB1*0401*	*DQA1*0401*	*DRB1*0801*	
		DQw4	*DQB1*0401*	*DQA1*0401*	*DRB1*0802*	
		DQw7(w3)	*DQB1*0301*	*DQA1*0601*	*DRB1*0803*	
		DQw6(w1)	*DQB1*0601*	*DQA1*0103*	*DRB1*0803*	
DRw53	*DR4*	*DQw8(w3)*	*DQB1*0302*	*DQA1*0301*	*DRB1*0401*	*DRB4*0101*
		DQw8(w3)	*DQB1*0302*	*DQA1*0301*	*DRB1*0402*	*DRB4*0101*
		DQw8(w3)	*DQB1*0302*	*DQA1*0301*	*DRB1*0404*	*DRB4*0101*
		DQw8(w3)	*DQB1*0302*	*DQA1*0301*	*DRB1*0404*	*DRB4*0101*
		DQw2	*DQB1*0202*	*DQA1*0301*	*DRB1*0405*	*DRB4*0101*
		DQw5(w1)	*DQB1*0501*	*DQA1*0301*	*DRB1*0405*	*DRB4*0101*
		DQw7(w3)	*DQB1*0301*	*DQA1*0301*	*DRB1*0401*	*DRB4*0101*
		DQw7(w3)	*DQB1*0301*	*DQA1*0301*	*DRB1*0403*	*DRB4*0101*
	DR7	*DQw2*	*DQB1*0201*	*DQA1*0201*	*DRB1*0701*	*DRB4*0101*
		DQw9(w3)	*DQB1*0303*	*DQA1*0201*	*DRB1*0701*	*DRB4*0101*
	DR9	*DQw9(w3)*	*DQB1*0303*	*DQA1*0301*	*DRB1*0901*	*DRB4*0101*

Not included in the Table are haplotypes carrying recently discovered *DRB1* alleles **0103*, **1202*, **1303*, **1403*, **1404*, **1405*, **0406*, **0407*, **0408*, **0409*, **0410*, **0411*, **0702*, **09012*, **08022*, **08032*, and **0804* (WHO Committee 1991).

The *DR2 group* consists of only one major haplotype, *DR2*, whose organization has been elucidated by Kawai and his co-workers (1989). The *DR2* haplotype contains at least three *DRB* genes (*DRB1, DRBVI,* and *DRB5*, arranged in this order) and one *DRA* gene. The *DRBVI* is a pseudogene, which seems to be transcribed (Corell et al. 1991) but not expressed on the cell surface. Both the *DRB1* and *DRB5* genes appear to be active (Kawai et al. 1989) and moderately polymorphic (Marsh and Bodmer 1991). The *DR2* haplotype was originally recognized as coding for a distinct serologically detectable antigen DRw2 (Bodmer et al. 1977) which was later split into two separate determinants (subtypes), DRw15 and DRw16 (Batchelor et al. 1989). These are believed to be encoded in different alleles at the *DRB1* locus -- *DRB1*15* and *DRB1*16* (WHO Committee 1991). Minor variants of these two subtypes were then recognized by DNA sequencing (Marsh and Bodmer 1991). Corresponding to the *DRB1* variants are *DRB5* variants which have been differentiated by DNA sequencing (Marsh and Bodmer 1991).

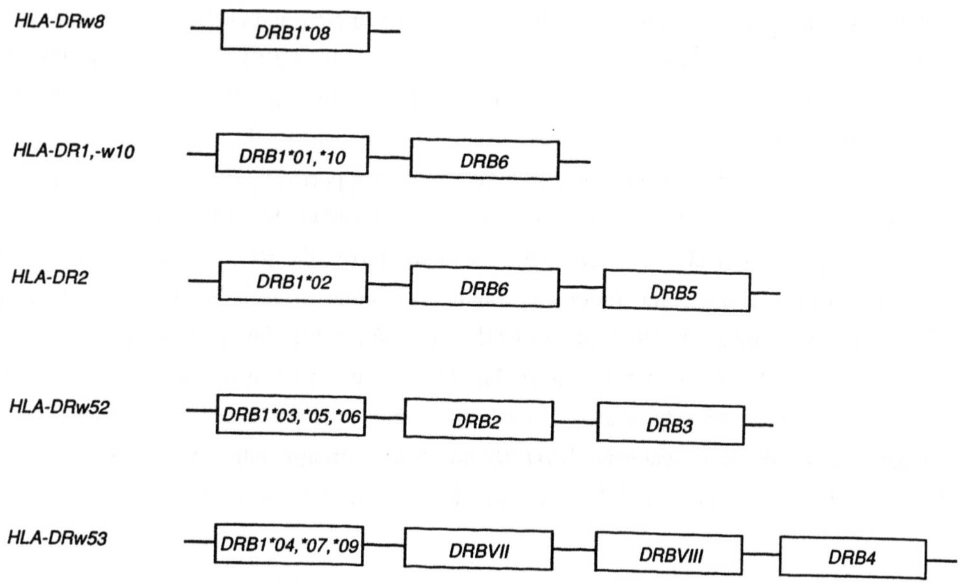

Fig. 1. The gene organization of the *HLA-DRB* haplotype groups.

The *DR1 group* consists of two haplotypes, *DR1* and *DRw10*. The gene organization of these two haplotypes is not known, but restriction fragment length polymorphism (RFLP) analysis suggests that they are the simplest of all the *DR* haplotypes, with the exception of *DRw8* (Bidwell 1988). Particularly informative are experiments in which the DNA was digested with the *Taq* I restriction enzyme and the Southern blots were hybridized with exon-specific *DRB* probes. The number of fragments produced by this method correlated well with the number of genes in those *DR* haplotypes for which the organization was known. The digestion of *DR1* and *DRw10* DNA produced, in each case, two fragments, one of which was shared between the two haplotypes. It is therefore assumed that each of the two haplotypes contains two genes. One of these genes must code for the serologically defined antigens DR1 in the *DR1* haplotype and DRw10 in the *DRw10* haplotype. These genes are designated *DRB1*01* and *DRB1*10*, respectively (WHO Committee 1991). The second gene in both haplotypes has recently been identified as corresponding to the *DRBVI* pseudogene of the *DR2* haplotype (Corell et al. 1991). The exon 2 sequences of the *DR1* and *DR2* haplotype pseudogenes differ at nine positions out of 200 compared. Assuming a substitution rate of 5 x 10^{-9} per site per year for *DRB* pseudogenes (O'hUigin and Klein 1991), it can be estimated that the *DRBVI* pseudogenes of *DR1* and *DR2* diverged more than 4.5 million years (my) ago. The limited sequence of the *DRBVI* gene in the *DRw10* haplotype is indistinguishable from that of the *DRBVI* sequence in *DR1* (Erlich et al. 1989).

The sharing of a closely related pseudogene between *DR1* and *DR2* haplotypes suggests that the two are related and this suggestion is supported by other data as well. First, in an RFLP analysis the two haplotypes cluster together but separately from other haplotypes (Simons et al. 1989). In spite of this similarity, the RFLP patterns of *DR1* and *DR2* are clearly distinct, and there are fragments unique to each haplotype. Second, in Caucasian populations, the *DRB* genes of *DR1* and *DR2* are in linkage disequilibrium with *DQB1*05* and *DQA1*01* genes; Table 1). Third, the *DRB1* genes of *DR1* and *DR2* haplotypes are more closely related to each other than they are to any other *DRB1* gene. Although in dendrogram analysis *DRB1*01* sometimes groups with other genes (Fig. 2), this clustering is due to a shared three-nucleotide substitution at the beginning of exon 3.

Fig. 2. Dendrogram depicting the relationships between known *DRB* alleles of humans (*HLA-*), chimpanzee (*Patr-*), gorilla (*Gogo-*), orang-utan (*Popy-*), pig-tail macaque (*Mane*), and cotton-top tamarin (*Saoe*). The tree was constructed by the neighbor-joining method of Saitou and Nei (1987) from genetic distances calculated by Kimura's two-parameter method (Kimura 1980). The human sequences are taken from Marsh and Bodmer (1991); the references for the remaining sequences are as follows: *Gogo*, Kasahara et al. (1990, 1991), D. Klein et al. (1991); *Patr*, Fan et al. (1989), Vincek et al. (1991), Figueroa et al. (1991); *Popy*, C. Schönbach and J. Klein, unpublished data; *Mane*, Zhu et al. (1991); *Saoe*, Grahovac et al. (1991). The coding sequences specifying the mature polypeptide chain were used. Numbers on branches indicate genetic distances.

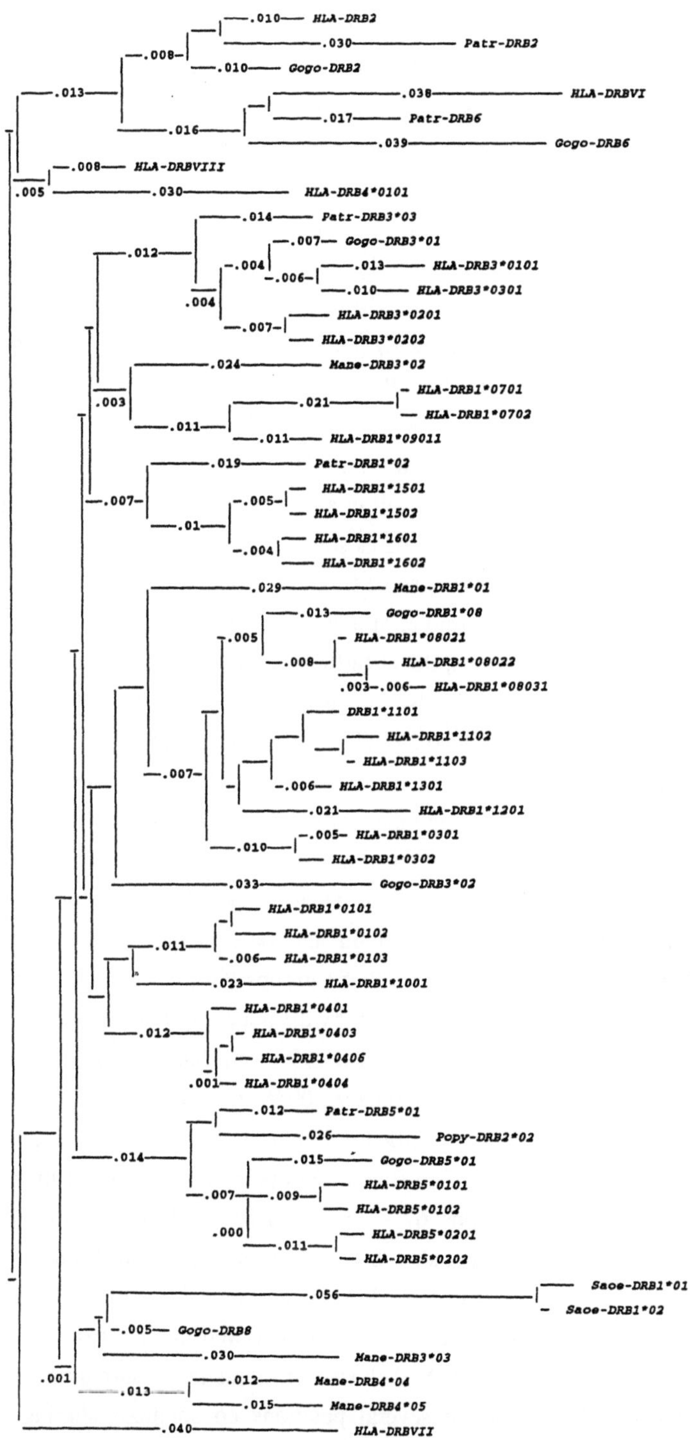
.010 HLA-DRB2
.030 Patr-DRB2
.008
.010 Gogo-DRB2
.013
.038 HLA-DRBVI
.017 Patr-DRB6
.016
.039 Gogo-DRB6
.008 HLA-DRBVIII
.005 .030 HLA-DRB4*0101
.014 Patr-DRB3*03
.007 Gogo-DRB3*01
.012
.004
.013 HLA-DRB3*0101
.006
.010 HLA-DRB3*0301
.004
HLA-DRB3*0201
.007
HLA-DRB3*0202
.024 Mane-DRB3*02
HLA-DRB1*0701
.003
.021
HLA-DRB1*0702
.011
.011 HLA-DRB1*09011
.019 Patr-DRB1*02
HLA-DRB1*1501
.005
HLA-DRB1*1502
.007
.01
HLA-DRB1*1601
.004
HLA-DRB1*1602
.029 Mane-DRB1*01
.013 Gogo-DRB1*08
.005
HLA-DRB1*08021
.008
HLA-DRB1*08022
.003 .006 HLA-DRB1*08031
DRB1*1101
HLA-DRB1*1102
HLA-DRB1*1103
.007
.006 HLA-DRB1*1301
.021 HLA-DRB1*1201
.005 HLA-DRB1*0301
.010
HLA-DRB1*0302
.033 Gogo-DRB3*02
HLA-DRB1*0101
.011
HLA-DRB1*0102
.006 HLA-DRB1*0103
.023 HLA-DRB1*1001
HLA-DRB1*0401
.012
HLA-DRB1*0403
HLA-DRB1*0406
.001 HLA-DRB1*0404
.012 Patr-DRB5*01
.026 Popy-DRB2*02
.014
.015 Gogo-DRB5*01
.007
.009
HLA-DRB5*0101
HLA-DRB5*0102
.000
HLA-DRB5*0201
.011
HLA-DRB5*0202
.056
Saoe-DRB1*01
Saoe-DRB1*02
.005 Gogo-DRB8
.030 Mane-DRB3*03
.001
.012 Mane-DRB4*04
.013
.015 Mane-DRB4*05
.040 HLA-DRBVII

When these three positions are excluded, the *DRB1* genes of *DR1* and *DR2* group together and an exon by exon analysis confirms this conclusion. Sharing sites with other *DRB* genes might reflect distant common ancestry. Assuming that synonymous sites evolve at the same rate as sites in pseudogenes, the divergence time of the *DRB1* genes in *DR1* and *DR2* haplotypes can be estimated as approximately 8.7 my. We can conclude, therefore, that the *DR1* and *DR2* haplotypes are related to each other; the only reason why they are not regarded as belonging to the same group is that they differ in gene organization, which is the primary criterion for haplotype grouping. However, it is very easy to imaginge how *DR1* arose from *DR2* (or the other way around), either by duplication of *DRB1* to give rise to *DRB5* in *DR2* or by deletion of *DRB5* in *DR1*.

A presumed similar organization and the sharing of the *DRBVI* gene groups the *DRw10* haplotype with *DR1* (and hence also with *DR2*). This conclusion is further supported by RFLP analysis (Lee et al. 1988; Bidwell et al. 1988; Riisome et al. 1988; Semana et al. 1988). The analysis shows that *DRw10* shares some fragments with *DR1* which are absent in other haplotypes, but that it also shares some fragments with the *DR2* and the *DRw52* group, which are absent in *DR1*; furthermore, it possesses fragments not found in any other haplotype. In Caucasian populations, the *DRw10* genes are found exclusively in association with the *DQB1*0501* allele which is also in linkage disequilibrium with the *DRB* genes of *DR1* and *DR2*. The *DRw10*-associated *DQA1*0301* allele is, however, otherwise in linkage disequilibrium with the *DRw53* group of haplotypes. The relationship of *DRB1*10* to *DRB1*01* is also not without ambiguities; however, the *DRB1*10* gene specifies the DRw10 antigen which has proved to be difficult to type serologically. Antisera monospecific for DRw10 are rare and no DRw10-specific monoclonal antibodies are available. Most anti-DRw10 reagents contain antibodies which crossreact with antigens specified by haplotypes of the *DR1* and *DRw53* groups (Grumet and Ness, 1980; Hurley et al. 1984). The problem is compounded further by the generally low frequencies of *DRw10* in most populations. The crossreactivity with *DRw53* group haplotypes can be explained by sharing of epitopes. The monoclonal antibody 109d6, for example, recognizes an epitope which is expressed on the ß chains specified not only by the *DRB1*10* gene but also by the *DRB1*09* and *DRB4*0101* genes. The shared epitope probably resides in the LLERRRA stretch of amino acids at positions 67 - 73 (Toguchi et al. 1984; Hurley et al. 1984; Sorrentino et al. 1985; Matsuyama et al. 1986, 1988; Merryman et al. 1988).

Analysis of the *DRB1*10* nucleotide sequence reveals its exon 1 to be identical to that of the *DRw52* group of haplotypes (Gorski 1989a) and distinct from exon 1 sequences of other haplotypes. Exon 2 of *DRB1*10* clusters in sequence dendrograms with that of the *DRw53* group (Fig. 2) although at several positions conspicuous sharing of nucleotide substitutions with *DRB1* of *DR1* occurs. The rest of the gene appears approximately equidistant to *DRB1*01* and the *DRB1* genes of the *DRw52* group. In the dendrogram constructed from the entire coding sequence, however, *DRB1*10* clearly clusters with *DRB1*01*. We conclude, therefore, that the *DRB1*10* gene is closest to *DRB1*01* (and hence also to the *DRB1* genes of the *DR2* haplotype), but that the distance between the two genes is considerable so that the affinity is difficult to recognize. The *DRB1*10* gene also appears to have some affinity with the *DRB1* and *DRB3* genes of the *DRw52* group of haplotypes, probably reflecting distant ancestral relationships. All in all, it seems justified to place *DRw10* together with *DR1* in a single group which in turn is related to the *DR2* group.

Haplotypes in the *DRw52 group* each contain three genes, *DRB1*, *DRB2*, and *DRB3* (Rollini et al. 1985; Figueroa et al. 1991). The first and last of these are expressed on the cell surface, whereas *DRB2* is a pseudogene. The product of the *DRB1* gene carries antigenic determinants which are recognized by both antibodies (DR determinants) and T lymphocytes (Dw determinants) and which differentiate the major (and also some of the minor) haplotypes belonging to this group. The major haplotypes, *DR3, DR5*, and *DRw6*, specify antigens of the same denomination. The DR5 antigen is split into two subtypes, DRw11 and DRw12, as is the DRw6 antigen (DRw13 and DRw14). Hence there are five main alleles at the *DRB1* locus of the DRw52 group, *DRB1*03, *11, *12, *13,* and **14*. The product of the *DRB3* gene carries a supertypic antigen, DRw52, which is shared by all the haplotypes in this group. Three main alleles are distinguished at the *DRB3* locus by sequencing, *DRB3*01, *02,* and **03*. In human populations, the *DRB1*03* alleles have been found in association with the *DRB3*01* and **02* alleles, but not with the *DRB3*03* alleles; the *DRB1*11* and **12* alleles (*DR5* haplotypes) have been found only in association with the *DRB3*02* alleles; the *DRB1*13* alleles occur in linkage with *DRB3*01, *02*, or **03* alleles; and the *DRB1*14* alleles are linked to either *DRB3*01* or **02* (see Table 1).

In addition to their linkage to *DRB3* genes encoding the supertypic DRw52 antigens and to their sharing the same organization and the same *DRB2* pseudogene, the *DR3, DR5*, and *DRw6* haplotypes are also united by several other features. First, the *DRB1*03, DRB1*11, DRB1*12, DRB1*13*, and *DRB1*14* genes, are related to one another more closely than they are to *DRB1* genes outside the *DRw52* group. This kinship is most conspicuous in the relatively conserved first and third exons (Gorski 1989a). In the first exon, all the *DRB1* genes of the *DRw52* group have an identical sequence which is distinguished from other sequences by several group-specific substitutions. Similar group-specific substitutions exist also in the third exon. The similarity extends, however, also to the variable exon 2: in dendrograms constructed from the exon 2 (first domain) sequences as well as from the entire coding sequences, the *DRB1* alleles of the *DRw52* group always cluster together, separately from other alleles (Fig. 2). Second, the differences between the *DRB3* alleles are relatively minor, indicating close relationships among them. Here again, exon 1 sequences of all the *DRB3* alleles are identical yet different from other exon 1 sequences (Gorski 1989a). Exon 3 sequences are characterized by group-specific substitutions, and exon 2 (as well as the entire coding) sequences cluster together in dendrograms. As pointed out by Gorski (1989a), the *DRB3*03* allele has a segment in exon 3 in which it differs from the other *DRB3* alleles but which it shares with the *DRB1*03* alleles. Such segmental sharing can be explained by recombination, by complex mutation, or by common ancestry (Klein and Schönbach 1991). The *DRB3* and the *DRw52*-group *DRB1* genes indeed seem to be derived from a relatively recent common ancestor (Rollini et al. 1985). Third, the *DRB* genes of the *DRw52* group are in linkage disequilibrium with the *DQA1*05* and *DQA1*01* genes, and to a lesser degree with the *DQB1*03* genes (Table 1). Fourth, in RFLP analysis, all the haplotypes of the DRw52 group cluster together (Simons et al. 1989).

All these similarities indicate that the *DRw52* haplotypes are derived from a common ancestral haplotype with the basic structure *DRB1...DRB2...DRB3*. In this haplotype, the *DRB1* and *DRB3* genes apparently arose by duplication from a common ancestral gene (Rollini et al. 1985). The ancestral haplotype then gave rise to three founder haplotypes from which the modern *DR3, DR5*, and *DRw6* haplotypes were produced by diversification of their *DRB1* genes. This diversification occurred primarily through mutations (Klein and Schönbach 1991). In the *DRw52* urhaplotype, the *DRB* loci were probably linked to the *DQA1*05* and *DQB1*03* genes, but recombination with other *DQ* genes later produced new *DQ-DR* combinations.

The *DRw8 group* is represented by a single major *DRw8* haplotype which apparently contains only one *DRB* gene (Böhme et al. 1985). According to several criteria, the haplotype is closely related to the *DRw52* group. First, the *DRB* gene of the *DRw8* haplotype codes for the supertypic antigen DRw52 characterizing the *DRw52* group. Second, in RFLP analyses, the *DRw8* haplotype clusters with the *DRw52*-group haplotypes (Simons et al. 1989). Third, in dendrograms based on nucleotide sequence, the *DRB1*08* gene (except for its most 3' part) clusters with the *DRB1* genes of the *DRw52*-group haplotypes (Fig. 2); its 3' part shows affinity to the *DRB3* gene. Fourth, the *DRB1*08* exon 3 sequences are characterized by the *DRw52*-group specific substitutions (Gorski 1989a). All these features have led Gorski (1989b) to propose that the *DRw8* haplotype arose from an ancestral *DRw52*-group haplotype by fusion of the *DRB1* and *DRB3* genes with the concomitant deletion of the entire inbetween sequence, including the *DRB2* pseudogene. As we shall discuss later, strong support for this explanation is now provided from studies of *DRB* haplotypes in the gorilla. Hence, the *DRw8* haplotype can be regarded as a member of the *DRw52* group, even though it has a different gene organization.

The *DRw53 group*, which consists of major haplotypes *DR4, DR7*, and *DR9*, is characterized by four *DRB* genes, the *DRB1* and *DRB4* expressed genes, and the *DRBVII, DRBVIII* pseudogenes (Spies et al. 1985; Andersson et al. 1987; Fig. 1). The *DRB1* gene codes for the serologically detectable antigens DR4, DR7, and DR9 (alleles *DRB1*04*, *1*07*, and *1*09*, respectively). Additional variation at the *DRB1* locus is identified by T-cell typing and two-dimensional gel electrophoresis of the protein products, as well as RFLP analysis and sequencing of the genes themselves (Dupont 1989; Marsh and Bodmer 1991). The *DRB4* locus is monomorphic; it is occupied by only one allele identified thus far, *DRB1*0401*, which codes for the supertypic antigen DRw53. The sharing of the *DRB1*0401* allele (and of the DRw53 antigen) by the *DR4, DR7*, and *DR9* haplotypes is one reason for their assignment into a single group, but there are others. First, in RFLP analysis, the *DR7* and *DR9* haplotypes cluster together and the two of them cluster with *DR4*, separately from all other *DR* haplotypes (Simons et al. 1989). Second, by pulsed field gel electrophoresis (PFGE) analysis, the *DR4*, *DR7*, and *DR9* haplotypes have the same length (which is different from that of the other haplotypes) and similar (though not identical) distribution of sites for rare cutting restriction enzymes (Dunham et al. 1987; Tokunaga et al. 1988, 1989; Niven et al. 1990; Kendall et al. 1991). Third, sequence analysis of the *DRB1*04, DRB1*07*, and *DRB1*09* alleles reveals them to be related to one another more closely than to other *DRB1* alleles (Fig. 2). Although there are only a few group-specific substitutions (mainly in exons 1 and 3, see Gorski 1989a), the overall intra-group similarity is significantly higher than the intergroup similarities. In dendrograms, however, especially those based on exon 2 sequences, the *DRB1*04* alleles often cluster with other genes, separately from *DRB1*07* and *DRB1*09*. Fourth, the *DRB* genes of the *DRw53* group are in linkage disequilibrium with *DQA1*03* (or **02*) and *DQB1*03* (or **02*) alleles. All these data indicate that *DR4, DR7*, and *DR9* are derived from a common ancestral haplotype which had the four-locus structure of modern *DRw53*-group haplotypes. The diversification occurred by mutational differentiation of the *DRB1* genes, with the *DRB1*04* gene diverging first, and the *DRB1*07* and *DRB1*09* diverging later.

Taken together, all these data provide reasons for the clustering of *HLA-DR* haplotypes into three major groups, *DR2, DRw52*, and *DRw53*. Since many of the haplotypes described above are found in different racial groups, they have apparently existed as separate entities and without major rearrangements for at least 50,000 years, which is the current minimum estimate for the separation of human races (Nei 1985). According to the calculations described earlier, haplotypes this old should by now have shown signs of progress toward linkage disequilibrium. No such progress, however, has been observed. Is it then possible that the haplotypes are actually *frozen* and hence do not recombine at all or do so extremely rarely? To answer this question, we need to go back more than 50,000 years and examine the *DR* haplotypes in our nearest relatives, the gorilla and the chimpanzee.

DRB haplotypes in apes

It is now firmly established that much of the allelic *Mhc* polymorphism predates speciation (Klein 1980; Arden and Klein 1982; Figueroa et al. 1988; Mayer et al. 1988; McConnell et al. 1988; Lawlor et al. 1988; Fan et al. 1989; Gyllensten and Erlich 1989; Gyllensten et al. 1990). In primates, virtually all alleles found thus far in apes have their counterparts in humans, and this is true for both class I (Mayer et al. 1988; Lawlor et al. 1988) and class II (Fan et al. 1989; Gyllensten and Erlich 1989; Gyllensten et al. 1990) alleles. Examples of this trans-species evolution of *Mhc* polymorphism are given in Figure 2, in which *DRB* alleles of chimpanzees, gorillas, and an orang-utan are seen clustering with those of humans. Clustering according to zoological species, as would have been expected had the *DRB* polymorphisms evolved in each species separately, does not occur. Trans-species evolution of *Mhc* polymorphism can be observed by comparing either entire genes or individual exons, indicating that it occurs not only at positions that are under positive selection pressure, but also at other positions which are either selectively neutral or are subject to negative selection (Klein et al. 1990). In view of the *DRB* linkage conservation in humans and the trans-specific character of *DRB* polymorphism, one could consider the possibility that some of the linkage relationships observed between human *DRB* alleles may have been established prior to the emergence of *Homo sapiens* or even prior to the emergence of hominids. We have therefore started a project aimed at elucidating the organization of some of the ape *DRB* haplotypes. Thus far we have succeeded in determining the organization of three ape haplotypes (Fig. 3), one from a chimpanzee (*Hugo*; Brändle et al. 1991) and two from a gorilla (*Sylvia 1* and *Sylvia 2*; Kasahara et al. 1990, 1991).

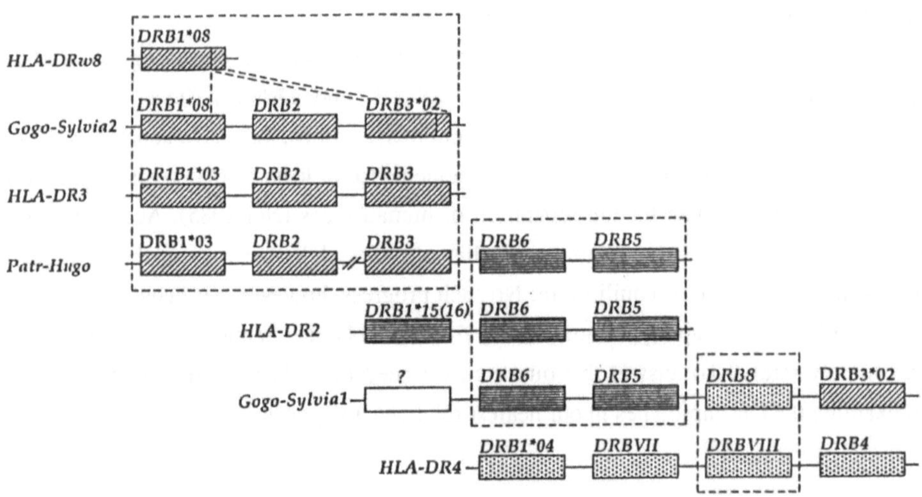

Fig. 3. The comparison of human (*HLA*), gorilla (*Gogo*), and chimpanzee (*Patr*) haplotypes. Homologous regions are indicated by broken-line rectangles. The presumed origin of *HLA-DRw8* from a haplotype similar to *Gogo-Sylvia 2* is also shown. For references, see text.

Several conclusions can be drawn from these results. First, none of the three ape haplotypes corresponds, in its entirety, to any of the known *HLA-DRB* haplotypes. At least some of the ape *DRB* haplotypes contain more genes than the longest human haplotype (i.e., *Sylvia 1* contains at least five *DRB* genes; the same is true for *Hugo*). A high number of *DRB* genes in chimpanzee haplotypes is also indicated by RFLP analysis of genomic DNA obtained from animals maintained at the ITRI-TNO Primate Center at Rijswijk, The Netherlands (Bontrop et al. 1990). It appears, therefore, that a considerable simplification of *DRB* haplotypes occurred during the evolution of the hominid lineage. Of course, the long length of ape *DRB* haplotypes makes the task of defining them more difficult than it would have been otherwise.

The second conclusion that can be drawn from the ape *DRB*-haplotype studies is that in spite of the differences between human and ape haplotypes, some of the allelic combinations appear to be the same. Thus, both *Hugo* and *Sylvia 1* contain apparent counterparts of the human *DRBVI* and *DRB5* genes in the same combination and orientation as they appear in the *HLA-DR2* haplotypes. The *Hugo* haplotype also contains the combination *DRB1*03...DRB2...DRB3*, which corresponds to the human haplotypes of the *DRw52* group. The *Hugo* haplotype in fact combines the entire *HLA-DR3*-like haplotype with most of the *HLA-DR2*-like haplotype on a single chromosome .

The third conclusion from these studies is that ancestral forms of some human *DRB* haplotypes can still be found among the apes. As mentioned earlier, the *HLA-DRw8* haplotype is believed to have originated by deletion and fusion from a *DR3*-like haplotype (Gorski 1989b). In fact, the *Sylvia 2* gorilla haplotype corresponds exactly to the postulated hypothetical *DR3*-like haplotype that may have given rise to *HLA-DRw8*. This finding has several implications, some of which will be discussed later.

The main conclusions from the ape *DRB* haplotype studies is, however, that some of the linkages between *DRB* genes have been preserved for over 5 - 10 my, which is the estimated time of separation of the hominid and ape lineages (Martin 1990). This is a much longer time than expected for the persistence of ordinary linkage disequilibrium and the haplotypes can therefore be regarded as "frozen" according to the definition given earlier. The preliminary data therefore support the notion (Klein and Takahata 1990) that not only allelic but also haplotype *Mhc* polymorphism evolves trans-specifically.

Why are *DRB* haplotypes frozen?

The results discussed in the earlier sections indicate that there are three main groups of human *DRB* haplotypes (the *DR2*, *DRw52*, and *DRw53* groups) which show no evidence of intergroup gene exchange, and that in portions of the *DRB* chromosomes recombination may have been suppressed for more than 5 my. One possible reason for the suppression could be the different length and organization of the haplotypes in the different groups. Mapping with overlapping cosmids (Spies et al. 1985; Rollini et al. 1985; Andersson et al. 1987; Kawai et al. 1989) indicates that the *DR* haplotypes in the different groups contain a different number of genes (Fig. 1). Similarly, PFGE studies (Tokunaga et al. 1988, 1989; Dunham et al. 1989; Niven et al. 1990; Kendall et al. 1991) reveal that compared to *DRw52* haplotypes, the *DR2*-group haplotypes contain a 130 kb deletion at one position and a 30 kb insertion at another position, and that the *DRw53*-group haplotypes are at least 140 kb longer than the *DRw52*-group haplotypes. Hence when chromosomes bearing haplotypes from different groups come together at meiosis, they can align only by looping out portions of the *DRB* subregion in one of the chromosomes. The lack of proper alignment may hinder genetic exchanges between the homologous chromosomes and so keep the groups in the frozen state. If this explanation were correct, one would expect to find evidence for unhindered recombination within each group of *DRB* haplotypes. Analysis of the *DRw52*- and *DRw53*-group haplotypes suggests that this, indeed, might be the case.

In the *DRw52* group, the *DRB1*03* alleles occur in combination with either the *DRB3*01* or *DRB3*02* alleles; the *DRB1*11* and *DRB1*12* alleles occur in combination with the *DRB3*02* allele; and the *DRB1*13* and *DRB1*14* alleles occur in combination with either *DRB3*01, DRB3*02*, or *DRB3*03* (Table 1). (The *DRw8* haplotype is a special case which will be dealt with separately.) Although no crossing-over between *DRB* genes has ever been documented in family studies, the simplest explanation for the occurrence of the various *DRB1-DRB3* allelic combinations in human populations is that they arose by rare intra-*DRB* subregion recombination.

In the *DRw53* group, the various *DRB1* alleles appear to be associated with the same *DRB4*0101* allele. This observation, too, can be interpreted as indirect evidence for intra-*DRB* subregion recombination. The argument is, however, in this case somewhat oblique. We can assume that the diversification of the ancestral *DRB1* gene into the major *DRB1*04*, **07*, and **09* alleles, as well as a host of minor alleles, was accompanied by a corresponding diversification of the ancestral *DRB4* gene in the *DRw53* urhaplotype. Should the *DRB4* gene not have been functional, or only poorly functional, it may not have been subjected to positive selection and hence it might have diversified at a slower rate than the *DRB1* gene, but the long separation times between the *DRB1*04, *07, *09* alleles (suggested by large genetic distances between them), should have nevertheless provided an opportunity for at least some diversification of the *DRB4* gene. The fact that all haplotypes in the *DRw53* group contain the same *DRB4*0101* allele must therefore mean that an intra-group homogenization of haplotypes by recombination occurred. One can envision, for example, that three major haplotypes, *DRB1*04...DRB4*01, DRB1*07...DRB4*02*, and *DRB1*09... DRB4*03* existed at some point in time, but that the *DR7* and *DR9* haplotypes subsequently acquired the *DRB4*01* gene from the *DR4* haplotype by recombination. This hypothetical homogenization must have occurred relatively recently, not giving the *DRB4*01* gene time to diversify. On the other hand, it could not have occurred *very* recently (e.g., less than 50,000 years ago), otherwise the *DRw53*-group haplotypes isolated from different racial groups would not bear the same *DRB4*0101* allele, as they seem to do.

It seems reasonable to assume that the recombination occurred only once for the *DR7* and once for the *DR9* haplotypes and so it must have represented a bottleneck phase for the *DRB1* gene pools. If so, all the polymorphism that had accumulated at the *DRB1*07* and *DRB1*09* genes up to that point in time must have been lost and the diversification of the genes must have started anew to produce new minor haplotypes. Hence the degree of divergence between the new minor haplotypes could be used to estimate the time homogenization occurred, which is approximately 4 my ago. Since the *DRB4* gene has not diversified appreciably during this period, this estimate also provides a measure of the effect of positive selection on the *DRB1* locus. Furthermore, if homogenization affected only two of the three major *DRw53* haplotypes (*DR7* and *DR9* but not *DR4*) in our hypothetical scheme, then in one haplotype more *DRB1* polymorphism should have accumulated than in the other two haplotypes (the one haplotype would retain both the pre- and post-homogenization polymorphism, whereas the other two would have only the post-homogenization polymorphism). This supposition seems to be supported by the observation that there are now 11 known *DRB1*04* alleles but only two *DRB1*07* and two *DRB1*09* alleles (WHO Committee 1991).

The difference in chromosome length probably does not fully explain why *DRB* haplotypes remain frozen. Although Bridges and his co-workers demonstrated as long as 55 years ago that the frequency of crossing-over can be reduced in the immediate neighborhood of a deletion (Bridges et al. 1936), this is not a general finding and the reduction, when found, is often small. Other factors may therefore either contribute to or provide an alternative explanation of inter-group suppression of crossing-over in the *DRB* subregion. Evidence is now accumulating for the existence of specialized recombination signals which promote crossing-over in certain regions (recombinational hot spots) of the chromosome (reviewed by Steinmetz et al. 1987 and Jarman and Well 1989). Recombination often appears to be initiated by a class of middle-repetitive nucleotide sequences, some 10^5 copies of which are interspersed in the genome. Such repeats have indeed been found in the segment between the *DQB1* and *DQA1* genes, in which recombination is known to occur (Satyanarayana and Strominger 1991). The absence of recombination signals in the *DRB* subregion would, however, not explain why the suppression of crossing-over is apparently restricted to inter-group combinations. To explain the available data one would have to postulate that different haplotype groups have different recombination signals and that homozygosity for the signal sequences is necessary for the initiation of crossing-over. Another possibility is that the *DRB* subregions of different haplotypes may contain a series of short inversions. Whatever the explanation, the inter-group suppression of crossing-over cannot be absolute, otherwise it would be difficult to understand how the haplotype groups arose in the first place. The suppression is, however, strong enough to freeze haplotypes for periods much longer than those predicted by the classical theory of linkage disequilibrium.

The strange case of the *DRw8* haplotype

One haplotype, *HLA-DRw8*, requires special consideration because it may hold clues to understanding how the haplotypes relate to one another and how they arose from other, more ancestral haplotypes. As was discussed earlier, the human single-gene *DRw8* haplotype is

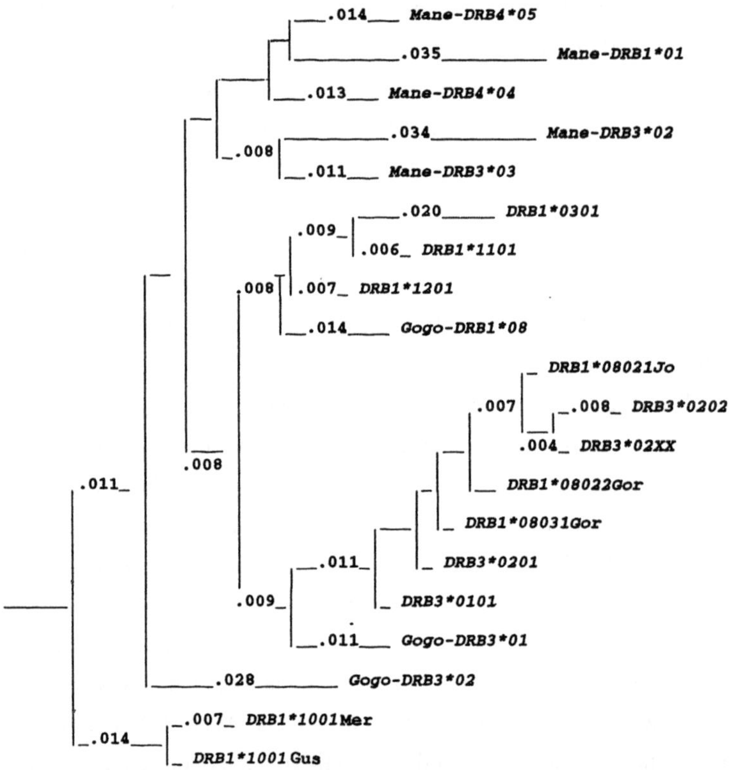

Fig. 4. Portion of a dendrogram depicting the relationship between *DRB1* genes of the *DRw52* group, *DRB1*08*, and *DRB3*, based on the sequence of the 3'UT region. The tree was constructed by the neighbor-joining method of Saitou and Nei (1987) from genetic distances (numbers on branches) calculated by Kimura's two-parameter method (Kimura 1980). The references for the individual sequences are as follows (from top to bottom): all *Mane* sequences, Zhu et al. (1991); *DRB1*0301*, Gustafsson et al. (1984); *DRB1*1101*, Tieber et al. (1986); *DRB1*1201*, Navarrete et al. (1989); *Gogo-DRB1*08*, Kasahara et al. (1991); *DRB1*08021Jo*, Jonsson et al. (1989); *DRB3*0202*, Didier et al. (1986); *DRB3*02XX*, Didier et al. (1986); *DRB1*08031*, Gorski (1989b); *DRB1*0822*, Gorski (1989b); *DRB3*0201*, Long et al. (1983); *DRB3*0101*, Gorski and Mach (1986); *Gogo-DRB3*01*, Kasahara et al. (1991); *Gogo-DRB3*02*, Kasahara et al. (1991); *DRB1*1001*, Merryman et al. (1988), Gustafsson et al. (1984). Species abbreviations are as in Figure 2.

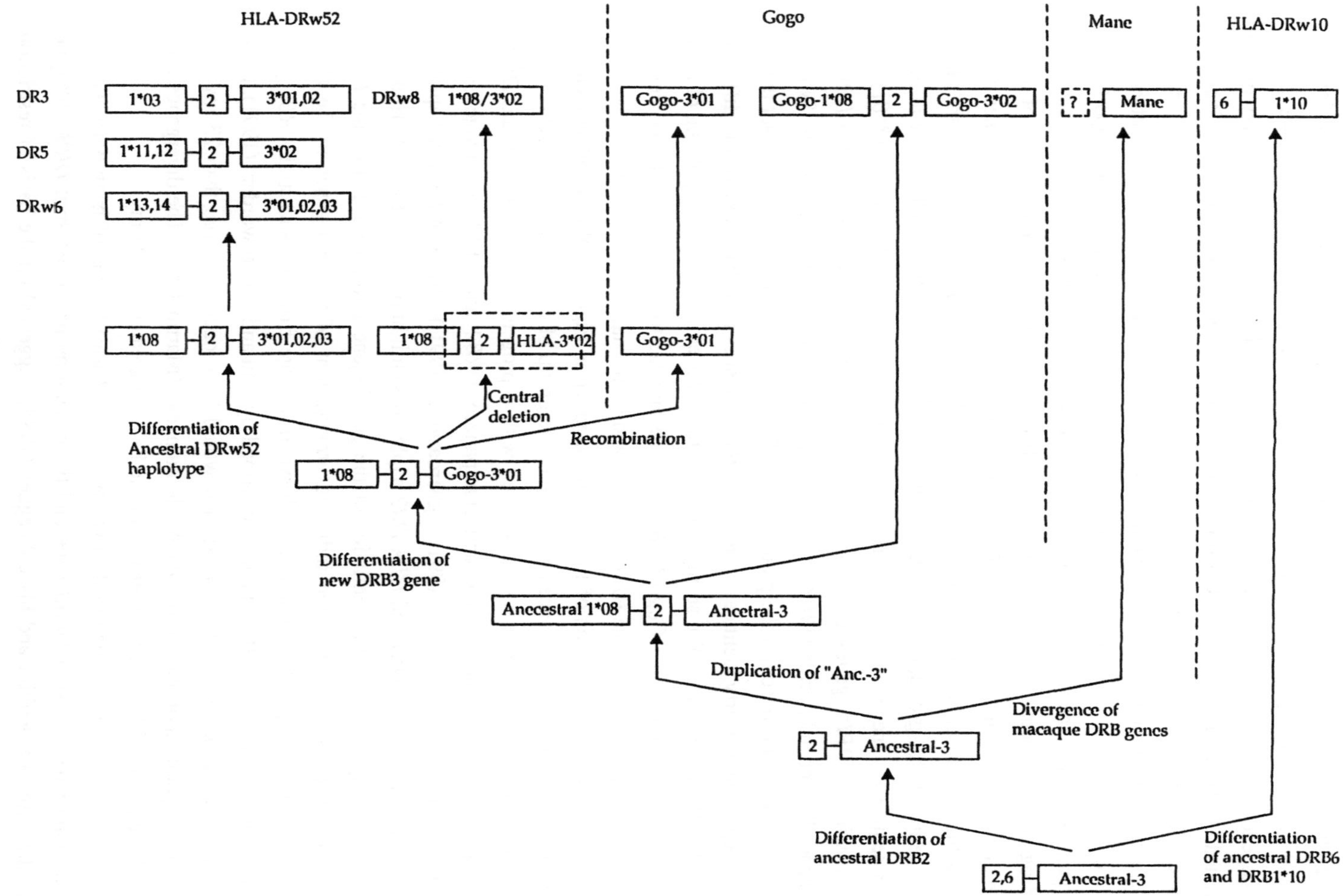

Fig. 5. The postulated origin of the *HLA-DRw52* group and related haplotypes. Explanation in the text.

believed to have originated from an *HLA-DR3*-like three-gene haplotype by the deletion of a segment that included the 3' untranslated (UT) region of *DRB1*, the entire *DRB2* pseudogene, and exons 1 through 5 of the *DRB3* gene, followed by the fusion of the truncated *DRB1* and *DRB3* genes (Gorski 1989b). (From the sequence of *DRB1*08* it cannot be determined whether exons 4 and 5 derive from *DRB1* or *DRB3* genes, so the exact point of exchange cannot be established; it could have been anywhere between the end of exon 3 and the beginning of exon 6.) The analysis of the gorilla *DRB* haplotypes supports this interpretation but at the same time complicates it somewhat. Figure 4 depicts a portion of a dendrogram constructed on the basis of genetic distances between *DRB* sequences of the 3' UT region. As expected, the *Gogo-DRB1*08* sequence clusters with the *DRB1* genes of the *HLA-DRw52* group of haplotypes, and the *HLA-DRB1*08* sequences clusters with the *HLA-DRB3* genes of that group, confirming that the 3' UT region of *HLA-DRB1*08 is* in fact a *DRB3* sequence. One of the *DRB* genes of the *Sylvia 1* haplotype (*Gogo-DRB3*01*) also clusters with the *DRB3* genes, even though it is apparently not linked to any other *DRw52*-like gene. The second gorilla *DRB3* gene (the *Gogo-DRB3*02* pseudogene from the *Sylvia-2* haplotype), however, obscures the picture; it *is* a *DRB3* gene as indicated by the fact that its exon 1 sequence is nearly identical with the *DRB3* sequences and characteristically different from other *DRB* sequences. The strong similarity to *DRB3* extends to the beginning of exon 2, but beyond this point, the *Gogo-DRB3*02* gene appears to have diverged considerably and displays some affinity to the *HLA-DRB1*10* and to the macaque *DRB* genes.

A possible interpretation of these results is given in Figure 5. We postulate a *Gogo-DRB3*02*-like sequence to be ancestral to the entire group of contemporary sequences. We postulate further that the gene was associated with a progenitor of the *DRB2* and *DRB6* pseudogenes. (The human, gorilla, and chimpanzee *DRB2* and *DRB6* pseudogenes are clearly related; see Ỵincek et al. 1991; Figueroa et al. 1991.) The *HLA-DRw10* haplotype (and hence also the *HLA-DR1* and *HLA-DR2* haplotypes) might be derived directly from this ancestral haplotype by the differentiation of the pseudogene into *DRB6* and of the *DRB3* gene into the *DRB1*10* gene (the indicated arrangement of the two genes in the *HLA-DRw10* haplotype is purely hypothetical.) Next to diverge from the ancestral *DRB3* gene was a group of macaque *DRB* sequences described by Zhu and her co-worders (1991). If the proposed scheme is correct, this divergence would have occurred more than 23 my ago, before the separation of apes from Old World monkeys (Martin 1990). The divergence time of *DRB3* and *DRB1*10* would therefore have to be even greater than 23 my.

The postulated next step in the evolution of the *DRw52* group of haplotypes was the duplication of the ancestral *DRB3* gene and the generation of an ancestral *DRB1*08* gene from the duplicate. Rollini and his colleagues (1985) suggested on the basis of restriction map similarity that the *DRB1* and *DRB3* genes of the *DRw52* haplotypes may have arisen by duplication from a common ancestral gene. The genetic distance dendrogram in Figure 4 is consistent with this suggestion. In contrast to Rollini and his colleagues, however, we do not postulate that the *DRB3* gene arose from the *DRB1* gene but vice versa. In our scheme, *DRB3* is an old gene that might have once been highly active but its function was later taken over by the *DRB1* gene.

The duplication may have generated an ancestral *DRB* haplotype whose direct descendant is the extant *Sylvia 2* haplotype. Further differentiation of the *DRB3* gene in this haplotype produced the modern *Gogo-DRB3*01* allele and the founding alleles of the *HLA-DRB1* and *-DRB3* loci in the *DRw52* haplotypes. We presume that the deletion generating the *HLA-DRw8* haplotype occurred close to or (more likely) shortly after the separation of the hominid and gorilla lineages more than 5 my ago.

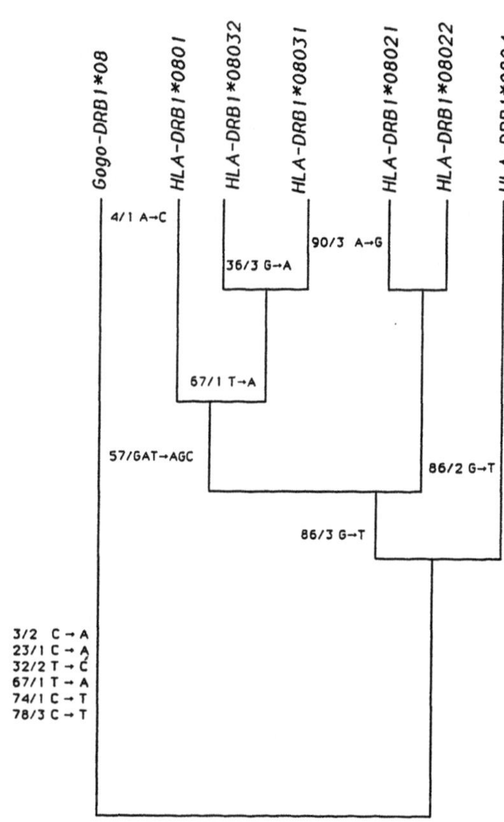

Fig. 6. Postulated diversification of human (*HLA*) and gorilla (*Gogo*) *DRB1*08* alleles. A maximum parsimony tree (Felsenstein 1988) was constructed on the basis of nucleotide sequences coding for the ß1 domain. Substitutions on the individual branches are indicated by arrows; (the first number indicates the codon and the second number the nucleotide position in this codon in which the substitution has occurred. Human sequences are from Marsh and Bodmer (1991); the gorilla sequence is from Kasahara et al. (1991).

The proposed scheme explains an observation that is otherwise rather puzzling. As already mentioned, the products of the *HLA-DRB3* genes are characterized by the DRw52 antigen. Since the hybrid *HLA-DRB1*08* gene acquired only the 3' UT region from *DRB3* (or at the most, also the regions coding for the cytoplasmic and transmembrane domains), the DRw52 antigen could not have been contributed to *HLA-DRB1*08* by *DRB3*. In our scheme, the *DRB1*08* gene, which is ancestral to the other *HLA-DRB1* genes of the *DRw52* group, may have coded for the DRw52 antigen from the beginning and only when this ancestral gene differentiated, did some of the *DRB1* genes of the DRw52 lose the information coding for the antigen (although certain other genes may have retained the information for certain components of the complex DRw52 antigen; see Taylor et al. 1989).

If, as seems likely, the *DRw8* haplotype was created in the hominid lineage after its separation from the great ape lineage, then all the extant *HLA-DRB1*08* polymorphism must have been generated in the last 6 my or so. Figure 6 depicts how the diversification of the *HLA-DRB1*08* alleles might have proceeded in the region coding for the ß1 domain of the DR molecule. It seems that the variants arose by point mutations substituting nucleotides at different positions. The only exception might be codon 57. The consensus sequence at this position is GAC; however, several *DRB* genes, some of them only distantly related to one another, have AGC instead. The *HLA-DRB1*08021, *08022, *0804*, and *Gogo-DRB1*08* have GAT, all the others have AGC at this position. This nucleotide triplet is therefore an example of segmental sharing between distantly related alleles. The repeated occurrence of this complex change may be the result of information transfer by recombination or of a complex mutation of the sort depicted in Figure 7.

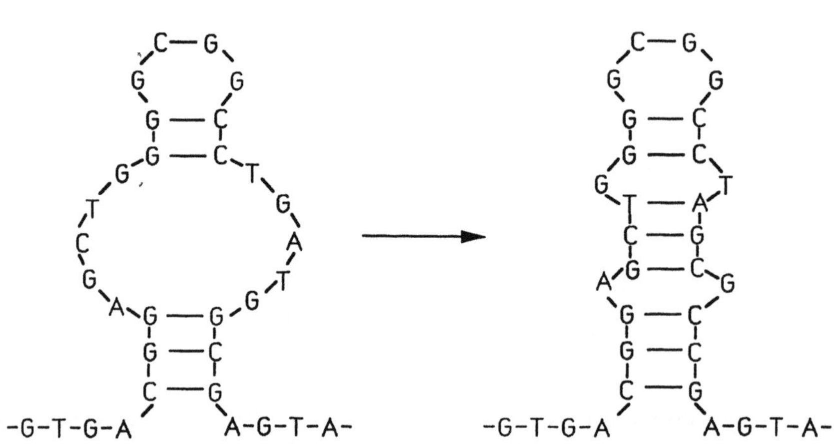

Fig. 7. A possible explanation for the GAC -> AGC change in codon 57 of *DRB1*08* alleles as a single event through the formation of a stem-loop structure and correction to improve complementary pairing.

The low divergence of the *HLA-DRB1*08* alleles illustrates how slowly the wheels of the process generating *Mhc* polymorphism turn. (The slow rate of *Mhc* divergence is discussed in more detail elsewhere in this volume; see Satta et al. 1991.)

Conclusion: Why frozen *DRB* haplotypes?

There are two possibilities as far as the meaning and significance of frozen *DRB* haplotypes are concerned, and they both depend on how widespread frozen haplotypes will turn out to be. If the frozen haplotypes are a unique feature of the *DRB* region, then it is probably safe to conclude that they are the result of an evolutionary accident and hence that they do not have any special functional significance. If, on the other hand, similar frozen haplotype polymorphisms exist in other regions and in other species, they may play some specific part in the evolution of the *Mhc*. As was already mentioned, the *DQ* and *DP* regions of the *HLA* complex do not show any structural haplotype variation and the same may be true for *Mhc*s of some other primates. The class I region of the *Mhc*, however, displays extensive variation in the number of genes and general length, at least in the mouse (Klein 1986). Whether any of the variant class I haplotypes are evolutionarily frozen remains to be seen.

Regardless of their functional significance, however, frozen *DRB* haplotypes provide a unique set of markers for the study of human evolution. They give another dimension to the trans-species polymorphism and to all its implications concerning the founding population of the human species. It should be possible to use the existence of different *DRB* haplotypes and their frequencies to make inferences about the origin not only of the different racial groups, but also of the different species of *Homo* and indeed, the manner of separation of the hominids from the African apes. Above all, however, *DRB* haplotype polymorphism seems to be uniquely suitable for resolving the so-called African hominoid trichotomy problem (Groves 1986; Mellars and Stringer 1989). For decades, evolutionary biologists have been debating whether the closest relative of humans is the chimpanzee or the gorilla (a small minority even claims it to be the orang-utan; see Schwartz 1987). The methods applied thus far, including protein and DNA sequencing, have failed to resolve the controversy (Holmes et al. 1989). The *DRB* haplotype polymorphism may enable us to take a fresh look at this old problem with a set of new, unique markers. The demonstration of sharing particular structural features between any two species of the trio and their absence in the third species may help to sway the argument in favor of these two species being the closest relatives. The haplotype polymorphism may, therefore, prove to be as useful for the study of human evolution as allelic polymorphism.

Acknowledgments. We thank Ms. Lynne Yakes for editorial help. The work was supported in part by grants R01 AI23667 from the National Institutes of Health, Bethesda, MD, and 163 258 from the Fonds der Chemischen Industrie, Frankfurt, Germany.

References

Andersson, G., Larhammar, D., Widmark, E., Servenius, B., Peterson, P.A., and Rask, L.A.: Class II genes of the human major histocompatibility complex: Organization and evolutionary relationship of the *DRβ* genes. *J Biol Chem 262*:8748-8758, 1987

Arden, B. and Klein, J.: Biochemical comparison of major histocompatibility complex molecules from different subspecies of *Mus musculus*: Evidence for trans-specific evolution of alleles. *Proc Natl Acad Sci USA 79*:2342-2346, 1982

Awdeh, Z.L., Raum, D., Yunis, E.J., and Alper, C.A.: Extended HLA complement allele haplotypes. Evidence for T/t like complex in man. *Proc Natl Acad Sci USA 80*:259-263, 1983

Batchelor, J.R., Curtoni, S., Mervart, H., Noreen, H., Davey, N.J., Moses, J.H., and Amoroso, A. Antigen Society No. 22 Report (DR2). *In* B. Dupont (ed.) *Immunobiology of HLA*. Vol. I. *Histocompatibility Testing 1987*, pp. 240-242. Springer-Verlag, New York 1989

Bidwell, J.: DNA-RFLP analysis and genotyping of HLA-DR and DQ antigens. *Immunol Today 9*:18-23, 1988

Bidwell, J.F., Bidwell, E.A., Savage, D.A., Middleton, D., Klouda, P.T., and Bradley, B.A.: A DNA-RFLP typing system that positively identifies serologically well-defined and ill-defined HLA-DR and DQ alleles, including DRw10. *Transplantation 45*:640-646, 1988

Blanck, G. and Strominger, J.L.: Molecular organization of the DQ subregion (DO-DX-DV-DQ) of the human MHC and its evolutionary implications. *J Immunol 141*:1734-1737, 1988

Bodmer, W.F. and Cavalli-Sforza, L.L.: *Genetics, Evolution, and Man*. W.H. Freeman, San Francisco 1976

Bodmer, J.G., Pickbourne, P., and Richards, S.: Ia Serology. *In* W.F. Bodmer et al. (eds.) *Histocompatibility Testing 1977*, pp. 35-84, Munksgaard, Copenhagen 1977

Böhme, J., Andersson, M., Andersson, G., Möller, E., Peterson, P.A., and Rask, L.: HLA-DR ß genes vary in number between different DR specificities whereas the number of DQ ß genes is constant. *J Immunol 135*:2149-2155, 1985

Bontrop, R.E., Broos, L.A.M., Pham, K., Bakas, R.M., Otting, N., and Jonker, M.: The chimpanzee major histocompatibility complex class II *DR* subregion contains an unexpectedly high number of beta chain genes. *Immunogenetics 32*:272-280, 1990

Brändle, U., Ono, H., Figueroa, F., and Klein, J.: Organization of major histocompatibility complex *DRB* genes in the chimpanzee. Submitted

Bridges, C.B., Skoog, E.N., and Li, J.-C.: Genetical and cytological studies of a deficiency (*Notopleural*) in the second chromosome of *Drosophila melanogaster*. *Genetics 21*:788-795, 1936

Cepellini, R., Curtoni, E.S., Mattiuz, P.L., Miggiano, V., Scudeller, G., and Serra, A.: Genetics of leukocyte antigens: A family study of segregation and linkage. *In* E.S. Curtoni, P.L. Mattiuz, and R.M. Tosi (eds.): *Histocompatibility Testing 1967*, pp. 149-184, Munksgaard, Copenhagen 1967

Corell, A., Martin-Villa, J.M., Morales, P., De Juan, M.D., Varela, P., Vicario, J.L., Martinez-Laso, A., and Arnaiz-Villena, A.: Exon-2 nucleotide sequences, polymorphism and haplotype distribution of a new HLA-DRB gene: HLA-DRB6. *Mol Immunol 28*:533-543, 1991

Dausset, J. and Pla, M. *HLA complexe majeur d'histocompatibilité de l'hommes*. Flammarion, Paris 1989

Deevey, E.S., Jr.: The human population. *Sci Am 203* (3):195-204, 1960

Didier, D.K., Schiffenbauer, J., Shuman, S., Abruzzini, L.F., Gorski, J., Watling, D.L., Tieber, V.L., and Schwartz, B.D.: Characterization of two distinct DRß chain alleles at the $ß_{III}$ locus of the DR5 haplotype: $ß_{III}$ alleles are highly conserved. *J Immunol 137*:2627-2631, 1986

Dupont, B. (ed.): *Immunobiology of HLA*, Vol. 2: *Immunogenetics and Histocompatibility*. Springer-Verlag, New York 1989

Dunham, I., Sargent, C.A., Dawkins, R.L., and Duncan Campbell, R.: An analysis of variation in the long-range genomic organization of the human major histocompatibility complex class II region by pulsed-field gel electrophoresis. *Genomics 5*:787-796, 1989

Erlich, H.A., Scharf, S.J., Long, C.M., and Horn, C.T.: Analysis of isotypic and allotypic sequence variation in the HLA-DRß region using the in vitro enzymatic amplification of specific DNA segments. *In* B. Dupont (ed.): *Immunobiology of HLA*. Vol. II, pp. 181-185, Springer-Verlag, New York 1989

Fan, W., Kasahara, M., Gutknecht, J., Klein, D., Mayer, W.E., Jonker, M., and Klein, J.: Shared class II Mhc polymorphisms between humans and chimpanzees. *Hum Immunol 26*:107-121, 1989

Felsenstein, J.: Phylogenies from molecular sequences: Inference and reliability. *Annu Rev Genet 22*:521-565, 1988

Figueroa, F., Günther, E., and Klein, J.: *Mhc* polymorphism pre-dating speciation. *Nature 335*:265-267, 1988

Figueroa, F., O'hUigin, C., Inoko, H., and Klein, J.: Primate *DRB6* pseudogenes -- clue to the evolutionary origin of the *HLA-DR2* haplotype. *Immunogenetics*, in press 1991

Gorski, J.: HLA-DR ß-chain polymorphism. Second domain polymorphism reflects evolutionary relatedness of alleles and may explain public serologic epitopes. *J Immunol 143*:329-333, 1989a

Gorski, J.: The HLA-DRw8 lineage was generated by a deletion in the DRB region followed by first domain diversification. *J Immunol 142*:4041-4045, 1989

Gorski, J. and Mach, B.: Polymorphism of human Ia antigens: gene conversion between two DRß loci results in a new HLA-D/DR specificity. *Nature 322*:67-70, 1986

Grahovac, B., Mayer, W., Vincek, V., Figueroa, F., O'hUigin, C., Tichy, H., and Klein, J.: Major histocompatibility complex *DRB* genes of a new world monkey, the cotton-top tamarin (*Saguinus oedipus*). *Mol Biol Evol*, in press 1991

Groves, C.P.: Systematics of great apes. *In* D.R. Swindler and J. Erwin (eds.): *Comparative Primate Biology*. Vol. I. *Systematics, Evolution, and Anatomy*, pp. 187-217, Alan R. Liss, New York 1986

Grumet, F.C. and Ness, D.B.: Joint report: DRw10. *In* P. Terasaki (ed.): *Histocompatibility Testing 1980*, p. 542-545, Springer-Verlag, New York 1980

Gustafsson, K., Wiman, K., Emmoth, E., Larhammar, D., Böhme, J., Hyldig-Nielsen, J.J., Ronnne, H., Peterson, P.A., and Rask, L.: Mutations and selection in the generation of class II histocompatibility antigen polymorphism. *EMBO J 3*:1655-1661, 1984

Gyllensten, U.B. and Erlich, H.A.: Ancient roots for polymorphism at the *HLA-DQ* alpha locus in primates. *Proc Natl Acad Sci USA 86*:9986-9990, 1989

Gyllensten, U.B., Lashkari, D., and Erlich, H.A.: Allelic diversification at the class II *DQB* locus of the mammalian major histocompatibility complex. *Proc Natl Acad Sci USA 87*:1835-1839, 1990

Hartl, D.L. and Clark, A.G.: *Principles of Population Genetics*. 2nd. ed. Sinauer, Sunderland, MA 1989

Holmes, E.C., Pesole, G., and Saccone, C.: Stochastic models of molecular evolution and the estimation of phylogeny and rates of nucleotide substitution in the hominoid primates. *J Hum Evol 18*:775-794, 1989

Hurley, C.K., Giles, R.C., Nunez, G., DeMars, R., Nadler, L., Winchester, R., Stastny, P., and Capra, D.C.: Molecular localization of human class II MT2 and MT3 determinants. *J Exp Med 160*:472-493, 1984

Jarman, A.P. and Well, R.A.: Hypervariable ministatellites: recombinators or innocent bystanders? *Trends Genet 5*:367-371, 1989

Jonsson, A.-K., Andersson, L., and Rask, L.: A cellular and functional split in the *DRw8* haplotype is due to a single amino acid replacement (DRß$^{ser57\text{-}asp57}$). *Immunogenetics 29*:308-316, 1989

Kasahara, M., Klein, D., Fan, W., and Gutknecht, J.: Evolution of the class II major histocompatibility complex alleles in higher primates. *Immunol Rev. 113*:65-82, 1990

Kasahara, M., Klein, D., Vincek, V., Sarapata, D.E., and Klein, J.: The *DRB* subregion of the gorilla major histocompatibility complex: Evidence for linkage disequilibrium conserved across species. *J Immunol.*, submitted 1991

Kawai, J. Ando, A., Sato, T., Nakatsuji, T., Tsuji, K., and Inoko, H.: Analysis of gene structure and antigen determinants of DR2 antigens using DR gene transfer into mouse L cells. *J Immunol 142*:312-317, 1989

Kendall, E., Todd, J.A., and Campbell, R.D.: Molecular analysis of the MHC class II region in *DR4, DR7*, and *DR9* haplotypes. *Immunogenetics*, in press 1991

Kimura, M.: Estimation of evolutionary distances between homologous nucleotide sequences. *Proc Natl Acad Sci USA 78*:454-458, 1981

Kimura, M.: *The Neutral Theory of Molecular Evolution.* Cambridge University Press, Cambridge 1983

Klein, J.: Generation of diversify at MHC loci: Implications for T cell receptor repertoires. *In* M. Fougereau and J. Dausset (eds.): *Immunology 80: Progress in Immunology IV*, pp. 239-253. Academic Press, London 1980

Klein, J. and Schönbach, C.: Origins of *Mhc* diversity. *In* B.G. Solheim, E. Möller, and S. Ferrone (eds.): *The HLA System. Basic Concepts and Clinical Importance in Transplantation Immunology.* Springer-Verlag, Heidelberg, in press 1991

Klein, J. and Takahata, N.: The major histocompatibility complex and the quest for origins. *Immunol Rev 113*:5-25, 1990

Klein, J., Kasahara, M., Gutknecht, J., and Figueroa, F.: Origin and function of *Mhc* polymorphism. *Chem Immunol 49*:35-50, 1990

Klein, D., Vincek, V., Kasahara, M., Schönbach, C., O'hUigin, C., and Klein, J.: Gorilla *Mhc-DRB* pseudogene orthologous to *HLA-DRBVIII. Hum Immunol*, in press 1991

Lawlor, D.A., Ward, F.E., Ennis, P.D., Jackson, A.P., and Parham, P.: *HLA-A* and *B* polymorphisms predate the divergence of humans and chimpanzees. *Nature 335*:268-271, 1988

Lee, S., Nunez-Roldan, A., Dwyer, E., Pompeo, L., and Winchester, R.: Definition of DRw10 by restriction fragment length polymorphism. *Tissue Antigens 33*:466-474, 1989

Long, E.O., Wake, C.T., Gorski, J., and Mach, B.: Complete sequence of an HLA-DR ß chain deduced from a cDNA clone and identification of multiple non-allelic DR ß chain genes. *EMBO J 2*:389-394, 1983

Marsh, S.G.E. and Bodmer, J.G.: *HLA* class II nucleotide sequences, 1991. *Immunogenetics 33*:321-334 1991

Martin, R.D.: *Primate Origins and Evolution: A Phylogenetic Reconstruction.* Chapman and Hall, London 1990

Matsuyama, T., Schwenzer, J., Silver, J., and Winchester, R.: Structural relationships between the DRß$_1$ and DRß$_2$ subunits in DR4, 7, and w9 haplotypes and the DRw53 (MT3) specificity. *J Immunol 137*:934-940, 1986

Matsuyama, T., Winchester, R., Lee, S., Shookster, L., and Nunez-Roldan, A.: Identification of the DRw10 DRß$_1$-chain allele as encoding a polymorphic class II major histocompatibility complex epitope otherwise restricted to DRß$_2$ molecules of the DRw53 type. *J Immunol 140*:537-543, 1988

O'hUigin, C. and Klein, J. The age and evolution of the *HLA-DRB* pseudogene. *In* J. Klein and D. Klein (eds.): *Molecular Evolution of the Major Histocompatibility Complex.* Springer-Verlag, Heidelberg 1991

Mayer, W.E., Jonker, M., Klein, D., Ivanyi, P., van Seventer, G., and Klein, J.: Nucleotide sequences of chimpanzee MHC class I alleles: Evidence for trans-species mode of evolution. *EMBO J 7*:2765-2774, 1988

McConnell, T.J., Talbot, W.S., McIndoe, R.A., and Wakeland, E.K.: The origin of *Mhc* class II gene polymorphism within the genus *Mus. Nature 332*:651-654, 1988

Mellars, P. and Stringer, C. (eds.): *The Human Revolution.* Princeton University Press, Princeton, New Jersey 1989

Merryman, P., Gregersen, P.K., Lee, S., Silver, J., Nunez-Roldan, N., Crapper, R., and Winchester, R.: Nucleotide sequence of a DRw10 ß chain cDNA clone. Identity of the third D region with that of the DRw53 allele of the $ß_2$ locus and as the probable site encoding a polymorphic MHC class II epitope. *J Immunol 140*:2447-2452, 1988

Nei, M.: Human evolution at the molecular level. *In* T. Ohta and K. Aoki (eds.): *Population Genetics and Molecular Evolution*, pp. 41-64, Springer-Verlag, Berlin 1985

Navarette, C., Seki, T., Miranda, A., Winchester, R., and Gregersen, P.K.: DNA sequence analysis of the HLA-DRw12 allele. *Hum Immunol 25*:51-58, 1989

Niven, M.J., Hitman, G.A., Pearce, H., Marshal, B., and Sachs, J.A.: Large haplotype-specific differences in inter-genic distances in human MHC shown by pulsed field electrophoresis mapping of healthy and Type 1 diabetic subjects. *Tissue Antigens 35*:19-24, 1990

Riisome, K., Sorensen, I.J., Moller, B., Kruse, T.A., Grangaard, B., and Lamm, L.U.: HLA-DR typing using restriction fragment length polymorphism (RFLP) with one enzyme. *Tissue Antigens 31*:141-150, 1988

Rollini, P., Mach, B., and Gorski, J.: Linkage map of three HLA-DR ß chain genes: Evidence for a recent duplication event. *Proc Natl Acad Sci USA 82*:7197-7201, 1985

Saitou, N. and Nei, M.: The neighbor-joining method. A new method for reconstructing phylogenetic trees. *Mol Biol Evol 4*:406-425, 1987

Satta, Y., Takahata, N., Schönbach, C., Gutknecht, J., and Klein, J.: Calibrating evolutionary rates at major histocompatibility loci. *In* J. Klein and D. Klein (eds.): *Molecular Evolution of the Major Histocompatibility Complex.* Springer-Verlag, Heidelberg, 1991

Satyanarayana, K. and Strominger, J.L.: DNA sequences near a meiotic recombination breakpoint within the human *HLA-DQ* region. *Immunogenetics*, in press 1991

Schwartz, J.H.: *The Red Ape. Orang-utans and Human Origins.* Houghton Mifflin Co., Boston 1987

Semana, G., Bignon, J.D., Quillivic, F., Cheneau, M.L., Hernion, E., Muller, J.Y., Genetet, B., and Fauchet, R.: Definition of DRw10 specificity by restriction fragment length polymorphism. *Tissue Antigens 32*:113-120, 1988

Simons, M.J., Wheeler, R., Cohen, D., Lalonel, J.-M., and Dupont, B.: Restriction fragment length polymorphism of HLA genes: Summary of the Tenth International Workshop Southern Blot Analysis. *In* B. Dupont (ed.) *Immunobiology of HLA*. Vol. I. *Histocompatibility Testing 1987*, pp. 959-1023. Springer-Verlag, New York 1989

Sorrentino, R., Lillie, J., and Strominger, J.L.: Molecular characterization of MT3 antigens by two-dimensional gel electrophoresis, NH_2-terminal amino acid sequence analysis and Southern-blot analysis. *Proc Natl Acad Sci USA 82*:3794-3798, 1985

Spies, T., Sorrentino, R., Boss, J.M., Okada, K., and Strominger, J.L.: Structural organization of the *DR* subregion of the human major histocompatibility complex. *Proc Natl Acad Sci USA 82*:5165-5169, 1985

Steinmetz, M., Uematsu, Y., and Fischer Lindahl, K.: Hotspots of homologous recombination in mammalian genomes. *Trends Genet 3*:7-10, 1987

Stormont, C.: Linked genes, pseudoalleles, and blood groups. *Amer Nat 89*:105-116, 1955

Taylor, C.J., Ugozzoli, L., Tanigaki, N., Tosi, R., Bunce, M., Ting, A., and Ferrara, G.B.: Antigen Society No. 29 Report (DRw52). *In* B. DuPont (ed.): *Immunobiology of HLA*. Vol. 2: *Immunogenetics and Histocompatibility*. Springer-Verlag, New York 1989

Tieber, V.L., Abruzzini, L.F., Didier, D.K., Schwartz, B.D., and Rotwein, P.: Complete characterization and sequence of an HLA class II DRß chain cDNA from the DR5 haplotype. *J Biol Chem 261*:2738-2742, 1986

Toguchi, T., Burmester, G., Nunez-Roldan, A., Gregersen, P., Seremetis, S., Lee, S., Szer, I., and Winchester, R.: Evidence for the separate molecular expression of four distinct polymorphic Ia epitopes on cells of DR4 homozygous individuals. *Hum Immunol 10*:69-81, 1984

Tokunaga, K., Saueracker, G., Kay, P.H., Christiansen, F.T., Anand, R., and Dawkins, R.L.: Extensive deletions and insertions in different MHC supratypes detected by pulsed field gel electrophoresis. *J Exp Med 168*:933-940, 1988

Tokunaga, K., Kay, P.H., Christiansen, F.T., Saueracker, G., and Dawkins, R.L.: Comparative mapping of the human major histocompatibility complex in different racial groups by pulsed-field gel electrophoresis. *Hum Immunol 26*:99-106, 1989

Vincek, V., Klein, D., Figueroa, F., Hauptfeld, V., Kasahara, M., O'hUigin, C., Mach, B., and Klein, J.: The evolutionary origin of the *HLA-DR3* haplotype. *Immunogenetics*, in press 1991

WHO Nomenclature Committee for Factors of the *HLA* System: Nomenclature for factors of the *HLA* system, 1990 *Immunogenetics 33*:301-309, 1991

Zhu, Z., Vincek, V., Figueroa, F., Schönbach, C., and Klein, J.: *Mhc-DRB* genes of the pigtail macaque (*Macaca nemestrina*): Implications for the evolution of human *DRB* genes. *Mol Biol Evol*, in press 1991

THE AGE AND EVOLUTION OF THE *DRB* PSEUDOGENES

Colm O'hUigin and Jan Klein*.
Max-Planck-Institut für Biologie, Abteilung für immungenetik, Corrensstraße 42, 7400 Tübingen, Germany
*Department of Microbiology and Immunology, University of Miami School of Medicine, Miami, FL 33101, USA

The evolutionary origins of DRB pseudogenes have been examined using molecular evolutionary techniques. Data for several primate DRB genes, both functional and pseudogenes were used. Careful consideration was given in determining suitable analytical methods to avoid two confounding effects, namely positive selection for variation within functional loci and the loss of functional constraint in pseudogenes, while allowing the use of data from both coding and non-coding regions. A core data set of exon sequences was used to determine the position of the pseudogenes within the major DRB phylogenetic groupings. The branching order within the pseudogenes themselves was then inferred using supplementary data from introns. The ages of the DRB pseudogenes were estimated using 2 methods which date the origin of the lineage giving rise to DRB4-like pseudogenes at 20-30 mya. Minimum age estimates show that the DRB8 and DRB7 lineages branched from that of DRB2 at least 18-20 mya, followed by DRB6 at about 8-12 mya and DRB4 at about 8 mya.

Since their discovery, pseudogenes have proved useful in demonstrating many of the processes which underlie molecular evolution. Substitution rates and biases in the absence of selection have been assessed and times since various gene duplication and inactivation events inferred (Li et al. 1984, Graur et al. 1989). At the same time, the major histocompatability complex (*Mhc*), with its extreme polymorphisms and structural diversity, presents opportunities to further our understanding of molecular evolution, while presenting new problems which must be interpreted within the present framework. By analyzing pseudogenes and introns in the *Mhc* we can hope to characterize the essentially neutral molecular processes which provide the raw material for subsequent selection in this region.

Sequences of protions of the class II *DR* region in humans have been reported by Andersson and colleagues (1987) and Rollini and colleagues (1985). Additional sequence data have become available recently (Figueroa et al. 1991, Klein et al. 1991, Vincek et al. 1991) extending our knowledge of this region not only in humans but also in other primates. We now have sequence available from several haplotypes covering several functional genes, pseudogenes, and their introns. In addition, cDNA sequences of many functional alleles are available, notably through the compilations of Marsh and Bodmer (1991). The coding regions of class II genes are comparatively short, less than 1 kilobase (kb) of sequence

NATO ASI Series, Vol. H 59
Molecular Evolution of the
Major Histocompatibility Complex
Edited by J. Klein and D. Klein

spread over six exons. However, the genomic region covered by a class II locus is approximately ten times larger and contains considerably more informative substitutions, outlining relationships among *DRB* alleles.

The human DR region contains one *A* and from one to four *B* loci. Haplotypes with more than one functional *DRB* gene will often include pseudogenes. While the functional *DRB1* locus accounts for most of the allelic polymorphism, considerable haplotype polymorphism is attributable to the arrangement of additional *B* loci (Klein et al., these proceedings). These haplotype arrangements appear to be conserved for relatively long periods of time, with apparently orthologous groupings being found in humans, chimps and gorillas. The pseudogenes are constitutent parts of these haplotype groupings and can be characterized on the basis of the haplotype from which they are derived.

The pseudogenes used in this study are the *DRB2* alleles (Vincek et al 1991) of human (*HLA-DRB2*), chimpanzee (*Patr-DRB2*) and gorilla (*Gogo-DRB2*); the *DRB6* alleles (Figueroa et al. 1991) of human and chimpanzee (*HLA-DRBVI* and *Patr-DRB6* respectively) and the *DRB8* alleles (Klein et al. 1991) of human and gorilla (*HLA-DRBVIII* and *Gogo-DRB8* respectively) as well as the human *HLA-DRB7* pseudogene (Andersson et al. 1987). The functional *DRB* genes contain six exons. The pseudogenes however may have fewer exons (the DRB6 pseudogenes lack exon 1 while DRB8 pseudogenes lack exons 1 and 2) in addition to alterations at splicing junctions, premature stop signals, and frameshifting insertions and deletions.

It is initially necessary to determine the broad relationships among *DRB* alleles. Although cDNA sequences are available for many *DRB* alleles, two considerations confound the phylogenetic analysis in the coding regions. Firstly, selection favoring non-synonymous substitution is known to occur for functional alleles (Hughes and Nei 1989). Such selection will enhance the rate of substitution, in a fashion which reflects the functional importance of particular alleles and thereby masks phylogenetic relationships. Secondly, pseudogenes will not have been subjected to functional constraint since the time of their inactivation; their evolutionary rate will reflect the underlying mutation pattern (Gojobori et al. 1982, Li et al 1984). Thus, phylogenetic analysis based on all sites in the coding region will engender problems associated with the comparisons based on widely varying substitution rates. We avoid these problems by analyzing selectively neutral (or nearly neutral) sites only. By limiting the analysis to synonymous sites in the coding region and 3' untranslated (UT) region sites, we avoid variable rate problems at the expense of a smaller data set.

Twenty-three *DRB* alleles were chosen and their coding and conserved 3'UT regions aligned. Substitutional distances between sequences were calculated using the method of Li, Wu, and Luo (1985) for synonymous sites in coding regions, and the method of Kimura (1981) for the 3'UT region. The overall substitutional distances between sequences were taken from weighed average of synonymous and 3'UT substitutional distances. The neighbor-joining method (Saitou and Nei 1987) was used to construct dendrograms from the distance data.

Figure 1. shows a tree constructed using this method. The 23 alleles from five species appear to fall into five major groups in the dendrogram. The tamarins genes are an outgroup; we have elsewhere (Grahovac et al. 1991, these proceedings) used a canine *DRB* sequence to show that this rooting holds. First to branch from the root is the *DRB1*04* and *DRB1*07* pair of sequences followed by the *DRB5* group. The final split occurs between the *DRB1* locus alleles and most of the pseudogenes, including *DRB4*. One of the pseudogenes, the DRB7, groups with the DRB1 alleles. The method of Nei and Jin (1989) was used to test the significance of the four major branching patterns, all of which are statistically significant at the 5% level.

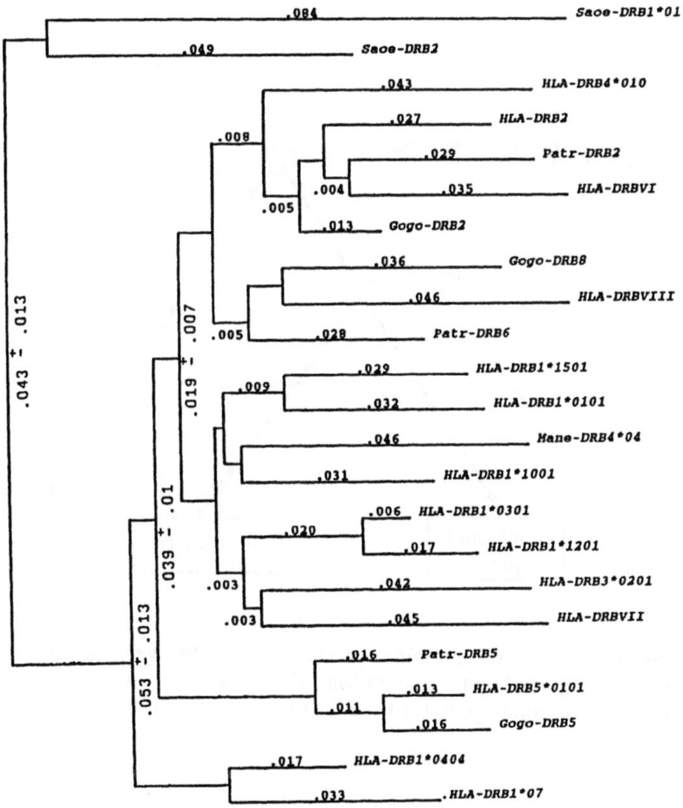

Fig. 1. Dendrogram depicting the relationship of the *DRB* pseudogenes to the major *DRB* groups of other primates. The tree was constructed using the neighbor-joining method (Saitou and Nei 1987) and distances were calculated using the methods of Li, Wu, and Luo (1985) and of Kimura (1981). For references of the original sequences see Grahovac et al. (1991).

In using the dendrogram to date the events leading to the generation of these pseudogenes, it is necessary to fix a time standard for the lineages. This standard is provided by the New-World, Old-World monkey split. The tamarin *DRB* sequences branched from the lineages that subsequently gave rise to the multiple *DRB* pseudogenes at about the time of the New-World, Old-World monkey split some 37 million years ago (mya).

A general scheme outlining the major events in pseudogene production is shown in Figure 2. At time T_o, the outgroup diverged from the lineage leading to the pseudogene and its functional counterpart. At time T_d, the pseudogene and its functional counterpart diverge and after a possible period of activity, the pseudogene becomes inactive at time T_n. If the pseudogene is inactivated at the time of its production then T_n equals T_d.

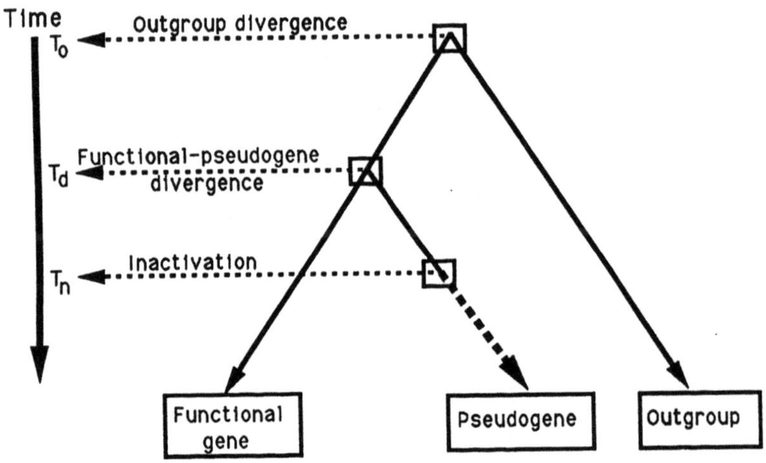

Fig. 2. Major events in pseudogene production. The vertical axis represents time. Significant events in pseudogene generation, divergences from the outgroup (at time To), duplication of the functional gene (at time Td) and inactivation of the pseudogene (at time Tn) are noted.

Methods used to determine T_d have been described by several groups (Li et al. 1981; Miyata and Yasunaga 1981; Matsuo and Yokoyama 1990). They are based upon comparison of non-synonymous substitution rates in the functional lineages with substitution rates at similar positions in the non-functional lineages. Given the short segment of DNA available, its low substitution rate and the relatively short time scale over which this study focuses, pooling of data into the groups shown in the dendrogram is essential. Table 1. shows the outcome obtained using this method. Alleles were grouped according to their positions in the dendrogram (Figure 1). The groupings are then defined to be either outgroups, functional genes or non-functional genes. The HLA-DRB4 and HLA-DRB7 sequences are omitted from this analysis since their placement is uncertain. Although, it appears to be a close relative of the pseudogenes, it is not itself know to be non-functional and contains none of the pseudogene attributes previously outlined.

Substitutional distances were measured between the various groups and are shown in Table 1. Also shown are the estimated substitution rates for functional and non-functional lineages. On the basis of an estimated divergence time of 37 million years between New and Old world monkeys, we can estimate the divergence times between these major groups using either the method of Li et al. (1981) or of Matsuo and Yokoyama (1990). There are slight methodological differences in the outcome between the two with the latter giving generally higher estimates than the former. However, combined estimates indicate that the *DRB5* group branched first from the pseudogene lineages at about 29-35 mya. This divergence was followed by the seperation of the *DRB1*04* and *DRB1* groups at approximately 27 to 32 million years ago.

Table 1. Estimated divergence times between major DRB allele groupings and the DRB pseudogenes

Functional Loci			Non-Functional	Outgroup
DRB1 GROUP	DRB1*04 GROUP	DRB5 GROUP	PSEUDOGENES	TAMARIN-DRB
*HLA-DRB1*0301*	*HLA-DRB1*0401*	*Patr-DRB5*	*HLA-DRB2*	*Saoe-DRB1*01*
*HLA-DRB1*0201*	*HLA-DRB1*07*	*HLA-DRB5*0101*	*HLA-DRBVI*	*Saoe-DRB2*
*HLA-DRB3*0201*		*Gogo-DRB5*	*HLA-DRBVIII*	
*HLA-DRB1*1001*			*Gogo-DRB8*	
*HLA-DRB1*1501*			*Gogo-DRB2*	
*HLA-DRB1*0101*			*Patr-DRB6*	
*Mane-DRB4*04*			*Patr-DRB2*	

FUNCTIONAL	NON-FUNCTIONAL	D(OF)	D(OP)	D(FP)	A	B	T_1	T_2
DRB1 group	Pseudogenes	0.065	0.075	0.066	1.08	1.46	32.2	27.3
DRB1*04 group	Pseudogenes	0.065	0.075	0.067	1.09	1.45	32.3	26.5
DRB5 group	Pseudogenes	0.067	0.075	0.074	1.12	1.37	35.5	29.4

D(XY)= distance from X to Y in terms of nonsynonymous subsitutions per site.
A = Rate of nonsynonymous substitution in functional lineages. (/site/year x 10^{-9})
B = Rate of nonsynonymous substitution in non-functional lineages. (/site/year x 10^{-9})
T_1 = Divergence time (million years) estimated by method of Matsuo and Yokoyama (1981)
T_2 = Divergence time (million years) estimated by method of Li et al. (1981)

The dating of inactivation times, T_n, is more problematic. Both Miyata and Yasunaga (1981) and Li et al. (1981) have described methods to determine T_n. These methods involve comparison of synonymous and non-synonymous rates in functional lineages with the rates in similar positions in pseudogene lineages. Subsequent steps involve division by the difference of two small numbers. This analysis does not provide reliable estimates for the sequences used here (data not shown), because of a limited number of synonymous positions available and the relatively low number of substitutions.

Having assessed the information available from exons, we proceed to study the pseudogenes by incorporating the available intron data. In addition to the pseudogenes, genomic sequences are available for the *HLA-DRB4* and *HLA-DRB1*0401* (Andersson et al. 1987) genes. Data were gathered for up to 5 kb of sequence from all these *DRB* alleles. Some sequences are complete over almost the entire length (*HLA-DRB4*, *HLA-DRB1*0401* and *HLA-DRB7*), while others lack sizeable segments (*HLA-DRB8* and *Gogo-DRB8*). Since the degree of divergence between the *DRB* loci is not great, alignment by eye is possible. Any poorly aligned regions or regions containing insertion sequences were excluded from the analysis.

It was then possible to test the substitution patterns across the sequenced regions, both to check our alignments and to detect any peculiarities in the evolutionary histories of the genes. This was done by taking a rolling average of substitution rates across a short window of 30 nucleotides and then by taking a second rolling average of these primary rolling averages (O'hUigin et al. 1990). Two rolling averages are necessary for longer sequences, in order to enhance large scale substitutional trends at the expense of minor local detail. To faciliatate the analysis, a majority consensus sequence was determined for the various *DRB2*, *DRB6* and *DRB8* sequences. Where no consensus was found, the human sequence was preferred to that of the chimpanzee or gorilla.

Figure 3 shows the pattern found when comparing the *HLA-DRB1*0401* allele with the *HLA-DRB4*01* gene and the consensus sequences of the pseudogenes, *DRB2* and *DRB6* (Fig.3, upper part) The pattern found when comparing *HLA-DRB1*0401* with *DRB7* and *DRB8* (Fig. 3, lower part) is also shown. The horizontal axis values correspond to the position in the aligned sequences and vertical axis values show the percentage substitution per nucleotide site, determined using Kimura's two-parameter method (Kimura 1981). The position of exons is also shown. Note that the relative exonic positions depend on the final alignment, following removal of insertion sequences, such as the *Alu* repeats, and of poorly aligned regions. It therefore does not correspond to the structure of any particular allele.

The absolute substitution rate seen in Figure 3 varies from position to position within the alignment. Strong divergences (substitutional distances of about 20%) are seen for all sequences in the region of nucleotides 200 to 300 of the alignment. The consensus *DRB6* sequences diverges strongly from *HLA-DRB1*04* around nucleotide 500, and examination

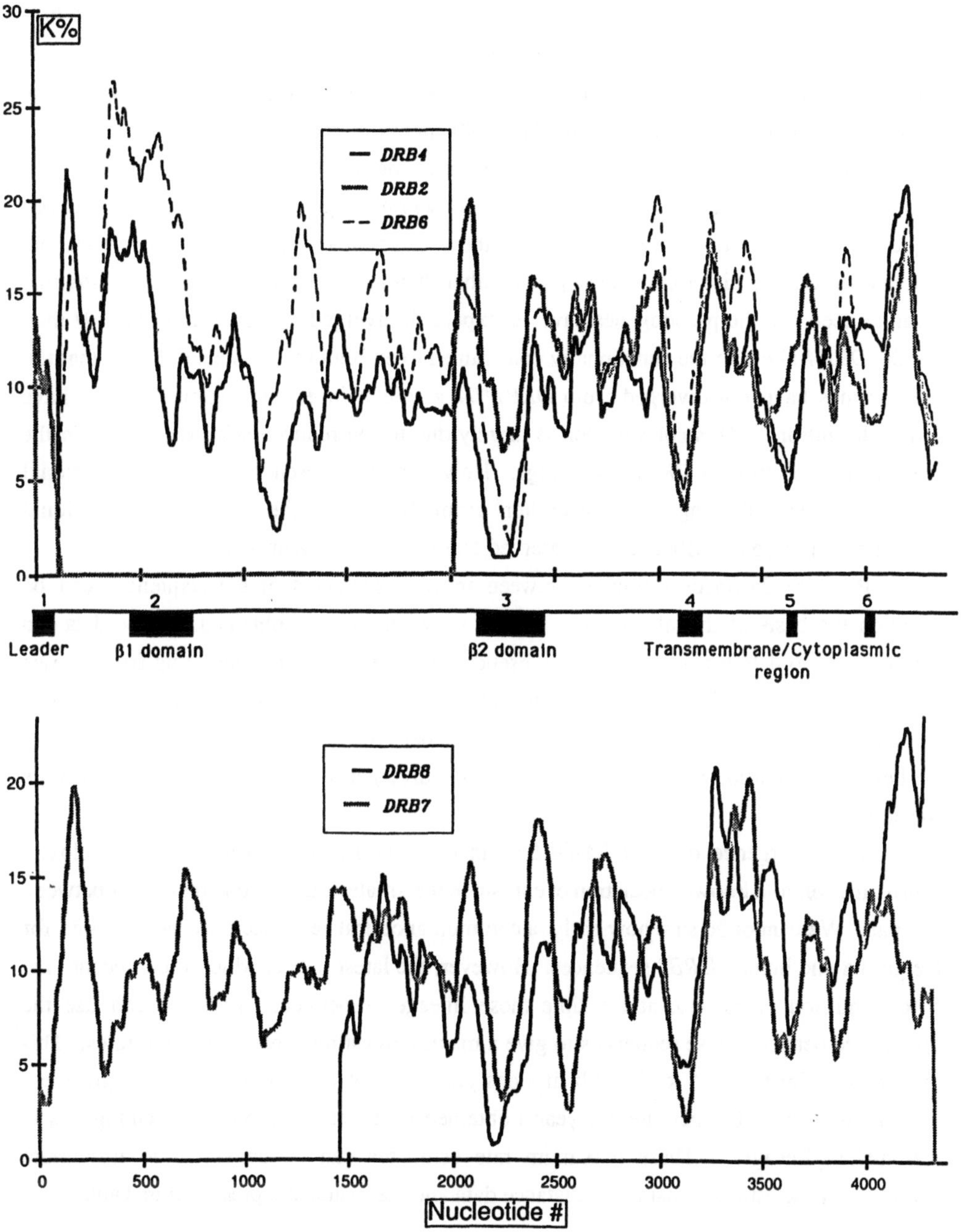

Fig. 3. Substitutional patterns seen in *DRB* pseudogenes when compared to *HLA-DRB1*0401*. The vertical axes show the number of substitutions per 100 nucleotides between selected sequences. The horizontal axis shows the nucleotide position in the final alignment of sequences, starting with the first nucleotide of exon 1 at position 1. The *HLA-DRB1*0401* sequence provides a reference against which the *DRB2*, *DRB6* and *DRB4* sequences (upper part) are compared or the *DRB7* and *DRB8* sequences (lower part). Between upper and lower sections is inserted a map showing the positions of exons (numbered) and the corresponding domains of the protein.

of component *DRB6* sequences shows this to be true of both the *HLA-DRB6* and *Patr-DRB6* alleles. Areas of conservation (substitutional distances of about 5%) are also seen particularly around the β2-domain (nucleotides 2100 to 2400 approximately) and around the tramsmembrane region (nucleotides 3080 to 3190 approximately).

While there are differences between sequences, the overall pattern for each comparison is strikingly similar. Areas of conservation generally correspond between sequences as do areas of divergence from the *DRB1*0401* sequence. This reflects the shared evolutionary histories of these genes rather than any localized functional constraints. It can be seen that the pseudogenes have, as expected, diverged further from *DRB1*04* than has the *DRB4* allele in the functionally constrained exon regions. It is also of note that the pseudogenes have not diverged from *DRB1*04* as much in exonic regions as they have across the introns. This conservation is largely due to the reduced substitution rate in the functionally constrained *DRB1*04* lineage. However, if the pseudogenes were functional for some time following their seperation from *DRB1*04*, the consequent functional constraint will also contribute to the observed conservation between sequences.

Pairwise distances estimates between intronic segments of the sequences can be made on the basis of this alignment. Table 2 shows the results obtained. These data can then be used to date branchpoints in the pseudogene lineages by extrapolating from known time points in the hominoid lineages. Since all our data come from either gorilla, human or chimpanzee, the smallest distance we can measure between alleles in different species gives the best approximation to the degree of nucleotide divergence which has occurred since the speciation event.

The divergence of the *HLA-DRB2* and *Gogo-DRB2* pseudogenes give the closest approximation to a known speciation event, since the smallest distance is measured between this pair. We cannot be sure how early a common ancestral sequence may have existed for the gorilla and human *DRB2* sequences. However, the latest date at which it can occur is at 5 to 6 million years ago, at the time these lineages seperated. Thus we can use the divergence between these sequences to give a maximum estimate of substitution rates. This maximum estimate is 3.5×10^{-9} substitutions per sites per year, in reasonable agreement with a value (5×10^{-9} per site per year) obtained by Li et al. (1985) for pseudogenes in mammalian lineages. These maximum rate estimates can be converted to mimimum divergence time estimates using the distance data and the results are presented in Table 2.

Table 2. Substitutional distances between selected sequences.

Pairwise comparison	D	s.e.	T_d (mya)
Gogo-DRB2 to *HLA-DRB2*	3.65	(0.62)	5.5
Gogo-DRB2 to *Patr-DRB2*	4.77	(0.62)	7.2
HLA-DRB2 to *Patr-DRB2*	4.55	(0.62)	6.9
DRB6			
Gogo-DRB2 to *HLA-DRBVI*	5.59	(0.64)	8.4
HLA-DRB2 to *HLA-DRBVI*	4.27	(0.68)	6.4
Gogo-DRB2 to *Patr-DRB6*	8.22	(0.77)	12.4
HLA-DRB2 to *Patr-DRB6*	6.81	(0.85)	10.3
DRB7			
Gogo-DRB2 to *HLA-DRB7*	11.97	(0.95)	18.0
HLA-DRB2 to *HLA-DRB7*	10.84	(1.09)	16.3
DRB8			
Gogo-DRB2 to *HLA-DRBVIII*	13.76	(1.02)	20.7
HLA-DRB2 to *HLA-DRBVIII*	11.71	(0.90)	17.7

The table shows estimated substitutional distances (D), estimated using the two-parameter method of Kimura (1981), and its standard error (in brackets) for selected paired sequence comparisons. Distances are estimated for non-coding segements of the alignments used in figure 3. and are calculated by the method of Kimura (1981). Also shown are estimated divergence times (T_d) between sequences based on a minimum estimated divergence time between the *Gogo-DRB2* and *HLA-DRB2* sequences of 5.5 million years.

On the basis of intron sequences, we can date the split of the *DRB2* pseudogene lineage from the *DRB1*0401* lineage at a minimum 19-21 million years, which can be compared with the independently estimated 27-32 mya on the basis of exonic sequences (Table 1). The difference between these two estimates can be ascribed to uncertainty in dating the divergence of tamarins from Old-World monkeys and in dating the divergence of *Gogo-DRB2* and *HLA-DRB2* alleles. Among pseudogenes, the *DRB2* sequences are most distant from the *DRB8* group, the average distance dates the split at a minimum 18-21 mya. This is followed by a split with the *DRB7* pseudogene at a minimum 16-18 mya.

The *DRB6* pseudogenes show some disagreement in estimated divergence dates from the *DRB2* sequences. *Patr-DRB6* appears to have diverged from the gorilla and human *DRB2* alleles some 10-12 mya, whereas the *HLA-DRBVI* appears much more closely related to *DRB2*, having apparently diverged from it some 6-8 mya. A possible explanations for this phenomenon is that *Patr-DRB6* and *HLA-DRBVI* are not orthologous genes. They are originally classified on the basis of being found in similar haplotypes in human and chimpanzee, of sharing certain sequence motifs in coding regions, and of the common loss of the exon I (Figueroa et al. 1991). Although both contain in-frame stop codons in the coding region, these codons occur in different positions. Alternatively, genetic exchange mechanisms may have partially altered one or the other sequence. The conclusion will have to await more detailed analysis of these haplotypes.

Finally, we date the split of the *DRB2* pseudogenes from their closest functional relative, the *DRB4*01* gene, at about 7-8 mya which is quite close to the time of divergence of the human, chimp and gorilla lineages. The divergence of the *Patr-DRB2* pseudogene from the human and gorilla *DRB2* sequences also dates to about this time. Consequently, it is unlikely that pseudogenes orthologous to *DRB2* will be found in non-hominoid primate lineages, although orthologues of *DRB7*, *DRB8* and perhaps *DRB6* may occur in the orang-utan (divergence time ca. 12 mya).

Acknowledgments. We thank Ms. Lynne Yakes for editorial help. The work was supported in part by grants R01 AI23667 from the National Institutes of Health, Bethesda, MD, and 163 258 from the Fonds der Chemischen Industrie, Frankfurt, Germany.

References

Andersson, G., Larhammar, D., Widmark, E., Servenius, B., Peterson, P.A., and Rask, L.A.: Class II genes of the human major histocompatibility complex: Organization and evolutionary relationship of the *DRβ* genes. *J Biol Chem 262*:8748-8758, 1987

Figueroa, F., O'hUigin, C., Inoko, H., and Klein, J.: Primate *DRB6* pseudogenes -- clue to the evolutionary origin of the *HLA-DR2* haplotype. *Immunogenetics*, in press 1991

Gojobori, T., Li, W.-H., and Graur, D.: Patterns of nucleotide substitution in pseudogenes and functional genes. *J Mol Evol 18*:360-369, 1982

Grahovac, B., Mayer, W., Vincek, V., Figueroa, F., O'hUigin, C., Tichy, H., and Klein, J.: Major histocompatibility complex *DRB* genes of a new world monkey, the cotton-top tamarin (*Saguinus oedipus*). *Mol Biol Evol*, in press 1991

Graur, D., Shuali, Y., and Li, W.-H.: Deletions in processed pseudogenes accumulate faster in rodents than in humans. *J Mol Evol 28*:279-285, 1989

Hughes, A. and Nei, M. Nucleotide substitution at major histocompatability complex class II loci: evidence for overdominant selection. *Proc Natl Acad Sci USA 86*:958-962, 1989

Kimura, M.: Estimation of evolutionary distances between homologous nucleotide sequences. *Proc Natl Acad Sci USA 78*:454-458, 1981

Klein, D., Vincek, V., Kasahara, M., Schönbach, C., O'hUigin, C., and Klein, J.: Gorilla *Mhc-DRB* pseudogene orthologous to *HLA-DRBVIII*. *Hum Immunol*, in press 1991

Li, W.-H., Gojobori, T., and Nei, M.: Pseudogenes as a paradigm of neutral evolution. *Nature 292*:237-239, 1981

Li, W.-H., Wu, C.-I., and Luo, C.-C.: Nonrandomness of point mutation as reflected in nucleotide substitutions in pseudogenes and its evolutionary implications. *J Mol Evol 21*:58-71 1984

Li, W.-H., Wu, C.-I. and Luo, C.-C.: A new method for estimating synonymous and nonsynonymous rates of nucleotide substitution considering the relative likelihood of nucleotide and codon changes. *Mol Biol Evol*, *2*:150-174, 1985.

Matsuo, Y. and Yokoyama, S.: Cloning and sequencing of a processed pseudogene derived from a human class III alcohol dehydrogenase gene. *Am J Hum Genet 46*:85-91, 1990

Marsh, S.G.E. and Bodmer, J.G.: *HLA* class II nucleotide sequences, 1991. *Immunogenetics 33*:321-334, 1991

Miyata, T. and Yasunaga, T.: Rapidy evolving mouse α-globin-related pseudogene and its evolutionary history. *Proc Natl Acad Sci USA 78*:450-453, 1981

Nei, M. and Jin, L.: Variances of the average numbers of nucleotide substitutions within and between populations. *Mol Biol Evol 6*:290-300, 1989

O'hUigin, C., Chan, L., and Li, W.-H. Cloning and sequencing of bovine apolipoprotein A-I cDNA and molecular evolution of apolipoproteins A-I and B-100. *Mol Biol Evol 7*:327-339, 1990

Rollini, P., Mach, B., and Gorski, J.: Linkage map of three HLA-DR ß chain genes: Evidence for a recent duplication event. *Proc Natl Acad Sci USA 82*:7197-7201, 1985

Saitou, N. and Nei, M.: The neighbor-joining method. A new method for reconstructing phylogenetic trees. *Mol Biol Evol 4*:406-425, 1987

Vincek, V., Klein, D., Figueroa, F., Hauptfeld, V., Kasahara, M., O'hUigin, C., Mach, B., and Klein, J.: The evolutionary origin of the *HLA-DR3* haplotype. *Immunogenetics*, in press, 1991

ORGANIZATION AND EVOLUTION OF THE HLA-DRB GENES

Göran Andersson* Leif Andersson°, Dan Larhammar*, and Lars Rask+

*Department of Medical Genetics, Uppsala University, Biomedical Center, Box 589, S-751 23 Uppsala, Sweden

°Department of Animal Breeding and Genetics, Swedish University of Agricultural Sciences, Biomedical Center, Box 596, S-751 24 Uppsala, Sweden

+Department of Cell Research, Swedish University of Agricultural Sciences, Biomedical Center, Box 596, S-751 24 Uppsala, Sweden.

The DRB genes of the human major histocompatibility complex display extensive polymorphism and vary in number from one to four between haplotypes. Recently, the DRB genes have been classified into five allelic series. We review here the molecular organization of the various DR haplotypes and analyze the evolutionary relationships of the DRB genes. The variation in the number of DRB genes between haplotypes along with the complex similarities between allelic as well as nonallelic sequences makes definitive locus assignments difficult. We propose a simplistic view of the locus assignments of the DRB genes, which is based on 1) the relative chromosomal localization in the DR region, 2) which genes are linked on the same haplotype and, 3) the level of gene expression. These criteria can be met by a model system with only two expressed DRB loci, and two loci for the DRB pseudogenes. This model accepts the extensive variation between the presently recognized DRB alleles as being due to genetic exchange with nonallelic DRB genes.

HLA class II *DRB* genes of the human major histocompatibility complex are highly polymorphic. A striking feature of this genetic polymorphism is the large number of amino acid replacements between sequences. Furthermore, the number of nonallelic genes varies from one to four between different *DR* haplotypes. Based on common restriction fragment length polymorphism (RFLP) patterns, and common number of *DRB* genes, cells carrying distinct DR specificities could be organized into different haplotypic groups, confirming previously noted serological similarities (Böhme et al. 1985). The organization of the *DRB* genes from different *DR* haplotypes has been investigated using genomic cloning, RFLP analyses, and pulse field gel electrophoresis (PFGE; reviewed by Kappes and Strominger 1988). The present nomenclature of the *DRB* genes is based on this information and nucleotide sequences of the genes from the *DR2, DR3*, and *DR4* haplotypes (WHO Nomenclature Committee 1989). This nomenclature recognizes five loci with the *DRB1* locus representing the *DRB* genes expressing the polymorphic *DR* specificity. More than 30 alleles have been assigned to the *DRB1* locus (Marsh and Bodmer 1989, 1990). *DRB2* is a pseudogene found in the *DR3* haplotype. The loci *DRB3* and *DRB4* correspond to the *DRB*

NATO ASI Series, Vol. H 59
Molecular Evolution of the
Major Histocompatibility Complex
Edited by J. Klein and D. Klein

genes that express the DRw52 and the DRw53 specificities, respectively. These two loci are less polymorphic than the *DRB1* locus, and only four *DRB3* alleles and three *DRB4* alleles have been observed. The two *DRB* genes in the *DR2* haplotype were given the designations, *DRB1* and *DRB5*, respectively. Two additional genes in the *DR4* haplotype, *DRBΨ1* and *DRBΨ2*, with unique features apparent from their nucleotide sequences (Larhammar et al. 1985; Andersson et al. 1987), were not included in the present nomenclature nor was the pseudogene in the *DR2* haplotype (Kawai et al. 1989).

Genomic Organization of the DRB Genes.

DR regions of different haplotypes exhibit variation in the number of *DRB* genes and express different allelic forms of DRB molecules. *DRB* genes from the *DR2*, *DR3*, and *DR4* haplotypes have been cloned in overlapping cosmids (Rollini et al. 1985; Larhammar et al. 1985; Spies et al. 1985; Andersson et al. 1987; Kawai et al. 1989). The *DRB* genes are organized in a head-to-tail orientation, but are oriented in opposite orientation to the *DRA* gene (Fig. 1). The four *DRB* genes from the *DR4* haplotype extend over approximately 200 kb, whereas the three *DRB* genes from the *DR3* haplotype are contained within approximately 100 kb (Fig. 1). In contrast to the *DRB* genes, the number of other class II genes does not vary between haplotypes (Böhme et al. 1985).

Most *DR* haplotypes express two *DRB* genes, which exhibit distinct levels of expression. DRB genes that carry the serological specificities DRw52 and DRw53 are expressed in lower quantities as compared to *DRB* genes that carry the so called DR specificities. The DRw52 specificity is expressed in haplotypes expressing the DR3, 5, w6, and w8 DR specificities. Accordingly, DRw52 is called a supertypic specificity. The DR3 specificity has recently been split into two new specificities called DRw17 and DRw18. The DR5 specificity has been split into DRw11 and DRw12 and the DRw6 specificity into DRw13, and DRw14 (WHO Nomenclature Committee 1989). In this report we will use the old designations DR3, DR5, and DRw6, since the cells used in our study have not been retyped. The *DRB* genes, in the *DR3, 5*, and *w6* haplotypes, expressing the DR specificities and the DRw52 supertypic specificity have been named *DRB1* and *DRB3*, respectively. In addition to the two expressed *DRB* genes, these haplotypes contain one *DRB* pseudogene, denoted *DRB2*. The *DRw8* haplotype contains a single *DRB* gene that is closely related to DRB genes in the DRw52 supertypic group of haplotypes (Andersson et al. 1988; Gorski 1989; Jonsson et al. 1989). The *DR1* haplotype contains two *DRB* genes of which only the *DRB1* gene has been assigned to a locus. In the *DR2* haplotype there are three *DRB* genes of which one is a pseudogene which lacks locus assignment. The two expressed genes are denoted *DRB1* and *DRB5*, respectively (Fig. 1).

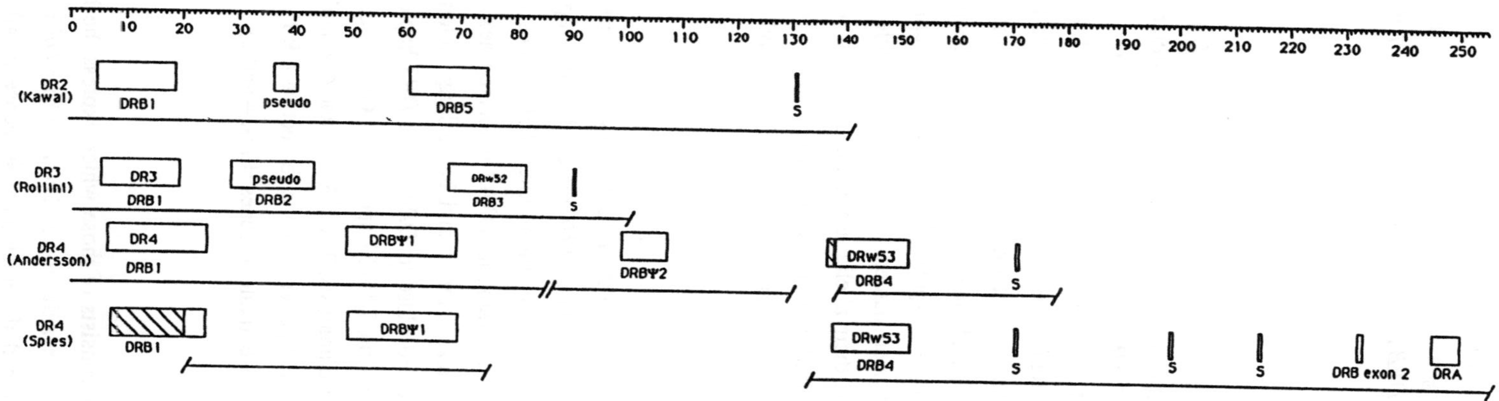

Fig. 1. Comparison of the *DR2*, *DR3*, and *DR4* regions. Schematic representation of the *DR2* region (Kawai et al. 1989), the *DR3* Rollini et al. 1985), and the *DR4* regions (Andersson et al. 1987; Spies et al. 1985). Genes are shown as open boxes. *DRB* genes are named according to the WHO Nomenclature Committee for Factors of the HLA system 1989. Hatched boxes in the DR4 libraries correspond to regions of the genes that were not cloned from the respective DR4 contigs. S denotes sequences which share nucleotide sequence similarities with exons encoding signal sequences. *DRB* exon 2 denotes a divergent exon 2 detected in all haplotypes tested (Haas et al. 1987), and *DRA* is the designation for the monomorphic gene encoding the α chain.

Haplotypes expressing the *DR4, 7,* and *9* specificities constitute another group containing four *DRB* genes with similar genomic organization (Böhme et al. 1985). These haplotypes contain two expressed genes, *DRB1* and *DRB4* (*DRw53β*) as well as two *DRB* pseudogenes, *DRBΨ1* and *DRBΨ2*.

The *DR1* and *-2* haplotypes

Previous RFLP analyses did not indicate a close relationship for the *DR1* nor for the *DR2* haplotypes with any of the other haplotypes (Böhme et al. 1985). Currently, no cloning data is available for the *DRB* genes from the *DR1* haplotype. The *DRB* genes from the *DR2* region were recently cloned by cosmid walking (Kawai et al. 1989). Like in the *DR3* haplotype (see below) this haplotype contains three *DRB* genes. However, the three *DRB* genes were proposed to be organized in an opposite order as compared to the *DRB* genes in other haplotypes (Kawai et al. 1989; Fig. 1). These expressed genes have been named *DRB1* and *DRB5*, respectively. In addition to the two expressed genes, the *DR2* haplotype contains a *DRB* pseudogene, which has not yet been named. Awaiting the nucleotide sequence determination of these three *DRB* genes, their allelic assignments remain tentative. Thus, both the RFLP analyses (Böhme et al. 1985) and the recent molecular cloning data (Kawai et al. 1989) clearly demonstrate that the *DR2* haplotype is distinct from all other haplotypes.

DR3, 5, and *w6* haplotypes

The genomic organization of the *DRB* genes from the *DR3* haplotype was established using cosmid cloning (Rollini et al. 1985, 1987). The *DRB* region in this haplotype covers approximately 100 kb of the chromosome and contains three *DRB* genes (Fig. 1). The polymorphic *DRB1* gene is located most centromeric. *DRB2* is located next to DRB1 and shows several pseudogene criteria. It lacks exon 2 and has a large deletion in exon 3 (Rollini et al. 1987). The *DRB* gene located close to *DRA* in the *DR3* haplotype is called DRB3. It is an expressed gene and encodes the DRw52 specificity. Only three *DRB3* alleles have been detected by nucleotide sequence analysis. The *DR3, 5,* and *w6* haplotypes can be grouped together based on the similar RFLP patterns of the respective *DRB* regions as well as a high degree of sequence similarity between allelic genes (Figs. 4 and 5).

DR4, 7, and *9* haplotypes

Another supertypic group of haplotypes consists of those which express the DR4, 7, and 9 specificities. They have a similar genomic organization and contain four *DRB* genes (Fig. 1). We have previously cloned and determined the nucleotide sequence of all the four *DRB* genes isolated from an HLA homozygous individual, serologically typed as DR4, Dw4, and DRw53 (Larhammar et al. 1985; Andersson et al. 1987). In addition to *DRB1* which expresses the *DR* specificity, these haplotypes contain one gene, *DRB4*, expressing the DRw53 supertypic specificity. This gene, as the *DRB3* gene from the *DR3, 5*, and *w6* haplotypes, displays limited allelic polymorphism within the supertypic group of haplotypes. In addition to the two expressed genes, the *DR4* haplotype contains two pseudogenes, denoted *DRBΨ1* and *DRBΨ2*. *DRBΨ1* has several deleterious mutations. Four of the splice junctions deviate from the G-T/A-G rule, and two premature termination codons as well as one frame-shift mutation are present (Larhammar et al. 1985). *DRBΨ2* is truncated and lacks exon 1 and 2. Moreover, it exhibits an unusually high ratio of replacement to silent substitutions as compared with expressed *DRB* genes (Andersson et al. 1987). It is very likely that *DRBΨ1* and *DRBΨ2* are also present in the *DR7* and *DR9* haplotypes. Previously, we have shown by RFLP analysis using genomic DNA prepared from *DR* homozygous typing cells that the hybridizing *DRB* fragments consistently have the same lengths in the *DR4, 7*, and *9* haplotypes. Some of these fragments correlate to *DRBΨ1* in the *DR4* haplotype as deduced from the nucleotide sequence information obtained from the cloned gene (Böhme et al. 1985, Andersson et al. 1987 and 1988). To gain further support for this view we performed Southern hybridizations to genomic DNA prepared from HLA homozygous individuals. The genomic DNA was probed with a *DRBΨ1*-specific probe corresponding to a unique retroviral insertion in intron 2 (Larhammar et al. 1985). With this probe we could detect a 6.5 kb *Bam* HI band exclusively in the *DR4, 7*, and *9* haplotypes (Fig. 2). The size of this band corresponds perfectly with the *Bam* HI fragment of *DRBΨ1*. The absence of RFLP using this probe indeed indicates that the strongly hybridizing band corresponds to *DRBΨ1*. Given the facts mentioned above, and the fact that haplotypes except *DR4, 7*, and *9* have fewer *DRB* genes, it is conceivable that *DRBΨ1* is unique for these three haplotypes. In addition to the strongly hybridizing band, two weaker bands were seen in all DR haplotypes (indicated by asterisks in Fig. 2). These bands may correspond to crosshybridizing endogenous copies of retroviral-like sequences present in the human genome.

The second pseudogene isolated in the *DR4* haplotype is called *DRBΨ2*. It lacks exons 1 and 2, corresponding to the signal sequence and the polymorphic extracellular domain of the *DRβ* chain, respectively. We have previously speculated that *DRB2* (Rollini et al.1987) in the *DR3* haplotype might be allelic to *DRBΨ2* based on the genomic localization of these genes (Andersson et al. 1987). However, this hypothesis was not supported by nucleotide sequence comparison of these two genes. Additional nucleotide sequence information of non-coding sequences from *DRB* genes might help to assign allelic relationships between these genes (see below).

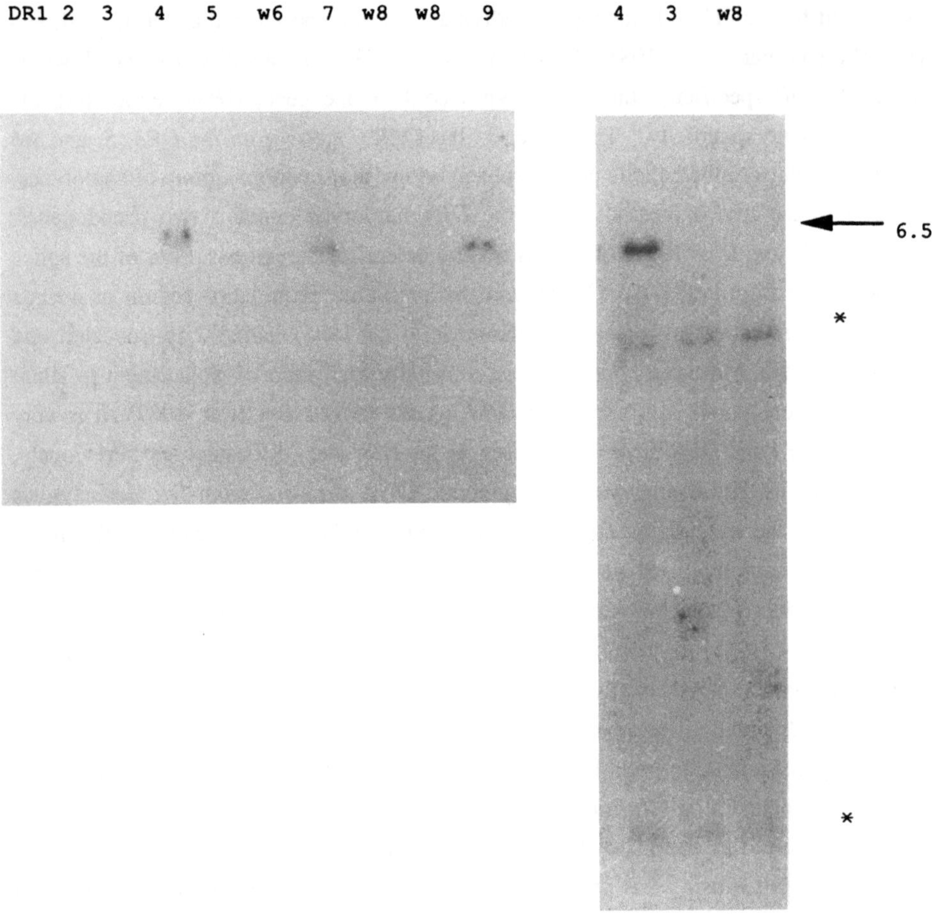

Fig. 2. Hybridization with the *DRBΨ1* probe. Hybridization with the *DRBΨ1*-specific probe to *Bam* HI digested human genomic DNA corresponding to DR specificities 1 to 9. The sizes of restriction fragments are given in kilobases. Asterisks denote two cross-hybridizing fragments detected in all *DR* haplotypes. The probe was a 300 bp *Stu* I fragment prepared from p801-2 (Larhammar et al. 1985). Southern blots were done using standard procedures.

The *DRw8* haplotype

The single DRB gene in the DRw8 haplotype expresses both the DRw8 specificity and the DRw52 specificity. RFLP analysis suggested that this DRB gene is most closely related to the *DRB3* gene in the *DR3, 5,* and *w6* haplotypes (Andersson et al. 1988). In contrast, the nucleotide sequence of a *DRB* cDNA clone isolated from a *DRw8* haplotype is clearly more similar to the *DRB1* sequences in that haplotypic group (Jonsson et al. 1989; Fig. 4). Based on these findings an hypothesis was put forward suggesting that the *DRw8* haplotype was generated by a gene contraction event from a *DRw52* haplotype containing both a *DRB1* and a *DRB3* gene (Jonsson et al. 1989), as also suggested by Gorski (1989).

Evolution of the *DR4 DRB* Genes

The nucleotide sequence analysis of the four nonallelic *DRB* genes from the *DR4* haplotype described previously (Larhammar et al. 1985; Andersson et al. 1987), permitted us to evaluate the evolutionary relationship of these genes. The nucleotide sequences revealed a high degree of similarities in exons as well as introns. Average intron differences between the genes are around 10 %. This is similar to the differences found in comparisons between exons corresponding to the 3'-untranslated region and the polymorphic first domain, which are 12% and 10%, respectively. Comparisons of exons 3, 4, and 5 gave very low average differences of 5%, 4%, and 2%, respectively. The short exon 5 is in fact identical between the *DRB1, DRB4*, and *DRBΨ2* genes (Andersson et al. 1987).

The common origin of the *DRB* genes in the *DR4* haplotype is supported by the presence of a 76 bp repeat, corresponding to the last 49 bp of exon 3 and 27 bp of intron 3, which is found in intron 2 of all four genes (Fig. 3). Most likely, this insertion was created prior to the duplication events that gave rise to the present day *DRB* genes in the DR4 haplotype. This is corroborated by the fact that each copy is more similar to the corresponding sequence in another gene than it is to the other copy within the same gene. Nucleotide sequence comparison of the four nonallelic *DRB* genes revealed a patchwork pattern of similarity, indicating that exchange of sequence information has occurred between these genes. For example, *DRB1* and *DRBΨ1* are very similar in the 5' end whereas in the 3' end, *DRB1* is more similar to *DRB4* and *DRBΨ1* to *DRBΨ2*. A conspicuous result of the sequence comparisons of the *DRB* genes is that no two *DRB* genes are more homologous than any other pair. Instead, a pattern of patchwork similarities was found. Thus, in different parts of the genes, pairwise higher degrees of nucleotide sequence similarities could be detected. Almost complete identity in long stretches of exon as well as intron sequences between different pairs of the genes indicate that genetic exchange mechanisms have homogenized these genes, making deductions of evolutionary relationship difficult. In summary, no pair of *DRB* genes in the *DR4* haplotype is more similar than any other pair.

76 bp DUPLICATION

```
        1       10       20        30         40         50
DRB1-i2 TGCCAAGTGGAGAACCCAAGTGTGACGAGCCCTCTCA**CAGTAGAATGGAG
DRB4-i2          C              T  C          G
DRB7-i2 GC                      T    A         G  G
DRB8-i2          T              T   **         G

DRB1-e3             C      CC                     G
DRB4-e3             C T    CA   T              G  GC
DRB7-e3   T   C     C      CA   T     T     TG    G
DRB8-e3             C T    CA   T              G  GC


                60        70
DRB1-i2 TGAGAAGCTTTCTGACCTCATAAATT
DRB4-i2    C        A
DRB7-i2     C               C   G
DRB8-i2                        AG

DRB1-e3 GT  C          T  G
DRB4-e3  T  C          T  C
DRB7-e3     C          T     C
DRB8-e3  T  C          T
```

Fig. 3. A 76 bp duplication in the *DR4 DRB* genes. Alignment of nucleotide sequences of a 76 bp repeat present in intron 2 in all *DRB* genes of the *DR4* haplotype, corresponding to the last 49 bp of exon 3 and 27 bp of intron 3. Nucleotide mismatches are indicated. Stars indicate gaps introduced to maximize similarity.

Separate *DRB* Exons

In all known *DR* haplotypes, a single *DRB* first domain exon has been detected by genomic hybridization (Haas et al. 1987). Nucleotide sequence determination of this exon from the *DR4* and *DR7* haplotypes showed that it is a highly divergent exon with multiple pseudo criteria (Meunier et al. 1986; Haas et al. 1987; Jonsson et al., manuscript in preparation). This exon is the most divergent member of the *DRB* gene family found so far. The extreme divergence of this exon indicates that it has escaped the homogenization process that is acting on other *DRB* genes (see above). Using a phylogenetic analysis, Hughes and Nei (1990) noted that this *DRB* exon branches together with other primate *DRB* genes. However, the branch for this sequence is very long reflecting its high divergence. Sequence analysis of this 1st domain exon from additional haplotypes should be informative regarding evolutionary relationships among *DR* haplotypes. In addition to the separate first domain exon, three separate highly divergent signal sequence exons have been detected in the *DR4* haplotype (Spies et al. 1985; Jonsson et al., manuscript in preparation; Fig. 1).

Comparison of *DRB* Allelic Nucleotide Sequences

A dendrogram was constructed by the parsimony method using the PAUP computer program (Dr. D.L. Swofford, Illinois Natural History Survey, Champaign, Illinois), to illustrate existing sequence similarities among the *DRB1, DRB3, DRB4, DRB5*, and *DRBΨ1* genes (Fig. 4). The most striking observation is that the DRB1 sequences do not cluster in the same branch. On the contrary, these presumed allelic sequences intermingle with the nonallelic sequences. This analysis clearly shows that the current locus assignments of human *DRB* genes does not have a clearcut genetic basis where allelic genes are more similar to each other than nonallelic are. The parsimony analysis also supports previous findings of close genetic relationships between the *DRB4* genes of the DRw53 supertypic group of haplotypes on the one hand, and the *DRB3* genes of the DRw52 supertypic group of haplotypes on the other hand (Rollini et al. 1985, 1987; Andersson et al. 1987).

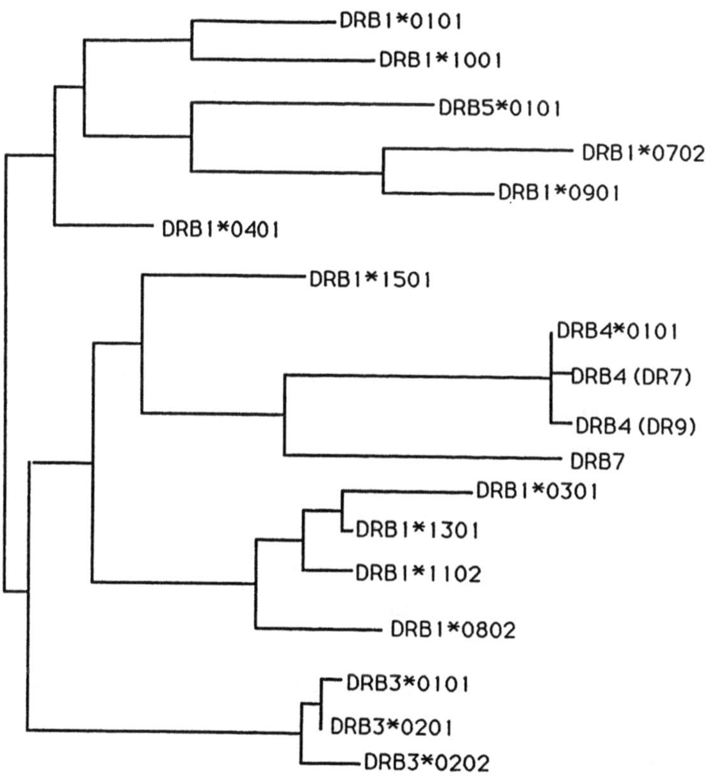

Fig. 4. Parsimony tree based on *DRB* nucleotide sequences. The tree is adapted from Jonsson et al. (1989), and was built using the computer program PAUP. For nucleotide sequence references see Jonsson et al. (1989).

Consistent with previous RFLP data, this analysis groups the *DRw8B* sequence together with the *DRB1* sequences of the DRw52 supertypic group of haplotypes. Finally, *DRB* sequences from the *DR1, DR2*, and *DRw10* haplotypes do not show any close relationship with any other *DRB* sequences using this type of analysis. This is in agreement with the distinct RFLP patterns of these haplotypes (Böhme et al. 1985).

To further study the allelic relationships of expressed human *DRB* genes we have calculated the relative frequencies of silent and replacement substitutions in pairwise comparisons of *DRB1, DRB3, DRB4*, and *DRB5* sequences. This approach was recently used to illustrate the relationships of bovine *DQB* genes (Sigurdardottir et al. submitted). The analysis was carried out separately for the first and second domain exons (Fig. 5A and 5B, respectively). The most striking observation is the great overlap in the data between non-allelic comparisons and comparisons of presumed *DRB1* sequences. This analysis clearly shows that the human expressed *DRB* genes do not exhibit any distinct locus characteristics in their coding sequences. Three clusters of closely related sequences are easily recognized in agreement with the parsimony analyses (Fig. 4). These are 1) the *DRB1* sequences in the *DR3, 5*, and *w6* haplotypes, 2) the *DRB3* alleles, and 3) the *DRB4* sequences. The clusters of observations close to the origo in Figures 5A and B all represent pairwise comparisons within these three groups.

Discussion

Our studies concerning the evolutionary relationships of the *DRB* genes suggest a complex evolutionary history involving duplications, contractions and exchange of genetic information between these genes. A conspicuous finding is the pattern of patchwork similarities among the four nonallelic *DRB* genes from the *DR4* haplotype as well as among human *DRB* sequences in general. The patchwork similarities support the notion that genetic exchange between these genes have occurred, thereby obscuring the allelic relationships. The variability in the number of genes among haplotypes and the fact that the nonallelic genes do not appear to evolve independently make unambiguous locus assignments very difficult. According to the current nomenclature, one gene in each haplotype has been assigned to the *DRB1* locus on the basis of its map location and high level of expression. Thus, in spite of the sequence heterogeneity, it has been considered operationally appropriate to regard the *DRB1* sequences as alleles of a single locus. The second expressed *DRB* gene in each haplotype (except *DRw8*), often expressed at lower level compared with *DRB1* has been assigned to one of the *DRB3, DRB4*, or *DRB5* loci depending on haplotype. To the best of our knowledge the *DRB3, DRB4*, or *DRB5* genes are never inherited on the same chromosome and occur in equivalent map positions on their respective chromosomes.

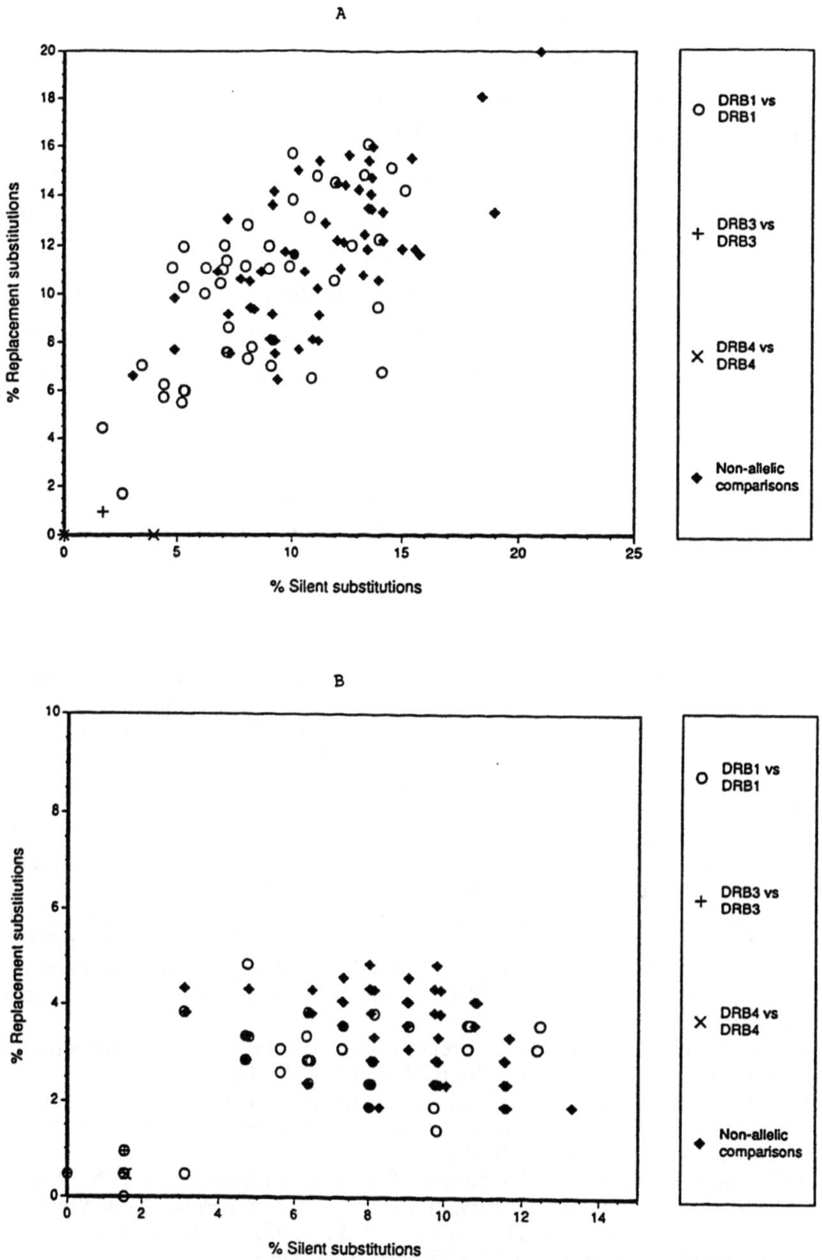

Fig. 5. Plot of the relative frequencies of silent and replacement substitutions in *DRB* sequences. The relative frequencies of silent and replacement substitutions in pairwise comparisons of human *DRB1*, *DRB3*, *DRB4*, and *DRB5* sequences. The frequencies of silent and replacement substitutions were estimated according to the method of Nei and Gojobori (1986) applying Jukes and Cantor's (1969) correction for multiple hits. The sequences used are those given in Fig. 4 excluding the DRBΨ1 sequence. **A**, first domain exon; **B**, second domain exon.

In analogy with the *DRB1* allelic series, *DRB3, DRB4*, and *DRB5* may be treated as alleles of a second locus. This "locus" would not be more heterogenous than the present *DRB1* locus. The two sequenced pseudogenes found in the *DR4* haplotype clearly demand separate locus assignments. The pseudogenes in the *DR2* and *DR3* haplotypes seem as divergent from each other as from the two *DR4* pseudogenes and all expressed genes. Thus, sequence analyses and exon organization comparisons (Fig. 1) would seem to warrant four *DRB* pseudogene loci. However, as all of these are located between the two expressed genes, two pseudogene loci would suffice in analogy with the *DRB1* allelic series. Cloning, sequence analysis and comparison of noncoding sequences of additional *DRB* genes from more haplotypes is expected to add to our understanding of the evolutionary relationship in this complex genetic locus.

Acknowledgment. This work was supported by grants from the Swedish Natural Science Research Council, Swedish Cancer Society and the Swedish Medical Research Council.

References

Andersson, G., Larhammar, D., Widmark, E., Servenius, B., Peterson, P. A., and Rask, L.: Class II genes of the human major histocompatibility complex. Organization and evolutionary relationship of the DRß genes. *J Biol Chem 262*: 8748-8758, 1987

Andersson, G., Lindblom, B., Andersson, L., Gorski, J., Mach, B., and Rask, L.: The single *DRβ* gene of the *DRw8* haplotype is closely related to the *DRβ3III* gene encoding DRw52. *Immunogenetics 28*: 1-5, 1988

Böhme, J., Andersson, M., Andersson, G., Müller, E., Peterson, P.A., and Rask, L.: HLA DRß genes vary in number between different DR specificities, whereas the number of DQß genes is constant. *J Immunol 135*: 2149-2155, 1985

Gorski, J.: The HLA-DRw8 lineage was generated by a deletion in the DR B region followed by first domain diversification. *J Immunol 142*: 4041-4045, 1989

Haas, D., Boss, J.M,, Strominger, J.L., and Spies, T.: A highly diverged ß1 exon in the DR region of the human MHC: sequence and evolutionary implications. *Immunogenetics 25*: 15-20, 1987

Hughes, A. and Nei, M.: Evolutionary relationships of class II major histocompatibility-complex genes in mammals. *Mol Biol Evol 7*: 491-514, 1990

Jonsson, A.-K., Andersson, L., and Rask, L.: A cellular and functional split in the DRw8 haplotype is due to a single amino acid replacement (DRß ser 57- asp 57) *Immunogenetics 29*: 308-316, 1989

Jukes, T.H. and Cantor, C.R.: Evolution of protein molecules. *In* H.N. Munro (ed.): *Mammalian Protein Metabolism*, pp 21-132, Academic Press, New York 1969

Kappes, D. and Strominger, J.L.: Human class II major histocompatibility complex genes and proteins. *Annu Rev Biochem 57*: 991-1028, 1988

Kawai, J., Ando, A., Sato, T., Nakatsuji, T., Tsuji, K., and Inoko, H.: Analysis of gene structure and antigen determinants of DR2 antigens using DR gene transfer into mouse L cells. *J Immunol 142*: 312-317, 1989

Larhammar, D., Servenius, B., Rask, L., and Peterson, P.A.: Characterization of an HLA DRß pseudogene. *Proc Natl Acad Sci USA 82*:1475-1479, 1985

Marsh, S.G.E. and Bodmer, J.G.: HLA-DR and -DQ epitopes and monoclonal antibody specificity. *Immunol Today 10*: 305-312, 1989

Marsh, S.G.E. and Bodmer, J.G.: HLA-DRB nucleotide sequences. *Immunogenetics 31*: 141, 1990

Meunier, H.F., Carson, S., Bodmer, W.F., and Trowsdale, J.: An isolated ß1 exon next to the DRα gene in the HLA-D region. *Immunogenetics 23*: 172-180, 1986

Nei, M. and Gojobori, T.: Simple methods for estimating the numbers of synonymous and nonsynonymous nucleotide substitutions. *Mol Biol Evol 3*: 418, 1986

Rollini, P., Mach, B., and Gorski, J.: Linkage map of three HLA-DRß chain genes: evidence for a recent duplication event. *Proc Natl Acad Sci USA 82*: 7197-7201, 1985

Rollini, P., Mach, B., and Gorski, J.: Characterization of an *HLA-DRβ* pseudogene in the DRw52 supertypic group. *Immunogenetics 25*: 336-342, 1987

Spies, T., Sorrentino, R., Boss, J., Okada, K., and Strominger, J. Structural organization of the DR subregion of the human major histocompatibility complex. *Proc Natl Acad Sci USA 82*: 5165-5169, 1985

WHO Nomenclature Committee: Nomenclature for factors of the HLA system, 1989. *Immunogenetics 31*: 131-140, 1990

THE MHC OF *PEROMYSCUS LEUCOPUS* (*Mhc-Pele*) ILLUSTRATES LARGE- AND SMALL-SCALE EXPANSION IN THE PHYLOGENY OF MHC LOCI

M.D. Crew, G.S. Smith, and R.L. Walford
Department of Pathology, Center for Health Sciences, UCLA School of Medicine, Los Angeles, CA 90024-1732, USA

Our current understanding of the Peromyscus leucopus *major histocompatibility complex (*MhcPele*) genes is reviewed. Several Pele class I loci appear to have undergone duplications relative to class I genes of the* H-2. *In addition, short-tract sequence reiteration in the transmembrane domain-encoding exon of* Pele *class I genes and several otehr mammalian class I genes is observed. Thus, class I genes of* Pele *reflect evolutionary expansion of class I loci at widely disparate levels.*

Members of the genus *Peromyscus* (family Cricetidae) are among the most successful rodents in terms of habitat range and number of species (Baker 1968; Carlton 1989; MacMillen and Garland 1989). From the Arctic circle to the Panama canal, an indigenous *Peromyscus* can be trapped. Owing to this, they have the distinction of being perhaps the most intensely studied mammal by field biologists (Kaufman and Kaufman 1989). *Peromyscus* has another noteworthy attribute - long lifespan. *P. leucopus* lives 7-8 years versus 3-4 years for *Mus* and *Rattus* (Sacher and Hart 1978; Smith et al. 1990). Thus, in cross-species assessments of aging parameters, the deer mouse (*P. maniculatis*) or the white-footed mouse (*P. leucopus*) frequently provides a point on gerontologists' biomarker versus lifespan plots (c.f. Sacher and Hart 1978; Wilson et al. 1987).

Operating under the premises that (i) the relationships between MHC genes from other species will prove insightful to structure-function relationships in present day human genes and (ii) the study of a long-lived rodent species will shed some light on the role of the MHC in regulating the rate of aging in *Mus* (Smith and Walford 1977; Walford 1990), we have begun an exploration of the MHC of *P. leucopus* (*MhcPele*). Deep in the bowels of the UCLA Center for Health Sciences is probably the world's largest colony of *Peromyscus* leucopus and, as far as is known, the only colony of inbred *Peromyscus*. Employing these animals, *MhcPele* has recently taken shape at the molecular level.

This essay is divided into two parts. First, we will review our current knowledge of *Pele* and how it may play in interpreting the origins of modern day MHCs. Secondly, we will discuss transmembrane domain (TM) length variation in the evolution of mammalian MHC class I genes, a feature that became apparent through the analysis of *Pele* class I genes. The first part emphasizes what is becoming a leitmotif in mammalian MHC evolution -

NATO ASI Series, Vol. H 59
Molecular Evolution of the
Major Histocompatibility Complex
Edited by J. Klein and D. Klein

expansion of class I and II loci. While the second part may be thought of as a microcosm of the first, we stress that the underlying mechanisms are not necessarily similar.

Genes of *MhcPele* - expansion of gene families

Variation in the number and types of MHC loci among mammals (especially those encoding class I molecules) are two of several levels of diversity exhibited by MHCs. Relative to the *HLA* and *H-2* systems, both *Pele* class I and II loci provide examples of intra- and inter-specifies variation of loci type and number.

MhcPele class I genes. Initial cloning and characterization of *Pele* class I genes has been presented elsewhere (Crew et al. 1990). A few major points from this and more recent data (unpublished) are summarized in Table 1 and Figure 1 and deserve elaboration. First, *Pele* class I loci constitute a large multi-gene family which judging from Southern blot analysis and genomic cloning experiments consists of *at least* forty members. Second, and more importantly, *Pele* class I genes corresponding to *H-2* nonclassical class I genes (i.e. those in the *Qa*, *Tl*, and *Hmt* regions) can be identified based upon exon 5 (TM-encoding) sequences. (Classification of *H-2* genes by this method is both convenient and perhaps biologically relevant [Brorson et al. 1989]). Three subtypes of *Pele* class I genes are evident. *Pele-A* genes form the largest group and are most similar to BALB/c mouse *H-2D*, *K*, *L*, *Q*, and six *T* genes. The *Pele-B* genes are similar to several *Tl* region genes (BALB/c *T1-3*, *T6*, and *T16-18*). *Pele-C* genes are likely to be homologs of *H-2M1* or other mouse *Hmt* genes. Thus, these genes have been present since the divergence of Muridae and Cricetidae families - about forty million years ago (MYA; Brownell 1983; see below).

Table 1. *Pele* class I genes

subtype	*H-2* homologs	number of genes[1] SB	GC	cDNA	expression[2]	polymorphism[3]
Pele-A	*K,D,L,Q*	>25	21	21	+++	+
Pele-B	*T1-3,T6, T16-18*	3->10	13	2	+/-	+
Pele-C	*M1*	4	2	0	-	-
Pele-X	?	6	6	2	-	-

[1]Number of genes refers to the number estimated by Southern blots (SB), genomic cloning (GC), or by characterization of PCR-derived cDNAs (cDNA).
[2]Expression by Northern blot analysis. +++ indicates strong signal obtained; +/- indicates weak signal obtained with some probes; - indicates no hybridization detected.
[3]Allelic polymorphism detected by RFLP analysis. + indicates polymorphism detected with some gene-specific probes.

Hughes and Nei (1989) have proposed that nonclassical loci evolve rapidly in mammals and were derived independently in distantly related phylogenetic families. Our data does not discount this hypothesis but rather indicates that *H-2* nonclassical genes arose prior to the emergence of Muridae rodents. However, the often cited claim that nonclassical class I genes have not been evolutionarily conserved in closely related species and are thus nonfunctional (Rogers 1985) may need amendment.

```
            PPPSTVSNMVIIAVLVVLGAVIIIGAVVAFVMKR   RRNT
H-2Kd       L--------T---------A-VT----------M   ----
H-2Ld       -----D-Y---V---G----MA------------   ----
H-2Dd       --S--KT-T-----P------V-L---M------   ----
Pele-A11b   --Q-M-TI-GT-----L----- F-----V-KM-   ----
Pele-A52a   --Q---PI-A----P-L---G------W-V-WW-   ----
Pele-A42c   --Q----I-A----    V---- -----V-WW-   ----
H-2T3       --QTSMP-RTTVRA  L---M-TL-FMSGS--MWM  -K-N
H-2T2       --QPIIFIRT-VGT  ----GV-L-FIIGG-KMWMKKKRK
Pele-B40b   --QPSIPI-T-VCA  -----V-L-FIIGG--MWV  -K--
Pele-B11a   --Q SIPI-T-VCA  -----V-L-FII-G--MWM  -KK-
H-2M1       --EP-IPFLA-V-I- ----LLM ---MT-LI  TK--TR
Pele-C53    --QPS-PI-P-VTG- -----LM ----T-LI  WK--TK
```

Fig. 1. Examples of *Pele* class I genes corresponding to *H-2* classical and nonclassical genes based upon exon 5 sequences. Amino acid sequences from several *Pele* class I genes and *H-2* genes are compared to the consensus sequence of *H-2* "group 1" transmembrane domains (top line, Brorson et al. 1989). Dashes indicate identity to the consensus sequence. Gaps introduced for alignments are shown by spaces in the sequence. Residues in bold and italic type denote putative subtype-specific residues. Sequences were taken from Brorson et al. (1989), Crew et al. (1990, 1991), Singer et al. (1988), and our own unpublished data.

Southern blots using probes specific for each subtype suggest that number of both *Pele-A* and *Pele-B* loci differ among our *P. leucopus* lines (unpublished data). Of these two subtypes, variability in the number of *Pele-B* genes is more dramatic (from three genes to over ten). Preliminary results with *Pele* class I gene-specific probes suggests that only a small fraction of genes in each subtype are polymorphic among our inbred lines of *P. leucopus*.

The complexity of *Pele* class I loci is mirrored in the expression of *Pele* class I genes (Table 1, unpublished observations). Northern analysis using probes specific for each subtype indicates that members of the *Pele-A* group are far more abundantly expressed than *Pele-B* and *-C* genes. PCR-mediated cDNA cloning experiments suggest that at least 25 genes are transcribed - most are *Pele-A* genes, though two are *Pele-B* genes. Hybridization of *Pele* class I gene genomic clones with the most abundant PCR-derived cDNAs have identified *Pele* class I genes which we tentatively designate as the major transplantation antigen genes of *Peromyscus*.

Evidence that several *Pele* class I loci have expanded relative to their *H-2* counterparts comes from the following observations: *(i)* Every *Pele* class I nucleotide sequence obtained thus far is more homologous to another *Pele* gene than to MHC genes from any other species. (However, it is possible that this is due to sequence homogenization through gene conversion events [Rada et al. 1990]). *(ii)* Some *Pele-A* and *-B* genomics clones differ from 3-5 other similarily subtyped genomic clones by one or two restriction sites (8-10 different restriction enzymes tested per clone). As an outcome of this, Southern blots have consistently underestimated the number of *Pele* class I loci. *(iii)* Nearly identical short-tract duplications are seen in at least five *Pele-A* genes (see below). Either the duplications arose independently, or probably more likely, each *Pele-A* gene exhibiting this feature was derived from a common primordial gene.

Table 2. Cosegregation, gene number, and polymorphism of *Pele* class II genes

P. leucopus line	*Pele* class II gene(s)[1]			
	DQα-like	*DRα* (*Eα*)-like	*DQβ**DPβ*-like	*DRβ*-like
109	2,2	2,2	2,2	1,1
8	2,4	1,2	2,4	1,1
18	4,4	1,1	4,4	1,1
14	3,3	2,2	3,3	1,1
157	2,3	2,2	2,3	1,1
number of genes:	1	1	4	1
number of alleles or genotypes:	5	2	5	1

[1]Pairs of numbers in each column refer to the alleles or genotype contained in these individuals according to the RFLP signatures described in Crew et al. (1989).

MhcPele class II genes. An examination of *Pele* class II genes using mouse and human derived probes (Crew et al. 1989) is summarized in Table 2. Several features worth noting are *(i)* the *DQα*- and *DQß/DPß*-like alleles display a high degree of polymorphism, *(ii)* there is an additional *DQß/DPß*-like locus relative to the *H-2*, and *(iii)* a *DRα*(*Eα*)-like allele contains a deletion encompassing the α1 domain. Since the MHC of another Cricitidae rodent, *Spalax ehrenbergi* (*MhcSpeh*) also contains additional *DQß/DPß*-like loci (Nizetiç et al. 1987), it is likely that the expansion of class II *Q* and *P* loci occurred before the divergence of *Spalax* and *Peromyscus* species.

MhcPele class III genes. This catchall category includes genes encoding the cytokines lymphotoxin and tumor necrosis factor. So far, the chief value of our studies of *Pele* class III genes has been in deriving an estimate of the time of divergence of *P. leucopus* from *Mus* and *Rattus* by comparison of silent substitutions in mouse, rat and *P. leucopus* tumour necrosis factor genes. (This is difficult to ascertain using *Pele* class I and II genes since orthologous loci are difficult to unambigously define). This analysis puts the Muridae/Cricetidae split at 38-46 MYA. *Mus* and *Rattus*, by this same method, diverged about 25 MYA (unpublished data).

c-----...c	--t---------------	-------c-t--------	---g---tt-	*I*
c-----...g	t---------------a-	-a---c-t-a--ttt-a-	-ag--gaag-	*II*
t--c-ggcct	---------a-c--t---	-------...--------	-------gc-	*III*
CTGTTC----	**GGCTGTCCTTGGAGCTGT**	**GGTCATC-T-GGAGCTGT**	GGTCGCT--T	*IV*

Fig. 2. An early duplication event in the phylogeny of mammalian classical and nonclassical MHC class I genes. Exon 5 consensus sequences of *H-2* group 1 (*H-2K,D, L,* and most *Q*), *H-2* group 2 genes (*H-2T2, T12*) and HLA-B genes (drawn from Brorson et al. 1989; Parham et al. 1989), respectively denoted *I-IV* at the left, were aligned using the Needleman and Wunsch (1970) algorithm. Each unit of a direct repeat is boxed. The first forty and last thirty nucleotides of exon 5 were omitted.

Transmembrane domains of mammalian class I genes - expansion of gene segments

During the course of characterizing *Pele* class I genes, we noticed a previously unrecogonized facet of MHC class I gene evolution - namely, that a short-tract (18 base pairs, bps) duplication in the middle of (modern day) exon 5 was an early event in the phylogeny of mammalian class I molecules (Crew et al. 1991). Figure 2 shows that such an event preceded the divergence of *H-2D* and *Tl* genes and in fact, occured in a common ancestor of rodents and primates.

Even more intriguing is that short-tract duplications have occured again and again in both rodents and primates, and in roughly the same portion of exon 5 (Crew et al. 1991). Figure 3 depicts some examples. That these repeats have independent origins is supported by the following observations: *(i)* The same sequence duplicated in *Pele-A* genes is triplicated in the rat class I gene, RT.BM1. *(ii)* A duplication of 15 bps rather than 18 bps is observed in several rat class I genes (e.g. RT16 in Fig. 3). *(iii)* The 9 bp repeat in *ChLA-B1* described by Mayer et al. (1988) and shown in Fig. 3*f*, is observed in only one other *ChLA-B* allele. (iv) While 18 bp duplications of nearly the same sequence are observed in exon 5 of both *H-2Q1* (Fig. 3*c*) and *RT1.A*^{a} (not shown), they are clearly metalogous loci, being found telomeric and centromeric, respectively of class II loci (Rada et al. 1990). Taken together, it appears that the fifth exon of MHC class I genes is a hotspot for such mutations.

The *chi*-site of bacteriophage λ (GCTGGTGG) has been implicated in the recombination of class II loci (Steinmetz et al. 1986) and in micro-recombination (i.e. gene conversion) of class I loci (Wheeler et al. 1990). *chi*-like sequences (GCTGTGG) are in fact present in and around most of the exon 5 duplicated sequences (Fig. 4). In addition, as shown in Figure 4, tandem repeats observed in rodent exon 5 sequences are also very similar to tandem repeats of a human hypervariable minisatellite core sequence (Jeffreys et al. 1985). A causal role of hypervariable repeat-like motifs in generating exon 5 short-tract duplications may be inferred from the fact that extremely polymorphic nature of minisatellite loci in humans is predominantly due to the high frequency of tandem duplications and deletions of these DNAs (Jeffreys et al. 1988).

```
a.  55
    TTGGAGCTAT GGCCATCATTGGAGCTGT GGCCATCATTGGAGCTGT GGTGGCTGTT
    Pele-A24

b.  55
    TTGGAGCTGT GGCCATCATTGGAGCTGT GTCCATCATTGGAGCTGT GGCGGCTGTT
    Pele-A34c

c.  37
    TAGCTGTTCT GGTTGTCCTTGGAGCTGT GATCGTCATTGGAGCTGTGGTAATCATT
    H-2Q1

d.  46
    TTGGAGCTGT GGCCATCATTGGAGCTGT GGCCATCATTGGAGCTGT GGCCATCATTGGAGCTGTTGTGAGGGAG
    RT.BM1

e.  59
    AGCTGTGGCC ATCATTGCAGCTGTG ATCATTGCAGCTGTG GTGGCTGTTG
    RT16

f.  65
    CACCGTAGCT GTGGTCGCT GTGGTCGCT GCTGTGATGT
    ChLA-B1
```

Fig. 3. Examples of short-tract duplications in MHC class I genes. Portions of MHC class I exon 5 sequences are shown with arrows denoting each repeated unit. The numbers above each sequence indicates distance (in base pairs) to the start of exon 5. The source of sequences is given in Crew et al. (1991).

```
          ---------------------->----------------------->
ATTGGAGCTATGGCCATCATTGGAGCTGTGGCCATCATTGGAGCTGTGGT   Pele-A24
XX XXXX   XXX XX  XX XXXX   XXX XX  XX XXXX   XXX
AXGGGAGG--TGGGCAGGAXGGGAGG--TGGGCAGGAXGGGAGG--TGGG   minisatellite
```

Fig. 4. Homology of exon 5 direct repeats to the l *chi*-site and human hypervariable minisatellite sequences. l *chi*-site-like sequences are in bold type and repeats are denoted by arrows.

Conclusions and Prospectus

Pele class I and II loci illustrate an evolutionary theme of expansion at two widely disparate levels. The largest level is the duplication of chromosomal segments yielding gene families. It has been proposed that such duplicated genes rapidly undergo mutation giving rise to (unexpressed) psuedogenes (Hughes and Nei 1989). The best example of an expanded *Pele-A* gene family are those genes which have, in addition, undergone expansion at the lowest level - i.e. these genes have short-tract duplications in exon 5. Most members of this group are abundantly and ubiquitously expressed (Crew et al. 1991) suggesting that their reiteration was a relatively recent event. The functional relevance of both large- and small-scale expansions is of great interest. We expect that further studies of *Pele* genes will contribute in this regard. For example, the dubbing of H-2 nonclassical class I genes as "evolutionary junkyards" (Klein and Figueroa 1986) is an apropos analogy if one considers that these genes may not merely rust away but may supply variation for classical class I antigens through template directed mutational events (i.e. gene conversion). The greater number of nonclassical class I loci in mice relative to primates and the higher degree of inter-loci segmental exhange in H-2 class I genes may be taken as support for this hypothesis (Lawlor et al. 1990, Parham et al. 1989). One might therefore predict that *Pele* class I loci contain a high number of alleles that show allelic differences attributable to gene conversion because of the greater number of potential donor sequences.

As to the functional ramifications of TM length variation, it will be important to determine whether these represent allelic differences (as it apparently does for *ChLA-B* genes) or whether lengthened TMs are peculiar for all alleles of a particular locus (as is the case for *H-2Q1* genes). Clues about the selective pressures operating on the TM will be gained by such studies coupled with conventional mutate and transfect approaches.

Finally, Klein and colleagues (1990) has pointed out in regards to the trans-specific nature of MHC polymorphism, that the true age of MHC alleles will become apparent only when alleles from diverse rodents or primates are sequenced. In the case of *H-2* alleles, *Peromyscus* seems to be an appropriate paradigm. In many ways *Pele* seems to be fertile ground for the growth of MHC evolutionary concepts.

Acknowledgements. We would like to thank E. Zeller, M. Filipowsky, and M. Neshat for their excellent technical assistance. Support for this work was provided from U.S. Public Health Service Grants AG04419-06 and AG08936-01 from the National Institute on Aging and U.S. Public Health Service Training Grant T32 AG-00121-01A1.

References

Baker, R.H.: Habitats and distibution. *In* J.A. King (ed.): *Biology of Peromyscus (Rodentia)*, pp. 98-126, American Society for Mammals Special Publication No. 2, Stillwater, Oklahoma 1968

Brorson, K.A., Hunt, S.W., Hunkapiller, T., Sun, Y.H., Cheroutre, H., Nickerson, D.A., and Hood, L.: Comparison of exon 5 sequences from 35 class I genes of the BALB/c mouse. *J Exp Med 170:* 1837-1858, 1989

Brownell, E.: DNA/DNA hybridization studies of muriod rodents: symmetry and rates of molecular evolution. *Evolution 37:* 1034-1051, 1983

Carlton, M.D. Systematics and Evolution. *In* G.L. Kirkland and J.N. Layne (eds.): *Advances in the Study of Peromyscus (Rodentia)*, pp. 7-142, Texas Tech, Lubbock, Texas 1989

Crew, M.D., Smith, G.S., Zeller, E.C., and R.L. Walford: Polymorphism in the major histocompatibility complex class II genes of *Peromyscus leucopus. Immunogenetics 30:* 214-217, 1989

Crew, M.D., Filipowsky, M.E., Zeller, E.C., Smith, G.S., and R.L. Walford: Major histocompatibility complex class I genes of *Peromyscus leucopus. Immunogenetics 32:* 371-379, 1990

Crew, M.D., Filipowsky, M.E., Smith, G.S. and R.L. Walford: Transmembrane domain length variation in the evolution of major histocompatibiltiy complex class I genes. *Proc Natl Acad Sci USA 88*, in press 1991

Hughes, A.L. and Nei, M.: Evolution of the major histocompatibility complex: independent origin of nonclassical class I genes in different groups of mammals. *Mol Biol Evol 6:* 559-579, 1989

Jeffreys, A.J., Wilson, V., and Thein, S.L.: Hypervariable 'minisatellites' regions in human DNA. *Nature 314:*67-73, 1985

Jeffreys, A.J., Royle, N., Wilson, V., and Wong, Z.: Spontaneous mutation rates to new length alleles at tandem-repetitive hypervariable loci in human DNA. *Nature 332:*278-281, 1988

Kaufman, D.W. and Kaufman, G.A.: Population biology. *In* G.L. Kirkland and J.N. Layne (eds.): *Advances in the Study of Peromyscus (Rodentia)*, pp. 233-270, Texas Tech, Lubbock Texas 1989

Klein, J., Kasahara, M., Gutknecht, J., and Figueroa, F.: Origin and function of *Mhc* polymorphism. *In* B.H. Waksman (ed.): *1939-1989: Fifty Years Progress in Allergy*, pp. 35-50, Karger, Basel, 1990

Klein, J. and Figueròa, F.: Evolution of the major histocompatibility complex. *CRC Crit Rev Immunol 6:* 295-388, 1986

Lawlor, D.A., Zemmour, J., Ennis, P.D., and Parham, P.: Evolution of class I MHC genes and proteins: from natural selection to thymic selection. *Annu Rev Immunol 8*:23-63, 1990

MacMillen, R.E. and Garland, T.: Adaptive physiology. *In* G.L. Kirkland and J.N. Laynyne (eds.): *Advances in the Study of Peromyscus (Rodentia)*, pp. 143-168, Texas Tech, Lubbock Texas 1989

Needleman, S.B. and Wunsch, C.D.: A general method applicable to the search for similarities in the amino acid sequence of two proteins. *J Mol Biol 48:* 443-453, 1970

Nizetic, D., Figuero, F., Dembic, Z., Nevo, E., and Klein, J.: Major histocompatibility complex gene organization in the mole rat, *Spalax ehrenbergi:* evidence for transfer of function between class II genes. *Proc Natl Acad Sci USA 84*: 5828-5832, 1987

Parham, P., Lawlor, D.A., Lomen, C, and Ennis, P.D.: Diversity and diversification of *HLA-A,B,C* alleles. *J Immunol 142:*3937-3950, 1989

Rada, C., Lorenzi, R., Powis, S.J., van den Bogaerde, J., Parham, P. and Howard, J.C.: Concerted evoltuon of class I genes in the major histocompatibility complex of murine rodents. *Proc Natl Acad Sci USA 87:*2167-2171, 1990

Rogers, J.H.: Mouse histocompatibility-related genes are not conserved in other mammals. *EMBO J 4*:749-753, 1985

Sacher, G.A. and Hart, R.W.: Longevity, aging and comparitive cellular and molecular biology of the house mouse, *Mus musculus*, and the white-footed mouse, *Peromyscus leucopus. Birth defects 14:* 71-96, 1978

Singer, D.S., Hare, J., Golding, H., Flaherty, L., and Rudikoff, S.: Characterization of a new subfamily of class I genes in the H-2 complex of the mouse. *Immunogenetics 28:* 13-21, 1988

Smith, G.S. and Walford, R.L: Influence of the main histocompatibilty complex on aging in mice. *Nature 270:* 727-729, 1977

Smith, G.S., Crew, M.D., and R.L. Walford: *Peromyscus* as a gerontological animal: Aging and the MHC. *In* D. Harrison (ed.): *Genetics of Aging II*, pp. 457-472, Telford, Caldwell, New Jersey 1990

Walford, R.L.: The major histocompatibility complex and aging in mammals. *In* C. Finch and T. Johnson (eds.): *Molecular Biology of Aging* pp. 31-41, Liss, New York 1990

Wilson, V.L., Smith, R.A., Ma, S., and Cutler, R.G.: Genomic 5-methyldeoxycytidine decreases with age. *J Biol Chem 262:* 9948-9951, 1987

SEQUENCE AND EVOLUTION OF BOVINE MHC CLASS I GENES

Theodore L. Garber*, Austin L. Hughes°, Norman L. Letvin+, Joe W. Templeton*, and David I. Watkins+

*Texas A&M University, Department of Veterinary Pathobiology, College Station, TX 77843, USA

°Pennsylvania State University, Department of Biology, 208 Mueller Laboratory, University Park, PA 16802, USA

+Harvard Medical School, New England Regional Primate Research Center, One Pine Hill Drive, Southborough, MA 01772, USA

Bovine MHC class I cDNAs were cloned and sequenced from PBLs of a MHC class I heterozygous bull (serotype w2, w30). Multiple copies of four unique cDNAs were sequenced. These four cDNAs were separated into two groups based upon sequence comparisons. Each of the groups appear to represent alleles of two different loci. Comparisons of these bovine sequences with published sequences from other artiodactyls demonstrates that one of the bovine MHC class I genes shares more sequence similarity with a sheep MHC class I sequence than with the other bovine MHC class sequences. These data suggest that concerted evolution of bovine and bovine MHC class I genes either has not occurred, or has not occurred at a rate sufficient to cause extensive sequence divergence between the MHC class I loci of cattle and sheep.

Diversity of BoLA class I molecules have been detected by a variety of methods including microcytotoxicity (Bull et al. 1989) and one-dimensional isoelectric focusing (1-D IEF) (Joosten et al. 1988; Watkins et al. 1989) Reports from three international workshops have defined 31 serological specificities (Bull et al. 1989; Anonymous 1982; Spooner et al. 1979). In spite of these efforts BoLA remains poorly characterized. There is up to a 50% frequency of blank alleles in some breeds (Bull et al. 1989). Thus far only one BoLA class I locus has been formally identified. There is, however, limited but growing evidence for a second locus (Ennis et al. 1988; Bull et al. 1989; Toye et al. 1990; Stear et al. 1982; and Watkins et al. 1989). To establish the number of expressed MHC class I loci in cattle we cloned and

NATO ASI Series, Vol. H 59
Molecular Evolution of the
Major Histocompatibility Complex
Edited by J. Klein and D. Klein

sequenced MHC class I cDNAs from a heterozygous Angus Bull. We then compared these BoLA class I cDNA sequences with cDNA sequences from other artiodactyls to determine potential mechanisms operating during the evolution of BoLA class I genes.

Cloning and sequencing of BoLA class I cDNAs

We used the polymerase chain reaction to amplify cDNA for cloning and sequencing (Ennis et al. 1990). PCR primers, designed from published BoLA cDNAs of Ennis et al. (1988) and Brown et al. (1988), were located in the leader peptide and 3' untranslated region. Four different cDNAs were cloned from the heterozygous black Angus bull and both strands of multiple copies of cDNAs were sequenced.

The nucleotide composition of these four cDNAs suggested that they encode classical BoLA class I molecules. Classical MHC class I genes of other species have a high frequency of G and C nucleotides in the third codon position of exons 2 and 3 (Hughes and Nei 1989). The G and C content of these BoLA class I cDNAs was similar to values for the same regions of HLA classical MHC class I genes (88.4%) suggesting that the BoLA cDNAs probably encode classical, antigen (Ag) presenting molecules.

The BoLA class I molecules have been under selective pressures. Strong selection pressure operates upon classical MHC class I genes that result in a high rate of nucleotide substitutions, causing amino acid changes (nonsynonymous substitutions) in the Ag-recognition site (Hughes and Nei 1988). The rate of nonsynonymous substitutions exceeds the rate of synonymous substitutions in the Ag-recognition site of the four BoLA cDNAs. In the remainder of the gene the rate of synonymous substitution is greater than the rate of nonsynonymous substitution. Thus selection has increased variability at the Ag-recognition site of the BoLA class I genes.

There are at least two BoLA class I loci

The bull from which we cloned these four cDNAs was serologically typed to be a heterozygote w2/w30. The isolation of what appear to be four classical BoLA class I cDNAs establishes that there must be at least two BoLA class I loci. Since many locus-specific nucleotide substitutions are found in exons 4-8 in human and mouse MHC class I genes, we constructed gene trees (Saitou and Nei 1987) using these exons of the BoLA class I cDNAs. To compare only these exons, the bovine cDNAs separated into two clusters (cluster A and B). Previously, Ennis et al. (1988) had sequenced two bovine cDNAs and concluded, based upon sequence dissimilarity in the 3' untranslated region, that they had cloned the products of two loci. Our gene tree agrees with this analysis, separating the two previously cloned cDNA of Ennis et al. and clustering two of our cDNAs with each of them

1-D IEF of BoLA class I immunoprecipitates was used to study the transmission of one of the bull's haplotypes. He was mated with his daughter to create a backcross homozygote. We used two different antibodies to immunoprecipitate BoLA class I antigens from the bull and his backcross offspring. Using a broadly reacting anti beta-2 microglobulin antibody, there are at least four bands expressed by the bull. The homozygote progeny inherited only two of the bull's bands. There is a differential recognition of these two bands in the progeny using the anti-HLA monoclonal antibody W6/32 with one of the bands is not recognized by W6/32. This provides further evidence that these two bands may represent the products of two BoLA class I loci.

The gene tree analysis clustering the BoLA class I cDNA into two distinct groups is supported by examination of the nucleotide sequences of the four BoLA cDNAs. Sixteen locus-specific nucleotide substitutions were present in the exons encoding the alpha-3, transmembrane, and cytoplasmic domains. Comparison of predicted amino acid sequences encoded by these cDNAs shows that there are also ten apparent locus-specific amino acid differences in the transmembrane and cytoplasmic domains. The predicted amino acid sequences of one of the BoLA class I loci products (cluster A) have two more amino acids in the transmembrane domain than are found in the predicted amino acid sequences of the other locus products (cluster B).

Evidence for inter-locus recombination in BoLA class I loci

A recombination event, involving exons encoding part of the leader peptide, all of the alpha-1, and most of the alpha-2 domains appears to have taken place between a cluster A gene and a cluster B gene. We ruled out the possibility that this recombinant was a PCR artifact since we recovered numerous copies of this cDNA, and the cluster B cDNA to which it is most closely related came from a different laboratory.

One of the BoLA class I loci is related to an bovine MHC class I locus

To determine the relationship between bovine MHC class I genes and those of other artiodactyls, gene trees were constructed using cow, sheep and pig MHC class I genes. Using the conserved 3' domains of MHC class I cDNAs to construct these trees, the sheep MHC class I cDNA clustered with one of the BoLA A cluster cDNAs. Thus this particular cluster of bovine cDNAs was more closely related to the sheep cDNA than to the other cluster of bovine cDNAs. A gene tree that made use of the entire coding region gave similar results. Sheep therefore appear to have a locus homologous to the BoLA cluster A locus. This indicates that an MHC class I gene duplication event must have taken place prior to the bovine-ovine divergence which occured about 7 to 10 million years ago. Two porcine MHC class I cDNAs were also included in the analysis; neither cDNA was found to be closely related to either of the cattle clusters. This is not unexpected because pigs and cattle diverged perhaps as many as 40 million years ago.

Acknowledgments. This work was supported by Division of Research Resources grand RR00168 and DK36350 from the National Institutes of Health. Dr. Letvin is a recipient of an American Cancer Society Faculty Research Award. We wish to thank S. Kotlikoff for help in preparation of this manuscript.

References

Anonymous: Proceedings of the second international bovine lymphocyte antigen (BoLa) workshop. *Anim Blood Groups Biochem Genet 13*: 33-53, 1982

Brown, P., Spooner, R.L., and Clark, A.J.: Cloning and characterization of a BoLA class I cDNA clone. *Immunogenetics 29*: 58-60, 1989

Bull, R.W. et al.: Joint report of the third international bovine lymphocyte antigen (BoLa) workshop, Helsinki, Finland, 27 July 1986. *Anim Genet 20*: 109-132, 1989

Ennis, P.D., Jackson, A.P., and Parham, P.: Molecular cloning of bovine class I MHC cDNA. *J Immunol 141*: 642-651, 1988

Hughes, A.L. and Nei, M.: Pattern of nucleotide substitution at major histocompatibility complex loci reveals overdominant selection. *Nature 335*: 167-170, 1988

Hughes, A.L. and Nei, M.,: Evolution of the Major Histocompatibility Complex: Independent origins of nonclassical class I genes in different groups of mammals. *Mol Biol Evol 6*: 559-579, 1989

Joosten, I., Oliver, R.A., Spooner, R.L., Williams, J.L., Hepkema, B.G., Sanders, M.F., and Hensen, E.J.: Characterization of class I bovine lymphocyte antigens (BoLA) by one-dimensional isoelectricfocusing. *Anim Genet 19*: 103-113, 1988

Saitou, N. and Nei, M.: The neighbor-joining method: A new method for reconstructing phylogeneti trees. *Mol Biol Evol 4*: 406-425, 1987

Spooner, R.L. et al.: Analysis of alloantisera against bovine lymphocytes. Joint report of the 1st international bovine lynphocyte antigen (BoLa) workshop. *Anim Blood Groups Biochem Genet 15*: 63-86, 1979

Stear, M.J., Newman, M.J., and Nicholas, F.W.: Two closely linked loci and one apparently independent locus code for bovine lymphocyte antigens. *Tissue Antigens 20*: 289-299, 1982

Toye, P.G., Machugh, N.D., Bensaid, A.M., Alberti, S., Teale, A.J., and Morrison, W.I.: Transfection into mouse L cells of genes encoding two serologically and functionally distinct bovine class I MHC molecules from a MHC-homozygous animal: evidence for a second class I locus in cattle. *Immunology 70*: 20-26, 1990

Watkins, D.I., Shadduck, J.A., Stone, M.E., Lewin, H.A., and Letvin, N.L.: Isoelectric focusing of bovine major histocompatibility complex class I molecules. *J Immunogenet 16*: 233,245, 1989

EVOLUTION OF MHC MOLECULES IN NONMAMMALIAN VERTEBRATES

Jim Kaufman*, Jan Salomonsen*, and Karsten Skjødt+
*Basel Institute for Immunology[1], Grenzacherstrasse 487, Basel CH-4005, Switzerland
+Institute of Medical Microbiology, University of Odense, Campusvej 55, Odense C., DK-5230, Denmark

Some MHC molecules from nonmammalian vertebrates have been identified and characterized at the level of protein or DNA or both. We summarize comparisons between B-F sequences (chicken β_2-microglobulin and class I α chains) and the sequences of other MHC-like molecules. As we naively expected, many residues that are involved in interactions (within and between domains, and with peptide, T cell receptor and CD8) are conserved. However, we also found certain conserved surface patches with no known function as well as differential evolution and codon usage in different domains. A speculative model is presented.

The major histocompatibility complex (*Mhc*) is a large genetic region that encodes many different kinds of molecules and is associated with a variety of biological functions. Chief among the molecules and functions that we understand are the highly polymorphic class I and class II heterodimers that are involved in presentation of peptide antigen to T lymphocytes. However, there are many other genes and molecules that are structurally related to *Mhc* class I and class II genes and molecules (Klein 1987; Guillemot et al. 1988a, Kaufman et al. 1991a), which we will refer to as Mhc molecules, whether or not they are encoded in the *Mhc*.

The enormous complexity of the *Mhc* can only be explained by evolution, a process composed of accident and selection. As we understand evolution, there is variation that appears in the population by mutation of some kind; this may be selected directly, or it may become fixed in the population by some kind of accident, e.g., from some founder effect. Thus, some features of the *Mhc* genes, molecules and functions are due to continuing selection, others were under selection sometime in the past, and others are totally accidental. Even with the selected features, it is a matter of accident that one particular feature was originally present rather than another.

---

[1]The Basel Institute for Immunology was founded and is supported by Hoffmann-La Roche, Basel, Switzerland.

NATO ASI Series, Vol. H 59
Molecular Evolution of the
Major Histocompatibility Complex
Edited by J. Klein and D. Klein

For vertebrates, there is no easy way to do real-time experiments, so the best that we can do is to make observations, construct plausible scenarios, and then see if further observations continue to be consistent with those scenarios. A number of such scenarios have been discussed (Davis and Bjorkman 1988; Kaufman et al. 1984, 1988, 1990c, 1991a). Since membrane proteins and soft tissues do not leave good fossils, observation means understanding the phylogeny of the living vertebrates. There are at last two problems with this approach (Kaufman et al. 1990c, 1991a). First, there are many gaps in the tree of vertebrates as constructed by fossils: some important groups have disappeared altogether, others are represented by just a few species that may not be representative of the ancestral group, others have radiated to become highly overrepresented, and many of the connections are as yet simply not known. For example, the jawless fish, a key group that dominated the ancient oceans, are represented now by only a few species of parasitic lampreys and scavenger hagfish. Second, in all lineages there has been secondary evolution, which means that the extant animals may not resemble the distant shared ancestors. For example, different vertebrate taxa apparently have different strengths of immune response, which have evidently evolved independently in each lineage.

The details of the evolution of the *Mhc* are very complicated, but we have to wade through them, in order to understand what special underlying principles, if any, are at work. These underlying principles will be concise and clear and oversimplified, whereas the details will be complicated and messy and true.

We have tried two approaches to understanding the details of the evolution of *Mhc* molecules in the vertebrates. One is to use antibodies to identify and characterize *Mhc* molecules from nonmammalian vertebrates in an attempt to understand what molecules and epitopes are present in various vertebrate taxa (Kaufman et al. 1990a, 1990b). The other is to isolate *Mhc* genes and cDNA clones by various means from nonmammalian vertebrates in an attempt to determine which sequence features are conserved and which are allowed to vary in evolution (Kaufman et al. 1991c, unpublished data). In this short review, we will first consider what we have learned about Mhc molecules in nonmammalian vertebrates from these two appróaches, and then suggest a very speculative model of macroevolutionary change to explain some puzzling aspects of these results.

Naive expectations

Based on what is known about phylogeny, general molecular evolution and the structure and function of Mhc molecules in mammals (Young 1981; Benjamin et al. 1984; Chothia 1984; Creighton 1983; Klein 1987), we initially had certain naive expectations about the relationships between Mhc molecules in different vertebrate taxa. We expected that the mammals, birds and reptiles would form one group that was clearly separated from the amphibians and fish. We expected that the class II α chains and $ß_2$-microglobulin (ß2m) would be the most conserved chains, and that the Ig-like (immunoglobulin-like) domains (and the transmembrane region of class II molecules) would be the most conserved domains. We expected that those portions of the molecule involved in contacts would be more conserved in evolution than those that were not. In other words, the surface residues that simply hang in the solvent would be quite diverged, while those surface residues that interact with other molecules (peptide, T cell receptor, CD4 or CD8, intracellular kinases, assembly molecules like "invariant chain") as well as intradomain and interdomain contact sites would be more conserved. We expected that changes in one residue of a contact would be compensated by appropriate changes in the other residue of the contact. In addition, if there were strong rules about binding peptides and T cell receptors, then positions of those residues, both polymorphic and invariant, responsible for binding peptides and T cell receptors would be conserved.

As will rapidly become apparent, some of these naive expectations were fulfilled and others were not. In addition, some unexpected phenomena also were found.

Chains and Domains

Of the four polypeptides that make up the mammalian class I and class II heterodimers, ß2m and class II α chains are the least polymorphic. Within each chain, it is the Ig-like domain that is invariant or nearly so, with the non-Ig domain ranging from invariant to highly polymorphic. In addition, the nonpolymorphic domains (ß2m, class I α3, class II α2, and class II ß2 domains) are the most conserved beween different mammals. From what little we know thus far, the same situation is found between different avian Mhc molecules (Kaufman et al. 1991c, unpublished observations).

The situation is markedly different in comparisons between mammals and other vertebrate taxa. Monoçlonal antibodies and polyclonal antisera to human, mouse and chicken Mhc molecules detect class I α chain and class II ß chain in other vertebrates, but rarely ß2m and never class II α chains (Kaufman et al. 1990a,1990b). Thus it would appear that ß2m and class II α chain have diverged faster in evolution than class I α chain and class II ß chain.

These antibody cross-reaction results have two unexpected implications: the Ig-like domains diverge faster than the peptide/T cell receptor-binding domains and the two peptide/T cell receptor-binding domains also evolve at different rates. This is confirmed, as far as we have information, by comparison of chicken and mammalian *Mhc* sequences (Bourlet et al. 1988; Guillemot et al. 1988b; Kaufman et al. 1991c, unpublished data). First, the least homologous of the class I extracellular domains is the class I α3 domain (around

30% identical residues between chicken and mammalian classical class I sequences). In fact, the chicken α3 domain is just as similar to α3 of mammalian classical and nonclassical class I molecules as it is to class II ß2 and α2 domains, ß2m, and various Ig constant regions. Second, the class I α2 domain and the class II ß1 domain are the most homologous extracellular domains (more than 55% identical residues), significantly more homologous than their peptide/T cell receptor-binding partners, the class I α1 domain and the class II α1 domain (less than 50% identical residues). This is interesting, since both the class I α1 and α2 domains are clearly more similar to classical than nonclassical class I molecules in all comparisons (excluding the New World monkey sequences, which are like *HLA-E* and the Syrian Hamster sequence, which is more like *Qa*). In addition, the sequence comparisons show that ß2m is more similar between mammals and chickens than the α3 domain.

The sequences of three other vertebrate *Mhc* clones, a frog class I α chain, a carp class I α chain and a carp class II ß chain, have been reported (Hashimoto et al. 1990; Flajnik et al. 1991). Comparison of these sequences with mammalian and chicken sequences gives a different picture: the α3 domain remains at the same level (around 30% identical residues), but frog α1 and α2 drop to the same low level, with the carp α1 and α2 even lower (around 20 and 10% identical residues). This may reflect the possibility that these clones are not the major polymorphic Mhc proteins in the frog and carp, since none of these sequences have been shown to correspond to protein and in the case of carp it is not yet clear that the genes are transcribed. On the other hand, these levels of homology may reflect the long time period of divergence, in which the sequence has varied to the limits of structural integrity for the different domains. Interestingly, there are approximately 14 α3 residues which are absolutely or nearly invariant between the mammalian classical, chicken, frog and carp class I α sequences; most of these are typical for Ig constant region domains (Williams 1987). In contrast there are only 3-4 such residues in the α1 and α2 domains (as in comparisons with mammalian nonclassical class I α chains).

Assuming that these are all the correct (orthologous?) molecules to compare, then what does all of this mean? It would appear that there are fewer overall structural constraints (i.e., intradomain and maybe interdomain contacts) but more functional constraints (i.e., contacts with other molecules, particularly peptide and T cell receptor, as well as "invariant chain" in class II) on the peptide/T cell receptor-binding domains compared to the Ig-like domains. Thus, the Ig domains rapidly diverge to the structural limit; the α3 domains of chickens and mammals are already at this limit, with ß2m not yet there. In contrast, the peptide/T cell receptor-binding domains diverge more slowly in deference to the requirements of peptide and T cell receptor binding, but eventually drop to their structural limit, which is below that of the Ig domains. This makes some sense when the particular conserved residues are considered (as described below).

It is the surface residues that are responsible for the differences between α1 and α2 and between α3 and ß2m in the similarity of chicken and mammalian classical class I sequences (Kaufman et al. 1991c). Virtually the only α3 residues on the surface of the class I molecule (i.e., excluding those thought to be involved in intradomain and interdomain contacts) that are invariant between mammalian classical class I and chicken are two that have been implicated in CD8 binding (and one adjacent residue that is also involved in intrachain contacts). In comparison, there are several large epitopes at the exposed end of the ß2m barrel that are invariant between mammals and chickens (the "nosetopes"); the possible functions of these are unknown. Perhaps they are involved in CD8 binding, assembly of class I molecules, or in other functions of ß2m (eg., the proposed role in chemotaxis, see Brinckerhoff et al. 1989; Dargemont et al. 1989; Dunon et al. 1990). There are clearly many more invariant residues on the surface of α2 compared to α1, including large epitopes on the α helix that could presumably come in contact with the T cell receptor. This may imply that different portions of the T cell receptor have also diverged at different rates, for instance, the two chains of the T cell receptor coevolve with the two peptide/T cell receptor-binding domains of the class I molecule.

Surface and buried residues

Comparisons of sequence strings are clearly useful for gaining an idea of the relatedness of molecules, but in fact the evolutionary selection for proteins is on particular residues in the context of their three-dimensional environment. The sterile identities in aligned sequences come alive when the sequences are placed in a three dimensional model. Identities throughout evolution are explained by positions in the secondary structure or as contact sites. In addition, inexplicable identities that are distant in the sequence turn out to form interesting groups in the three-dimensional structure.

Based on our experiences with the antibody cross-reactions, we expected a difference in the rate of evolution between the surfaces and the buried portions of native Mhc molecules (Kaufman et al. 1990a, 1990b). We found that very few monoclonal antibodies to human, mouse and chicken Mhc molecules cross-reacted with the Mhc molecules of other vertebrate taxa, and that these cross-reactions were almost entirely with polymorphic linear determinants in the peptide/T cell receptor-binding domains. In contrast, the rabbit antisera to human and chicken Mhc molecules predominantly recognized nonlinear cryptic determinants that were exposed when the chains were separated but not denatured (with only a few antisera that reacted like the monoclonal antibodies). From this we concluded that, except for the polymorphic α-helices, the surfaces of Mhc molecules are composed of epitopes that either completely diverged during evolution of the vertebrates or are invariant in each triad of animals examined (source of immunogen, source of antibody and the tested vertebrate). We also concluded that the cryptic sites, presumably the interdomain contact sites, are conserved in evolution (but nevertheless were immunogenic since they are usually buried).

There are enough sequence identities and few enough gaps between human class I molecules (HLA-A2, HLA-Aw68, and ß2m) and chicken class I molecules (B-F19, B-F12, and ß2m) to attempt an evolutionary comparison using a template-modeling approach (Kaufman et al. 1991c). We found that the information from the 3.5 Å structure of HLA-A2 resolution (in which the mainchain atoms could be traced but the details of the sidechains were not clear, see Bjorkman et al. 1987a, 1987b) did not have high enough resolution for such comparisons. The 2.6 Å structure of HLA-Aw68 (in which many details of the sidechains were clear, see Saper et al. 1991) was adequate to draw certain conclusions, but even with this model, the details of the interactions were of primary importance. In particular, we thought that we could simply compare contact residues, but in fact the exact atoms involved were important, as will be described below. So we used a computer program to determine 4 Å contacts (all pairs of atoms within 4 Å of each other) in order to determine intradomain and interdomain contacts. We used a molecular modeling system (FRODO PS390) for visual inspection, and a surface dot program to visualize the surface residues of the molecule. (While the surface dot models were quite effective as slides for both detailed and large-scale pictures, the figures for the whole molecule left much to be desired. We recommend a CPK presentation (for instance, the MacImdad software package) for clarity.) The observations that follow refer to residues that are invariant (or nearly so) between 120 mammalian classical class I sequences, 6 mammalian ß2m sequences and the chicken sequences (Kaufman et al. 1991c).

We presumed that there would be some five types of intradomain contacts in the extracellular region of an Mhc molecule. In the Ig-like domains, these would be residues in the ß-strands (alternate and thus perhaps predictable) and residues in the bends (not likely to be predictable) that point inwards toward the center of the domain. In the peptide/T cell receptor-binding domains, these would be the residues pointing up from the ß-strands (alternate and thus perhaps predictable), residues pointing down from the α-helix (every third or fourth and thus perhaps predictable), and residues interacting between the α-helices (at the ends and not likely to be predictable). In fact, many of these residues are invariant or highly conserved. But there were also many exceptions, including highly charged residues that clearly had to point into the presumed hydrophobic interior of the domain. One possible explanation was the formation of buried salt bridges, as postulated for certain residues (Travers et al. 1984). We thought that we could use the three-dimensional structure to devise an easy definitive criterion for intrachain contacts.

We rediscovered what every crystallographer apparently already knows--that determining which residues are crucial for intradomain contacts is not a trivial task! We found that over 90% of the residues are in contact with other residues in HLA-Aw68, so this can not be used as a criterion for intradomain contacts. Also, over 90% of the residues were accessible to the solvent in some degree, so this is also not a very useful criterion for intradomain contacts (it should be noted that the second analysis was only done with the 3.5 Å structure). There are two reason for this, which became clear from a simple inspection of the Ig-like domains at 2.6 Å resolution. First, the Ig-like domains are composed of a pair of offset twisted ß-sheets that are farther apart at one end than at the other (with the Cα backbone of the ß-strands having a distinct lopsided wedge shape), filled in with broad turns. The result, which must be seen in three dimensions to be properly appreciated, is that many of the "inward-pointing" alternate residues of the ß-strands C, D, and E do not fill the hydrophobic interior of the domain at all, but are more-or-less solvent exposed. These residues need only have part of the sidechain be hydrophobic (eg Cß) and the rest can be hydrophilic or even charged and turn out into the solvent. An example of this is an arginine (R63) in chicken ß2m which is leucine or isoleucine (LI64) in mammals. Second, many of the amino acid sidechains have several chemical features which allow them to fulfill different roles in the context of the immediate environment, and the overall protein structure can accommodate significant changes in detailed structure with minor local adjustments. One example is a lysine (K243) in HLA-Aw68 (which is a threonine (T239) in chicken B-F) in which the 4 hydrophobic methylene groups of the lysine side-chain traverse the hydrophobic interior of the domain, with the terminal amino group shoving out between two ß-strands on the opposite sheet into the solvent.

It was less tempting to try to predict the interdomain contacts (since they need not be regular), but they are easier to define using the 4 Å proximity program. We found that in many cases (18/54 contact pairs), both residues of a contact site were invariant between mammalian classical class I and chicken class I molecules. We found no examples of compensatory changes in contact site residues; instead there were three classes of changes in the interdomain contact sites in the chicken class I heterodimer. In most contacts, one residue was invariant and the other was highly diverged. But in many of these cases (11/54 contact pairs), the contact was between the side-chain atoms of the invariant residue and the main-chain atoms of the highly diverged residue, so that the actual contact was invariant! A second class (12/54) consisted of one invariant residue in contact with a diverged residue in which the side-chain of the diverged residue could allow a contact to be maintained. The third class of contacts (13/54 contact pairs) consisted of one invariant residue and one residue which became much smaller, often becoming glycine in the chicken sequences. There were no other contacts that we could discern in the chicken molecule, but of course we are not in a good position to predict contacts that are specific to the chicken due to changed residues.

We identified the surface residues by visual inspection of the dot matrix model displayed on a Evans and Sutherland raster. Much of the surface of α3 and ß2m were diverged. Virtually the only invariant residues on the surface of α3 were three residues that formed a protuberant epitope: the glutamine and aspartic acid (Q226 and D227 in HLA-Aw68) implicated in CD8 binding, along with an adjacent tyrosine (Y257 in HLA-Aw68). In contrast, a nine residue concave epitope (the "nosetope") was found at the exposed end of the exposed ß-barrel of ß2m, along with adjacent smaller concave epitopes. For what purpose the "nosetopes" are conserved is still a mystery. There are a number of invariant residues on the surface of α1, and many more on the surface of α2, in particular a large clump of residues on the α2 helix.. Some of these invariant residues hang off the side of these domains, and the function, if any, is unknown. Other residues point up from the α-helix and may interact with the T cell receptor. Others point into the cleft from the T cell receptor or the ß-strands and these invariant residues have been implicated in hydrogen-bonding to the ends of the peptides in HLA-B27. Of 28 polymorphic residues identified by comparison of the presumed chicken class I alleles B-F19 and B-F12, two are in α3 and seven are in the bends of α1/α2. The remaining 19 are in the peptide/T cell receptor-binding site and are also polymorphic in class I molecules of mouse and/or man. Finally, the C-terminal cytoplasmic tail has very little homology to mammals, but a tyrosine in exon 6 and two serines in exon 7 that are used as kinase sites in mammals are also found in chicken B-F19.

From the analysis of the individual residues of the class I heterodimers of mammals and chickens in this template modeling approach, we can conclude that the residues involved in contacts (intradomain, interdomain, CD8, peptide, T cell receptor, and kinase) are more generally more conserved than those that simply hang out in the solvent. For such a conclusion, we have to exclude polymorphic residues in the peptide/T cell receptor-binding site (or take note of the fact that their "conservation" selects for sequence divergence) and we have to postulate that the big invariant patches in ß2m and α2 are actually involved in some, as yet undiscovered, interactions. Even so, the conclusion is not very strong, since multiple amino acids can perform the same task, and the structure can adjust even to quite radical changes with minimal distortion.

As mentioned in the last section, the frog and carp class I α sequences (Hashimoto et al. 1990; Flajnik et al. 1991) are quite diverged from the chicken and mammalian sequences. It is interesting that these sequences lack the invariant residues (between chicken and mammalian classical class I molecules) that are implicated in CD8 binding and in positioning the ends of the peptide antigen. Thus, either the class I molecules function in quite different ways in frog and carp, there have been many compensatory changes, or these are not the correct ("orthologous"?) molecules for comparison.

We have postulated that class II genes gave rise to class I α and ß2m genes by an inversion, such that certain class I domains are equivalent to certain class II domains (ie., class I α1, α2, and α3 equivalent to class II α1, ß1, and ß2; ß2m equivalent to class II α2) (Kaufman et al. 1984, 1990c, 1991a). This has recently received some support from the comparisons of carp class I α1 and mammalian class II α1 sequences (Hashimoto et al. 1990). Several chicken class II ß chain sequences have been reported (Bourlet et al. 1987; Xu et al. 1989) and we have just cloned a chicken class II α chain cDNA (unpublished data). It will be interesting to analyze the class II heterodimer by a similar approach, to see whether the different extracellular domains follow the same order of homology, whether the same surface residues are invariant (in particular, evidence for a CD4 interaction site in the α2 domain and evidence for an "invariant chain" interaction site), and whether the Ig-like domains interact as in class I molecules, as in antibody constant regions, or in some other fashion.

Surprises and speculations

We were surprised by two other observations made in the course of this work. In the absence of a good idea to explain either of these two phenomena, we proposed (Kaufman et al. 1991b) that a macroevolutionary event, the appearance of microchromosomes in reptiles and birds, might be responsible for both. This model became more interesting when we realized that it might also explain a spectacular oddity of the avian immune system. Of course, it is likely that we are completely wrong in these speculations, but at the worst they provide a basis for further experiments.

The first observation (Kaufman et al. 1990a, 1990b) is that antibodies against human Mhc molecules cross-reacted better with amphibian Mhc molecules than with chicken and reptile Mhc molecules, while antibodies against chicken Mhc molecules cross-reacted better with reptile Mhc molecules than with mammal and amphibian Mhc molecules. This was true for both rabbit antisera and monoclonal antibodies, for both class I and class II molecules, and for three reptiles and two amphibians. The observation is unexpected, since mammals and reptiles should be more closely related to each other than to amphibians. (In addition, it is not completely clear that birds should be more closely related to snakes and lizards and to turtles than to mammals.)

The second observation (Kaufman et al. 1991c) is that the nucleotide sequences of the chicken α3 domain and ß2m are extremely G+C rich (greater than 97% G+C in the wobble bases) compared to the α1 and α2 domains (86% and 69% G+C in the wobble bases, respectively). The same is true of the chicken class II ß2 domain compared to the ß1 domain. In contrast, the G+C contents are much lower in the *Mhc* DNA sequences from other taxa (approximately 80% G+C in the wobble bases for mammals and approximately

50% for frog and carp, see Aota and Ikemura 1986; Hashimoto et al. 1990; Flajnik et al. 1991). In fact, the chicken Mhc protein sequences are biased towards those amino acids with G+C rich codons (glycine, alanine, tryptophan, proline and arginine) in comparison to mammal, frog and carp molecules. (This is apparently the reason why so many contact sites disappear in chicken class I molecules.)

Both the origin of this drive to G+C richness and the reason why the Ig-like domains are preferentially affected are mysteries. However it is clear that evolutionary trees constructed from either nucleotide or amino acid sequences must take this bias into account, in order to be taken at all seriously.

Our speculative notion is based on the following points (Bernardi 1989; Bickmore and Sumner 1989). Mammals have large chromosomes that can be stained with dyes to show two kinds of bands. They can also be broken into large pieces of DNA with homogeneous properties called isochores. The Giemsa staining (G) bands correspond to A+T rich isochores of DNA, while the Reverse Giemsa staining (R) bands correspond to G+C rich isochores. These two kinds of DNA differ in a number of properties besides the G+C content, including the replication time, the kinds of repetitive elements, and the number or type of genes present. It is important for our argument that, at least some authorities maintain, the R bands (the G+C rich bands) contain most of the active genes. It should be pointed out that the mechanism of maintaining these isochores with separate properties is not known.

In contrast to mammals, birds have a few macrochromosomes that are roughly the size of mammalian chromosomes, and a large number of smaller microchromosomes. Banding studies indicate that the macrochromosomes have both A+T rich and G+C rich portions, whereas the microchromosomes are generally G+C rich (Auer et al. 1987; Holmquist 1989). The chicken *Mhc* is located on a microchromosome of roughly 8 million base pairs (apparently chromosome 16 or 17) and the genes that have been characterized thus far (class I α, class II ß and B-G) have some exons with high G+C content and have mostly short introns (Bloom and Bacon 1985; Bourlet et al. 1988; Guillemot et al. 1988b; 1989; Kroemer et al. 1990; Kaufman et al. 1989, 1991c). The chicken ß2m gene has a high G+C content (Kaufman et al. 1991c) and also has mostly short introns; moreover it is also located on a small chromosome different from the Mhc microchromosome (M. Dominguez-Steglich, P. Riegert and J. Kaufman, unpublished data). A final point is that reptiles also have some microchromosomes, while amphibians have big chromosomes.

Based on these points, we make the following speculative argument (Kaufman et al. 1991b). First, microchromosomes were produced during evolution from macrochromosomes by some unknown mechanism, that involved removal of much unnecessary DNA from the chromosome (truly the most fascinating part of the story). Thus, the introns, flanking regions and the (presumably gene-poor) A+T rich isochores were eliminated from microchromosomes, all in accordance with observation. Second, the general mechanism (DNA site?) responsible for maintaining the G+C rich isochore on the microchromosome was not eliminated, and was no longer opposed by the general mechanism (DNA site?) for maintaining the A+T rich isochore, which allowed the G+C content to rise above what is normally seen. This allowed particular genes or portions of genes to be driven to G+C richness.

If this is all true, then the amphibian and mammalian Mhc molecules might share epitopes that are lost in the reptiles and birds, because the reptiles and birds have evolved away quickly from the ancestor molecules in a particular direction, that is, towards proteins with amino acids that have G+C rich codons. We would therefore expect reptile *Mhc* genes to have a bias towards G+C richness, both at the nucleotide and protein level, and to be located on microchromosomes (and parenthetically to have short introns). This would then be a fascinating case of a molecular macroevolutionary event having a host of consequences that would otherwise be interpreted as microevolutionary events.

Another interesting feature of the chicken immune system is the mechanism for generating antibody diversity. For both the light and heavy chain loci, there is a single active *V* gene segment that rearranges with downstream *D* and/or *J* segments, and is then the target for multiple gene conversion events from upstream *V*-like pseudogenes. The whole light chain locus is about 4 kb and the heavy chain is about 60 kb, which is much smaller than the immunoglobulin loci in man and mouse (Reynaud et al. 1986, 1989). We have suggested that the chicken *Ig* loci are located on microchromosomes, and that during evolution of these microchromosomes, the pressure to eliminate excess DNA rapidly led to small *Ig* loci, which then selected for the use of gene conversion (already present as a minor pathway) in order to generate enough antibody diversity to insure survival of the animal (Kaufman et al. 1991b). In fact, the *Ig* genes have small introns and are mostly G+C rich (but not astonishingly so, as in the *Mhc* genes); as yet their chromosomal location is unknown.

References

Aota, S.-I. and Ikemura, T.: Diversity in G+C content at the third position of codons in vertebrate genes and its cause. *Nucl Acids Res 14*: 6345-66355, 1986

Auer, H., Mayr, B., Lambrou, M., and Schleger, W.: An extended chicken karyotype, including the NOR chromosome. *Cytogenet Cell Genet 45*: 218-221, 1987

Benjamin, D., Berzofsky, J., East, I., Gurd, F., Hannum, C., Leach, S., Margoliash, E., Michael, J., Miller, A., Prager, E., Reichlin, M., Sercaz, E., Smith-Gill, S., Todd, P., and Wilson, A.: The antigenic structure of proteins: a reappraisal. *Annu Rev Immunol 2*: 67-102, 1984

Bernardi, G.: The isochore organization of the human genome. *Annu Rev Genet 23*: 637-631, 1989

Bickmore, W. A. and Sumner, A. T.: Mammalian chromosome banding--an expression of genome organization. *Trends Genet 5*: 144-148, 1989

Bjorkman, P. J., Saper, M. A., Samraoui, B., Bennet, W. S., Strominger, J. L., and Wiley, D. C.: Structure of the human class I histocompatibility antigen, HLA-A2. *Nature 329*: 506-512, 1987a

Bjorkman, P. J., Saper, M. A., Samraoui, B., Bennet, W. S., Strominger, J. L., and Wiley, D. C.: The foreign antigen binding site and T cell recognition regions of class I histocompatibility antigens. *Nature 329*: 512-518, 1987b

Bloom, S. and Bacon, L.: Linkage of the major histocompatibility (B) complex and the nucleolar organizer in the chicken. *J Hered 76*: 146-154, 1985

Bourlet, Y., Behar, G., Guillemot, F., Frechin, N., Billault, A., Chausse, A.-M., Zoorob, R., and Auffray, C.: Isolation of chicken major histocompatibility complex class II (B-L) ß chain sequences: comparison with mammalian ß chain and expression in lymphoid organs. *EMBO J 7*: 1031-1039, 1988

Brinckerhoff, C., Mitchell, I., Karmilowicz, M., Kluve-Beckerman, B., and Benson, M.: Autocrine induction of collagenase by serum amyloid A-like and β_2-microglobulin-like proteins. *Science 243*: 655-657, 1989

Chothia, C.: Principles that determine the structure of proteins. *Annu Rev Biochem 53*: 537-572, 1984

Creighton, T. E.: *Proteins*. Freeman, New York 1983

Dargemont, C., Dunon, D., Deugnier, M., Denoyelle, M., Girault, J., Lederer, F., Le, K., Godeau, F., Thiery, J., and Imhof, B.: Thymotaxin, a chemotactic protein, is identical to β_2-microglobulin. *Science 246*: 803-806, 1989

Davis, M. and Bjorkman, P.: T-cell antigen receptor genes and T-cell recognition. *Nature 334*: 395-402, 1988

Dunon, D., Kaufman, J., Salomonsen, J., Skjodt, K., Vainio, O., Thiery, J.-P., and Imhof, B.: T cell precursor migration towards ß2-microglobulin is involved in thymus colonization of chicken embryos. *EMBO J 9*: 3315-3322, 1990

Flajnik, M., Canel, C., Kramer, J., and Kasahara, M.: Evolution of the major histocompatibility complex: molecular cloning of major histocompatibility complex class I from the amphibian *Xenopus*. *Proc Natl Acad Sci USA 88*: 537-541, 1991

Guillemot, F., Auffray, C., Orr, H., and Strominger, J. MHC antigen genes. *In* B. Hames and D. Glover (eds.): *Molecular Immunology*, pp. 81-144, Oxford IRL Press 1988a

Guillemot, F., Billault, A., Pourquie, O., Behar, G., Chausse, A.-M., Zoorob, R., Kreiblich, G., and Auffray, C.: A molecular map of the chicken major histocompatibility complex: the class II ß genes are closely-linked to the class I genes and the nucleolar organizer. *EMBO J 7*: 2775-2785, 1988b

Guillemot, F., Kaufman, J., Skjoedt, K., and Auffray, C.: The major histocompatibility complex in the chicken. *Trends Genet 57*: 300-304, 1989

Hashimoto, K., Nakanishi, T., and Kurosawa, Y.: Isolation of carp genes encoding major histocompatibility complex antigens. *Proc Natl Acad Sci. USA 87*: 6863-6867, 1990

Holmquist, G. P.: Evolution of chromosome bands: molecular ecology of noncoding DNA. *J Mol Evol 28*: 469-486, 1989

Kaufman, J. F., Auffray, C., Korman, A. J., Shackelford, D. A., and Strominger, J. L.: The class II molecules of the human and murine major histocompatibility complex. *Cell 36*: 1-9, 1984

Kaufman, J.: Vertebrates and the evolution of the MHC class I and class II molecules. *Verh Dtsch Zool Ges 81*: 131-144, 1988

Kaufman, J., Salomonsen, J., and Skjodt, K.: B-G cDNA clones have multiple small repeats and hybridize to both chicken MHC regions. *Immunogenetics 30*: 440-451, 1989

Kaufman, J., Skjoedt, K., Salomonsen, J., Simonsen, M., Du Pasquier, L., Parisot, R., and Riegert, P.: MHC-like molecules in some nonmammalian vertebrates can be detected by some crossreactive xenoantisera. *J Immunol 144*: 2273-2257, 1990a

Kaufman, J., Ferrone, S., Flajnik, M., Kilb, M., Volk, H., and Parisot, R.: MHC-like molecules in some nonmammalian vertebrates can be detected by some crossreactive monoclonal antibodies. *J Immunol 144*: 2257-2265, 1990b

Kaufman, J., Skjoedt, K., and Salomonsen, J.: The MHC molecules of nonmammalian vertebrates. *Immunol Rev 113*: 83-117, 1990c

Kaufman, J., Flajnik, M., and Du Pasquier, L.: The MHC molecules of ectothermic vertebrates. *In* G. W. Warr and N. Cohen (eds.): *Phylogenesis of Immune Functions*, pp. 125-150, CRC Press, Boca Raton, Florida 1991a

Kaufman, J., Salomonsen, J., Riegert, P., and Skjodt, K.: Using chicken class I sequences to understand how xenoantibodies crossreact with MHC-like molecules in nonmammalian vertebrates. *Am Zool 31*, in press 1991b

Kaufman, J., Andersen, R., Avila, D., Engberg, J., Lambris, J., Salomonsen, J., Welinder, K., and Skjodt, K.: Different features of class I molecules have evolved at different rates: chicken B-F and β_2-microglobulin sequences reveal invariant surface epitopes. *J Immunol*, in press 1991c

Klein, J. *Natural History of the Major Histocompatibility Complex*. Wiley, New York, 1987

Kroemer, G., Zoorob, R., and Auffray, C.: Structure and expression of a chicken MHC class I gene. *Immunogenetics 31*: 405-409, 1990

Reynaud, C.-A., Anquez, V., Dahan, A., Dixon, V., Grimal, H., and Weill, J.-C.: A single rearrangement event generates most of the chicken immunoglobulin light chain diversity. *Cell 40*: 283-291, 1986

Reynaud, C.-A., Dahan, A., Anquez, V., and Weill, J.-C.: Somatic hyperconversion diversifies the single V_H gene of the chicken with a high incidence in the D region. *Cell 59*: 171-183, 1989

Saper, M., Bjorkman, P., and Wiley, D.: Refined structure of the human histocompatibility antigen at 2.6 Å resolution. *J Mol Biol*, in press 1991

Travers, P., Blundell, T., Sternberg, M., and Bodmer, W.: Structural and evolutionary analysis of HLA-D region products. *Nature 310*: 235-238, 1984

Williams, A.: A year in the life of the Immunoglobulin superfamily. *Immunol Today 8*: 2998-3003, 1987

Xu, Y., Pitcovski, J., Peterson, L., Auffray, C., Bourlet, Y., Gerndt, B., Nordskog, A., Lamont, S., and Warner, C.: Isolation and characterization of three class II MHC genomic clones from the chicken. *J Immunol 142*: 2122-2132, 1989

Young, J.: *The Life of the Vertebrates*. Clarendon Press, Oxford 1981

THE POLYMORPHIC B-G ANTIGENS OF THE CHICKEN MHC - DO THE STRUCTURE AND TISSUE DISTRIBUTION SUGGEST A FUNCTION?

Jan Salomonsen*, Karsten Skjødt+, and Jim Kaufman*

*Basel Institute for Immunology[1], Grenzacherstrasse 487, Basel CH-4005, Switzerland
+Institute of Medical Microbiology, University of Odense, Campusvej 55, Odense C., DK-5230, Denmark

The chicken MHC, the B complex, encodes three polymorphic multigene families: B-F, B-L and B-G (Pink et al. 1977; Longenecker and Mosmann 1981; Crone and Simonsen 1987; Guillemot and Auffray 1989). B-F and B-L products are the avian homologues of classical mammalian class I and class II antigens. The basic structures of both these integral membrane glycoproteins are non-covalently associated hetero-dimers consisting of membrane-proximal immunoglobulin-like domains followed by non-Ig-like domains where most of the allelic variability is located. They also function as T-cell restriction elements and their tissue distributions also parallel the mammalian counterparts.

The third polymorphic MHC gene family, encoding B-G antigens, was until recently considered to be erythrocyte-specific in expression (Longenecker and Mosmann 1980a; Salomonsen et al. 1987). The high polymorphism and immunogenicity of B-G was useful in the production of alloantisera, but a role in the immune system was difficult to envisage due to the limited tissue distribution. Some people even considered the polymorphism to be due to the mere proximity of B-G genes to the classical MHC regions.

Several B-G related phenomena with immunological impact have been described, but they have been difficult to explain. One is the existence of "natural antibodies" that are directed to polymorphic B-G epitopes and are found in allogeneic chickens as well as in a variety of other species (Longenecker and Mosmann 1980b; Neu et al. 1984). It was demonstrated that non-immune sera could be absorbed on a panel of chicken erythrocytes of different haplotypes without losing much activity against other MHC haplotypes. Another example is that polymorphic B-G epitopes were found to elicit a preferential response when chicken erythrocytes were injected into mice (Longenecker et al.1979). Other experiments on tolerance induction and in vivo erythrocyte elimination (Schulmanova et al. 1978; Hala et al. 1980; Havele et al. 1982) also led to the conclusion that the presumed B-G antigens are

[1]The Basel Institute for Immunology was founded and is supported by Hoffmann-La Roche, Basel, Switzerland.

NATO ASI Series, Vol. H 59
Molecular Evolution of the
Major Histocompatibility Complex
Edited by J. Klein and D. Klein

more immunogenic and less tolerogenic than other blood group antigens. Finally, B-G antigens have been demonstrated to have an "adjuvant-effect" on the humoral response to other alloantigens on the same erythrocyte (Schierman and McBride 1967; Hala et al. 1981; Salomonsen et al. 1991b). A detailed account and discussion of these and other phenomena involving B-G antigens has been published recently (Kaufman et al. 1991).

We now know more about the biochemical structure of the B-G antigens (which can be placed in the immunoglobulin superfamily), the distribution of B-G epitopes (which are located on many cell types and tissues of prime importance for the immune system), and the genomic organisation of the *B-G* region (which includes many genes). The many genes found give room for involvement in a multitude of immunological as well as non-immunological functions. All in all, B-G is as complex in structure and distribution as class I and class II, but thus far no biological function has been proven.

We have suggested a role for some B-G antigens in the development of B cells, based mainly on the polymorphism of B-G, the wider tissue distribution and the above mentioned phenomena (Salomonsen et al. 1991a,b; Kaufman et al. 1991). This specula-tive model has yet to pass a critical test.

Structure of proteins

The polymorphisms detected by serology are paralleled when erythrocyte B-G antigens are studied by electrophoresis. Highly variable haplotype-specific patterns are obtained by IEF- or SDS-electrophoresis after immunoprecipitation with some monoclonal antibodies, rabbit antisera or alloantisera (Miller et al. 1984; Salomonsen et al. 1987; Kline et al. 1988; Kaufman et al. 1990a). The B-G proteins derived from erythrocytes are not glycosylated, are mostly disulphide-linked dimers and bear intrachain disulphide bonds. Both homo- and heterodimers are found which resolve into 3-6 bands of 30-55 kD apparent molecular mass when analyzed under reducing conditions. The combination of the monomers in the dimeric structures is not random. In a heterozygous animal, the bands from different haplotypes do not form heterodimers, as assessed by immunoprecipitation with some haplotype specific antibodies, although both kinds of antigens are expressed on all erythrocytes (Fig. 1).

The extracellular region of the differently sized molecules was found to be very similar as assessed by partial proteolysis of bands precipitated from cell-surface iodinated erythrocytes. Proteolysis of antigen iodinated after isolation generates additional bands and thus indicates that the size variation between B-G molecules primarily is due to different lengths of the cytoplasmic regions. Analysis of the DNA sequences from the 3' end of erythrocyte B-G genes showed many 21 nucleotide repeats, which could give rise to size variation by means of alternative splicing of transcripts from the same gene.

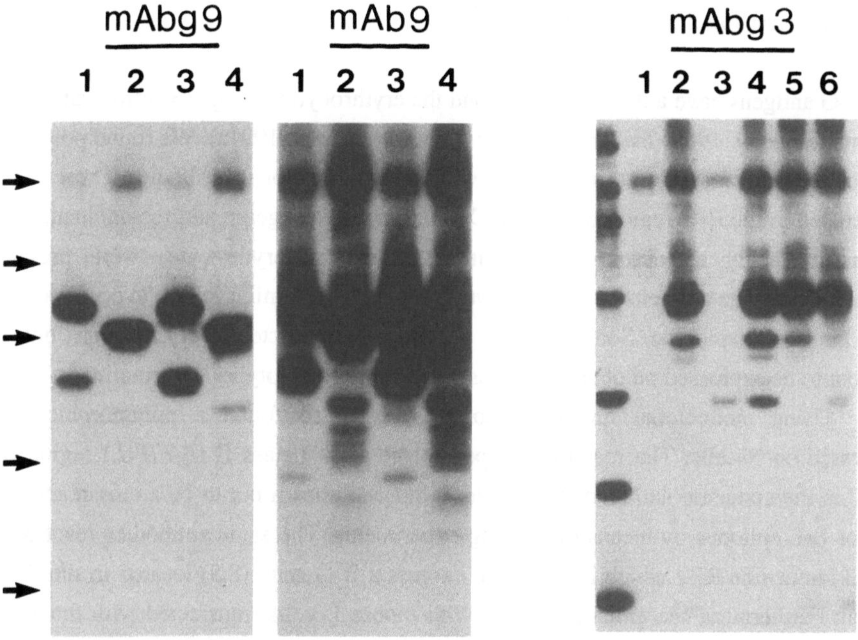

Fig.1. Precipitation of B-G antigens from surface iodinated erythrocytes by means of monoclonal antibodies to B-G antigen (mAbg3 and mAbg9). Samples were analysed by SDS-PAGE under reducing conditions. The erythrocytes were derived from homozygous chicken lines B15, B21, R4, and R5 (*lanes 1-4*), a mixture of B15 and B21 erythrocytes (*lane 5*), and erythrocytes from a heterozygous chicken B15 x B21 (*lane 6*). The recombinant R5 shares the *B-G* region with B21 but has the *B-F/B-L* region of B15. The recombinant R4 has the *B-G* region of B15 and the *B-F/B-L* region of B21.
Left panel: Demonstration of the size variability of erythrocyte B-G antigens within and between haplotypes by means of short and long exposure of mAbg9 precipitates. The mAbg9 precipitates from both B15 and B21 haplotypes and all bands segregate with the *B-G* regions. Arrowheads indicate standards: 95, 70, 45, 30, and 20 kDa.
Right panel: The reaction with mAbg3 is specific for B-G from the B21 haplotype. No co-precipitation of bands from the B15 haplotype can be observed in the heterozygous sample (*lane 6*), showing that the pairing of monomers in the B-G dimers is not random. First lane shows standards: 200, 95, 70, 45, 30, 20, and 14 kDa.
Flow cytometry analyses show that mAbg3 only reacts with erythrocytes with B-G21 antigen, but with a monomorphic epitope on B-cells. The mAbg9 reacts with a cytoplasmic B-G epitope on erythrocytes from all haplotypes. (Salomonsen et al. 1991a).

The translated repeats indicate an α-helical coiled-coil structure (Fig. 2) which may be formed by the two chains of the the dimer and which might interact with cytoskeletal components (Kaufman et al. 1990).

Tissue distribution and ontogeny

The B-G antigens have a distribution beyond the erythrocytic lineage (Kaufman et al. 1989a; Salomonsen et al. 1989; Miller et al.1990; Salomonsen et al. 1991a). We found polymorphic B-G epitopes expressed on thrombocytes with many monoclonal antibodies. These reactions all mapped to the *B-G* region of the MHC as defined by congenic and recombinant chicken strains, but only a subset of the epitopes found on erythrocytes were present on thrombocytes. The antigens on thrombocytes seemed to be similar in size to erythrocyte B-G from the same haplotype. Some epitopes appear to be restricted to erythrocytes, but might turn out to be expressed on other cell types when further haplotypes are examined.

Using monoclonal antibodies to B-G, we also found a polymorphic epitope expressed on B-cells. The reaction mapped to the class I/class II (*B-F*/*B-L*) region of the MHC in the congenic and recombinant lines, but was shown not to be a crossreaction with B-F or B-L epitopes by means of capping experiments. The same antibodies react with the B-cell tumor line RP9 which is known to express a B-G gene (8.5) located in the B-F/B-L region. Furthermore the antibodies react with mouse L cells transfected with the 8.5 gene (Kaufman et al. 1991; Salomonsen et al. 1991a).

Some antibodies reacted with polymorphic B-G on erythrocytes but with monomorphic epitopes on peripheral B-cells or B- and T-cells. Interestingly, the B-G antigens precipitated from erythrocytes by these antibodies are a different subset compared to the antigens precipitated by antibodies which only react with erythrocytes and thrombocytes. These monomorphic antibodies gave a spectacular staining pattern in the bursa of Fabricius, some reacting with stromal cells and basal lamina as well as epithelial cells at the cortico-medullary border. Other antibodies stained mainly the medulla. (Salomonsen et al. 1991b).

The same monoclonal antibodies stained strongly in the caecal tonsils and also the medulla of the thymus. The section staining reactions were verified by Northern blot analyses with B-G cDNA probes, showing signals in both the lymphoid and stromal compartments of bursa and thymus as well as a very strong signal in the caecal tonsil area of the intestine. A wider distribution of B-G antigens has also been found by Miller's group, which demonstrated B-G expression in the intestine and maybe also in liver (Miller et al. 1990). They also found, consistent with our results, that some B-G antigens in these non-erythroid tissues are not disulfide-linked to other chains and that some are larger than those on erythrocytes. We found the larger size to be due, at least in part, to glycosylation, which is not found on erythrocyte B-G molecules.

We hoped that a study of the expression of these monomorphic epitopes through the ontogeny of the chicken would provide us with some clues as to the biology of B-G antigens. We used flow cytometry to study bursa, spleen and thymus cells from day 10 of incubation, as well as blood and bone-marrow cells from day 16. Section staining was done from day 10 on the same organs as well as caecal tonsils and liver. The tissues have also been tested in Northern blot analyses using B-G cDNA probes.

Preliminary flow cytometry data has shown that peripheral blood lymphocytes and spleen cells only began to be positive for B-G staining shortly before hatching (which is day 21 of incubation) and then only with one of three monoclonal antibodies. In contrast, the class II positive cells in the bone-marrow and on bursal lymphocytes were stained from day 16. Section staining was positive in the bursa and intestine at day 10 and from day 11 also weakly positive in the thymus. Strong staining was obtained at day 14 in bursa and day 13 in the intestine. The staining pattern of the three monoclonals were different and correlated, in the intestine, with different sizes of message. B-G epitopes in bursa appeared before expression of immunoglobulin and Bu-1 antigen. Thus, the results so far show a great complexity of B-G (or crossreacting) epitopes, which are present on lymphoid cells and tissues early in embryogenesis. On erythrocytes, the B-G antigens seems to be expressed in quantity from the earliest stage a embryogenesis. By immunofluorescense with monoclonal antibodies, the B-G antigen has been detected on erythrocytes in the blood islands at day two of incubation (Lassila and Skjødt, unpublished data). Furthermore, using flow cytometry we found no major difference in the level of expression of B-G antigen on erythrocytes from day 10 to day 21 of incubation, compared with adult levels. Likewise, precipitation and electrophoresis of B-G antigen from surface labelled erythrocytes from day 10 to day 21 of embryogenesis also did not reveal any difference of embryonal and adult antigen.

The B-G multigene family

Several B-G cDNAs and genes have been isolated but only partial sequences have yet been published (Kaufman et al 1989b; Miller 1991). The first cDNAs were obtained by expression cloning and were not full length. Our first clones contained the 3' ends of several different sequences, some of which contained unspliced introns. The 3' end of the isolated cDNA had a high identity between clones, whereas the 5' end contained multiple short 21 nucleotide repeats which varied between the clones. By Southern blot analyses, all the sequences were found to map in the MHC, (based on the congenic lines CB and CC). Most map into the *B-G* region, but some map into the *B-F/B-L* region based on the haplotypes *B15*, *B21* and their apparently reciprocal recombinants B21r3 (R4) and B15r1 (R5).

The sequences of B-G now offer a general picture of a *B-G* gene: the translated part of the gene begins with an exon containing an apparent signal sequence, but since the exon contains at least 40 amino acids and the cleavage site has not been determined, there might be additional sequence belonging to the protein. The next exon encodes an Ig V-like domain. The sequence of this exon includes the invariant cysteines and tryptophan of Ig V-like regions, another cysteine present which might be involved in the intrachain disulfide bond, but no site for asparagine-linked carbohydrate. This is followed by an exon encoding a hydrophobic transmembrane segment. The cytoplasmic region is encoded by multiple exons, although the precise number varies between different genes. Many of these cytoplasmic exons are composed of poorly conserved 21 nucleotide repeats, some of which are apparently subject to alternative splicing. The 21 bp repeats encode 7 amino acid repeats with hydrophobic residues in the first and fourth positions, corresponding to an α-helical coiled coil (Fig. 2). This picture is pieced together from erythrocyte B-G protein and cDNA data and the sequence of the 8.5 *B-G* gene (Fig 3). Other genes are bound to vary since precipitation data of B-G molecules from lymphoid tissues has demonstrated glycosylated antigen.

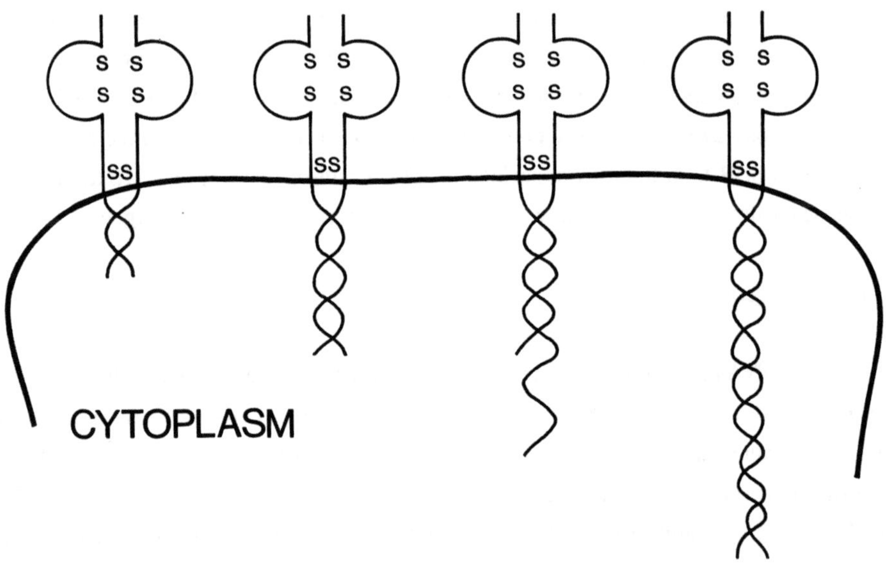

Fig. 2. Biochemical structure of B-G molecules. Transmembrane dimers of different size are shown. The extracellular Ig V-like regions contains intrachain disulfide bonds and also the interchain disulfide bonds are shown. The cytoplasmic tails contain heptad repeat regions that presumably form α-helical coiled coils. Size differences are mainly due to variation in the length of the cytoplasmic tails. (Kaufman et al. 1990).

The number of *B-G* genes in the MHC is high when studied by Southern blot analysis (Goto et al. 1988; Kaufman et al. 1989b) and evidence has also been obtained for recombination within the *B-G* genes expressed on erythrocytes (Miller et al. 1988b). We are currently mapping the *B-G* genes derived from a B12 (CB) cosmid library (Guillemot et al. 1988). So far, approximately 18 *B-G* genes have been mapped in the *B-G* region and few B-G hybridizing fragments have been found in the *B-F/B-L* region of the MHC. One of the latter fragments is due to the polymorphic 8.5 gene, which is expressed on B-cells.

To study the products of single genes in isolation, we are transfecting the B-G genes into mouse L cells and analysing reactions of the transfectants with antibodies to B-G. In some cases, we have developed new monoclonal antibodies to the transfected cells in order to study the distribution of the epitopes encoded by that particular gene. Eventually, this should provide a good estimate of which genes are active in which cells and tissues.

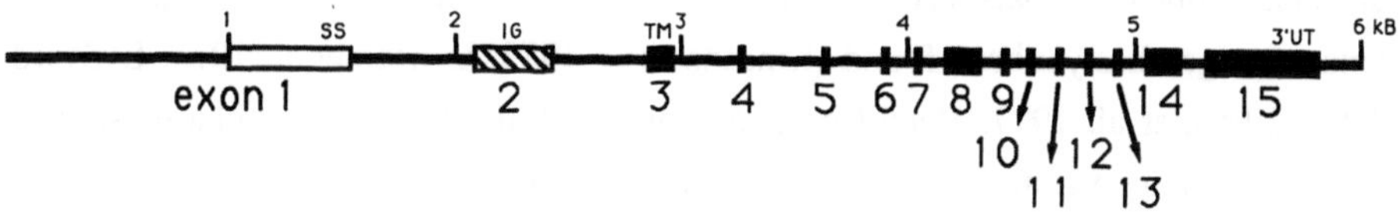

Fig. 3. Intron-exon structure of the 8.5 gene. This is a preliminary picture, since the 5' end of the gene has not been mapped and the extent of possible alternative splicing is not known.

Speculations on the function of B-G

The distribution and structure of B-G antigens have now been refined to a level such that their involvement in cell recognition or receptor interaction seems plausible, as has been proven for many other molecules of the Ig-superfamily. Since many *B-G* genes are now known to exist and are expressed in several diverse lymphoid and stromal compartments, it is likely that these antigens are involved in several different tasks. Such pleiomorphism is known from studies of class I gene function in rodents. At this time, it is not possible to present a unified scheme that accounts for all of the various phenomena that we have observed, although we can speculate about some likely functions of B-G.

We have already suggested that the selection process of the thymus could have a counterpart in the bursa (Kaufman et al. 1990b). We will further argue that B-cells are selected on the basis of binding to a self antigen in the bursa and propose that it is polymorphic antigens from the *B-G* multigene family that are involved in the selection process. This proposal is mainly based on the polymorphism of *B-G*, the reaction of antibodies to *B-G* on bursa stroma, the preferential humoral response to B-G and the preimmune level of "natural antibodies" to allogeneic *B-G*. These findings clearly suggest an involvement of B-G antigens in the development of the chicken B-cell repertoire.

The chicken B-cell system is derived from a limited number of ontogenically-restricted extra-bursal stem cells, which depend on the bursal stromal microenviroment to develop. The first hemopoetic cells to enter the bursal epithelium are of the macrophage-dendritic lineage. These cells localize to form small epithelial buds, which are then colonized by lymphoid precursors (Houssaint 1987). Individual follicles are populated by few stem cells and antibody diversity is created by rounds of gene conversion on the single rearranged gene (Ratcliffe 1985; Pink et al. 1987; Weill and Reynaud 1987). The rapid proliferation of cells in the bursal follicles is balanced by extensive cell death but no absolute requirement of exogenous antigen has yet been found (Kincade et al. 1973; Lawton et al. 1975; Lydyard et al. 1976; Eskola et al. 1977; Ratcliffe 1985; Lassila et al. 1989; Mansikka et al. 1990). Uptake of environmental antigens via the cloaca to the bursa follicles can occur (Sorvari et al. 1975), but will be considered a secondary event for expansion of previously selected B-cells. The large number of cells dying in the bursa is indicative of a selection process or a high proportion of abortive gene-conversion events. Furthermore, a rather restricted antibody repertoire dominated by germline specificity would be anticipated, unless the bursa can specifically preselect mutant clones (Metcalf and Moore 1971).

The follicles differentiate to form a cortex and medulla which are separated by a basal membrane and a layer of epithelial cells. Immunoglobulin positive cells appear first in the medulla (Metcalf and Moore 1971; Lawton et al. 1975). Around hatching, the cortex develops strongly, consisting of rapidly dividing large lymphocytes. The traffic of cells from the medulla through the basal membrane into the cortex has often been described (Bosch 1968; Ackerman and Knouff 1959; Grossi et al. 1976). The route by which B-cells traffic from the cortex to the periphery is not known in detail, although a network of lymph vessels outside the follicles (Ekino et al. 1979) and blood capillaries within the cortex are the obvious probable conduits. The reactivity of some of our monoclonal antibodies, which show strong staining in the medullary stroma and on the basal membrane epithelial cells, suggests that B-G antigen could function as selection ligands for developing B cells within the medulla of the bursa or as they traffic to the cortex.

If B-G antigens are involved in selecting the B-cells, it could happen by a negative- or a positive selection model (Fig. 4). The theory of Jerne, linking generation of diversity with establishment of tolerance to self antigens by having antibody germline specificity to histocompatibility antigens (Jerne 1971), is tempting to apply in the case of B-G. Negative selection would involve B-G as an anergizing antigen and only B-cells which have mutated away from germline specificity would be allowed to migrate to the periphery. Positive selection would mean that those B-cells that still bind B-G after gene-conversion would be stimulated to further conversion and those B-cells that don't bind could escape. Tolerance induction of cells which initially have self B-G specificity in the germ line could account for the level of "natural antibodies" reactive with allogeneic B-G (Longenecker and Mosmann 1980), the preferential response to allogeneic B-G antigen (Longenecker et al. 1979) and the "adjuvant effect" (Hala et al. 1981).

The previous models focus on an interaction of B-cell receptors with antigen expressed on the bursa stroma, which we find particularly intriguing. However, the distribution of B-G on other cell types also needs to be discussed.

The high levels of polymorphic B-G found on erythrocytes might be involved in B-cell development by anergizing the non-mutated B-cells that leak through the bursal selection process. Alternatively, as for B-F antigen, the expression of B-G on erythrocytes could have no functional role in the immune system. The presence of MHC antigens on erythrocytes may reflect a phylogenetically distant use of B-G as a restriction element on erythroblasts. This will not be possible to appreciate until homologues are studied in taxa with metabolically active erythrocytes. A further issue is the glycosylation of non-erythroid B-G, which could be imagined to play a role for the function or recognition of B-G. All in all, the high expression of the polymorphic non-glycosylated B-G antigens on erythrocytes remains a separate mystery that might not be relevant for understanding the immunological function of B-G antigens.

The presence of B-G on thrombocytes also needs to be studied in greater detail. Thrombocytes are the major phagocytic cell type in the chicken and the main type that clear bacteria (Chang and Hamilton 1979). They synthesize a subset of the polymorphic B-G molecules found on erythrocytes (Salomonsen et al. 1991a). In addition, they express cytoplasmic class II antigen. Thus, thrombocytes, whose lineage relationship to other cells is unclear, could have a potential role as antigen-presenting cells. Recently, resistance to fowl cholera has been reported to be linked to the *B-G* region of the MHC (Lamont et al. 1987). Although the effect of differences in the *B-G* region was moderate, it is an interesting possibility that B-G on thrombocytes is involved in effects on bacterial clearing.

The 8.5 gene, which is located in the class I/class II region of the MHC and is expressed on B-cells, also needs further attention. 8.5 is the first and so far the only known polymorphic B-G gene on lymphocytes (Salomonsen et al. 1991a). Whether or not the 8.5 gene contributes to some of the many effects of the *B-F/B-L* region in immunological assays is unknown. However, we have found that some antibodies reacting with lymphocyte B-G epitopes activate bursa B-cells as measured as Ca^{2+}-flux. This further suggests that B-G on B-cells might be involved in signaling and could even play a role similar to that proposed for B-G on bursa stroma. The antibody staining and high level of B-G mRNA in the intestine (Miller et al. 1990; Salomonsen et al. 1991a) could reflect a functional, selecting role similar to the bursa. However, there is no separation of the cellular architecture of the caecal tonsil germinal centers into cortex and medulla. Furthermore, the B-cells which become seeded from the bursa should already have been through the previously proposed selection process.

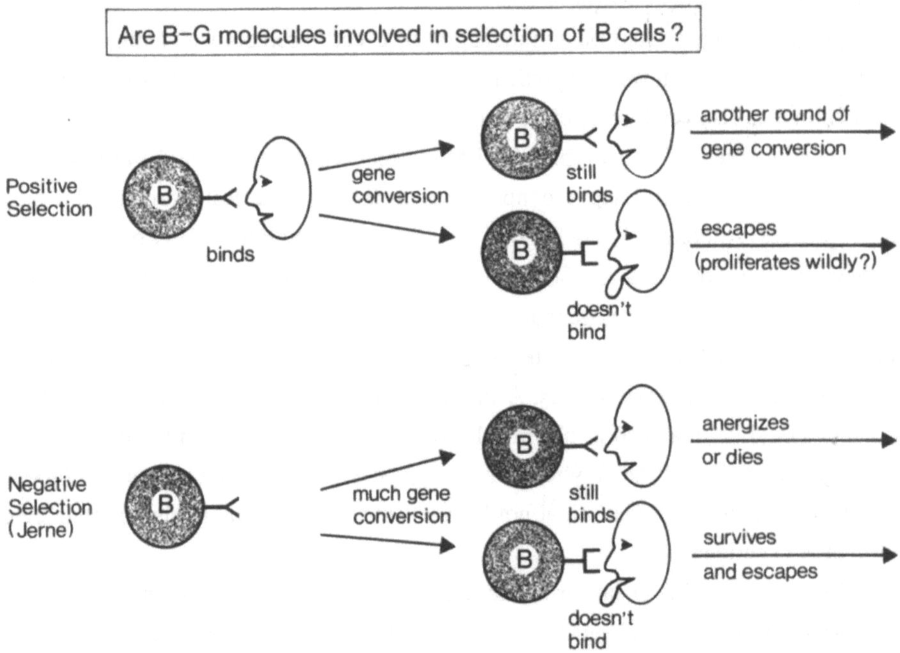

Fig. 4. A hypothezised scheme of selection of B-cells in the bursa of Fabricius.

Another possibility is that the strong B-G expression detected in the intestinal lymphoid compartment could mean, that the active genes code for molecules involved in trapping enviromental debris or bacteria (as suggested for thrombocytes above) to enhance antigen dependent B-cell expansion.

The staining on the thymus also suggests a possible role for B-G in T-cell maturation. We have not yet been able to demonstrate clear cut T-cell cytotoxicity or proliferative responses to allogeneic erythrocyte and thrombocyte B-G. However, in the light of the differences between the antigens expressed in these cell types and the antigens from lymphoid compartments, a new study of T-cell responses to lymphoid B-G might be rewarding.

Clearly, these various speculations leave many issues unresolved. We have only found monomorphic expression of B-G epitopes on the sections but that might be due to a limited number of reagents. Furthermore, the high expression of some B-G epitopes on peripheral B-cells and their low expression on bursa B-cells, together with the fact that some polymorphic epitopes are expressed on erythrocytes and crossreacting monomorphic epitopes are expressed on lymphocytes, suggests a complicated developmentally regulated and tissue specific expression of the *B-G* multigene family.

Our suggestion that B-G antigens are involved in central immunological processes has the implication that one would expect to find B-G homologues in other species. The indications that this is so include the presence of "natural antibodies" to polymorphic B-G molecules on chicken erythrocytes in all the species investigated and the preferential antibody response to B-G. Southern blot analyses have demonstrated B-G hybridizing bands in some other species of birds (Miller et al. 1988a). We have also obtained hybridizing fragments in Southern blots with some other bird species and have found some monoclonal antibody reactions on turkey erythrocytes. The putative B-G antigen was affinity-purified from turkey erythrocyte lysate and injected mice produced anti-sera that reacted with chicken B-G antigens when tested in western blot on chicken erythrocytes. The genes and antigens involved in these reactions have still to be proven formally as B-G homologues by family studies. In any event, the high expression on erythrocytes may be an avian peculiarity and the relevant epitopes might be found only on non-erythroid cells in other taxa. We have recently found that some of our antibodies, which stain the bursa stroma, also stain stromal cells in the sheep Ileal Peyers Patches (IPP). The sheep generates its B-cell repertoire by a somatic hypermutation mechanism (Reynaud et al. 1991) and arguments for the necessity of a selection process in this species could be made in analogy to the bursa. The cross-reaction of some antibodies specific for B-G with sheep IPP is intriguing, but either molecular cloning or a polymorphic reaction which can be shown to segregate with class I and class II is necessary before the B-G antigens enter the mammalian immune system.

Acknowledgments. We thank Drs. Shigeo Ekino and Wayne Hein for critical reading.

References

Ackerman, G.A. and Knouff, R.A. : Lymphocytopoiesis in the Bursa of Fabricius. *Am J Anat 104*: 163-205, 1959

Bosch, J.: Morphogenetische, histochemiche und experimentelle Untersuchungen uber die Bursa Fabricii. *Zool Jb Anat 85*: 327-385, 1968

Chang, C.F. and Hamilton, P.B.: The thrombocyte as the primary circulating phagocyte in chickens. *J Reticuloendothel Soc 25*: 585-590, 1979

Crone, M. and Simonsen, M. : Avian major histocompatibility complex. *In* A. Toivanen and P. Toivanen (eds.): *Avian Immunology: Basis and Practice.* Vol. 2: 25-41, CRC, Boca Raton, Florida 1987

Ekino, S., Matsuno, K. and Kotani M.: Distribution and role of lymph vessels of the Bursa of the Bursa Fabricii. *Lymphology 12*: 247-252, 1979

Eskola, J., Ruuskanen, O., Fraki, J.E., Viljanen, M.K. and Toivanen, A.: Alkaline phosphatase in the developing Bursa of Fabricius. *Scand J Immunol 6*:185-194, 1977

Goto, R., Miyada, C.G., Young, S., Wallace, R.B., Abplanalp, H., Bloom, S.E., Briles, W.E., and Miller, M.M.: Isolation of a cDNA clone from the B-G subregion of the chicken histocompatibility (*B*) complex. *Immunogenetics 27*: 102-109, 1988.

Grossi, C.E., Lydyard, P.M., and Cooper M.D.: B-Cell ontogeny in the Chicken. *Ann Immunol (Inst. Pasteur) 127C*: 931-941, 1976

Guillemot, F., Billault, A., Pourquie, O., Behar, G., Chausse, A-M., Zoorob, R., Kreiblich, G., and Auffray, C.: A molecular map of the chicken major histocompatibility complex: the class II ß genes are closely-linked to the class I genes and the nucleolar organizer. *EMBO J 7*: 2775-2785, 1988

Guillemot, F. and Auffray, C.: Molecular biology of the chicken major histocompatibility complex. *CRC Crit Rev Poultry Biol 2*: 255-275, 1989

Hala, K., Schulmannova, J., and Nemeckova, S.: The role of primary lymphoid organs of chickens in the elimination of ^{51}Cr-labeled erythrocytes. *Folia Biol (Praha) 26*: 26-33, 1980

Hala, K., Plachy, J., and Schulmannova, J.: Role of the B-G-region antigen in the humoral immune response to the B-F-region antigen of chicken MHC. *Immunogenetics 14*: 393-401, 1981

Havele, C., Wegmann, T. G., and Longenecker, B. M.: Tolerance and autoimmunity to erythroid differentiation (B-G) major histocompatibility complex alloantigens of the chicken. *J Exp Med 156*: 321-336, 1982

Houssaint, E.: Cell lineage segregation during Bursa of Fabricius ontogeny. *J Immunol 138*: 3626-3634, 1987

Jerne, N.: The somatic generation of immune recognition. *Eur J Immunol 1*: 1-9, 1971

Kaufman,J., Salomonsen, J., Skjoedt, K., and Thorpe, D.: MHC proteins and genes in birds and reptiles. *Dev Comp Immunol 13*: 374-375, 1989a

Kaufman, J., Salomonsen, J., and Skjødt, K.: B-G cDNA clones have multiple small repeats and hydridize to both chicken MHC regions. *Immunogenetics* 30: 440-451, 1989b

Kaufman, J., Salomonsen, J., Skjødt, K., and Thorpe, D.: Size polymorphism of the chicken MHC-encoded B-G molecules is due to length variation in the cytoplasmic heptad repeat region. *Proc Natl Acad Sci USA 87*: 8277-8281, 1990a

Kaufman, J., Skjødt, K., and Salomonsen, J.: The MHC molecules of nonmammalian vertebrates. *Immunol Rev* 113: 83-117, 1990b

Kaufman, J., Skjødt, K., and Salomonsen, J.: The B-G multigene family of the chicken MHC. *CRC Crit Rev Immunol*, in press 1991

Kincade, P.W., Self, K.S., and Cooper, M.D.: Survival and function of Bursa-derived cells in bursectomized chickens. *Cell Immunol 8*: 93-102, 1973

Kline, K., Briles, W.E., Bacon, L., and Sanders, B.G.: Characterization of two distinct disulfide-linked B-G molecules in the chicken. *J Hered 79*: 249-256, 1988

Lamont, S.J., Bolin, C., and Cheville, N.: Genetic resistance to Fowl cholera is linked to the major histocompatibility complex. *Immunogenetics 25*: 284-289, 1987

Lassila, O., Lefkovits, I., and Alanen, A.: Immunoglobulin diversification in bursal duct-ligated chickens. *Eur J Immunol 19*:1343-1345, 1989

Lawton, A.R., Kincade, P.W., and Cooper, M.D.: Sequential expression of germ line genes in development of immunoglobulin class diversity. Gene Regulation in Differentiation and Development. *Fed Proc 34*: 33-39, 1975

Longenecker, B.M., Mosmann, T.R., and Shiozawa, C.: A strong, preferential response of mice to polymorphic antigenic determinants of the chicken MHC, analyzed with mouse hybridoma (monoclonal) antibodies. *Immunogenetics 9*: 137-147, 1979

Longenecker, B. M. and Mosmann, T. R.: Restricted expression of an MHC alloantigen in cells of the erythroid series: A specific marker for erythroid differentiation. *J Supramol Struct 13*: 395-400, 1980a

Longenecker, B.M. and Mosmann, T.R.: "Natural" antibodies to chicken MHC antigens are present in mice, rats, humans, alligators and allogeneic chickens. *Immunogenetics 11*: 293-302, 1980b

Longenecker, B.M. and Mosmann, T.R.: Structure and properties of the major histocompatibility complex of the chicken. Speculations on the advantages and evolution of polymorphism. *Immunogenetics 13*: 1-23, 1981

Lydyard,P.M., Grossi,C.E. and Cooper, M.D.: Ontogeny of B cells in the Chicken. *J Exp Med 144*: 79-97, 1976

Mansikka, A., Jalkanen, S., Sandberg, M., Granfors, K., Lassila, O., and Toivanen, P.: Bursectomy of chicken embryos at 60 hours of incubation leads to an oligoclonal B cell compartment and restricted Ig diversity. *J Immunol 145*: 3601-3609, 1990

Metcalf, D. and Moore, M.A.S. : Haemopoietic Cells. Frontiers of Biology Vol. 24. *In* A. Neuberger and E.L. Tatum (eds.): *North-Holland Research Monographs*, 1971

Miller, M.M., Goto, R., and Abplanalp, H.: Analysis of the B-G antigens of the chicken MHC by two-dimensional gel electrophoresis. *Immunogenetics 20*: 373-385, 1984

Miller, M.M., Abplanalp, H., and Goto, R.: Genotyping chickens for the B-G subregion of the major histocompatibility complex using restriction fragment length polymorphisms. *Immunogenetics 28*: 374-379, 1988a

Miller, M.M., Goto, R., and Briles, W.E.: Biochemical confirmation of recombination within the B-G subregion of the chicken major histocompatibility complex. *Immunogenetics 27*: 127-132, 1988b

Miller, M.M., Goto, R., Young, S., Liu, J., and Hardy, J.: Antigens similar to major histocompatibility complex B-G are expressed in the intestinal epithelium in the chicken. *Immunogenetics 32*: 45-50, 1990

Miller, M. M.: The major histocompatibility complex of the chicken. *In* : G. W. Warr and N. Cohen (eds.): *Phylogenesis of Immune Functions*, pp. 151-169, CRC Press, Boca Raton, Florida 1991

Neu, N., Hala, K., and Wick, G.: "Natural" chicken antibodies to red blood cells are mainly directed against the B-G antigen, and their occurence is independent of spontaneous autoimmune thyroiditis. *Immunogenetics* 19: 269-277, 1984

Pink, J.R.L., Droege, W., Hala, K., Miggiano, V.C., and Ziegler, A.: A three-locus model for the chicken major histocompatibility complex. *Immunogenetics 5*: 203-216, 1977

Pink, J.R.L, Lassila, O., and Vainio, O.: B-Lymphocytes and their self-renewal. *In* A. Toivanen and P. Toivanen. (eds): *Avian Immunology: basis and practise.* Vol. 1: 65-78, CRC Press, Boca Raton, Florida 1987

Ratcliffe, M. J. H.: The ontogeny and cloning of B cells in the bursa of Fabricius. *Immunol Today 6*: 223-227, 1985

Reynaud, C.-A., Mackay, C.R., Müller, R.G., and Weill, J.-C.: Somatic generation of diversity in a mammalian primary lymphoid organ: the sheep Illeal Peyers Patches. *Cell 64*: 995-1005, 1991

Salomonsen, J., Skjødt, K., Crone, M., and Simonsen, M.: The chicken erythrocyte-specific antigen. Characterization and purification of the B-G antigen by monoclonal antibodies. *Immunogenetics 25*: 373-382, 1987

Salomonsen, J., Skjødt, K., Thorpe, D. and Kaufman, J.: The chicken MHC-encoded B-G antigen: structure and tissue distribution. *7th International Congress of Immunology*, West Berlin. G. Fischer Verlag, Stuttgart 1989

Salomonsen, J., Dunon, D., Skjødt, K., Thorpe, D., Vainio, O., and Kaufman, J.: Chicken major histocompatibility complex-encoded B-G antigens are found on many cells that are important for the immune system. *Proc. Natl. Acad. Sci.* USA 88: 1359-1363, 1991a

Salomonsen, J., Eriksson, H., Skjødt, K., Lundgreen, L., Simonsen, M., and Kaufman, J.: The "adjuvant effect" of the polymorphic B-G antigens of the chicken MHC analyzed using purified molecules incorporated in liposomes. *Eur J Immunol 21*: 649-658, 1991b

Schierman, L. W. and McBride, R. A.: Adjuvant activity of erythrocyte isoantigens. *Science 156*: 658-659, 1967

Schulmanova, J., Hala, K., and Hraba, T.: Tolerance to alloantigens of the A and B systems in chickens. *Folia Biol (Praha) 24*: 270-280, 1978

Sorvari, T., Sorvari, R., Ruotsalainen, P., Toivanen, A., and Toivanen, P. : Uptake of enviromental antigens by the bursa of Fabricius. *Nature 253*: 217-219, 1975

Weill, J.-C. and Reynaud, C.-A.: The chicken B cell compartment. *Science 238*: 1094-1098, 1987

EVOLUTION OF PRIMATE *C4* AND *CYP21* GENES

Hiroshi Kawaguchi*, Colm O'hUigin, and Jan Klein+
Max-Planck-Institut für Biologie, Abteilung Immungenetik, 7400 Tübingen, Germany and +Department of Microbiology and Immunology, University of Miami School of Medicine, Miami, FL 33101, USA

The C4 and CYP21 genes of the major histocompatibility complex are part of a module which may also include other closely linked loci. In several species, the module has been found duplicated or triplicated and to undergo continuous expansion and contraction. We argue that the primigenial duplication of the module occurred only once in the history of placental mammals but that because of a continuous homogenization (concerted evolution), orthologous relationships between genes in different species are not discernible. We suggest further that the expansion and contraction process is fuelled by selection for duplicated C4 genes and the high frequency of congenital adrenal hyperplasia caused by defects in the CYP21 gene is the consequence of this selection. The only chance for escaping this consequence is to fix monomodular haplotypes. Certain species seem to have indeed opted for this evolutionary pathway by adjusting their single C4 genes to perform the functions that are carried out by **two** *genes in other species.*

The UC island

Intercalated between the class I loci on the telomeric side and the class II loci on the centromeric side of the *HLA* complex on the short arm of human chromosome 6 is an island of unrelated chromatin (UC; Spies et al. 1989a,b; Sargent et al. 1989). The island is inhabited by at least 20 loci referred to by some investigators as "class III loci"; however, since this denomination implies functional and evolutionary relationship of the UC island to the major histocompatibility complex (*Mhc*) loci, while in reality no such relationship exists (Klein and Figueroa 1986), we refrain from using it. We regard the inhabitants of the UC island as a farraginous assembly of elements brought together by chance and remaining in their current position partly because of the conservative character of chromosome evolution and partly because they are now entrapped between the two *Mhc* clusters containing loci that might be kept together by selection.

---

*Current address: Department of Dermatology, Yokohama City University School of Medicine, Yokohama 232, Japan

NATO ASI Series, Vol. H 59
Molecular Evolution of the
Major Histocompatibility Complex
Edited by J. Klein and D. Klein

Two prominent inhabitants of the UC island are the *C4* and *CYP21* loci (Carroll et al. 1984, 1985a; White et al. 1985). The *C4* locus codes for a serum protein, the complement component 4, which upon activation by the C1 component in turn activates another member of the complement cascade (C2) and at the same time grounds the cascade to a substrate, such as a cell or an antigen-antibody complex, by interacting with their active groups (Ross 1986; Reid 1985). In many mammalian species, two functional forms of the C4 component, A and B, have been found. The C4A component binds efficiently to proteins in antigen-antibody complexes and thus facilitates their solubilization and clearance. The C4B component binds preferentially to cell surfaces and thus participates in cell lysis. A simple way of distinguishing the two C4 forms functionally is by using the hemolytic assay: the C4B protein is four times more efficient than the C4A component in lysing red blood cells. The functional difference between C4A and C4B is attributed to amino acid residues in the so-called C4d fragment encoded in exons 23 through 30 (the *C4* gene has altogether 41 exons!). Presumably, certain residues in this region of the C4A molecule interact preferentially with NH_2 groups forming amide bonds with proteins of the antigen-antibody complexes, whereas certain residues in the same region of the C4B molecule interact preferentially with OH groups forming ester linkages with sugars and other components of the cell surface (Law et al. 1984; Yu et al. 1986). The C4A and C4B proteins are encoded in two separate loci. These are members of a family that includes two other complement-encoding loci, C3 and C5, as well as the α_2-macroglobulin gene coding for a noncomplement serum protein (Sottrup-Jensen et al. 1985).

The *CYP21* gene codes for a key enzyme in the synthesis of glucocorticoid and mineralocorticoid hormones (cortisol and aldosterone, respectively) -- the cytochrome 450, subfamily XXI (steroid 21-hydroxylase). A defect in CYP21 prevents the conversion of 17-hydroxyprogesterone (17-OHP) to 11-deoxycorticosterone (New and Levine 1984; Speiser and New 1985; White et al. 1987; Miller 1988b). The 17-OHP accumulates and is shunted into the pathway leading to the biosynthesis of androgen hormones. The resulting excessive production of androgens causes virilization in the affected individuals (ambiguous genitalia in females, sexual precocity in males) together with other defects. A complete inactivation or absence of the CYP21 enzyme also means that progesterone cannot be converted into 11-deoxycorticosterone and hence that no aldosterone is produced. The absence of aldosterone results in inability to conserve urinary sodium ions and is manifested in the salt-wasting syndrome, which may lead to the death of an untreated patient within the first few weeks after birth. Because aldosterone is normally produced at a rate 100 to 1000 times lower than cortisol, a low level of CYP21 activity may be sufficient to sustain aldosterone synthesis but insufficient to sustain cortisol synthesis. Virilization can therefore occur without the salt-wasting syndrome. Individuals with CYP21 activity as low as 20% of the normal level remain asymptomatic or, at worst, develop mild symptoms of androgen excess such as precocious adrenarche in girls. All these manifestations of the CYP21 deficiency are clubbed together into a single disease entity -- the congenital adrenal hyperplasia (CAH), which is classified into three groups: salt-wasting CAH (in which total absence of CYP21 activity prevents the synthesis of both cortisol and aldosterone), simple virilizing CAH (in which very low CYP21 activity sustains aldosterone synthesis but prevents cortisol synthesis), and nonclassical CAH (in which moderate CYP21 activity causes mild manifestations of androgen excess); the salt-wasting and the simple virilizing CAH represents the "classical" forms of the disease. Most humans have one functional *CYP21* gene and one pseudogene (*CYP21P*; the former is frequently referred to as the *CYP21B* and the latter as the *CYP21A* gene).

The C4-CYP21 module

The *C4* and *CYP21* genes are organized into modules, each module consisting of one *C4* and one *CYP21* gene, transcriptionally oriented in the same direction (Carroll et al. 1985a, 1987; White et al. 1985; Dunham et al. 1987; Figs. 1 and 2).

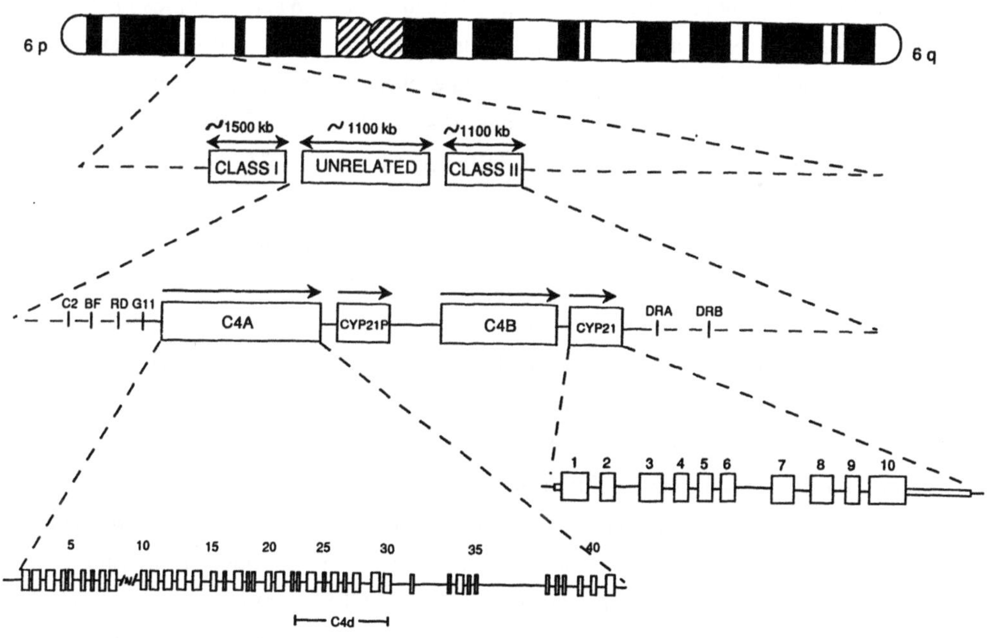

Fig. 1. Position and organization of human *C4* and *CYP21* genes on chromosome 6. The final blow-ups in the sequence show exon-intron arrangement in the two genes.

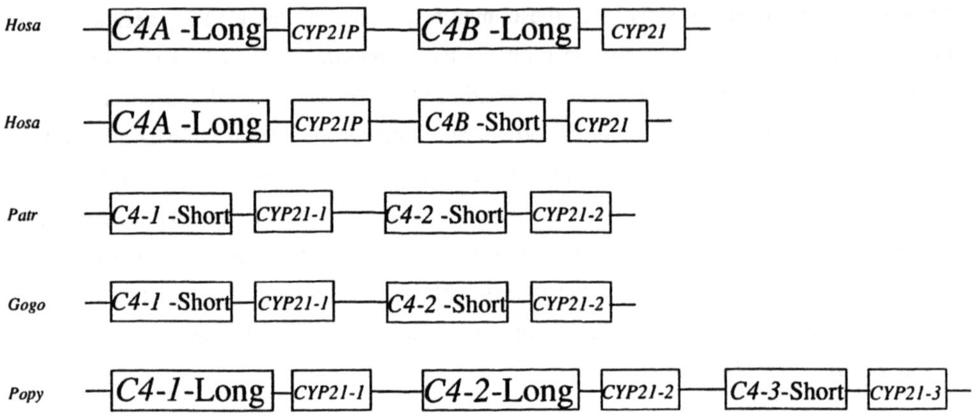

Fig. 2. Common *C4-CYP21* haplotypes of human (*Hosa*), chimpanzee (*Patr*), gorilla (*Gogo*), and orang-utan (*Popy*). For details on ape haplotypes see Kawaguchi et al. (1990) and Kawaguchi and Klein (1991).

One human *C4-CYP21* module is about 35 kb long and the distance between its two genes is approximately 3 kb (Dunham et al. 1987; Carroll et al. 1987). When multiples of *C4* and *CYP21* genes occur, they are always present as multiples of the basic module. For example, when two *C4* and *CYP21* genes occur on a single chromosome, as is the case in most humans, the *C4* genes alternate with *CYP21* genes (*C4...CYP21...C4...CYP21*) and one never finds a situation in which two *C4* genes are followed by two *CYP21* genes. The modular organization is retained in spite of extensive variation in length and gene copy number of the *C4-CYP21* region. The form with two modules appears to be the standard, at least in humans and mice, but chromosomes with one, three, or four modules are not at all rare (in Caucasians, for example, chromosomes with one module or with three modules occur at frequencies of 16% and 18%, respectively; Teisberg et al. 1988). The modular polymorphism is believed to arise by unequal crossing-over occurring between misaligned modules (e.g., when a *C4* gene of the first module on one chromosome pairs with a *C4* gene of a second module on the second chromosome; Raum et al. 1984; Carroll et al. 1985a,b; Donohoue et al. 1986a; Rodrigues et al. 1987; Fig. 3), The crossing-over theory is supported by pedigree analysis of at least one family (Sinnott et al. 1990). Unequal crossing-over between two chromosomes each carrying two *C4-CYP21* modules can produce chromosomes with one module (= deletion of one module) or with three modules (= duplication of one module; see Fig. 2). Deletions involving the functional *CYP21* gene are the cause of CAH in 20% to 30% of patients with the classical form of CAH (for a summary of deletion frequencies, see Mornet et al. 1991). Originally it was thought that unequal crossing-over occurs between the *C4* and *CYP21* genes (White et al. 1984a), but more recent studies have revealed that it often occurs between exons *within* genes (Donohoue et al. 1989). In the gene, however, the recombination breaks can occur at different positions.

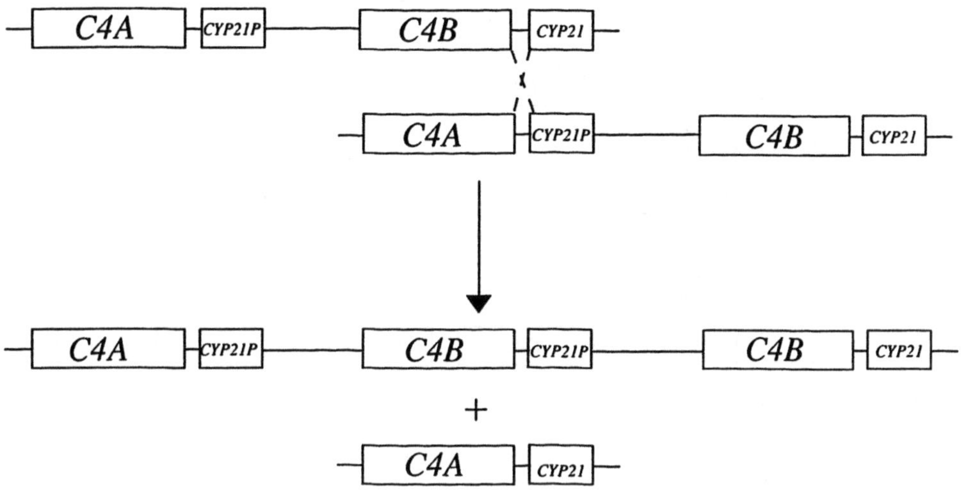

Fig. 3. Unequal homologous crossing-over between misaligned *C4-CYP21* regions (indicated by broken lines) and the trimodular and monomodular haplotypes it produces.

Taking the organization *C4A...CYP21P...C4B...CYP21* for a standard human haplotype, we can see from Figure 4 that the misalignment of two such haplotypes can theoretically produce six derived haplotypes (ignoring the fact that the unequal crossing-over is mostly intragenic) -- three haplotypes with a modular deletion and three with a duplication. The arrangement of *C4A* vs. *C4B* and *CYP21P* vs. *CYP21* genes in the derived haplotypes can differ depending on where the crossing-over took place. The derived haplotypes themselves can subsequently participate in unequal crossing-over events when paired with the standard haplotype or with other derived haplotypes. Such secondary unequal crossing-over can produce additional variation in the length and gene composition in the *C4-CYP21* region, but no matter how they rearrange, the *C4-CYP21* genes retain their modular organization. Unequal crossing-over cannot, in fact, disrupt the modular structure and so, turning the argument around, the observed retention of modular organization must indicate that most of the *C4-CYP21* haplotype length polymorphism has been generated by unequal crossing-over. How the modularity arose in the first place, however, is a different question.

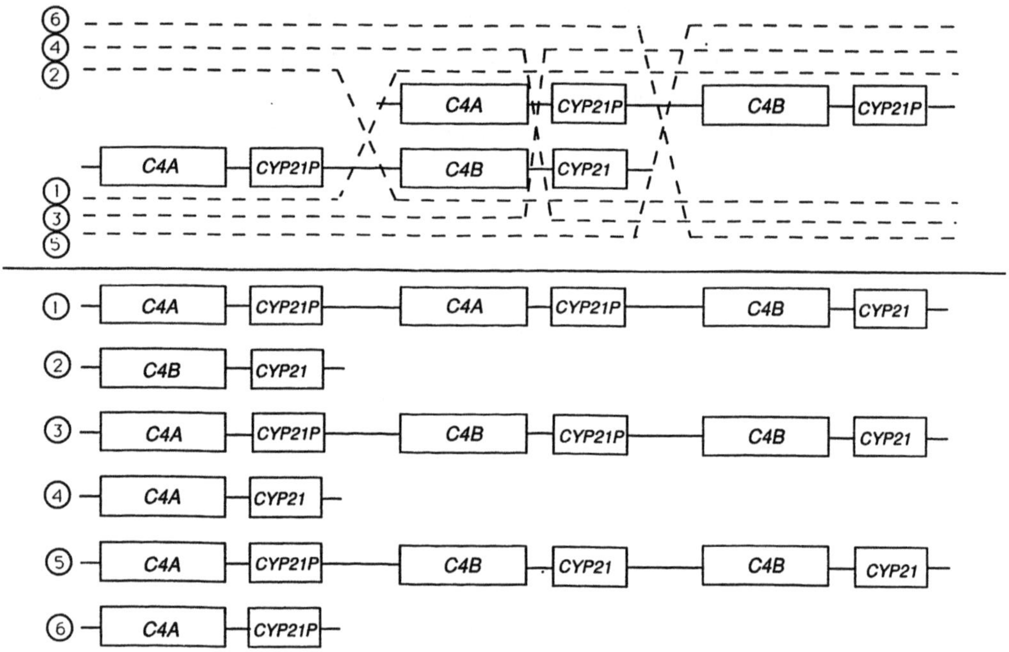

Fig. 4. The six possible types of unequal crossing-over between bimodular haplotypes, and the resulting haplotypes. Broken lines indicate the site of exchange and the genes involved in it.

The primigenial duplication

Before a chromosome with two *C4* and two *CYP21* genes appeared, there must have been a stage when all the chromosomes in the population had only one *C4* and one *CYP21* gene. While it is easy to contract two genes into one and then expand again to two or more genes, when haplotypes with duplicated genes are present in the population already, it is more difficult to explain how one gene can give rise to two when there are no duplicated genes in the population. Such initial, *primigenial* duplication cannot arise by the usual form of unequal crossing-over because there are no extra gene copies available for mispairing. The problem of the primigenious duplication is of course not limited to the *C4* and *CYP21* genes; it concerns *any* gene.

There are two ways of solving the problem. One way (Fig. 5) is to assume that as the sequences flanking a gene change by mutations, occasionally a short noncoding stretch of nucleotides at the 5' flank by chance acquires the same or very similar sequences as another short stretch at the 3' flank. This provides the potential for mispairing between the two short stretches and for a breakage-and-reunion process similar to that occurring in crossing-over.

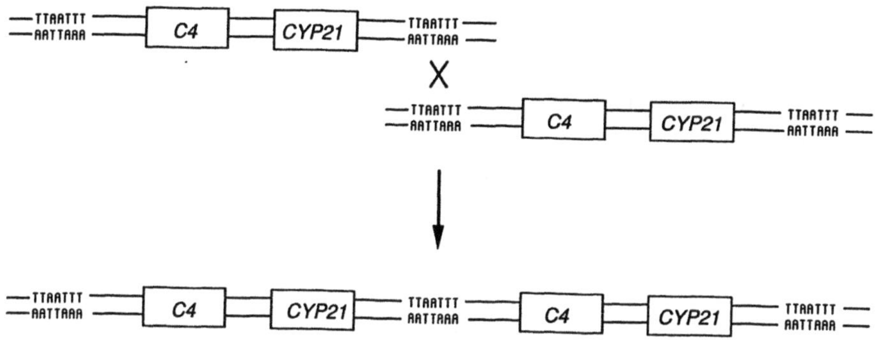

Fig. 5. A postulated mechanism of primigenial duplication in the *C4-CYP21* region by crossing-over between short, misaligned flanking sequences which happen to resemble each other by chance. The flanking repeats are purely hypothetical.

The exchange can generate one chromosome in which the gene has been deleted (a lethal or at least unfavorable condition) and another chromosome in which the gene has been duplicated. For the *C4* and *CYP21* genes we have to assume that the stretches of similar sequence bracketing the module occurred 5' of the original *C4* gene and 3' of the original *CYP21* gene (Fig. 5). The mispairing would have to be accompanied by the looping out of some 30 kb of chromatin. Whether under these conditions a short misaligned stretch of DNA can remain paired long enough for the breakage-and-reunion process to take place is doubtful. It is possible, however, that the *C4-CYP21* genes reside in a "natural" chromatin loop attached to the nuclear membrane by anchoring proteins. Involvement of such loops in the breakage-and-reunion process has been postulated by Vanin et al. (1983) as an explanation for length regularity of deletions in the ß-globin genes.

The second way of generating conditions for a primigenial duplication is by the insertion of movable repetitive elements epitomized by the *Alu* repeats. An insertion of one *Alu* element 5' of the *C4* gene and another 3' of the *CYP21* gene would create a situation in which, sooner or later, the two elements could be expected to mispair and through a breakage-and-reunion process produce some chromosomes with a deletion and others with a duplication of the *C4-CYP21* genes. Because an *Alu* element is approximately 300 base pairs (bp) long, whereas a stretch of nucleotides accidentally acquiring sequence similarity through mutations realistically cannot be expected to be much longer than 10 bp, the probability of exchange between mispaired *Alu* elements is much higher than that between the short stretches. Indeed, frequent occurrence of recombination leading to changes in gene copy numbers has been associated with the insertion of *Alu* elements in several genetic systems (Ottolenghi and Giglioni 1982; Lehrman et al. 1985, 1986, 1987; Rouyer et al. 1987; Myerowitz and Hogikyan 1987; Huang et al. 1989; Miura et al. 1989).

original ancestral haplotype. If one of the newly duplicated haplotypes became fixed in the rodent lineage and the other in the primate lineage and the genes in each haplotype then diverged genetically, dendrograms constructed for the genes would show the primate genes grouping separately from the rodent genes (Fig. 6B). In other words, unequal crossing-over would have *homogenized* the genes to look more alike within each of the two orders than between them.

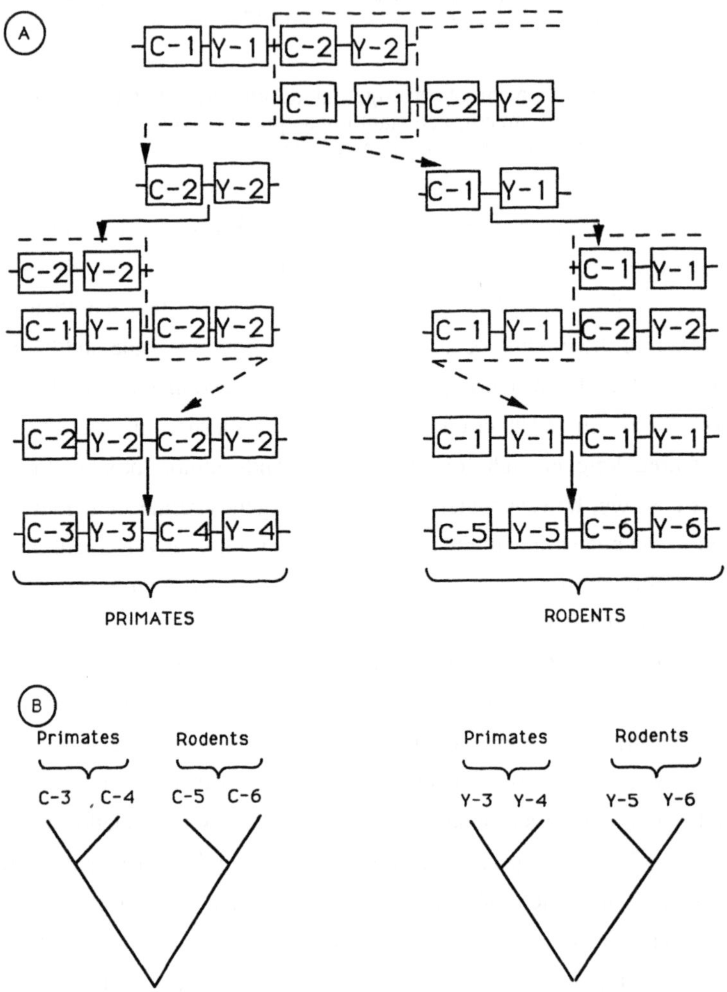

Fig. 6. Postulated homogenization of *C4* (= C) and *CYP21* (= Y) genes by unequal crossing-over during the evolution of primates and rodents (A). As a result of this process the primate genes are more similar to each other than they are to rodent genes and vice versa (B).

Such homogenization (concerted evolution) could have, and most likely has occurred repeatedly even within the same order (Fig. 7). The dendrogram in Figure 7 is based on partial sequences of human, gorilla, chimpanzee, and orang-utan genes (Kawaguchi et al. 1991a). It reveals that in the orang-utan lineage at least, the three *C4* genes borne by a single chromosome are derived from a single ancestral gene which was different from that of the gorilla, chimpanzee, and human ancestral *C4* genes. The same is true also for the *CYP21* genes of these primate species (Kawaguchi et al. 1991b). The homogenization of the orang-utan genes probably occurred by the process depicted in Figure 6A.

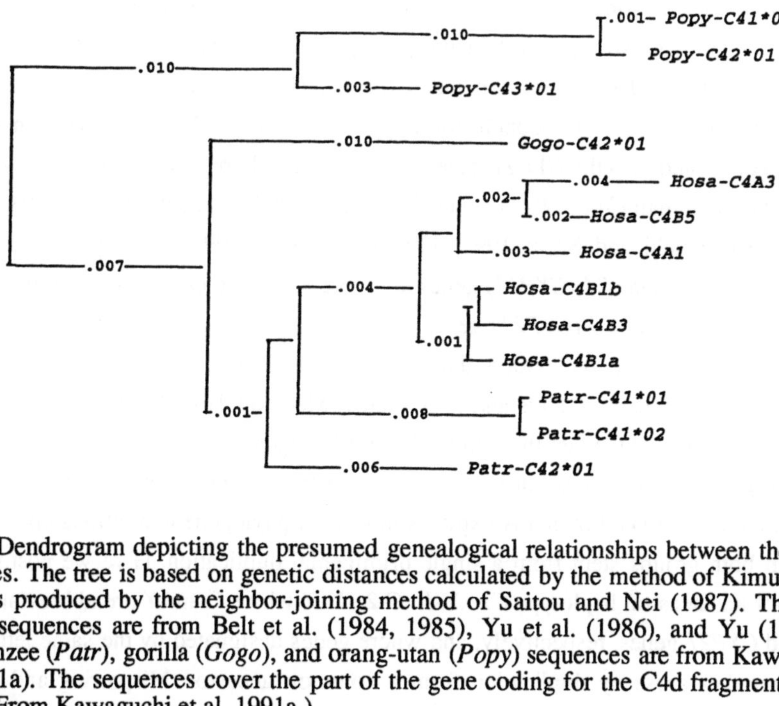

Fig. 7. Dendrogram depicting the presumed genealogical relationships between the primate *C4* genes. The tree is based on genetic distances calculated by the method of Kimura (1980) and was produced by the neighbor-joining method of Saitou and Nei (1987). The human (*Hosa*) sequences are from Belt et al. (1984, 1985), Yu et al. (1986), and Yu (1991); the chimpanzee (*Patr*), gorilla (*Gogo*), and orang-utan (*Popy*) sequences are from Kawaguchi et al. (1991a). The sequences cover the part of the gene coding for the C4d fragment of the α chain. (From Kawaguchi et al. 1991a.)

Homogenization by unequal crossing-over is also required to explain certain other features of the primate *C4-CYP21* modules. First, although in each species the *C4* genes are more similar to one another than they are to genes in other species, the differentiation into *A*- and *B*-like genes is retained in all four species. The *A* vs. *B* difference is determined by a short nucleotide stretch in the *C4d* region of the gene and may essentially depend on a single codon specifying the amino acid residue 1106 (Law et al. 1984; Yu et al. 1986). It is nevertheless unlikely that the *A*- and *B*-specifying codons arose in each species independently. Thus, different regions of the *C4* gene seem to display different phylogenies: The *A*- and *B*-specifying region appears to have diverged before the divergence of the orang-

Because neither *Alu* nor any other repetitive elements have thus far been found in the vicinity of the *C4* and *CYP21* genes in any of the species studied (although, admittedly, the search for them has not been thorough), we assume that the primigenial *C4-CYP21* module was generated by the first of the two described mechanisms. A search for 5' and 3' flanking stretches of homology may not reveal any good candidates because the original homologous regions may have long since diverged beyond recognition by the same process that created them in the first place.

Assuming that short homology regions 5' of *C4* and 3' of *CYP21* were indeed responsible for the primigenial duplication, we can assume further that this duplication occurred only once*. Moreover, because the *C4-CYP21* module is found in different orders of placental mammals, we assume that the primigenial duplication of the *C4* and *CYP21* genes occurred more than 75 million years (my) ago, before the main radiation of the placentals (Carroll 1988). This conclusion appears to clash with the observation that the mouse *C4* genes (and also the *CYP21* genes) are more closely related to each other than they are to any of the human *C4* (*CYP21*) genes (Tosi et al. 1984, 1985; White et al. 1984b). Our own data (Kawaguchi and Klein 1991; Kawaguchi et al. 1991a,b) suggest that even in primates, the orang-utan (*C4 (CYP21)* genes are more similar to each other than any of them are to any of the human *C4 (CYP21)* genes (see below). These relationships have led some investigators to advocate the notion that the *C4-CYP21* modules arose by independent primigenial duplications in rodents and in primates (Meo and Tosi 1985). We would like to point out, however, that the fact that mouse *C4 (CYP21)* genes form one phylogenetic group and human genes another does not necessarily mean that the two groups originated by two independent duplications. Figure 6A explains how such phylogenetic relationships can arise even when both groups stem from a single primigenial duplication. Here we postulate the existence of an ancestral haplotype *C1...Y-1...C-2...Y-2*, where *C* and *Y* are abbreviations for the *C4* and *CYP21* genes, respectively, and where genes designated by the same letter but a different number arose by duplication from a common ancestor. Mispairing and unequal crossing-over between two such ancestral haplotypes could have produced two different single-module haplotypes, *C-1...Y-1* and *C-2...Y-2*, which, in turn, could have produced *C-1...Y-1...C-1...Y-1* and *C-2...Y-2...C-2...Y-2* duplicated haplotypes by recombination with the

*It has been estimated that the rate of duplication in a vertebrate genome containing some 50,000 expressed genes is 0.005 per year or one duplication every 200 years (Li 1982). After 10 my, one would therefore expect all the genes to have duplicated once on average. The probability of duplication for a particular gene is given by the Poisson distribution. The probability of *not* duplicating is e^{-z}, where z is the mean number of duplications. The probability that a particular gene *is* duplicated is $1-p$, where p is the probability of it being duplicated. When $z = 3$, there is 95% probability that a particular gene will be duplicated after 30 my.

utan and the African-ape lineages more than 12 my ago, whereas the rest of the *C4d* region of the gene diverged separately in each of the four species after their formation. These contradictory phylogenies can be reconciled by assuming that most of the *C4d* region, but not its *A*- and *B*-specifying part, has been homogenized.

Second, the *C4* genes occur in two versions, long and short, differentiated by the presence or absence of a 7 kb insert in intron 9 (Yu et al. 1986; Prentice et al. 1986). In humans, all the *C4A* genes are of the long variety, whereas the *C4B* genes can be either long or short (Palsdottir et al. 1987; Fig. 2). In the chimpanzee (Kawaguchi et al. 1990) and the gorilla (Kawaguchi and Klein 1991), both the *C4A* and *C4B* genes are short; in the orang-utan (Kawaguchi and Klein 1991), the *C4* genes are long, whereas the *C4B* genes are short. Since the insert is a repetitive element (Yu 1991), the original *C4* gene must have been of the short type and the insertion must have occurred before the divergence of the orang-utan and the African-ape lineages but after the primigenial duplication in one of the two *C4* genes. The transfer of the insert from one copy to its duplicate must again have occurred by unequal crossing-over.

Third, the human and chimpanzee *CYP21P* genes contain a characteristic 8 bp deletion in exon 3 which might have been the original mutation silencing them (White et al. 1986; Higashi et al. 1986; Rodrigues et al. 1987; Kawaguchi et al. 1990, 1991b). This deletion is absent in the gorilla and orang-utan *CYP21* genes (Kawaguchi et al. 1991b) signaling that it occurred after the separation of the gorilla/orang-utan from the chimpanzee/human lineages, but before the separation of the human from the chimpanzee lineage. One would therefore expect the *Hosa-CYP21P* and *Patr-CYP21P* genes to be derived from a common ancestor. In fact, this is actually demonstrated by the dendrogram based on the overall *CYP21* sequence similarity (Kawaguchi et al. 1991b). Yet, the *C4* genes borne by the same haplotypes as the *CYP21* genes show a different phylogeny (Fig. 7). Once again, these contradictory findings are reconciled by the postulate of unequal crossing-over and the resulting homogenization of the *C4* genes.

A possible way these features could have evolved in the great ape phylogeny is depicted in Figure 8. One to two unequal crossing-overs in each evolutionary lineage are required to explain the current constitution of the *C4-CYP21* haplotypes in the different species. Additional unequal crossing-overs are required to account for the concerted evolution of the *C4* sequences (i.e., for the fact that over most of the sequence, the interspecific distances between the *C4* genes are greater than the intraspecific differences).

It is thus important to differentiate between the primigenial duplication, which is very rare and may have, in fact, occurred only once in the history of a particular taxon, and subsequent secondary duplications (and deletions), which are relatively common and may lead to repeated homogenization of sequences within a given taxon. We suggest that the *C4*-

CYP21 module arose only once but that it subsequently underwent many rounds of homogenization which have obscured the primigenial event.

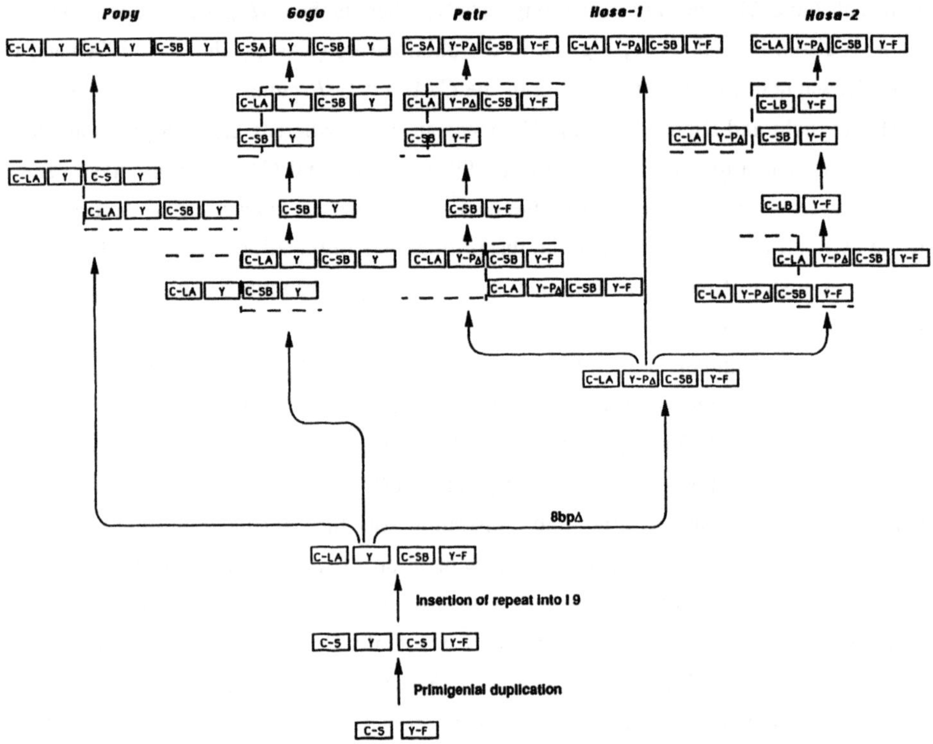

Fig. 8. Postulated evolution of the *C4-CYP21* region in humans and the great apes. Rectangles indicate genes, broken lines crossing-over between misaligned homologous chromosomes, arrows the direction of evolution. C-S, C4 undifferentiated with respect to *A* or *B* type; C-LA, long *C4A*; C-LB, long *C4B*; C-SB, short *C4B*; Y-PΔ, CYP21P (with 8 bp deletion); Y-F, *CYP21* (functional); Y, *CYP21* whose functional status is unknown; Y-P, *CYP21* pseudogene inactivated by defects different from those present in Y-PΔ; I, intron; *Gogo*, gorilla; *Hosa*, human; (-1 and -2 are two different haplotypes); *Patr*, chimpanzee; *Popy*, orang-utan. (From Kawaguchi and Klein 1991.)

How many genes per module?

In the preceding discussion, we assumed that the *C4-CYP21* module consists of these two genes only. The actual situation is, however, probably more complex. The first evidence for the presence of additional loci in the module was provided by Morel and his colleagues (1989) who found a human cDNA clone in a library prepared from adrenal gland RNA and showed that its sequence overlapped in part with that of *CYP21*. The corresponding gene, which we will refer to as *Hosa-X* (for "identity unknown"), lies on the DNA strand complementary to that containing the *CYP21* and *C4* genes. The 3' untranslated (UT) region and part of the last intron of *Hosa-X* overlaps with most of the 3'UT of the *CYP21* gene (the two 3'UT regions are, of course, in opposite orientation to each other). The coding sequence of the *Hosa-X* gene contains three repeats, each specifying about 100 amino acid residues.

In an effort to find the borders of the *C4-CYP21* module, we have embarked on sequencing the flanking regions of the *C4A...CYP21P...C4B...CYP21* segment, as well as the region between the two modules isolated from a chimpanzee cosmid clone. Preliminary results from this effort are summarized in Figure 9, which shows that a sequence corresponding to *Hosa-X* is present not only 3' of the *CYP21* (fragment F), which is as expected, but also 5' of the *C4B* gene (fragment D), between the two modules. Hence at least a portion of the *Hosa-X* gene must have been duplicated in the primigenial event and thus has become part of the module.

Two other genes, *G11* (Sargent et al. 1989) and *RD* (Lévi-Strauss et al. 1988) are located very close to the *C4A* and it would therefore not be totally unexpected if one or both of them were also included in the *C4-CYP21* module. Interestingly, the coding region of the *RD* gene contains eight imperfect repeats, each of approximately 15 bp. There are thus apparently several genes in or near the *C4-CYP21* region which contain repeats, some of which may have participated in or contributed to the primigenial duplication and the subsequent expansion-contraction cycles.

The conversions of gene conversion

In some of the patients with CAH, haplotypes are found in which the *CYP21* gene has acquired some of the defects normally present in the *CYP21P* gene (Donohoue et al. 1986b; Jospe et al. 1987; Matteson et al. 1987; Harada et al. 1987; Higashi et al. 1988). Two explanations have been offered for the apparent sequence transfer from *CYP21P* to *CYP21* -- gene conversion (Donohoue et al. 1986b; Matteson et al. 1987; Higashi et al. 1988; Miller et al. 1988a) and unequal crossing-over (Raum et al. 1984; Carroll et al. 1985a,b; Donohoue et al. 1986a; Rodrigues et al. 1987). At different times in the recent past, one explanation was favored over the other depending on the changing estimate of the frequency of deletions in the *C4-CYP21* region. In the original restriction fragment length polymorphism (RFLP) studies, the two *CYP21* loci were distinguished by characteristic fragments on Southern blots of DNA from CAH patients. When a fragment characterizing the *CYP21P* or the *CYP21* gene was absent on the blot, it was concluded that the corresponding gene was deleted in the particular patient and that the deletion was, most likely, the result of an unequal crossing-over. This conclusion was strengthened by the observation that when one *CYP21* gene was found missing, one *C4* gene segment was absent, too. The frequency of deletions in the *C4-CYP21* region, estimated in this way, was quite high -- up to 30% of CAH patients were believed to suffer from the disease because of a missing *CYP21* gene (Schneider et al. 1986; Werkmeister et al. 1986; Jospe et al. 1987; Rumsby et al. 1986; White et al. 1988). Later, when short-range restriction mapping (Donohoue et al. 1986; Jospe et al. 1987; Mattesson et al. 1987; Rumsby et al. 1986; White et al. 1988), mapping based on the use of specific oligonucleotide probes (Higashi et al. 1988), and sequencing (Harada et al. 1987; Higashi et al. 1988) were applied to the *CYP21-C4* region and evidence of interlocus transfer of sequences was brought to light, some investigators (Matteson et al. 1987; Miller 1988b) began to argue that the apparent deletions were in fact transmogrifications of one gene into another by gene conversion. More recently, however, pulsed-field gel electrophoresis (PFGE) analyses have authenticated deletions in the *C4-CYP21* region by demonstrating that portions of chromatin are missing in some of the CAH patients (Collier et al. 1989), and the unequal crossing-over hypothesis has enjoyed a certain comeback. A unique solution to the controversy has been proposed by some authors. Thus, we find the following statement in a recent paper by Morel and his colleagues (1989): "The term gene conversion refers to a change in part or all of the P450c21B gene, converting part or all of its structure to a P450c21A pseudogene ... Gene conversion can be most easily conceptualized as resulting from two unequal crossover events, but many other mechanisms may also be involved ..."

In this latest redefinition, the 60-year old history of the conversions of "gene conversion" have finally completed a full circle and it may be worth digressing at this point in order to recapitulate briefly the main stages of this history. Indeed, some authors seem to think that gene conversion was invented by Baltimore in 1981. In fact, the problem the gene conversion hypothesis attempts to resolve was encountered at least as early as 1926 in the studies of fungi.

Fig. 9. Search for the limits of the *C4-CYP21* module. The upper part of the drawing depicts the organization of the chimpanzee (*Patr*) module and indicates sequenced DNA segments (A through F; arrows indicate the direction of sequencing). The sequence of these segments is given in the lower part (each segment being numbered separately). Segments C, E, and F are aligned with the sequence of the *Hosa-X* gene described by Morel et al. (1989). Here, dashes (-) indicate identity with the chimpanzee sequence and dots (.) absence of sequence information. The dotted segments presumably correspond to introns, since they are missing from the cDNA sequence. The depicted exon/intron organization of the *X* gene is deduced on the basis of this presumption. Homology of segment C with part of segment F is also indicated.

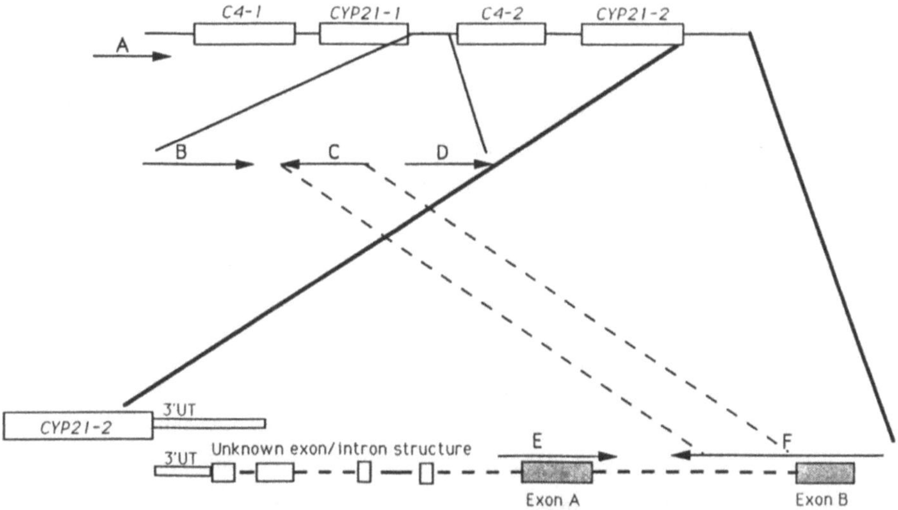


```
           1          11         21         31         41         51         61         71         81         91
Segment E  CCTGTGGTGC CCTCAGCTGA GAGGGGCCCC AGGCGCTTCC CTTCATGGAG GCCATAGAGG AGGAACCTGT AGGGGGTGCT GGGCTCCAGG CCTGAGATGA
Hosa-XA    ---------- ---------- ---------- ---------- ---------- ---------- ---------- ---------- ---------- ----------

           101        111        121        131        141        151        161        171        181        191
Segment E  GGATCTTGCT CTGGTCGCCG TCCACGAGCA AGGCCTGGGG CTGCCCGTTC GTGTCCTCAT ACTGGACCAC GAAGGAATCA AAGGGGCCCT GGGC.ACGCT
Hosa-XA    ---------- ---------- ---------- ---------- ---------- ---------- ---------- ---------- ---------- ----C-----

           201        211        221        231        241        251        261        271        281        291
Segment E  CCACGAGAGG CGCATGGAGT CTGGGGTTGT GTCGGTCACG GTCAGCACTC CTAGGCGGGG CTCTTCAGGA GGCTCAGGGG CCTCTGGGGC TAACTCTGGG
Hosa-XA    ---------- ---------- ---------- ---------- ---------- ---------- ---------- ---------- ---------- ----------

           301        311        321        331        341        351        361        371        381        391
Segment E  GCTGGTGTGT CCTCTTTCTG GGGCTGCGTG GGAGAAGCCC AGGGAAGAAT CTGAGTGAGG GGCGCCATGG GGTGCTCCAT TTTTTATCTT CCAGGCTTGG
Hosa-XA    ---------- ------.--- ------.... .......... .......... .......... .......... .......... .......... ..........

           401        411        421        431        441        451
Segment E  CCAAGGCTGA GGTGGAAGTT ACAGGTCCAG GCCAGTCAGA CAATGAAGTC GCTGTG

           1          11         21         31         41         51         61         71         81         91
Segment F  TCTGCTTCCC TCACCCCGTG TCTGCGAGCC ACCTCGAGAT ATCACAGCTC TGCCC.ACCA TGCCGGATAC ATTCACTGAG CTTGAGGAGT GTGGTGCTCC
Segment C  .......... .......... ......CTG- -GG------- ---------- -----C---- ---------- ---------- ---------- ----------

           101        111        121        131        141        151        161        171        181        191
Segment F  CTTCTGAGAG AAGCTGAGGG TGGAACTGGC TGGTTGAGGT GACTGGCAAA TCCCACCAGC CGAGCCGTGG TCAGGCCTGT CTGAGGTGGG CATCAGCGAG
Segment C  ---------- ---------- ---------- ---------- ---------- ---------- --T------- ---------- ---------- ----------

           201        211        221        231        241        251        261        271        281        291
Segment F  CTCTGGAAGA GGAGCCTGTA CCACAAATGC AGCCACTGCT GTTGGTTTCT GTGTCCCCGC TCATTTTGTT TTCCAGTGAT GTTCCTCTTA AGAAAATGCT
Segment C  ---------- ---------- ---------- ---------- ---------- ---------- ---------- ---------- ---------- ----------

           301        311        321        331        341        351        361        371        381        391
Segment F  CCTGACTCAT CCATGGCAGG GAGGTTTGCC GCTATCTGGA CAAGGCCAC. TTCGGGGAGG CGACAGCAGC CCCAGCGAGT AATGAGGAGC AGCGGGCAGT
Segment C  ---------- ---------- ---------- ---------- ---------C ---------- -------T-- ---------- ---------- -----.----

           401        411        421        431        441        451        461        471        481        491
Segment F  GATGGGGCAG AGTTGGGGCT GGGAGATTAG AGAGCCCCTG CCAGGGCCTT TCCCTCCCGC CTGGCCTGGC TCCTGCTCTG GACTCCTTGA TGGATGTTGA
Segment C  ---------- ---------- ---------- ---CG----- ---------- ---------- ---------- ---------- ---------- ----------

           501        511        521        531        541        551        561        571        581        591
Segment F  CACGACAGGG CTGCAGATCT CTCTTCTGGA CATTGCCAGG CGCCCACTCG CCTTCCCACT CCTCGAGTCG TCTTGTCTTC AGCCCAAATG CACATGAAAC
Segment C  ---------- ------.... .......... .......... .......... .......... .......... .......... .......... ..........

           601        611        621        631        641        651        661        671        681        691
Segment F  CCACACAAGC TGGCTTGCAT TAGCCAGCAC AGCAGCCTCA CCTGTCATTC CCAGGGCAGA GACCGGGCCC AGGCGCTTTC CCCCGAGGAG CCCGTAGAGC
Hosa-XB    .......... .......... .......... .......... ....------ ---------- ---------- ---------- ----A----- ----------

           701        711        721        731        741        751        761        771        781        791
Segment F  AGAAACTTGT ATTTCTTGCC AGGCTTCAGG TCCTCTACGG TGACTGTGCG CTGGTCTGCG GCCACAG.CA CTGCCCTGGG CTGCCCGTCC GTGTCCCTGT
Hosa-XB    ---------- ---------- -----C---- ---------- ---------- ---------- -------G-- ---------- ---------- ----------

           801        811        821        831        841        851        861        871        881        891
Segment F  ACTGGACCAC GAAGGAGTCA AAGGG.CCCT GGGCCACCGT CCAGGACAGG CGCAGAGAGC TGGAGGTCTC CTCAGCTACG GTCAGTTCCC CCAGGTGGGG
Hosa-XB    ---------- ---------- -----G---- ----T----- ---------- ---------- ---------- ------C--- ---------- ----------

           901        911        921        931        941        951        961        971        981        991
Segment F  AGGTAGCTCC TTCTCCAGGG GAGCTGTGCA GAGGGAGGAG GGAAAGCTCT TAGTCACATG CTGCCTTTGC CTAAGCCCTG GCAGCCTCCC GGAGGTGTGA
Hosa-XB    ---------- ---------- --------.. .......... .......... .......... .......... .......... .......... ..........

           1001       1011
Segment F  GGTTCTGGGA AAGGGTACC

Segment A    1 GGCTTATCAA AAGCACACAG GGGCCTGGCA GGAACCGTAA AAGCTTGATG TTAATCATAC TGGGACTTAG AGGATAGAGA ATGGTAGGAG CTGGGATACC
Segment A  101 CCTAAACATT CACATTAAAA CAAAAAAAAA CCCAAAGCTA AAAAACAACT GGGCAGGAGC TAAATAAAAA TCTAATTTTG AGAGGCTGTA TCTGGCTCAG
Segment A  201 GCATCCTACT TTGTAACCCA TGGAATATGT GAAAGCATTT GAAAAACTAT AGCACTGGTC TCACATGGGC AGACACACTC TCAGAGAGAT GTGGTGGGAG
Segment A  301 CCATGGCGCA GTCTGCCTAG GGCAGTGGCA GGAGCGCAGA AGACCCTGAT TCCTCTCCTC GGTCCTAAGA CTGAATGTGT GTCAGGACAT GTGTCAGGG

Segment B    1 GGATCCAAGA GAGGTTAGAT CCGTCTGTCT GTCTGCTACC TTCTGGCCAA GCTAGGGCTC GGGGCAGAGT TGACTCTGGA CCTTGCTCCT CCCCCAGCCC
Segment B  101 AGCTAAGCTG GAAACCACAT GACAGCCAAG AAGTGCAACT GGCCTCGGGC CAGCGTTGTG GGGGCACCCC GGACACCTGG GTGTCATGGG AAACTCTGAA
Segment B  201 TCTTGGCCCA GAAATAACAC TGTCCTTCCC CGTTAACTCC TCTCATTCCA GTGCCTGAAC CACTCCCCCT ACCAGGTATA TCTGTATCAG GGTATAGGTG
Segment B  301 CACACAAGCT CATGTATACC TGGGTGTACA AGGG

Segment D    1 CTGCAGCCCC TCACCTTTTG GGTGGCTCCT CAGGCCTCTA GGCTTATTCA CCTTTCCCCT TTCCTGTGCC ACTTCTCCTC TAGGGCGCCA GGCTGTCCTT
Segment D  101 AGCATGGTCC GGAAGGCAAA GTACCGGGAA CTGCTCCTAT CAGAGCTCCT GGGCCGGCGG GCACCTGTCG TGGTGCGGCT TGGCCTCACC TACCATGTGC
Segment D  201 ACGACCTCAT TGGGGCCCAG CTAGTGGACT GGTGAGTCTT TCCCTGGCCT CTGGCAGATT ATGGAGCAAT GACCCAAAGT GGGATTTCCT CCCAGCTCAC
Segment D  301 GCTTAGTTTC CTAGTGAAGG CCAGTGGCTC TCATTCTTCT CTGGAACCCG GGAGCACCCC TTCCAAGTTC TAAGTTCTCC TCACAGCTTG AGCCTAGGGC
Segment D  401 TCTGGCTCCA GCTT
```

One of the main advantages of using fungi in genetics studies is that the four cells (a tetrad) produced by the two meiotic divisions stay together (often after one subsequent mitotic division which increases their number to eight) in an arrangement that reflects the orientation of the meiotic spindles. If the cell undergoing meiosis is a heterozygote and the two alleles have a different visible effect on the resulting tetrads, one can deduce whether the alleles have recombined or not by recording the ratios and the distribution of the four cells. Both the recombined and nonrecombined tetrads will have two cells of one type and two cells of the second type (segregation ratio of 2:2 or 1:1) but the cells will be arranged differently, the 2:2 ratio of the recombined tetrads attesting to the reciprocity of the recombination process. In 1926, Brunswick may have been the first investigator to notice that the segregation ratio is sometimes not 2:2 but 3:1, suggesting that recombination may not always be reciprocal. Similar observations were subsequently made repeatedly by other investigators. In 1930 Winkler summarized all the then known cases of non-Mendelian segregation and deduced that recombination of characters was not the result of physical crossing-over but of a physiological process in which one allele influences its counterpart on the homologous chromosome and physiologically converts it to its own kind (e.g., allele *A* converts *a* to *A*). This was the original meaning of "gene conversion". Winkler's hypothesis was subsequently discredited experimentally but the term "gene conversion" was resurrected in 1949 by Lindgren as a synonym for nonreciprocal genetic recombination and it has been retained in this form by fungal geneticists to this day. The mechanism of nonreciprocal genetic recombination, however, remained controversial for a long time. For a while, a popular explanation had been the copy-choice hypothesis (Belling 1933), according to which conversion is the result of switching the replication mechanism from one gene to its allele on the homologous chromosome and then back again to the original gene. When, however, the true mechanism of gene replication was finally elucidated, Belling's hypothesis had to be abandoned. The current genetic theory explains fungal nonreciprocal recombination (gene conversion) as being the result of DNA repair following intragenic crossing-over (Holliday 1964). The theory postulates that crossing-over leads to the formation of DNA heteroduplexes -- molecules containing one strand from one chromosome (chromatid) and the other from the homologous chromosome (chromatid). As the point of exchange migrates along the gene and reaches the sites where the two alleles differ in their nucleotide sequence, it leaves the noncomplementary nucleotides unpaired. Proofreading enzymes then recognize the unpaired region, excise one of the nucleotides, and replace it with a complementary nucleotide, thus converting one of the alleles to the other.

In mammals and other vertebrates, the gametes produced by the meiotic divisions do not stay together as they do in fungi so that here a "tetrad analysis" of the kind carried out by fungal geneticists is not possible. In 1980, however, Slightom and colleagues published a paper in which they invoked a gene conversion-like mechanism for data not generated by tetrade analysis but by sequencing of mammalian genes (in their case the $G\gamma$ and $A\gamma$-globin genes). The principal observation made by these investigators was that in addition to genes in which the sequence of markers is *ABCDEF* and *abcdef*, a third kind of homologous gene sequence is occasionally found, in which the markers are arranged in the order *ABcdEF* (here "markers" are particular sequence motifs). In other words, the genes of the third kind look like the genes of the first kind into which information has been transferred from the genes of the second kind. Slightom and associates designated this information transfer as "gene conversion" and proposed that at the molecular level the transfer occurs by an invasion of one strand of one DNA molecule into the corresponding region of the second DNA molecule and subsequent strand cleavage. From this point on, virtually all investigators who have found the arrangement *ABcdEF* in the genes they happened to study, followed Slightom and his colleagues in invoking gene conversion as an explanation for their finding. There is, however, an essential difference between gene conversion as used by fungal geneticists and that used by mammalian geneticists. Fungal geneticists examine the direct product of meiosis, whereas mammalian geneticists study products of many cell generations, and in the overwhelming majority of cases also many offspring generations, removed from the meiosis that created the *ABcdE*F variant. In these generations, even low-probability events have a definitive chance of taking place so that the observed sequence may in fact be the outcome of two or more rare occurrences. Instead of being the product of gene conversion, the *ABcdEF* sequence could be the result of two independent, sequential, reciprocal crossing-overs. More than that, under most circumstances it is virtually impossible to differentiate between the products of gene conversion and those of double crossing-over. The preference for gene conversion over reciprocal crossing-over then becomes nothing more than a case for special pleading by the defense. Ojectively, therefore, most of the published papers invoking gene conversion as an explanation for the observed sharing of sequence stretches and not paying equal attention to alternative explanations must be regarded as being unscientifically biased.

There is only one situation in which gene conversion can be discriminated from unequal crossing-over with some degree of security. If the occurrence of the *ABcdEF* sequence is associated with deletion or duplication of chromatin, then it is more likely to be the outcome of unequal crossing-over than of gene conversion. In the case of *CYP21* (and *C4*), most of the sequences attributed to gene conversion have, in fact, not been examined for the presence of accompanying duplications and deletions. Those that have been analyzed in this way *have* been found associated with deletions. These findings therefore favor unequal crossing-over as an explanation for the observed sharing of defects between *CYP21P* and *CYP21* in CAH patients. Also, in the only instance in which the recombinant haplotype has been caught *in statu nascendi* (Sinnott et al. 1990), it was generated by unequal crossing-over and not by gene conversion. We hasten to add that these comments should not be interpreted as denying the occurrence of gene conversion between *CYP21* (and *C4*) genes, but rather as championing more rigorous criteria for accepting gene conversion as an explanation for the origin of *ABcdEF* sequences.

The *C4-CYP21* between Scylla and Charybdis?

Bimodular *C4-CYP21* haplotypes are common in humans (McLean et al. 1988; and others), in the chimpanzee (Kawaguchi et al. 1990; Bontrop et al. 1991; Christiansen et al. 1991), the gorilla (Kawaguchi and Klein 1991), the mouse (Roos et al. 1978), the rat (Tosi et al. 1985; Wurst et al. 1988), the pig (Kirszenbaum et al. 1985), the horse (Kay et al. 1987), and the macaque (Mevag et al. 1983). Monomodular haplotypes have been reported in the Syrian hamster (Lévi-Strauss et al. 1985), the dog (Kay and Dawkins 1984; Doxiadis et al. 1985), the cat (De Kroon et al. 1986), the guinea-pig (Bitter-Suermann et al. 1977), and several species of whale (Spilliaert et al. 1990); it is, however, not always clear how prevalent the monomodular haplotypes in these species are. Trimodular haplotypes have been reported in humans (Bruun-Peterson et al. 1982; Fielder et al. 1983; and others) and in the orang-utan (Kawaguchi and Klein 1991). The presence of multimodular haplotypes in a population provides the potential for unequal crossing-over and hence for the instability of the *C4-CYP21* region. Since this instability often means that haplotypes with deletions encompassing parts of or entire functional genes are produced (obviously an adverse situation) one might wonder why all the species have not returned to monomodular haplotypes. Whenever multimodular haplotypes are present in the population, monomodular haplotypes are always produced with an appreciable frequency, and so one would expect the latter to be favored by selection over the former and become fixed in all populations and all species. The fact that in many species the *C4-CYP21* haplotypes keep expanding and contracting could therefore mean that there is a certain selective advantage to possessing multimodular haplotypes. The advantage could apply to only one member of the module while the others expand and contract willy-nilly simply because they cannot withdraw from the module once a primigenial duplication has bound them to it. It is unlikely that there is any advantage in having two or more *CYP21* genes in a haploid genome; on the contrary, the fact that whenever extra *CYP21* copies are present, all but one of them are apparently silenced, suggests that having more than one active *CYP21* gene is an undesirable condition.

When not *CYP21*, then it would have to be either *C4* or an as yet unidentified other member of the module which is responsible for the maintenance of multimodular haplotypes in the population. As it is difficult to consider the advantages of an unknown gene, we must limit our discussion to *C4*. Although a number of cases of complete C4 deficiency have been reported since 1970, when the first case was described by Torisu and his colleagues (reviewed by Hauptmann et al. 1986), their overall frequency is very low. Most C4-deficient patients develop systemic lupus erythematosus (SLE) and other autoimmune diseases, presumably because of decreased ability to clear immune complexes from circulation. The unusual characteristic of the SLE in C4-deficient patients is that it develops very early, sometimes before six years of age, and that it is associated with frequent cutaneous lesions, photosensitivity, mild kidney disease, and recurrent bacterial and viral infections (pneumonia, upper respiratory tract infections, recurrent otitis, viral meningitis, and others).

Individuals with partial C4-deficiency are somewhat more frequent: In the Caucasian populations, the frequency of C4A-deficient individuals is 2.5% - 3.5% and that of C4B-deficient individuals 1%-2% (Hauptmann et al. 1986). Approximately 10% of Caucasians have only two *C4* genes expressed in their diploid genomes, 13-24% have three genes expressed, and 69%-76% have four genes expressed. The effects of partial C4-deficiency are far less drastic than those of complete deficiency, but are nevertheless demonstrable as increased susceptibility to the development of SLE, insulin-dependent diabetes mellitus (IDDM), Grave's disease, and various other autoimmune diseases.

It appears therefore that although the lack of either *C4A* or *C4B* genes is compatible with survival, it results in a significant decrease in the state of health. Humans with two *C4* genes in their haploid genomes fare better than those with only one copy and individuals with *C4A* and *C4B* genes apparently have an advantage over those with *C4A* or *C4B* genes only. Hence there *is* selection for bimodular haplotypes and its target is the *C4* gene. (This observation, however, does not exclude that another, as yet unidentified constituent of the module might also be the target of selection.)

How do the species with monomodular haplotypes therefore manage to thrive? There are two possible answers to this question. The first is a trivial, though by no means unlikely one: Since one-module status has not been confirmed in any of these species by cloning, the possibility remains that they are actually bimodular but the second *C4* copy has been overlooked until now. This explanation, however, does not apply to the mouse, which -- although a bimodular species -- apparently has only one functional *C4* gene (for the other, the *C4-Slp* gene, no complement activity could be demonstrated thus far, see Shreffler et al. 1984). The second possibility is that the monomodular species have somehow remodelled the one *C4* gene in such a way that it carries out functions executed in other species by the *C4A* and *C4B* genes.

Like the ancient mariner in the straight between Scylla and Charybdis, the bimodular haplotypes might come under two counteracting influences -- selection for two active *C4* genes and selection against two active *CYP21* genes. In some species, the negative selection gains the upper hand and these then become monomodular. In others, positive selection exerts a stronger influence, thereby establishing bimodularity (and with it the ceaseless expansion-contraction cycling). It may seem that the counteracting influences could be dissociated by decoupling of the *C4* and *CYP21* genes. If the *CYP21* did leave the module, it could remain a single copy gene while the *C4* remained duplicated. However, as discussed earlier, it probably is not possible to decouple the evolution of the two genes once the primigenial duplication tied them together. Only in monomodular species, if such do indeed exist, could a lucky accident bring about a new primigenial duplication in which only the *C4* and no other gene in that region has been duplicated. Given the crowding of the genes in this region and the presence of elements with repetitive structure, such a restricted duplication may be highly unlikely to occur.

If the *CYP21* gene is indeed involved in an undesired partnership with the *C4* gene, it may have made the best out of this situation by evolving mechanisms of speedily silencing the second active copy, once it arises by recombination. The rate of silencing of duplicated genes is a complex issue that has been the subject of many theoretical and experimental studies (reviewed by Li 1982). It depends on several parameters, especially on the population size, *N*. In small populations, it is largely determined by μ, the rate of inactivating mutations. If, for example, *N* is less than 100 and $\mu = 10^{-5}$, it takes approximately 50,000 generations to silence a duplicate gene (or 1 my for a species with a generation time of 20 years). In a large population, the duplicate gene may remain expressed for a long time. Since the population size in the hominid lineage may have rarely, if ever, dropped under 10,000 in the last 5 - 10 my of evolution (Klein et al. 1990; Satta et al. 1991), conditions for rapid silencing of the duplicate *CYP21* genes may not have existed. Yet, in humans (chimpanzee), gorilla, and orang-utan (as well as the mouse), the silencing occurred repeatedly through different defects. It seems therefore that the *CYP21* gene is more prone to silencing of duplicate copies than most other genes. Since the mutations silencing the *CYP21* duplicate often occur in regions of short repeats (Kawaguchi et al. 1991b) and might be the result of DNA polymerase slippage (Streisinger et al. 1966), one could postulate that such repeats might have been favored by selection to provide a faster way of silencing supernumerous *CYP21* genes. The undesired but perhaps inadvertent consequence of such selection could have been the proness to inactivating *all CYP21* genes on a given chromosome and thus the high risk of defects leading to CAH. This may be an explanation to satisfy a biologist, but try telling a patient that his CAH is good for the evolution!

References

Baltimore, D.: Gene conversion: some implications for immunoglobulin genes *Cell 24*:592-594, 1981

Belling, J.: Crossing-over and gene arrangement in flowering plants. *Genetics 18*:388-413, 1933

Belt, K.T., Carroll, M.C., Porter, R.R.: The structural basis of the multiple forms of human complement component C4. *Cell 36*:907-914, 1984

Belt, K.T., Yu, C.Y., Carroll, M.C., and Porter, R.R.: Polymorphism of human complement component C4. *Immunogenetics 21*:173-180, 1985

Bitter-Suermann, D., Kroenke, M., Brade, V., and Hadding, U.: Inherited polymorphism of guinea-pig factor B and C4: Evidence for genetic linkage between the Bf and C4 loci. *J Immunol 118*:1822-1826, 1977

Bontrop, R.E., Broos, L.A.M., Otting, N., and Jonker, M.J.: Polymorphism of *C4* and *CYP21* genes in various primate species. *Tissue Antigens 37*:145-151, 1990

Brunswick, J.: Die Reduktionsteilung bei den Basidiomyceten. *Z Bot 18*:481-498, 1926

Bruun-Peterson, G., Lamm, L.U., Jacobsen, B.K., and Kristensen, T.: Genetics of complement C4. Two homo duplication haplotypes C4SC4S and C4FC4F in a family. *Hum Genet 61*:36-38, 1982

Carroll, R.L.: *Vertebrate Paleontology and Evolution*. W.H. Freeman, New York 1988

Carroll, M.C., Belt, T., Palsdottir, A., and Porter, R.R.: Structure and organization of the *C4* genes. *Phil Trans Roy Soc B 306*:379-388, 1984

Carroll, M.C., Campbell, R.D., and Porter, R.R.: Mapping of steroid 21-hydroxylase genes adjacent to complement component C4 genes in HLA, the major histocompatibility complex in man. *Proc Natl Acad Sci USA 82*:521-525, 1985a

Carroll, M.C., Palsdottir, A., Belt, K.T., and Porter, R.R.: Deletion of complement C4 and steroid 21-hydroxylase genes in the HLA class III region. *EMBO J 4*:2547-2552, 1985b

Carroll, M.C., Katzman, P., Alicot, E.M., Koller, B.H., Geraghty, D.E., Orr, H.T., Strominger, J.L., and Spies, T.: Linkage map of the human major histocompatibility complex including the tumour necrosis factor genes. *Proc Natl Acad Sci USA 64*:8525-8539, 1987

Christiansen, F.T., Bontrop, R.E., Giphart, M., Cameron, P.U., Zhang, W.J., Townsend, D., Jonker, M., and Dawkins, R.L.: Major histocompatibility complex haplotypes in the chimpanzee: Identification using C4 allotyping. *Hum Immunol 31*:34-39, 1991

Collier, S., Sinnott, P.J., Dyer, P.A., Price, D.A., Rodney, H., and Strachan, T.: Pulsed field gel electrophoresis identifies a high degree of variability in the number of tandem 21-hydroxylase and complement C4 genes repeats in 21-hydroxylase deficiency haplotypes. *EMBO J 8*:1393-1402, 1989

De Kroon, A.I.P.M., Doxiadis, I., and Hensen, E.J.: Structure and polymorphism of the feline complement component C4. *Immunogenetics 24*:202-205, 1986

Donohoue, P.A., Jospe, N., Migeon, C.J., McLean, R.H., Bias, W.B., White, P.C., and Van Dop, C.: Restriction maps and restriction fragment length polymorphisms of the human 21-hydroxylase genes. *Biochem Biophys Res Commun 136*:722-729, 1986a

Donohoue, P.A., Van Dop, C., McLean, R.H., White, P.C., Jospe, N., and Migeon, C.J.: Gene conversion in salt-losing congenital adrenal hyperplasia with absent complement C4B protein. *J Clin Endocrin Metab 62*:995-1002, 1986b

Donohoue, P.A., Jospe, N., Migeon, C.J., and Van Dop, C.: Two distinct areas of unequal crossing-over within the steroid 21-hydroxylase genes produce absence of *CYP21B Genomics 5*:397-406, 1989

Doxiadis, G., Rebmann, V., Doxiadis, I., Krumbacher, K., Vriesendorp, H.M., and Grosse-Wilde, H.: Polymorphism of the fourth complement in the dog. *Immunobiology 169*:563-569, 1985

Dunham, I., Sargent, C.A., Trowsdale, J., and Campbell, R.D.: Molecular mapping of the human major histocompatibility complex by pulsed-field gel electrophoresis. *Proc Natl Acad Sci USA 84*:7237-7241, 1987

Fielder, A.H.L., Walport, M.J., Batchelor, J.R., Rynes, R.I., Black, C.M., Dodi, I.A., and Hughes, G.R.V.: Family studies of the major histocompatibility complex in patients with systemic lupus erythematosus: Importance of null alleles of C4A and C4B in determining disease susceptibility. *Brit Med J 286*:425-428, 1983

Harada, F., Kimura, A., Iwanaga, T., Shimozawa, K., Yata J., and Sasazuki, T.: Gene conversion-like events cause steroid 21-hydroxylase deficiency in congeital adrenal hyperplasia. *Proc Natl Acad Sci USA 84*:8091-8094, 1987

Hauptmann, G., Goetz, J., Uring-Lambert, B., and Grosshans, E.: Component deficiencies 2: The fourth component. *Progr Allergy 39*:232-249, 1986

Higashi, Y., Yoshioka, H., Yanae, M., Gotch, O., and Fujii-Kuriyama, Y.: Complete nucleotide sequence of two steroid 21-hydroxylase genes tandemly arranged in human chromosome: A pseudogene and a genuine gene. *Proc Natl Acad Sci USA 83*:2841-2845, 1986

Higashi, Y., Tanae, A., Inoue, H., Hiromasa, T., and Fujii-Kurijama, Y.: Aberrant splicing and missense mutation causing steroid 21-hydroxylase (P-450c21) deficiency in humans: possible gene conversion products. *Proc Natl Acad Sci USA 85*:7486-7490, 1988

Holliday, R.: A mechanism for gene conversion in fungi. *Genet Res 5*:282-304, 1964

Huang, L.-S., Ripps, M.E., Korman, S.H., Deckelbaum, R.J., and Breslow, J.L.: Hypobetalipoprotein B gene exon 21 deletion derived by Alu-Alu recombination. *J Biol Chem 264*:11394-11400, 1989

Jospe, N., Donohoue, P.A., Van Dop, C., McLean, R.H., Bias, W.B., and Migeon, C.J.: Prevalence of polymorphic 21-hydroxylase gene (CA21HB) mutations in salt-losing congenital adrenal hyperplasia. *Biochem Biophys Res Commun 142*:798-804, 1987

Kawaguchi, H. and Klein, J.: Organization of *C4* and *CYP21* loci in gorilla and orang-utan. *Human Immunol*, submitted 1991

Kawaguchi, H., Golubic, M., Figueroa, F., and Klein, J.: Organization of the chimpanzee *C4-CYP21* region: Implications for the evolution of human genes. *Eur J Immunol 20*:739-749, 1990

Kawaguchi, H., Zaleska-Rutczynska, Z., Figueroa, F., O'hUigin, C., and Klein, J.: *C4* Genes of the chimpanzee, gorilla, and orang-utan: Evidence for extensive homogenization. *Immunogenetics*, in press 1991a

Kawaguchi, H., O'hUigin, C., and Klein, J.: Evolutionary origin of mutations in the human cytochrome *P450c21* gene. *Amer J Hum Genet*, submitted 1991b

Kay, P.A. and Dawkins, R.L.: Genetic polymorphism of complement C4 in the dog. *Tissue Antigens 23*:151-155, 1984

Kay, P.H., Dawkins, R.L., Bowling, A.T., and Bernoco, D.: Heterogeneity and linkage of equine C4 and steroid 21-hydroxylase genes. *J Immunogenet 14*:247-253, 1987

Kimura, M.: A simple method for estimating evolutionary rates of base substitutions through comparative studies of nucleotide sequences. *J Mol Evol 16*:111-120, 1980

Kirszenbaum, M., Renard, C., Geffrotin, C., Chardon, P., and Vaiman, M.: Evidence for mapping pig *C4* gene(s) within the pig major histocompatibility complex (SLA). *Anim Blood Groups Biochem Genet 16*:65-68, 1985

Klein, J. and Figueroa, F.: Evolution of the major histocompatibility complex. *CRC Crit Rev Immunol 6*:295-386, 1986

Klein, J., Gutknecht, J., and Fischer, N.: The major histocompatibility complex and human evolution. *Trends Genet 6*:7-11, 1990

Law, S.K.A., Dodds, A.W., and Porter, R.R.: A comparison of the properties of two classes, C4A and C4B, of the human complement component C4. *EMBO J 3*:1819-1823, 1984

Lehrman, M.A., Schneider, W.J., Südhof, T.C., Brown, M.S:, Goldstein, J.L., and Russell, D.W.: Mutation in LDL receptor: Alu-Alu recombination deletes exons encoding transmembrane and cytoplasmic domains. *Science 227*:140-146, 1985

Lehrman, M.A., Russell, D.W., Goldstein, J.L., and Brown, M.S.: Exon-*Alu* recombination deletes 5 kilobases from the low density lipoprotein receptor gene, producing a null phenotype in familial hypercholesterolemia. *Proc Natl Acad Sci USA 83*:3679-3683, 1986

Lehrman, M.A., Russell, D.W., Goldstein, J.L., and Brown, M.S.: Alu-Alu recombination deletes splice acceptor sites and produces secreted low density lipoprotein receptor in a subject with familial hypercholesterolemia. *J Biol Chem 262*:3354-3361, 1987

Lévi-Strauss, M., Tosi, M., Steinmetz, M., Klein, J., and Meo, T.: Multiple duplications of complement *C4* gene correlate with *H-2*-controlled testosterone-independent expression of its sex-limited isoform, C4-Slp. *Proc Natl Acad Sci USA 82*:1746-1750, 1985

Lévi-Strauss, M., Carroll, M.C., Steinmetz, M., and Meo, T.: A previously undetected *MHC* gene with an unusual periodic structure. *Science 240*:201-204, 1988

Li, W.-H.: Evolutionary change of duplicate genes. *In* M.C. Rattazzi, J.G. Scandalias, and G.S. Whitt (eds.): *Isozymes: Current Topics in Biological and Medical Research.* Vol. 6, pp. 55-92, Alan R. Liss, New York 1982

Lindgren, C.C.: *The Yeast Cell -- Its Genetics and Cytology*. Education Publishers, St. Louis 1949

Matteson, K.J., Phillips, J.A. III, Miller, W.L., Chung, B., Orlando, P.J., Frisch, H., Ferrandez, A., and Burr, I.M.: P450XXI (steroid 21-hydroxylase) gene deletions are not found in family studies of congenital adrenal hyperplasia. *Proc Natl Acad Sci USA 84*:5858-5862, 1987

McLean, R.H., Donohoue, P.A., Jospe, N., Biase, W.B., van Dop, C., and Migeon, C.J.: Restriction fragment analysis of duplication of the fourth component of complement (C4A). *Genomics 2*:76-85, 1988

Meo, T. and Tosi, M.: Molecular genetics of the *S* region of the murine *H-2* major histocompatibility complex. *Ann Inst Past 136C*:225-243, 1985

Mevag, B., Olaisen, B., Teisberg, P., and Smith, D.G.: Two *C4* loci in macaca monkeys. Tenth International Complement Workshop, Mainz. *Immunobiology 164*: Abstract 1, 1983

Miller, W.L.: Gene conversions, deletions, and polymorphisms in congenital adrenal hyperplasia. *Am J Hum Genet 42*:4-7, 1988a

Miller, W.L.: Molecular biology of steroid hormone synthesis. *Endocrin Rev 9*:295-318, 1988b

Miura, O., Sugahara, Y., Nakamura, Y., Hirosawa, S., and Aoki, N.: Restriction fragment length polymorphism caused by a deletion involving *Alu* sequences within the human α_2-plasmin inhibitor gene. *Biochemistry 28*:4934-4938, 1989

Morel, Y., Andre, J., Uring-Lambert, B., Hauptmann, G., Betuel, H., Tossi, M., Forest, M.G. David, M., Bertrand, J., and Miller, W.L.: Rearrangements and point mutations of P450c21 genes are distinguished by five restriction endonuclease haplotypes identified by a new probing strategy in 57 families with congenital adrenal hyperplasia. *J Clin Invest 83*:527-536, 1989

Mornet, E., Crété, P., Kuttenn, F., Raux-Demay, M.-C., Boué, J., White, P.C., and Boué, A.: Distribution of deletions and seven point mutations on CYP21B genes in three clinical forms of steroid 21-hydroxylase deficiency. *Am J Hum Genet 48*:79-88, 1991

Myerowitz, R. and Hogikyan, N.D.: A deletion involving Alu sequences in the ß-hexosaminidase α-chain gene of French Canadians with Tay-Sachs disease. *J Biol Chem 262*:15396-15399, 1987

New, M.I. and Levine, L.S.: Recent advances in 21-hydroxylase deficiency. *Annu Rev Med 35*:649-663, 1984

Ottolenghi, S. and Giglioni, B.: The deletion in a type of d^0-β^0-thalassaemia begins in an inverted *Alu* I repeat. *Nature 300*:770-771, 1982

Palsdottir, A., Fossdal, R., Arnason, A., Edwards, J.H., and Jensen, O.: Heterogeneity of human *C4* gene size. A large intron (6.5 kb) is present in all *C4A* genes and some *C4B* genes. *Immunogenetics 25*:299-304, 1987

Prentice, H.L., Schneider, P.M., and Strominger, J.L.: C4B gene polymorphism detected in a human cosmid clone. *Immunogenetics 23*:274-276, 1986

Raum, D., Awdeh, Z., Anderson, J., Strong, L., Granados, J., Teran, L., Giblett, E., Yunis, E.J., and Alper, C.A.: Human C4 haplotypes with duplicated C4A or C4B. *Amer J. Hum Genet 36*:72-79, 1984

Reid, K.B.M.: Application of molecular cloning to studies of the complement system. *Immunology 55*:185-196, 1985

Rodrigues, N.R., Dunham, I., Yung Yu, C., Carroll, M.C., Porter, R.R., and Campbell, R.D.: Molecular characterization of the HLA-linked steroid 21-hydroxylase B gene from an individual with congenital adrenal hyperplasia. *EMBO J 6*:1753-1661, 1987

Roos, M.H., Atkinson, J.P., and Shreffler, D.C.: Molecular characterization of the Ss and Slp (C4) proteins of the mouse H-2 complex: Subunit composition, chain size, polymorphism an intracellular (pro-SS) precursor. *J Immunol 121*:1106-1115, 1978

Ross, G.D. (ed.): *Immunobiology of the Complement System. An Introduction for Research and Clinical Medicine*. Academic Press, Orlando 1986

Rouyer, F., Simmler, M.-C., Page, D.C., and Weissenbach, J.: A sex chromosome rearrangement in a human XX male caused by Alu-Alu recombination. *Cell 51*:417-425, 1987

Rumsby, G., Carroll, M.C., Porter, R.R., Grant, D.B., and Hjelm, M.: Deletion of the steroid 21-hydroxylase and complement C4 genes in congenital adrenal hyperplasia. *J Med Genet 23*:204-209, 1986

Saitou, N. and Nei, M.: The neighbor-joining method. A new method for reconstructing phylogenetic trees. *Mol Biol Evol 4*:406-425, 1987

Sargent, C.A., Dunham, I., and Campbell R.D.: Identification of multiple HTF-island associated genes in the human major histocompatibility complex class II region. *EMBO J 8*:2305-2312, 1989

Satta, Y., Takahata, N., Schönbach, C., Gutknecht, J., and Klein, J.: Calibrating evolutionary rates at major histocompatibility complex loci. *In* J. Klein and D. Klein (eds.): *Molecular Evolution of the Major Histocompatibility Complex*. Springer-Verlag, Heidelberg, in press 1991

Schneider, P.M., Carroll, M.C., Alper, C.A., Rittner, C., Whitehead, A.S., Yunis, E.J., and Colton, H.R.: Polymorphism of the human complement C4 and steroid 21-hydroxylase genes. *J Clin Invest 78*:650-657, 1986

Shreffler, D.C., Atkinson, J.P., Karp, D.R., Killion, C.:, Ogata, R.T., and Rosa, P.A.: The C4 and Slp genes of the complement region of the murine H-2 major histocompatibility complex. *Phil Trans Roy Soc London 306*:395-403, 1984

Sinnott, P., Collier, S., Costigan, C., Dyer, P.A., Harris, R., and Strachan, T.: Genesis by meiotic unequal crossing-over of a *de novo* deletion that contributes to steroid 21-hydroxylase deficiency. *Proc Natl Acad Sci USA 87*:2107-2111, 1990

Slightom, J.L., Blechl, A.E., and Smithies, O.: Human fetal $^{G}\gamma$- and $^{A}\gamma$-globin genes: Complete nucleotide sequences suggest that DNA can be exchanged between these duplicated genes. *Cell 21*:627-638, 1980

Sottrup-Jensen, L., Stepanik, T.M., Kristensen, T., Loenblad, P.B., Jones, C.M., Wierzbicki, D.M., Magnussen, S., Domdey, H., Wetsel, R.A., Lundwall, A., Tack, B.F., and Fey, G.H.: Common evolutionary origin of α_2-macroglobulin and complement components C3 and C4. *Proc Natl Acad Sci USA 82*:9-13, 1985

Speiser, P.W. and New, M.I.: Genetics of steroid 21-hydroxylase deficiency. *Trends Genet 1*:275-278, 1985

Spies, T., Bresnahan, M., and Strominger, J.L.: Human major histocompatibility complex contains a minimum of 19 genes between the complement cluster and HLA-B. *Proc Natl Acad Sci USA 86*:8955-8958, 1989a

Spies, T., Blanck, G., Bresnahan, M., Sands, J., and Strominger, J.L.: *Science 243*:214-217, 1989b

Spilliaert, R., Palsdottir, A., and Arnason, A.: Analysis of the *C4* genes in baleen whales using a human cDNA probe. *Immunogenetics 32*:73-76, 1990

Streisinger, C.Y., Okada, Y., Emrich, J., Newton, J., Tsugita, A., Terzahgi, E., and Inouye, M.: Frameshift mutations and the genetic code. *Cold Spring Harbor Quant Biol 31*:77-84, 1966

Teisberg, P., Jonasson, R., Mevag, B., Gedde-Dahl, T., Jr., and Olaisen, B.: Restriction fragment length polymorphisms of the complement component C4 loci on chromosome 6: Studies with emphasis on the determination of gene number. *Ann Hum Genet 52*:77-84, 1988

Torisu, M., Sonozaki, H., Inai, S., and Arata, M.: Deficiency of the fourth component of complement in man. *J Immunol 104*:728-737, 1970

Tosi, M., Lévi-Strauss, M., Duponchel, C., and Meo, T.: Sequence heterogeneity of murine complementary DNA clones related to the C4 and C4-Slp isoforms of the fourth complement component. *Phil Trans R Soc Lond B 306*:389-394, 1984

Tosi, M., Lévi-Strauss, M., Georgatson, E., Amor, M., and Meo, T.: Duplications of complement and noncomplement genes of the H-2S region: evolutionary aspects of the C4 isotypes and molecular analysis of their expression variants. *Immunol Rev 87*:151-183, 1985

Vanin, E.F., Henthorn, P.S., Kioussis, D., Grosveld, F., and Smithies, O.: Unexpected relationships between four large deletions in the human ß-globin gene cluster. *Cell 35*:701-709, 1983

Werkmeister, J.W., New, M.I., Dupont, B., and White, P.C.: Frequent deletion and duplication of the steroid 21-hydroxylase gene. *Am J Hum Genet 39*:461-469, 1986

White, P.C., New, M.I., and Dupont, B.: HLA-linked congenital adrenal hyperplasia results from a defective gene encoding a cytochrome P-450 specific for steroid 21-hydroxylation. *Proc Natl Acad Sci USA 81*:7505-7509, 1984a

White, P.C., Chaplin, D.D., Weiss, H.H., Dupont, B., New, M.I., and Seidman, J.G.: Two steroid 21-hydroxylase genes are located in the murine *S* region. *Nature 312*:465-467, 1984b

White, P.C., Grossberger, D., Onufer, B.J., Chaplin, D.D., New, M.I., Dupont, B., and Strominger, J.L.: Two genes encoding steroid 21-hydroxylase are located near the fourth component of complement. *Proc Natl Acad Sci USA 82*:1089-1093, 1985

White, P.C., New, M.I., and Dupont, B.: Stucture of human 21-hydroxylase genes. *Proc Natl Acad Sci USA 83*:5111-5115, 1986

White, P.C., New, M.I., and Dupont, B.: Congenital adrenal hyperplasia. *N Engl J Med 316*:1519-1524, 1987

White, P.C., Vitek, A., Dupont, B., and New, M.I.: Characterization of frequent deletions causing steroid 21-hydroxylase deficiency. *Proc Natl Acad Sci USA 85*:4436-4440, 1988

Winkler, H.: *Die Konversion der Gene*. Fischer, Jena 1930

Wurst, W., Rothermel, E., and Günther, E.: Genetic mapping of *C4* and *Bf* complement genes in the rat major histocompatibility complex. *Immunogenetics 28*:57-60, 1988

Yu, C.Y.: The complete exon-intron sequence of a human complement component C4A gene. DNA sequences, polymorphism, and linkage to the 21-hydroxylase gene. *J Immunol 146*:1057-1066, 1991

Yu, C.Y., Belt, K.T., Giles, C.M., Campbell, R.D., and Porter, R.R.: Stuctural basis of the polymorphism of human complement components C4A and C4B: gene size, reactivity, and antigenicity. *EMBO J 5*:2873-2881, 1986

MAPPING OF A HOT SPOT IN THE MAJOR RECOMBINATION AREA OF THE MOUSE *H-2* COMPLEX

M. Snoek*, P.C. Groot*, R.D. Campbell°, T. Spies+, and P. Demant*

*Department of Molecular Genetics, The Netherlands Cancer Institute, Plesmanlaan 121, 1066 CX Amsterdam, The Netherlands

°MRC Immunochemistry Unit, Dept. of Biochemistry, University of Oxford, England

+Division of Tumor Virology, Dana Farber Cancer Institute, Boston, MA 02115, USA

Analysis of the class III region of the mouse, using human probes, showed that all human class III genes tested thusfar seem to have homologous counterparts in the mouse genome. Furthermore, the relative gene order of these genes appears to be conserved also. The fine mapping of the C4-H-2D segment shows that the recombination in this interval is not random, but that recombination occurs preferentially in the Hsp70-Bat-5 segment, indicating the existence of a "hot spot". The determination of crossovers at the molecular level will be helpful in the identification of functionally defined genes, since our structural data gives more information of their localisation of genes controlling disease susceptibility.

The mouse Major Histocompatibility Complex, the *H-2* complex, can be divided in the *K* and *D* regions containing the class I genes, the *I* region containing the class II genes and the *S* or class III region containing the complement class III genes. The relative positions of the various regions have been determined genetically, through the study of intra-*H-2* recombinant haplotypes. Recombination in the mouse *H-2* region, does not occur at random but is localised frequently at specific sites or hot spots. Examples of recombination "hot spots" in the *H-2* region are the class II *Eb* and *Ea* genes (Steinmetz et al. 1982; Lafuse and David 1986).

The *H-2* complex has been found to affect many traits, for instance susceptibility to various diseases. Since mice with recombinant haplotypes have been used for the genetic mapping of this kind of genes, the crossover points at the molecular level are important. In our laboratory it was found that tumor susceptibility genes for chemically induced lung tumors and for hormonally induced mammary tumors are located in the *I-D* segment (Oomen et al. 1991; Röpcke et al. 1990). In order to map the responsible genes we need to understand the

NATO ASI Series, Vol. H 59
Molecular Evolution of the
Major Histocompatibility Complex
Edited by J. Klein and D. Klein

molecular organisation of this segment of the class II and class III region. In this study we analysed the telomeric part of the class III region. The class III region is conserved, since several genes (*Tnf, Hsp70*) have been mapped to this region and were found to be located in the same order in several mammals (Müller et al. 1987a; Dunham et al. 1987; Sargent et al. 1989a; Cameron et al. 1990; Wurst et al. 1989; Gaskins et al. 1990). Recently 15 transcribed genes have been identified in the *HLA* complex in the interval corresponding to the mouse *C4-D* segment. Those genes are designated *G 1- G 11* (Sargent et al. 1989b) and *BAT 1-BAT 9* (Spies et al. 1989), respectively. Some of the *G* and *BAT* genes are in fact identical. Chromosomal walking has confirmed a high degree of interspecific conservation, since the mouse homologue of the *BAT1* gene has been localised at the same position relative to *Tnf* in the mouse (Wroblewski et al. 1990). Interspecific homology has also been shown for several other *BAT* and *G* genes (Snoek et al. 1991).

To investigate the homology between mouse and man and to map the recombination sites occurring in the *C4-H-2D* interval, we undertook restriction fragment length polymorphism (RFLP) analysis of mouse strains with crossovers in this segment using the human *HSP70* gene and *G* and *BAT* cDNA's (Sargent et al. 1989b; Spies et al. 1989) as probes as well as oligonucleotides flanking polymorphic $(TA)_n$, $(CA)_n$ and $(TC)_n$ microsatellite regions of mouse *Hsp70*, *Tnfa* and *Tnfb*, respectively (Love et al. 1990) (Table 1). Using this approach, we recently analysed ten recombinants with crossover sites in the *C4-D* interval, which, with one exception, all have a recombination site in the *Hsp70-Bat-5* segment (Snoek et al. 1991), indicating the existence of a "hot spot" in this strech of DNA. Here we show the results with 4 additional recombinants (Table 1). From the 14 recombinant mouse strains tested, 11 show recombination within the *Hsp70-Bat-5* interval; for C3H.OH we could not determine whether the recombination site lies between *C4* and *Hsp70* or between *Hsp70* and *Bat-5*. The other two recombinations occurred at different sites, e.g. 1 within the *Bat-5-Bat-4* interval (A.AL) and 1 within the *Tnf-D* interval (B10.P(40R)).

The list of *H-2* haplotypes (Klein 1989) shows that a disproportionally high number of recombinations between *H-2K* and *H-2D* has occurred between *C4* and *H-2D*. We calculated that in at least 45% of 266 recombinants, the recombination is in this part of the class III (*S*) region. 6% of the recombinants show recombination between the class I and class II region, 18% has a crossover site within the class II region, concentrated in two hot spots at a distance of < 50 kb, whereas recombination between *Ea* and *C4* occurs in ± 17% of the cases, probably preferentially in a 300 kb segment next to *C4* (Lafuse et al. 1989). For ± 12% of the recombinants the crossover site is not defined.

Table 1. *H-2* alleles of *C4-H-2D* recombinant mice

Class		III	III	III	III	III	III	III	I
Gene		*C4*	*Hsp 70*	*Bat-5*	*G5 Bat-4*	*G2 Bat-2*	*Tnf*	*Bat-1*	*D,L*
strain	A.AL	*k*	.*[1]	*k*------	*d*	*d*	*d**	*d*	*d*
	A.TH	*s*	*s**--------	*d*	*d*	*d*	*d**	*d*	*d*
	O20.Q(R15)	*pz*	*pz*--------	*q*	*q*	*q*	*q**	*q*	*q*
	O20.Q(R26)	*pz*	*pz*--------	*q*	*q*	*q*	*q**	*q*	*q*
	O20.Q(R314)	*pz*	*p*---------	*q*	*q*	*q*	*q**	*q*	*q*
	B10.P(27R)	*b*	*b**--------	*p*	*p*	*p*	*p**	.	*p*
	B10.P(40R)	*b*	*b**	*b*	*b*	*b*	*b**---------	---------	*p*
	B10.A(1R)	*d*	*d**--------	*b*	.[1]	*b*	*b**	*b*	*b*
	B10.A(2R)	*d*	*d**--------	*b*	.[1]	*b*	*b**	*b*	*b*
	B10.A(18R)	*b*	*b**--------	*d*	.[1]	*d*	*d**	*d*	*d*
	B10.BAR5	*b*	*b**--------	*d*	.[1]	*d*	*d**	*d*	*d*
	B10.BAR12	*b*	*b**--------	*d*	.[1]	*d*	*d**	*d*	*d*
	B10.AKM	*k*	*k**--------	*q*	.[1]	*q*	*q**	*q*	*q*
	C3H.OH	*d*-------	.*[1]--------	*k*	*k*	*k*	*k**	*k*	*k*

RFLP was performed as described before (Oudshoorn-Snoek and Demant 1990)

* microsatellite (Love et al. 1990), for details of the method see Snoek et al. (1991)

.[1] no polymorphism between the parental haplotypes

. not tested

Our data indicates that the recombinations in the *C4-D* segment occur preferentially in the *Hsp70-Bat-5* interval. Obviously this region harbours the major recombination site(s) used in inbred mice. The parental haplotypes involved in the recombination appear to be relevant. Haplotype dependent use of hot spots has been described by Zimmerer and Passmore (1991). These authors observed that the $H\text{-}2^p$ haplotype is frequently involved in recombination at the *Ea* hot spot, but no case of *Eb* recombination involving this haplotype has been reported. They suggest that the deletion present in the Eb^p gene suppress recombination at the *Eb* hotspot. From the recombinations occurred in the *C4-D* region, 40% concerns *b/d* or *d/b* crossover sites. Hence the high number of recombinant haplotypes involving the *d* and *b* haplotype might bias our calculations. It has been suggested that homology facilitates recombination (Carpenter 1984). As a matter of fact we do find fewer polymorphisms in the putative hot spot region between *d* and *b* haplotypes than between others.

Although the *C4-D* interval is quite long, ≥ 600 kb (Müller et al. 1987b), our data indicate that recombination occurs at a relatively small piece of DNA. The length of the *Hsp70-Bat-5* interval in human is ± 100 kb, investigations are in progress to estimate the length of the homologous interval in the mouse. Whether the preference for recombinaton to take place in the relatively small *Hsp70-Bat-5* segment is due to one super hot spot or to more than one hot spot remains to be clarified. The situation might be similar to that in the class II region, where two hot spots have been identified within a stretch of less than 50 kb (*Eb* and *Ea*). The phenotypes observed for *Csp-1*, a gene controlling susceptibility for cleft palate (Gasser et al, 1988), in the *b/d* recombinant haplotypes suggest that more than one recombination site exists in the *Hsp70-Bat-5* segment (Table 2).

Several genes controlling disease susceptibility have been mapped to the class III region of the mouse. Our data together with those from the literature (Table 2) leads to a better mapping of those genes. For instance, more precise mapping was possible for the *Hh-1* gene (Rembecki et al. 1988; Table 2). C3H.OH maps the *Hh-1* gene telomeric of *C4*, B10.AKM maps it telomeric of *Hsp70* and finely A.AL maps it telomeric of *BAT-5*. The *Cps-1* gene (Gasser et al. 1988) controlling the susceptibility for cleft palate is localised in the *Hsp70-Bat-5* segment on the basis of the recombinants between the $H\text{-}2^d$ and $H\text{-}2^b$ haplotypes. Furthermore, a gene involved in susceptibility to autoimmune orchitis, *Orch-1* (Teuscher et al. 1990) maps to the same segment. The data of *Orch-1* combined with that of *Csp-1*, however, is conflicting with respect to the localisation of the crossover sites of B10.A(1R) and B10.A(2R). More data is needed to exclude false interpretations which may be caused by unidentified deletions or other chromosomal irregularities occurring during the recombination event. The genes influencing the susceptibility for both chemically induced lung tumors (Oomen et al. 1991) and hormonally induced mammary cancer (Röpcke et al. 1990) were mapped on the basis of phenotypic differences between strains B10.A(2R) and B10.A(4R). The latter strain is recombined in the *Eb* gene; the two strains differ therefore in the *Eb-Bat-5* segment and the tumor susceptibility genes map there.

Table 2. Fine mapping of genes involved in disease susceptibility

Class		III	III	III	III	III	III	I			
Gene		*C4*	*Hsp 70*	*Bat-5*	*G5 Bat-4*	*G2 Bat-2*	*Bat-1*	*D,L*	*Hh-1*[1]	*Csp*[2]	*Orch-1*[3]
strain	A.AL	*k*	.	*k----*	*----d*	*d*	*d*	*d*	*d*		
	A.TH	*s*	*s----*	*----d*	*d*	*d*	*d*	*d*	*d*		
	B10.A(1R)	*d*	*d----*	*----b*	.	*b*	*b*	*b*		*b*	*d*
	B10.A(2R)	*d*	*d----*	*----b*	.	*b*	*b*	*b*		*d*	*b*
	B10.A(18R)	*b*	*b----*	*----d*	.	*d*	*d*	*d*		*b*	
	B10.BAR5	*b*	*b----*	*----d*	.	*d*	*d*	*d*		*d*	
	B10.BAR12	*b*	*b----*	*----d*	.	*d*	*d*	*d*		*d*	
	B10.AKM	*k*	*k----*	*----q*	.	*q*	*q*	*q*	*q*		
	C3H.OH	*d----*	*----.----*	*----k*	*k*	*k*	*k*	*k*	*k*		

1 Rembecki et al. (1988)
2 Gasser et al. (1988)
3 Teuscher et al. (1990)
no polymorphism between the parental haplotypes

In conclusion, we found that the class III region of the mouse is homologous to that of man. All human class III genes tested thusfar seem to have homologous counterparts in the mouse genome. Furthermore, the relative gene order of these genes appears to be conserved also; no data contradicting this assumption was found as yet. The fine mapping of the *C4-D* segment shows that the recombination in this interval is not random, but that the recombination occurs preferentially in the *Hsp70-BAT-5* segment. Experiments are in progress to analyse this segment using the three genes present in the human 100 kb homologous region (*G7a/BAT6, G7, G6*). Polymorphism of these genes will help to unravel the molecular basis of the hot spot(s) in the major recombination area of the *H-2* complex of inbred mouse strains.

Acknowledgments. We thank Dr. J.A. Todd for the gift of the *Hsp70* and *Tnf* microsatellite probes; Drs. D.C. Shreffler and P. Wettstein for the gift of livers of recombinant mouse strains; and H. van Vugt, R. Klompmaker and D. Schipper for skillfull technical assistence.

References

Cameron, P.U., Tabarias, H.A., Pulendran, B., Robinson, W., and Dawkins, R.L.: Conservation of the central MHC genome: PFGE mapping and RFLP analysis of complement, HSP70, and TNF genes in the goat. *Immunogenetics 31*: 253-264, 1990

Carpenter, A.T.C.: Meiotic roles of crossing-over and of gene conversion. *Cold Spring Harbor Symp Quant Biol 49*:23-29, 1984

Dunham, I., Sargent, C.A., Trowsdale, J., and Campbell, R.D.: Molecular mapping of the human major histocompatibility complex by pulsed-field gel electrophoresis. *Proc Natl Acad Sci USA 84:* 7237-7241, 1987

Demant, P.: Corticosteroid-induced cleft palate: Cis interaction of MHC genes and hybrid resistance. *Immunogenetics 22*: 183-188, 1985

Gaskins, H.R., Prochazka, M., Nadeau, J.H., Henson, V.W., and Leiter, E.H.: Localization of a mouse heat shock Hsp70 gene within the H-2 complex. *Immunogenetics 32*: 286-289, 1990

Gasser, D.L., Yadvish, K.N., Trammell, M.A., and Goldman, A.S.: Recombinants in the H-2S/H-2D interval of mouse chromosome 17 define the map position of a gene for cleft palate susceptibility. *Teratology 38*: 571-577, 1988

Klein, J.: Congenic and segregating inbred strains. *In* M.F. Lyon and A.G. Searle (eds.): *Genetic variants and strains of the laboratory mouse*, pp. 797-825. Oxford University Press, Oxford 1989

Lafuse, W.P. and David, C.S.: Recombination hot spots within the I region of the mouse H-2 complex map to the Eß and Eα genes. *Immunogenetics 24*: 352-360, 1986

Lafuse, W.P., Lee, S.T., Castle, L., and David, S.C.: Restriction fragment analysis of H-2 recombinant mouse strains with crosovers between Eα and C4 genes. *Immunogenetics 30*: 387-389, 1989

Love, J.M., Knight, A.M., McAleer, M.A., and Todd, J.A.: Towards construction of a high resolution map of the mouse genome using PCR-analysed microsatellites. *Nucleic Acids Res 18*:,4123-4130, 1990

Müller, U., Jongeneel, C.V., Nedaspasov, S.A., Fischer Lindahl, K., and Steinmetz, M.: Tumour necrosis factor and lymphotoxin genes map close to H-2D in the mouse major histocompatibility complex. *Nature 325*: 265-267, 1987a

Müller, U., Stephan, D., Philippsen, P., and Steinmetz, M.: Orientation and molecular map position of the complement genes in the mouse MHC. *EMBO J 6*: 368-373, 1987b

Oomen, L.C.J.M., Van der Valk, M.A., and Demant, P.: MHC and non-MHC genes in lung tumor susceptibility in the mouse: Implications for the study of different lung tumor types and their cell of origin. *Exp Lung Res* (in press)

Oudshoorn-Snoek, M. and Demant, P.: Identification and expression of the Tla region gene $T11^b$ and its Qa-like product. *J Immunol 145*: 1270-1277, 1990

Rembecki, R.M., Kumar, V., David, C.S., and Bennet, M.: Polymorphism of Hh-1, the mouse hemopoietic histocompatibility locus. *Immunogenetics 28*: 158-170, 1988

Röpcke, G., Moen, C.J.A., Hart, A.A.M., and Demant, P.: Effects of the MHC on hormonal induction of mammary tumors and function of hypophyseal isografts in the mouse. *Immunogenetics 31*: 347-355, 1990

Sargent, C.A., Dunham, I., Trowsdale, J., and Campbell, R.D.: Human major histocompatibility complex contains genes from the major heat shock protein HSP70. *Proc Natl Acad Sci USA 80*: 1968-1972, 1989a

Sargent, C.A., Dunham, I., and Campbell, R.D.: Identification of multiple HTF-island associated genes in the human major histocompatibility complex class III region. *EMBO J 8*: 2305-2312, 1989b

Snoek, M., Groot, P.C., Spies, T., Campbell, R.D., and Demant,P.: Fine mapping of the crossover-sites in the C4-H-2D region of H-2 recombinant mouse strains. *Immunogenetics* (submitted 1991)

Spies, T., Bresnahan, M., and Strominger, J.L.: Human major histocompatibility complex contains a minimum of 19 genes between the complement cluster and HLA-B. *Proc Natl Acad Sci USA 86*: 8955-8958, 1989

Steinmetz, M., Minard, S., Horvath, S., McNicholas, J., Frelinger, J., Wake, C., Long, E., Mach, B., and Hood, L.: A molecular map of the immune response region from the major histocompatibility complex of the mouse. *Nature 300*: 35-42, 1982

Teuscher, C., Gasser, D.L., Woodward, S.R., and Hickey, W.F.: Experimental allergic orchitis in mice. VI. Recombinations within the H-2S/H-2D interval define the map position of the H-2-associated locus controlling disease susceptibility. *Immunogenetics 2*: 337-344, 1990

Wroblewski, J.M., Kaminsky, S.G., Milisauskas, V.K., Pittmann, A.M., Chaplin, D.D., Spies, T., and Nakamura, I.: The B114-H-2Db interval and the location of a mouse homologue of the human D6S81E locus. *Immunogenetics 32*: 200-204, 1990

Wurst, W., Benesch, C., Drabent, B., Rothermel, E., Benecke, B.-J., and Günther, E.: Localization of heat shock protein 70 genes inside the rat major histocompatibility complex close to class III genes. *Immunogenetics 30*: 46-49, 1989

Zimmerer, E.J. and Passmore, H.C.: Structural and genetic properties of the Eb recombinational hot spot in the mouse. *Immunogenetics 33*: 132-140, 1991

CONSERVATION VERSUS POLYMORPHISM OF THE MHC IN RELATION TO TRANSPLANTATION, IMMUNE RESPONSES AND AUTOIMMUNE DISEASE

Roger L. Dawkins[1], Mariapia P. Degli-Esposti, Lawrence J. Abraham, Wenjie Zhang, and Frank T. Christiansen

Department of Clinical Immunology, Royal Perth Hospital, GPO Box X2213, Perth, Western Australia 6001, Australia

The MHC comprises three megabases and includes both HLA and central polymorphic non HLA genes which are relevant to the immune response. Alleles at all MHC genes occur together in specific combinations or haplotypes inherited within families. The vast majority of these nuclear haplotypes are either conserved ancestral haplotypes (AHs) or recombinants between them. In our population approximately twenty such AHs occur with frequency of greater than 1% (some as frequent as 13%) and at least 70% of the population have at least 500 kilobases of one or more of these AHs. Many MHC alleles are specific for one AH whilst other alleles apparently shared between AHs can be split. It appears that within the MHC there are at least three conserved regions containing at least two or more contiguous loci which exhibit extensive polymorphism. Recombination between these regions can be identified but is rare within them. However, these frozen regions seem to be non randomly associated, perhaps through a cascade of protein peptide interaction.

The Major Histocompatibility Complex (MHC) can be discussed and perceived in several quite different ways. Some believe the MHC simply contains genes which encode products which are critical in transplantation. For others, immune responses and antigen presentation are of more fundamental importance. Autoimmune responses and diseases provide a basis for others attempting to understand the Mhc genes.

Depending upon the vantage point, it is possible to see the MHC only in terms of HLA Class

---

[1]Publication No.9110 of the Departments of Clinical Immunology, Royal Perth Hospital and Sir Charles Gairdner Hospital and the University of Western Australia, Perth, Western Australia

NATO ASI Series, Vol. H 59
Molecular Evolution of the
Major Histocompatibility Complex
Edited by J. Klein and D. Klein

I and Class II genes and their products. According to this view there are Class I and Class II haplotypes which are separated by 1 Megabase of irrelevant DNA. By contrast, it is our contention that the central non HLA region contains genes which are relevant not only to autoimmune responses but also to antigen presentation and transplantation rejection (Christiansen and Dawkins 1991). Thus, we see the MHC as a continuous sequence of at least three megabases which encodes non HLA as well as HLA genes.

Three facts are critical to any debate on the structure and function of the MHC.

1. The central MHC contains numerous genes which are structurally unrelated to HLA but *functionally relevant* to the immune response.
2. Many, if not all of these genes, are *polymorphic*. For C4, at least, the degree of polymorphism is similar to that of the most polymorphic HLA Class I and Class II genes.
3. Alleles at all MHC genes, HLA and non HLA, tend to occur together in specific combinations or *"haplotypes"*. These haplotypes can be seen as specific multi-megabase sequences encoding multiple linked alleles.

Three conclusions appear inescapable:

1. There is a genomic region (the MHC) which contains non HLA as well as HLA genes.
2. There is a *prĭma-facie* case for considering the function of non HLA genes in attempting to understand transplantation rejection, antigen presentation and auto-immune responses.
3. It has been extremely difficult to distinguish between the effects of HLA and non HLA genes because of the existence of haplotypes which result in specific non HLA alleles being present whenever particular HLA alleles are studied.

In the first instance it would seem useful to determine the genes which are present within the MHC. Recently there has been relatively rapid progress largely through the work of Dupont, Spies, Howard, Trowsdale, Campbell and others. As shown in Table 1, the catalogue includes genes known to be expressed and sequences or transcripts which require further study. There is apparent heterogeneity but inspection suggests the possibility of some functional groupings. Given that genes such as C2, Bf, C4 and TNF are most certainly relevant to immune responses we have suggested that there may be a "central cascade" such that central MHC gene products modulate those immune responses which are possible given appropriate HLA class I and II alleles (French and Dawkins 1990).

Table 1. Possible Functions of MHC genes and Transcripts

MHC genes/transcripts	Possible functions	References
Collagen	Cartilage structural component	Hanson, et al. (1989)
HLA-DP	Class II Molecules: Antigen Presentation	See Dupont, et al., 1987
HLA-DQ	"	"
HLA-DR	"	"
ABC Transporters	Peptide transport from cytosol to ER	Spies et al. (1990)
		Trowsdale et al. (1990)
C4	C3 Convertase	Carroll et al. (1984)
21OH	Steroid 21-Hydroxylase	White et al. (1985)
C2	Classical pathway C3 Convertase	Carroll et al. (1984)
Factor B	Alternate pathway C3 Convertase	"
HSP70	Chaperone Molecules	Sargent et al. (1989)
Bat 2	? Integrin-like	Banerji et al. (1990)
Bat 3	Chaperone?	"
B144	Modulation?	Spies et al. (1989)
TNF	Immune regulation	Carroll et al. (1987)
X, Y, V	?	Spies et al. (1989)
HLA-B	Class I Molecules: Antigen Presentation	See Dupont et al., 1987
HLA-C	"	"
HLA-A	"	"

Whilst attempting to determine the gene content of the central MHC, it is clearly critical to determine the polymorphism of coding and non coding sequences. Non polymorphic regions can be of only limited interest in the present context. Genes or sub regions which are polymorphic are inherently more interesting since transplantation rejection, antigen presentation or autoimmune responses are associated with *particular* HLA and C4 alleles; it is possible that some other polymorphic gene is important given that alleles at all genes, known and unknown, will be associated ie in linkage disequilibrium.

Methods for defining polymorphism have advanced in recent years and it may be possible to screen large sequences for haplospecific differences without the need for global sequencing. Some of these approaches are listed in Table 2. It should be emphasised that to date all polymorphisms within the MHC are in linkage disequilibrium and this is true whether the variations occur in coding or non coding sequence. The significance of polymorphism of non coding regions should not be underestimated; as one example it appears that polymorphisms in non coding regions around the TNF genes may be important in controlling production irrespective of whether or not there are qualitative differences in the protein (Messer et al. 1991).

Table 2. Screening methods for detection of polymorphism

Method	Advantages	Disadvantages
Sequencing	Type & position of polymorphism defined	Time consuming, expensive
Denaturing Gradient Gel Electrophoresis (1)		Lengthy set-up, special equipment needed
Chemical Cleavage (2)	Position of polymorphism defined	Technically demanding
RFLP Analysis	Established methods used	Limited information
Heteroduplex Detection (3)	Simple to perform	No information about nature of polymorphism
S1/RNase Methods		Difficult to perform, may not detect all differences

* See Cotton (1989) for review of screening methods.

(1) See Collins and Myers (1987) for review.

(2) Cotton, et al (1988).

(3) Bidwell and Hui (1990).

Having defined some of the genes and their polymorphisms it becomes possible to recognise conserved or ancestral MHC haplotypes. A current list is shown in Table 3 which provides a convenient summary of data derived locally and from the past and present International Workshop studies. The list began when it became clear that apparently unrelated subjects shared a whole series of alleles at multiple loci scattered over several megabases. Initially two mutually exclusive alternatives appeared possible. Firstly, these alleles could occur together without regard to the intervening sequence. If this were the case it might be suggested that the alleles were associated by virtue of some interaction between them. Secondly, it was possible that the alleles occurred together simply because they reflected conserved megabase sequences. In this case several explanations could be considered including the possibility that the haplotypes were identical by descent from some common ancestor. In the event, tne evidence was overwhelming that ancestral haplotypes are a function of conserved continuous sequence.

For reasons which remain uncertain, there has been considerable confusion surrounding the term *haplotype*. When introduced by Ceppellini, the term referred to serological patterns which segregated within nuclear families. Obviously these patterns reflected closely linked alleles.

Such "nuclear" haplotypes could only be defined by family studies. On the other hand, it was apparent that certain combinations or "population" haplotypes occurred rather frequently and more commonly than would be expected by the frequencies of the independent alleles i.e. certain alleles were in *linkage disequilibrium*. "Haplotype" frequencies were calculated and it became clear that similar results were obtained whether the calculations were performed on phenotypes or nuclear haplotypes derived from family studies. Thus most, but not all, nuclear haplotypes carried the combinations of alleles which would be expected from calculating linkage disequilibrium parameters within the population. This finding was not surprising - for any combination or "haplotype" to be maintained in the population it would seem obvious that the same would be transmitted from generation to generation as a nuclear haplotype. Nevertheless, the terminological confusion lead to some curious distractions. If haplotypes could only be defined by studies of nuclear families, haplotypes could only be conserved over several generations. Clearly there was a need to study apparently unrelated subjects in an attempt to derive ancestral families whilst recognising that recombination would occur in at least a few percent of successive generations.

Table 3. Provisional list of ancestral haplotypes

NAME	A	Cw	B	TNF	BAT3	C2	Bf	C4A	C4B	DR	Dw	DQw	RACE
7.1	3	7	7	L	L	C	S	3	1	15	2	6	C
7.2	24	7	7	?	L	C	S	3+3	1	1	1	5	M
8.1	1	7	8	S	S	C	S	Q0	1	3	3	2	C
13.1	30	6	13	L	S	C	S	3	1	7	?	2	C/M
18.1	25	?	18	L	L	Q0	S	4	2	15	2	6	C
18.2	30	5	18	L	S	C	F1	3	Q0	3	3	2	C
18.3	?	?	18	L	?	?	S	3	1	11	?	7	C
35.1 t	?	4	35	S	L	?	S	3	1	11	?	7	C
35.2	3	4	35	L	L	C	F	3+2	Q0	1	1	5	C
35.3	11	4	35	L	?	?	S	3	Q0	1	?	5	C
37.1	1	6	37	?	?	?	F	3	1	10	?	?	C
38.1	26	?	38	L	L	?	S	2	1	4	10	8	C
42.1	?	2	42	S	?	C	F	1	Q0	3	?	4	N
44.1	2	5	44	S	L	C	S	3	Q0	4	4	7	C
44.2	29	4	44	L	L	C	F	3	1	7	17	2	C
44.3	33	?	44	?	S	?	S	Q0	1	7	17	2	M
44.4	33	?	44	?	?	C	F	3	1	13	?	6	M
46.1	2	11	46	S	L	C	S	4	2	9	?	9	M
46.2	2	11	46	S	L	C	S	4	2	8	?	6	M
47.1	3	6	47	L	L	C	F	1	Q0	7	?	2	C
50.1	?	?	50	S	?	C	S07	2	1+12	7	?	?	C
51.1	?	?	51	?	?	C	F	3	Q0	4	14	3	C
52.1	24	?	52	?	?	C	S	3+2	Q0	15	?	6	M
54.1	?	1	54	L	L	C	S	3	5	4	15	4	M
55.1	?	3	55	L	L	B	S	4	5	14	?	?	C
57.1	1	6	57	L	L	C	S	6	1	7	11	9	C/M/N
58.1	33	?	58	S	S	?	S	3	Q0	3	?	2	M
58.2	33	?	58	?	?	?	F	Q0	1	13	?	?	M
59.1	?	1	59	?	?	C	S	3	5	9	?	?	M
60.1 t	?	3	60	?	L	C	S	3	1	4	?	3	C
60.2	?	3	60	L	L	C	S	3	Q0	8	8.1	4	C
60.3	2	3	60	S	S	C	S	Q0	2	13	19	6	C
61.1	26	3	61	?	?	C	S	3	1	9	23	?	M
62.1	2	3	62	L	L	C	S	3	3	4	4	8	C
62.2	?	3	62	L	L	B	S	4	2	4	?	8	C
62.3	?	?	62	S	?	C	F	3	1	13	?	6	C
62.4 t	11	4	62	?	L	C	S	3	1	4	?	?	C/M
64.1	?	8	64	L	?	C	S	3	1	7	?	?	C
65.1	?	8	65	L	L	C	S	2	1+2	1	20	5	C
65.2	?	8	65	?	?	C	F	3	1	13	?	?	C

BAT3 and TNF determined by RFLP typing (S, short fragment; L, long fragment)

RACE: C, Caucasoid; M, Mongoloid; N, Negroid. t, tentative assignment; ?, not defined.

One approach was to identify combinations of alleles (or "supratypes" as defined by Newton Morton), and to seek indirect evidence for a common ancestor (Dawkins et al. 1983). We therefore began to compare the genomic structure of the MHC carried by unrelated subjects expressing the same supratypes (Tokunaga et al. 1988; Zhang et al. 1990). Of course, the critical assumption was that identity by sequence (especially at polymorphic non coding regions) implied identity by descent from a remote ancestor.

In the intervening period Chester Alper and his colleagues had introduced the concept of complotypes (Alper et al. 1983). This group showed quite unequivocally that unrelated subjects frequently shared the same alleles at C2, Bf, C4A and C4B. Indeed many of the possible combinations of these four loci were extremely rare suggesting that there must be a remarkable degree of conservation in this region. Accordingly, the recognition of complotypes effectively provided evidence for ancestral haplotypes maintained from very remote ancestors. As an alternative possibility only certain combinations or complotypes are compatible with survival with selection operating at fertilisation or soon after. Whilst not formally excluded, this hypothesis appears inherently unlikely unless the complement region of the MHC contains some absolutely critical gene. There are other genes in the region including some which could conceivably be relevant to intrauterine selection but further evidence is required. In the meanwhile complotypes provide a potential model for perceiving the whole MHC.

The recognition that 21-Hydroxylase deficiency was often associated with a particular complotype provided indirect evidence for the concept of conserved ancestral haplotypes. The subsequent finding that there was a specific allele at HLA B (Bw47) suggested the possibility that the conservation might be maintained over at least several hundred kilobases. There are now many more examples where the complotype has allowed the recognition of specific splits of HLA Class I and Class II alleles. It should not be surprising that ancestral haplotypes were often first seen in groups of subjects with disease including for example C2 deficiency, Myasthenia Gravis, Systemic Lupus Erythematosus, Rheumatoid Arthritis and Insulin Dependent Diabetes Mellitus (Dawkins et al. 1991; Christiansen et al. 1991).

At this stage, Alper and his colleagues introduced the notion of "extended" haplotypes (Awdeh et al. 1983). Some combinations of alleles at complement and HLA loci were apparently "extended" due to a lack of recombination. Other combinations were "non extended" perhaps because they tended to fragment. The precise distinction between these two classes has remained difficult to define.

Our own view was that most if not all supratypes shared by unrelated subjects must be conserved and that less frequent combinations would be shown to be due to either recombination between two ancestral haplotypes or intrinsic rarity due to recent racial admixture or perhaps poor survival if associated with specific disease genes. If correct, it should be possible to show that the vast majority of nuclear haplotypes are either conserved ancestral haplotypes or explicable in terms of recombination between two or more ancestral haplotypes. The current estimates of frequency *in our population* are shown in Table 4. Several points can be made:

1. At least 70% of the total population have at least 500 kilobases of one or two of the haplotypes listed in Table 3.
2. Some ancestral haplotypes occur with a gene (or haplotype) frequency as high as 13%.
3. Only approximately 20 ancestral haplotypes occur with a frequency of approximately 1% or greater.
4. Many MHC alleles appear to be entirely specific for one ancestral haplotype (e.g. B8 for 8.1, BfF1 for 18.2, C4A6 for 57.1).
5. Some alleles which appear to be shared by ancestral haplotypes have already been split eg many HLA B alleles which appear to be shared are known to be heterogeneous by serology or isoelectric focusing (eg HLA B44, B18, B35, Bw62) (Yang et al. 1989). It seems possible that each ancestral haplotype may have a specific and unique allele at the HLA B locus. Whilst a similar phenomenon may be seen at DRB1 it is quite clear that some loci (eg TNF) are less polymorphic (Dawkins et al. 1989; Abraham et al. 1991). There may be a continuum of polymorphism of the coding regions but quite obviously some extensive sequences are entirely non-polymorphic.

At this stage it would appear that there are at least three blocks or regions containing at least two or more contiguous loci exhibiting extensive polymorphism. Firstly, the HLA class II region contains closely linked and polymorphic genes. Secondly, the complement region similarly contains at least several highly polymorphic genes within approximately 100kb. Thirdly, the HLA C and B region may be similar with high degrees of polymorphism over at least several hundred kilobases towards TNF (Egea et al. 1991). Thus, ancestral haplotypes consist of potentially variable sequences which are relatively "fixed" or "frozen" (Klein 1990)

Table 4. Frequency of ancestral haplotypes and their recombinants in an Australian population

Ancestral Haplotype	HLA B	Bf	C4A	C4B	HLA DR	B—C4	B—DR	C4—DR	TOTAL	%
7.1	7	S	3	1	2	13	19	13	45	12.9
8.1	8	S	0	1	3	1	40	5	46	13.2
13.1	13	S	3	1	7	0	5	4	9	2.6
18.1	18	S	4	2	2	0	3	1	4	1.2
18.2	18	F1	3	0	3	0	3	2	5	1.4
18.3	18	S	3	1	5	2	4	2	8	2.3
35.1t	35	S	3	1	5	5	5	2	12	3.4
35.2	35	F	3+2	0	1	0	3	0	3	0.9
35.3	35	S	3	0	1	1	5	2	8	2.3
38.1	38	S	2	1	4	3	1	0	4	1.2
44.1	44	S	3	0	4	4	10	5	19	5.5
44.2	44	F	3	1	7	2	6	1	9	2.6
44.3	44	S	0	1	7	1	0	0	1	0.3
44.5p	44	S	3	1	5	2	10	4	16	4.6
50.1	50	S1	1	1+1.5	7	1	1	0	2	0.6
51.1	51	F	3	0	4	1	1	0	2	0.6
51.2p	51	S	3	1	5	2	6	2	10	2.9
55.1	55	S	4	5	6	1	1	1	3	0.9
57.1	57	S	6	1	7	2	6	1	9	2.6
60.1t	60	S	3	1	4	4	2	5	11	3.2
60.3	60	S	0	2	6	1	6	2	9	2.6
62.1	62	S	3	3	4	1	3	0	4	1.2
62.2	62	S	4	2	4	2	2	4	8	2.3
62.4t	62	S	3'	1	4	2	1	3	6	1.7
64.1	64	S	3	1	7	1	2	2	5	1.4
65.1	65	S	2	1+2	1	0	2	0	2	0.6
65.2	65	F	3	1	6	0	4	3	7	1.7
					TOTAL	52	151	64	267/348	76.7

Where t = tentative (require further characterization at the DNA level) and p = provisional

throughout the three or more highly polymorphic regions. Recombination *between* these regions can be identified readily but appears to be very rare *within* the polymorphic regions.

An inverse relationship between polymorphism and recombination frequency could possibly be explained by the potential difficulty in recombining polymorphic regions but, since alleles often differ by only a few bases, other possibilities need to be considered.

It should be emphasised that estimates of recombination frequencies are extremely difficult. Although we assume that conservation of polymorphic regions has been maintained over many thousands of generations it is not yet possible to be precise and certainly not possible to compare different haplotypes without an accurate estimate of their age. In addition, recombination may be more dynamic than sometimes appreciated. For example, we have seen situations where ancestral haplotypes are reformed implying that sequential sections may be associated preferentially possibly by virtue of some survival advantage. It may be erroneous to think of the MHC as being a fixed structure which is undergoing unidirectional decay; recombination may result in apparent conservation as well as fragmentation. Further difficulty arises from our ignorance of the state of MHC haplotypes in haploid cells and immediately after fertilisation.

In the absence of reliable estimates of recombination frequencies and other possible mechanisms for generating new haplotypes, it remains possible to consider explanations for the extraordinary conservation of highly polymorphic regions.

We propose that there must be interactions between the allelic forms of more or less contiguous genes within the three or more polymorphic regions. Thus, for example, the alpha and beta chains of the class II genes may need to be closely linked because they are so polymorphic. Conceivably different allelic forms of the complement components have some significance in relation to each other although direct evidence is lacking. Polymorphic genes in the vicinity of HLA B may be relevant to peptide protein interactions as possibly suggested by inspection of Table 1.

Although, highly speculative, the notion that polymorphic segments may be "frozen" by virtue of interactive polymorphic products appears amenable to further testing. Furthermore, a similar model could be proposed to explain the fact that these "frozen" regions are non randomly associated. Thus, for example, a central cascade could modulate immune responses which are dependent upon protein peptide interactions determined by the genes in the vicinity of HLA B and HLA DR.

References

Abraham, L.J., ChinDu, D., Zahedi, K., Dawkins, R.L., and Whitehead, A.S.: Haplotypic polymorphisms of the TNFB gene. *Immunogenetics 33*: 50-53, 1991

Alper, C.A., Raum, D., Karp, S., Awdeh, Z.L., and Yunis, E.J.: Serum complement 'Supergenes' of the major histocompatibility complex in man (complotypes). *Vox Sang 45*: 62-67, 1983

Awdeh, Z.L., Raum, L.D., Yunis, E.J., and Alper, C.A.: Extended HLA/complement allele haplotypes: evidence for T/t-like complex in man. *Proc Natl Acad Sci USA 80*: 259, 1983

Banerji, J., Sands, J., Strominger, J.L., and Spies, T.: Gene pair from the human major histocompatibility complex encodes large proline-rich proteins with multiple repeated motifs and a single ubiquitin-like domain. *Proc Natl Acad Sci USA 87*: 2374-2378, 1990

Bidwell, J.L. and Hui, K.M.: Human HLA-DR/Dw allotype matching by analysis of HLA-DRB gene PCR product polymorphism (PCR "fingerprints"). *Technique 2(2)*: 93-100, 1990

Carroll, M.C., Campbell, R.D., Bentley, D.R., and Porter, R.R.: A molecular map of the human major histocompatibility complex class III region linking complement genes C4, C2 and factor B. *Nature 307*: 237-241, 1984

Carroll, M.C., Katzman, P., Alicot, E.M., Koller, B.H., Geraghty, D.E., Orr, H.T., Strominger, J.L., and Spies, T.: Linkage map of the human major hisotcompatibility complex including the tumor necrosis factor genes. *Proc Natl Acad Sci USA 84*: 8535-8539, 1987

Christiansen, F.T. and Dawkins, R.L.: Questions in marrow matching: The implications of ancestral haplotypes for routine pracitice. *Bone Marrow Transpl* (submitted 1991)

Christiansen, F.T., Saueracker, G.C., Leaver, A.L., Tokunaga, K., Cameron, P.U., and Dawkins, R.L.: Characterization of MHC ancestral haplotypes in insulin-dependent diabetes mellitus: evidence for involvement of non-HLA genes. *Immunogenetics 17*: (submitted 1990)

Collins, M. and Myers, R.M.: Alterations in DNA helix stability due to base modifications can be evaluated using denaturing gradient gel electrophoresis. *J Mol Biol 198*: 737-744, 1987

Cotton, R.G.H.: Detection of single base changes in nucleic acids. *Biochem J 263*: 1-10, 1989

Cotton, R.G.H., Rodrigues, N.R., and Campbell, R.D.: Reactivity of cytosine and thymine in single-base-pair mismatches with hydroxylamine and osmium tetroxide and its application to the study of mutations. *Proc Natl Acad Sci USA 85*: 4397-4401, 1988

Dawkins, R.L., Christiansen, F.T., Kay, P.H., Garlepp, M.J., McCluskey, J., Hollingsworth, P.N., and Zilko, P.J.: Disease associations with complotypes, supratypes and haplotypes. *Immunol Revs 70*: 5-22, 1983

Dawkins, R.L., Leaver, A., Cameron, P.U., Martin, E., Kay, P.H., and Christiansen, F.T.: Some disease-associated ancestral haplotypes carry a polymorphism of TNF. *Human Immunol 26*: 91-97, 1989

Dawkins, R.L., Zhang, W.J., Degli-Esposti, M.A., Abraham, L., McCann, V., and Christiansen, F.T.: Studies of haplotypes by pulsed field gel electrophoresis. *In* Clinical Endocrinology and Metabolism. Bailliere Tindall Limited, London (submitted 1991)

Dupont, B.: Immunobiology of HLA. Histocompatibility Testing 1987. Springer-Verlag, New York 1989

Egea, G.E., Yunis, I., Spies, T., Strominger, J., Awdeh, Z.L., Alper, C.A., and Yunis, E.J.:

Association of polymorphisms in the HLA-B region with extended haplotypes. *Immunogenetics 33*: 4-11, 1991

French, M.A.H. and Dawkins, R.L.: Central MHC genes, IgA deficiency and autoimmune disease. *Immunol Today 11*: 271-274, 1990

Klein, J. and Takahata, N.: The Major Histocompatibility Complex and the Quest for Origins. *Immunol Revs 113*: 5-25, 1990

Messer, G., Spengler, U., Jung, M.C., Honold, G., Blomer, K., Pape, G.R., Riethmuller, G., and Weiss, E.H.: Polymorphic structure of the tumor necrosis factor (TNF) locus: An NcoI polymorphism in the first intron of the human TNF-B gene correlates with a variant amino acid in position 26 and a reduced level of TNF-B production. *J Exp Med 173*: 209-219, 1991

Sargent, C.A., Dunham, I., and Campbell, R.D.: Identification of multiple HTF-island associated genes in the human major histocompatibility complex class III region. *EMBO J 8*: 2305-2312, 1989

Sargent, C.A., Dunham, I., Trowsdale, J., and Campbell, R.D.: Human major histocompatibility complex contains genes for the major heat shock protein HSP70. *Proc Natl Acad Sci USA 86*: 1968-1972, 1989

Spies, T., Blanck, G., Bresnahan, M., Sands, J., and Strominger, J.L.: A new cluster of genes within the human major histocompatibility complex. *Science 243*: 214-217, 1989

Spies, T., Bresnahan, M., Bahram, S., Arnold, D., Blanck, G., Mellins, E., Pious, D., and DeMars, R.: A gene in the human major histocompatibility complex class II region controlling the class I antigen presentation pathway. *Nature 348*: 744-747, 1990

Tokunaga, K., Saueracker, G., Kay, P.H., Christiansen, F.T., Anand, R., and Dawkins, R.L.: Extensive deletions and insertions in different MHC supratypes detected by pulsed field gel electrophoresis. *J Exp Med 168*: 933-940, 1988

Trowsdale, J., Hanson, I., Mockridge, I., Beck, S., Townsend, A., and Kelly, A.: Sequences encoded in the class II region of the MHC related to the 'ABC' superfamily of transporters. *Nature 348*: 741-744, 1990

White, P.C., Grossberger, D., Onufer, B.J., Chaplin, D.D., New, M.I., Dupont, B., and Strominger, J.L.: Two genes encoding steroid 21-hydroxylase are located near the genes encoding the fourth component in man. *Proc Natl Acad Sci USA 82*: 1089-1093, 9185

Yang, S.Y.: Assisgnment of HLA-A and HLA-B antigens for the reference panel of B-lymphoblastoid cell lines determined by one-dimensional isoelectric focusing (1D-IEF) gel electrophoresis. *In* B. Dupont (ed.): Immunobiology of HLA, Volume 1: Histo-compatibility Testing 1987, p. 43. Springer-Verlag, New York 1989

Zhang, W.J., Degli-Esposti, M.A., Cobain, T.J., Cameron, P.U., Christiansen, F.T., and Dawkins, R.L.: Differences in gene copy number carried by different MHC ancestral haplotypes: quantitation after physical separation of haplotypes by Pulsed Field Gel Electrophoresis. *J Exp Med 171*: 2101-2114, 1990

HLA ASSOCIATIONS WITH MALARIA IN AFRICA: SOME IMPLICATIONS FOR MHC EVOLUTION

Adrian V. S. Hill

Molecular Immunology Group, Institute of Molecular Medicine, University of Oxford, John Radcliffe Hospital, Oxford OX3 9DU, U.K.

A recent large case-control study of malaria in West African children has demonstrated the first clear HLA associations with a fatal infectious disease of humans, supporting the theory that MHC polymorphism is pathogen driven. The time for the protective HLA alleles to rise in frequency as a result of natural selection by malaria, and the possible mechanisms of this selection, overdominant, frequency-dependent or fluctuating, are discussed.

Theories of MHC polymorphism

Over the last decade several explanations for the extraordinary polymorphism of MHC genes have been discussed (Table 1). Of these, theory 1, that of parasite-driven MHC diversity (Snell 1968; Doherty and Zinkernagel 1975) has been particularly attractive, because it invokes an explanation related to the primary function of MHC class I and II molecules: defence of the host against infectious pathogens. It appears that much of the perceived need for other explanations of MHC polymorphism derives from the failure of attempts to demonstrate the primary corollary of this theory: that individuals of different MHC type should vary in their susceptibility to at least some fatal infectious agents.

Over the last few years two large data sets have provided important new evidence helping us to distinguish between these explanations. The first derives from the work of Pamela Bjorkman and her colleagues (1987) who deduced the crystal structure of the HLA class I molecule, HLA-A2, which when taken together with the accumulated class I DNA sequences from several laboratories (Bjorkman and Parham 1990) showed that polymorphic amino acid residues were concentrated along a clear peptide binding groove. Subsequent statistical analysis confirmed that non-synonymous codon changes are more frequent along this groove than elsewhere in the molecule (Hughes and Nei 1988). This work provided strong

NATO ASI Series, Vol. H 59
Molecular Evolution of the
Major Histocompatibility Complex
Edited by J. Klein and D. Klein

evidence that selection has been acting in the past on this peptide-binding groove but, despite the title of Hughes and Nei's (1988) paper, provided no information on the mechanism (e.g. overdominant v frequency-dependent) or stage of action (e.g. differential viability v differential fertility) of the selection.

Table 1. Theories of MHC polymorphism

THEORIES OF MHC POLYMORPHISM	
1)	Parasite driven selection
	- overdominant
	- frequency dependent
	- fluctuating
2)	Disassortive mating
3)	Maternal - fetal incompatibility
4)	High mutation rate
5)	Ancient neutral polymorphism

The other very recent data set, on HLA polymorphism and susceptibility to malaria in African children (Hill et al. 1991), will be discussed here. This work provided the first clear evidence that HLA alleles are associated with altered susceptibility to fatal infectious pathogens in humans, and showed that natural selection by at least one infectious disease continues to exert an important influence on the evolution of HLA polymorphism.

Considering the other four theories of MHC polymorphism, the last two now appear particularly unlikely. The high mutation rate theory appears incompatible with the great age of many MHC alleles, demonstrated by inter-species comparisons of MHC sequences (Lawlor at al. 1988; Figueroa et al. 1988), and with the concentration of polymorphisms in the nucleotides specifying coding changes in the peptide binding groove. Formal simulations have now confirmed the belief that the great age of MHC alleles is evidence that some form of balancing selection is operating (Takahata and Nei 1990), rather than evidence of the selective neutrality of these alleles (Klein 1987; Serjeantson 1989). Hence the concept of ancient neutral polymorphism, theory 5, appears untenable. The ideas of diversity maintained by disassortive mating (Yamazaki et al. 1976) or relative incompatibility of MHC identical maternal-fetal pairs (Clarke and Kirby 1966), remain theoretical possibilities. The latter has been shown to provide a sufficient basis for the maintenance of MHC diversity over considerable time periods (Hedrick and Thomson 1988). However, it is difficult to see how it can operate in species without immunological interaction of fetal and maternal cells, such as chickens, which nonetheless show considerable MHC polymorphism. Recent evidence that HLA molecules are present in sweat (Zavazava et al. 1990), and that various alleles may be associated with detectable olfactory differences (Ferstl et al. 1989), provides a basis for investigating theory 2 in more detail (cf Potts et al., this volume). However, I believe that both of these phenomena (theories 2 and 3), if they can be confirmed to occur, have probably evolved as an additional diversity-enhancing mechanisms in a limited number of species, exploiting a polymorphic system that had evolved primarily to combat infectious pathogens. Perhaps the strongest reason for favouring the theory of parasite-driven diversity is that it is the only one to provide a plausible purposive as well as a causative explanation for the diversity of MHC class I and II molecules in most vertebrate species.

Detecting HLA associations with infectious disease

The lack of associations between particular HLA alleles and infectious diseases, in contrast to the many established associations with particular autoimmune diseases, has been seen as evidence against the theory of parasite-driven MHC diversity. No detailed review of this scattered literature has been published recently and I shall not attempt one here. It may be more useful to point out the reasons why it has been particularly difficult to find such HLA-infection associations and why most of the available data are difficult to interpret.

Essentially the reasons for not finding associations in HLA-infectious disease case-control studies may be 1) inadequate sample size, 2) inadequate HLA typing methodology, and 3) the actual absence of associations. I believe that the first explanation has been the principal difficulty and I shall discuss it first. Because of the ever-growing list of HLA allelic types and subtypes the sample size requirement for demonstrating even fairly strong associations with moderately common HLA types is of the order of several hundreds to thousands: this is a result of the need to correct probability values for the number of comparisons made. Protective alleles, perhaps of most interest for dissecting mechanisms of protective immunity, often require larger sample sizes for their detection than susceptibility alleles. The best way around this problem is to do a study twice: generate a hypothesis in the first study and then test this alone in the second. Hardly any investigators have adopted this two stage strategy for infectious diseases, perhaps because the lack of strong associations in preliminary studies has deterred further work. The difficulty in collecting suitable cases of severe acute infectious diseases, particularly in developed countries where many of these are increasingly uncommon, cannot be overemphasised: this contrasts with the relative ease of studing patients with many chronic autoimmune diseases, who often obliging cluster in specialist clinics. Even in developing countries many of the diseases one would most like to study from an evolutionary point of view either occur in epidemics, thereby increasing already challenging logistic problems, or predominantly affect young children who, until the recent advent of molecular typing methods, could not have been expected to donate sufficient blood for detailed serology. Hence, case-control studies totalling a hundred or so cases and a couple of hundred controls, with sparse, if any details of matching are the rule. Confounding factors are rarely mentioned, much less corrected for, and the detailed genetic typing of parasite as well as host remains an area for the future.

A second difficulty, that is rapidly being overcome, has been the inability of serology to distinguish all relevant HLA types and subtypes, particularly in non-caucasians. For example, in The Gambia over half of the population carry a HLA-DRw13 allele but four different DRw13 subtypes are prevalent, two of which are the commonest two DRB1 alleles. Most of the older HLA - infectious disease studies could only type HLA class I alleles and there has been almost no information on DP alleles or DQ subtypes.

Finally, the alternative explanation for the lack of convincing associations with infectious diseases is that none exist. The molecular basis for this might be that most infectious pathogens contain very many epitopes with a random assortment of HLA restrictions, so that individuals of any given HLA genotype are equally capable of mounting a protective immune response. Conversely, the demonstration of a clear HLA association with an infectious disease would suggest that there are relatively few 'immunodominant' epitopes in the pathogen capable of generating a protective immune response.

Simplifying enormously a great body of literature, it may be helpful to note that the clearest MHC-disease association with a fatal infection is Marek's disease in chickens (Briles et al. 1977), but even here the precise MHC gene involved remains unclear. In humans, leprosy has been most extensively studied with reports of HLA-DR2 and -DR3 associations in different populations. This remains a controversial area with the majority view being that HLA has a role in determining the clinical type of leprosy rather that susceptibility to the disease *per se*. Recent data from small studies of HIV infection in humans also suggest HLA associations, but it is clear that neither leprosy (a non-fatal infection) nor AIDS have played major roles in selecting for HLA polymorphism. De Vries et al.'s (1979; 1989) study of about 60 descendants in Surinam of Dutch emigrants who survived typhoid and yellow fever epidemics, led to the suggestion that HLA-determined differential mortality might have led to the differences in their HLA frequencies from the present-day Dutch population: but the likelihood that genetic drift, admixture amongst the settlers, and spatial variation in HLA frequencies in early 19th century Holland could have produced such a difference, seems very high.

In conclusion, it is unclear whether HLA associations with most infectious diseases exist but, if present, they are probably not as striking as those seen with some autoimmune diseases. However such modest selective effects may be of considerable evolutionary importance.

Why malaria?

We designed the study of HLA and malaria in 1987, a year in which the idea of parasite-driven MHC diversity was, to say the least, unfashionable (Klein 1987; Flaherty 1987). A principal reason for choosing malaria was because it was possible to study a large number of cases of severe disease over a reasonable period of time. Also, in practical terms, the renewed interest in malaria as a major global health problem and our ignorance of the nature of protective immunity to *P. falciparum* were of importance both to us and the sponsors of this study. There were, however, other advantages from the perspective of studying HLA evolution. Any HLA association with malaria was likely to be of relevance because of the enormous mortality of this disease, estimated at one to two million childhood deaths each year in sub-Saharan Africa alone (Greenwood et al. 1991). This implies that a strong association might be reflected in the known geographical distribution of HLA alleles, which *might* therefore provide supporting evidence. Most attractively, because of the numerous known genetic loci which it is believed may alter susceptibility to malaria, perhaps as many as ten, we had the opportunity to compare the effects of several of these with HLA.

On the other hand malaria could in some respects be viewed as a risky choice. For example, it was argued that we were exceedingly unlikely to find a common susceptibility allele because this would have been rapidly eliminated from an African population by malarial selection. Likewise, assuming for the moment simple directional selection and ignoring more complex (and realistic) types of selection, strongly protective alleles were, arguably, unlikely to be found because these should rapidly go to fixation. We banked on balancing selection pressures: other pathogens take considerable tolls. Finally, considerable genetic polymorphism in *P. falciparum* has been demonstrated, so that associations between HLA and particular parasite strains could have been obscured by looking at *falciparum* malaria as a whole. We did not, and still do not, know how much of this parasite polymorphism is important, but observations on the nature of polymorphism in at least one malarial antigen strongly suggest that selective pressure by the human immune system has been consequential (Good et al. 1988).

Protective HLA class I and class II associations

The substantive account of this malaria study is now in press (Hill et al. 1991), so I shall here briefly decribe the approach and summarise the main results. We studied children of average age 3 years in two hospitals on the Gambian coast in West Africa. Studies over the last 40 years by the U.K. Medical Research Council Laboratories have given us a particularly detailed knowledge of malaria in this setting. Almost all malaria deaths occur in children under 7 years from either severe malarial anaemia or, more frequently in The Gambia, cerebral malaria. We studied just over 600 children with severe malaria and compared them to control children attending the same hospitals or clinics, with matching for age and area of residence. Although every child growing up in The Gambia gets a malarial infection every year, this group of 'severe malaria' cases represents the very severe end of a spectrum of clinical disease. Without treatment, a large majority of these would have died: even with treatment 88 did. We also HLA typed cases of uncomplicated or 'mild' malaria and other healthy controls, a total of over 2000 individuals.

For HLA class I typing we initially typed about half of the severe malaria cases and a similar number of controls serologically, and saw a possible association with HLA-Bw53. We then retested this hypothesis, that HLA-Bw53 was associated with protection from both cerebral malaria and severe malarial anaemia, on the remaining cases and controls. For this we used a polymerase chain reaction (PCR)-based method which allows just a single antigen to be typed, eliminating the statistical problem of assessing the results of multiple comparisons. The combined data (Table 2) showed that possession of this antigen was associated with about a 40% reduction in the relative risk of developing severe malaria. HLA-Bw53 is a common allele in sub-Saharan Africans, the highest antigen frequency of 40% having been reported for Nigerians. It is rare (1%) in Europeans and extremely rare or absent in most Asian and Oceanic populations.

For HLA typing of class II (DR and DQ) antigens we used the now well established method of restriction fragment length polymorphism (RFLP) haplotype determination, supplemented by oligonucleotide hybridization following polymerase chain reaction amplification. Again, a single clear association was found (Fig. 1).

Table 2. HLA-Bw53 antigen frequencies determined by serology and by PCR. There is no overlap between the samples analyzed by the two methods (Hill et al. 1991).

	HLA-Bw53 association with severe malaria			
	Serology		PCR	
	Number	%HLA-Bw53	Number	%HLA-Bw53
Severe malaria	306	15.7		16.9
Mild controls	144	24.3	364	25.4
Mild malaria	-	-	353	22.6
Healthy adults	112	25.0	106	26.4

The DRw13.02 (i.e. the 02 subtype of DRw13) - DQw1 haplotype was associated with an approximate halving of a childs risk of developing severe malarial anaemia. Sequence analysis and subsequent DQ subtyping showed that, in The Gambia, this haplotype is mainly DRB1*1302-DRB3*0301-DQA1*0102-DQB1*0501, a haplotype seen almost exclusively in Africans or their derivative populations.

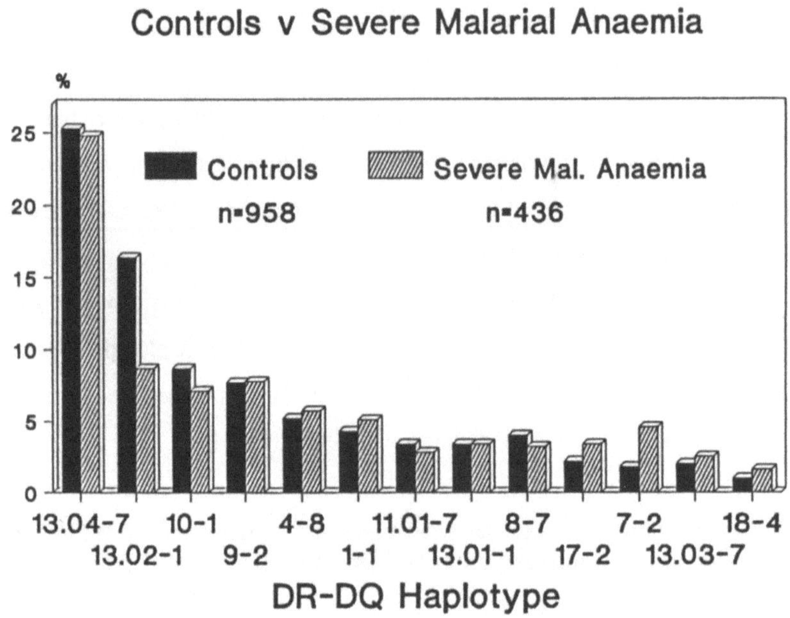

Fig. 1. HLA class II (DR-DQ) haplotype gametic frequencies in mild controls and cases of severe malarial anaemia in the Gambian study.

There was no clear association with cerebral malaria (Fig. 2) but homozygotes for this haplotype may enjoy significant protection (see below). An unusual feature of this HLA-disease association study was that the possible confounding effects of several other variables was considered in the final logistic regression analysis. None of these variables, in particular the ethnic group affiliation of the cases and controls, made any significant contribution to the observed associations.

In the case of both the HLA class I and class II associations the HLA effects were clearly seen only with cases of severe disease, although the mild cases always were intermediate in frequency between the severe cases and controls. Interestingly, a separate control group of children hospitalised with severe, mainly infectious, illnesses other than malaria also showed a small shift in the frequency of the protective alleles towards the severe malaria group frequencies. This fits very well with the evidence that some of the mortality caused by malaria is indirect: those with repeated malarial infections, perhaps through malaria-induced immunosuppression, are more likely to die from other severe infections (Marsh and Greenwood 1986).

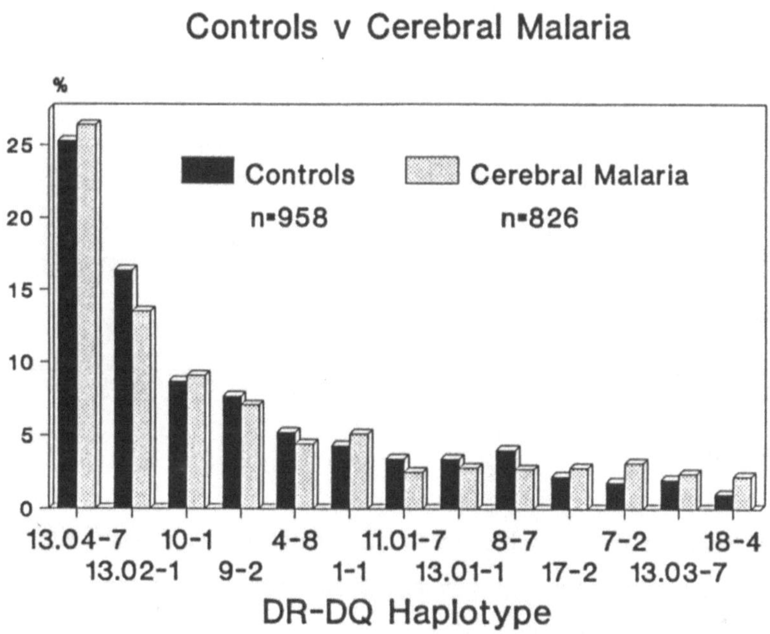

Fig. 2. HLA class II (DR-DQ) haplotype gametic frequencies in mild controls and cases of cerebral malaria in the Gambian study.

Comparison of the amount of protection associated with these HLA alleles to the effects of the African haemoglobinopathies that are associated with resistance to malaria (Table 3), shows that the haemoglobin S carrier enjoys more protection than individuals with either of these HLA alleles. Indeed the highest selection coefficient observed for any HLA allele, for HLA-Bw53, is just 0.028. However, because the HLA alleles are much more common in this population than haemoglobin S, a preventive fraction calculation (Bengtsson and Thomson 1981) indicates that at least as many cases of severe malaria are being prevented by these HLA antigens being in the population as are prevented by the presence of haemoglobin S (Hill et al. 1991).

Table 3. Degree of protection against severe malaria afforded by various alleles in The Gambia. The DRB1*1302 figure applies to protection against severe malarial anaemia. The α thalassaemia figure is an estimate from our unpublished data.

	Protection against severe malaria	
	Carrier rate	% Protection
Haemoglobin AS	13%	92%
HLA-DRB1*1302 haplotype	29%	55%
HLA-Bw53	25%	41%
α Thalassaemia	23%	<5%

Time for selection

One of the reservations about searching for an HLA association with malaria was that the high mortality form malaria produces such a strong selection pressure. Thus, no strongly protective alleles might be seen because these should rapidly go to fixation and, likewise, malaria susceptibility alleles should rapidly be eliminated. The degree of protection afforded to children with the HLA types we have identified allows some estimates to be made of the time it would take for these alleles to reach their present-day frequencies under natural selection by malaria, assuming, for the moment, constant directional selection. To maintain the balanced Hb S polymorphism at a carrier rate of 13% in The Gambia the fitness of carriers must be 7% greater than that of individuals with Hb A (Cavalli-Sforza and Bodmer 1971). Taking this difference in fitness as an indicator of the historical death rate from malaria, an estimate compatible with contemporary surveys (Greenwood et al. 1987), it can be calculated that to reach the current antigen frequency of HLA-Bw53 (25%) from an initial background allele frequency of, say, 0.001% would take about 7000 years (Cavalli-Sforza et al. 1971; Nei 1987) with directional selection and a selection coefficient of 0.028 (0.07 x [1-Relative Risk]). Because the allele is rare for most of this time, the estimates with dominance and semidominance are very similar (Nei 1987). Starting at the 0.5% allele frequency which is found in parts of Europe today would take approximately 2500 years. Although we feel that selection pressures are unlikely to be so constant, it is interesting that either of these figures is compatible with estimates of the length of time malaria has been a major cause of childhood mortality in Africa: it has been suggested that, prior to the start of farming in Africa about 4-7000 years ago, the population density may have been insufficient to maintain significant malarial transmission (Bruce-Chwatt 1988). A similar calculation for the DRB1*1302 haplotype is complicated by the lack of information on its frequency in different areas and uncertainty as to the relative importance of severe malarial anaemia and cerebral malaria historically, but it is interesting that HLA-DRB1*1302 has now reached about the same antigen frequency in The Gambia as HLA-Bw53. Although cerebral malaria has been the focus of more clinical research, because of its high in-hospital mortality and its importance in non-immune adults, it is possible, judging by the pattern of childhood illnesses in African countries with the greatest malarial endemicities, that more childhood deaths have been caused historically by severe malarial anaemia (Greenwood et al. 1991).

So, although this simple model of constant selection pressures is probably unrealistic, the example shows that even with the current intensity of malarial selection the time to fixation of beneficial alleles is not negligible.

Overdominant, frequency-dependent or fluctuating selection?

One type of evidence for the action of natural selection on HLA alleles is the observation that the number of observed alleles is too large to have been maintained in a finite population without some sort of balancing selection pressure. The nature of this balancing selection for HLA alleles has recently been of particular interest. As outlined above (Table 1) there are different types of selection mechanisms which can maintain a large number of alleles in finite populations. The best studied is overdominant selection, in which heterozygotes are fitter than any homozygote. Frequency-dependent selection, in which the fitness of an allele or genotype is dependent on its frequency may also maintain diversity when fitness increases with decreasing frequency (negative frequency-dependence). Finally, it has been shown that fluctuating selection pressures, at least in certain circumstances, can lead to increased genetic diversity in a population. This fluctuation may be related to either spatial or temporal variation in selection coefficients.

In their paper analyzing the striking excess of codon changes in the peptide binding site of HLA molecules Hughes and Nei (1988) interpreted this as evidence of overdominant selection. They favoured this mechanism apparently because an adequate model of the ability of frequency-dependent selection to maintain diversity in finite populations, incorporating the effects of mutation and genetic drift, was not available. Takahata and Nei (1990) addressed this problem, using computer simulations to assess the capacity of overdominant and three types of frequency-dependent selection to preserve a diversity of alleles over long time periods. Perhaps their most important finding is that the equations describing allele frequency dynamics under overdominant selection and under the most plausible type of frequency dependent selection, rare-allele advantage, are identical: a result generalised by Denniston and Crow (1990). Takahata and Nei (1990) found that the other two types of frequency-dependent selection, the 'molecular mimicry' and 'pathogen adaptation' models, (the latter, in any case, not really frequency-dependent) appeared less powerful at retaining allelic diversity. However, this conclusion must remain tentative until a larger range of parameters, particularly of effective population size, are explored.

In view of the identical allele frequency dynamics of the overdominant and rare-allele advantage models, Takahata and Nei (1990) concluded that experimental observations were required to distinguish between these. However, their particular suggestion, to assess the fitness of homozygotes for rare alleles, is almost certainly currently impractical. At least in a case-control study, the sample size required would be of the order of hundreds of thousands. Another possibility that may be generally applicable to case-control studies of HLA and infectious diseases is to compare the frequencies of HLA heterozygotes in the disease and control groups. If heterozygosity is advantageous then there should be a lower percentage of heterozygotes in the disease group. Applying this approach to the Gambian malaria study data we find that the opposite appears to be the case. Amongst the children with severe malarial anaemia there was a *higher* proportion of HLA class II heterozygotes than in the controls. Similarly, for the class I locus, HLA-B, there were *more* heterozygotes amongst the children with severe malaria (Table 4).

Another approach is to look at the children with protective alleles and to ask whether carriers of this allele or homozygotes for it enjoy more protection. With simple overdominant selection the heterozygous carriers should be more fit then the homozygotes. Yet, for the protective HLA class II allele, DRB1*1302, homozygotes had significantly lower relative risks (were more protected) than carriers of developing severe malarial anaemia or cerebral malaria (Table 5). For HLA-Bw53, with a smaller sample size, no significant difference was found. So, this first large data set on a protective HLA association with a parasitic disease does not appear to support the overdominance model.

Table 4. Heterozygosity rates for HLA DR-DQ class II haplotypes and HLA-B class I antigens in the various groups in the Gambian severe malaria study (Hill et al. 1991). For the DR-DQ haplotypes the heterozygosity rate is significantly higher in the severe malarial anaemia group compared to the mild controls, and for HLA-B antigens significantly higher in the severe malaria group than in the control children (both $p < 0.05$).

Heterozygosity rates			
DR - DQ Haplotypes			
SM Anaemics	Cerebrals	Mild controls	Adults
93.1%	85.5%	87.5%	86.8%

HLA - B Alleles		
Severe malaria	Control children	Adults
82.7%	74.5%	78.4%

Table 5. Relative risks of developing either severe malarial (S.M.) anaemia, cerebral malaria, or either of these manifestations of severe malaria, in children heterozygous and homozygous for the protective HLA class II (DRw13.02) and class I (Bw53) alleles (from Hill el al. 1991).

	Relative risks for heterozygotes and homozygotes		
	DRw13.02 in S.M. anaemia	DRw13.02 in cerebral malaria	HLA-Bw53 in all severe malarias
Heterozygotes	0.55	0.94	0.57
Homozygotes	0.23	0.36	1.03

On the other hand, it can be countered that the above observations apply only to malaria, and that a protective association with different alleles for different major infections would lead to overall overdominance. This is true, but remains to be demonstrated. Furthermore, the known genetic diversity of *P. falciparum*, which may be greatest in the immunodominant regions of major parasite surface proteins (Good et al. 1988), implies that we could well have observed an overdominant pattern for malaria alone. A potentially tractable approach to this general issue might be to compare HLA heterozygosity rates in children with major infections of all types to controls: the slightly lower class II heterozygosity amongst the severely ill non-malaria children (83% v 87% in controls) in the Gambian study (Hill et al. 1991) suggests that this might be worthwhile.

The third potential mechanism of maintaining allelic diversity, fluctuating selection, has not featured in discussions of HLA polymorphism. In view of the selective effect of malaria and, potentially, other major infectious pathogens on HLA alleles this may well be unjustified. Because infectious pathogens frequently appear in epidemics, often of novel antigenic type, temporally-varying selection coefficients are likely to be common. Seasonal variation in mortality from malaria and many other infections is found in much of the tropical world. The distribution of pathogens also, of course, varies spatially, so this factor also needs to be incorporated into realistic models. Dempster (1955) appears to have been the first to find that variable selection could maintain polymorphism without (arithmetic mean) overdominance, and Haldane and Jayakar (1963) demonstrated that temporally-varying selection coefficients could lead to a permanent polymorphism which, although not strictly stable in that allele frequencies vary all the time, is stable in the wider sense of being protected from allele extinction. This happens provided that the geometric mean fitness of the heterozygote is highest over time. Fluctuating selection pressures were incorporated as a central feature of Gillespie's (1978) general model to account for enzyme variation in natural populations.

The neglect in discussions of HLA polymorphism may relate to the difficulty of generating an adequate stochastic, as opposed to deterministic, model of fluctuating selection. Hedrick (1974) and Takahata (1981) showed that in finite populations rare alleles were lost more quickly when selection pressures fluctuated randomly, leading to *reduced* genetic diversity. However, there are certain situations in which the 'holding power' of fluctuating selection overcomes the tendency to lose rare alleles (Karlin and Lieberman 1974; Takahata

1981; Gillespie 1985), which seems to justify exploring models of HLA polymorphism and fluctuating selection by pathogens in more detail. In particular, there are a large number of fairly frequent HLA alleles, so that the tendency to lose rare ones may be less relevant, and cyclic variation, which may be more effective than random changes in selection coefficients in maintaining variation (Hedrick 1974), is probably a common feature of host parasite interactions.

When more than one genetic locus is subject to selection through parasite resistance new, and potentially powerful, security against allele loss comes into play (Wills 1978; Hamilton et al. 1978; Karlin and Campbell 1981; Hamilton et al. 1990). Polymorphism is preserved when pathology due to more than one infection is superadditive or, in what is almost an equivalent assumption, multiply susceptible persons tend to compose a truncated fraction. Such models, of multilocus resistance with numerous other infectious pathogens about, may be particularly relevant to malaria and selection in Africa. In effect, the idea here is that people carrying alleles which are currently bad for malaria may be carrying alleles which protect against some other infections. Due to the combination the bearer escapes the truncated section of the population and survives, and the allele is preserved. Permanence or continued *slow* decline in frequency depends on whether the selection is cyclic and reverses, or continues. In this view, the HLA alleles discussed here are probably seeing a long period of positive selection. So, these more complex, but plausible, models of host-parasite interactions offer a way around the difficulty that finding *common* protective alleles appears to pose for frequency-dependent selection.

Finally, it should be noted that these are not mutually exclusive mechanisms: all may operate in different circumstances.

Future directions

It will now be of interest to determine to what extent the HLA associations found in The Gambia are seen elsewhere. Considerable genetic diversity exists amongst wild isolates of *P. falciparum* (Creasey et al. 1990) and some of this diversity appears to have evolved in response to selection pressure by the host immune system (Good et al. 1988). Geographical variation in the allelic frequencies of immunodominant malarial antigens could lead to different HLA associations in different populations. With collaborators from Kenya and

Oxford a large case-control study of HLA and childhood malaria in a Kenyan coastal population is now in progress, to address this issue. Temporal variation, as predicted by models of frequency-dependent and fluctuating selection, may also exist but the available data argue against recent parasite antigenic shift in The Gambia (D. Conway and J.S. McBride, pers. comm.)

Recent progress in documenting genetic diversity and identifying important antigens in *P falciparum*, together with our growing understanding of the molecular basis of the host immune response, provide an exceptional opportunity for studying parasite-host coevolution at a molecular level. For example, genetic variation in the circumsporozoite protein of *P falciparum* appears to have been driven by immune selection (Good et al. 1988) and studies of the interaction of particular parasite variants with host HLA molecules should be fruitful.

As malaria is the first fatal infectious disease for which HLA antigens have been studied on this scale with molecular typing, it remains to be seen whether most other major pathogens of man have HLA associations. It is too late to look at smallpox and it may become impossible to study other major pathogens unless studies are undertaken soon. Given the potential these studies have for understanding the contribution infectious diseases have made to the evolution of HLA and other genetic loci, it would be a shame to miss the opportunity.

Acknowledgments. Some 40 collaborators, assistants and advisers who contributed to the study in The Gambia are listed in the recent manuscript: this paper reflects their work. In addition I am grateful to W. D. Hamilton and R. M. Harding for discussions and comments on this manuscript. AVSH is a Wellcome Trust Senior Clinical Fellow.

References

Bengtsson, B.O. and Thomson, G.: Measuring the strength of associations between HLA antigens and diseases. *Tissue Antigens 18:* 356-363, 1981

Bjorkman, P.J. and Parham, P.: Structure, function and diversity of class I major histocompatibility complex molecules. *Ann Rev Biochem 59:* 253-288, 1990

Bjorkman, P.J., Saper, M.A., Samraoui, B., Bennet, W.S., Strominger, J.L., and Wiley, D.C.: The foreign antigen binding site and T cell recognition regions of class I major histocompatibility antigens. *Nature 329:* 512-518, 1987

Briles, W.E., Stone, H.A., and Cole, R.K.: Marek's disease: effects of B histocompatibility alloalleles in resistant and susceptible chickens. *Science 195:* 193-195, 1977

Bruce-Chwatt, L.J. *In* W.H. Wernsdorfer and I.A. McGregor (eds.): *Malaria: Principles and*

Practice of Malariology, pp. 1-59. Churchill-Livingstone, Edinburgh 1988

Cavalli-Sforza, L.L. and Bodmer, W.F.: *The Genetics of Human Populations*, pp. 133-184. W.H. Freeman Co., San Francisco 1971

Clarke, B. and Kirby, D.R.S.: Maintenance of histocompatibility polymorphisms. *Nature 211:* 999-1000, 1966

Creasey, A., Fenton, B., Walker, A., Thaitong, S., Oliveira, S., Mutambu, S., and Walliker, D.: Genetic diversity of *Plasmodium falciparum* shows geographical variation. *Am J Trop Med Parasitol 42:* 403-413, 1990

de Vries, R.R., Khan, P.M., Bernini., L.F, van Loghem, E., and van Rood, J.J.: Genetic control of survival in epidemics. *J Immunogenet 6:* 271-287, 1979

de Vries, R.R., Schreuder, G.M.T., Naipal, A., D'Amaro, J., and van Rood, J.J.: Selection by typhoid and yellow fever epidemics witnessed by the HLA-DR locus. *In* B. Dupont (ed.): *Immunobiology of HLA*. Springer-Verlag, New York 1989

Dempster, E.R.: Maintenance of genetic heterogeneity. *Cold Spring Harbor Symp Quant Biol 20:* 25-32, 1955

Denniston, C. and Crow, J.F.: Alternative fitness models with the same allele frequency dynamics. *Genetics 125:* 201-205

Doherty, P.C. and Zinkernagel, R.M.: A biological role for the major histocompatibility antigens *Lancet i:* 1406-1409, 1975

Ferstl., R., Eggert, F., Pause, B. et al.: Immune system signalling to the others brain: MHC-specific body scents in humans. *6th International Symposium on Neuronal Control of Bodily Function: Basic and Clinical Aspects*. Hogrisi, Toronto 1989

Figueroa, F., Gunther E., and Klein, J.: MHC polymorphism pre-dating speciation. *Nature 335:* 265-267, 1988

Flaherty, L.: Major histocompatibility complex polymorphism: a nonimmune theory for selection. *Hum Immunol 21:* 3-13, 1987

Gillespie, J.H.: The interaction of genetic drift and mutation with selection in a fluctuating environment. *Theor Popul Biol 27:* 222-237, 1985

Gillespie, J.H.: A general model to account for enzyme variation in natural populations. V. The SAS-CFF model. *Theor Popul Biol 14:* 1-45, 1978

Good, M.F., Berzofsky, J.A., and Miller, L.H.: The T cell response to the malaria circumsporozoite protein: an immunological approach to vaccine development. *Ann Rev Immunol 6:* 663-688, 1988

Greenwood, B.M., Marsh, K., and Snow, R.W.: Why do some African children develop severe malaria? *Parasitol Today* (in press 1991)

Greenwood, B.M., Bradley, A.K., Greenwood, A.M., Byass, P., Jammeh, K., Marsh, K., Tulloch, S., Oldfield, F.S., and Hayes, R.: Mortality and morbidity from malaria among children in a rural area of The Gambia, West Africa. *Trans Roy Soc Trop Med Hyg 81:* 478-486, 1987

Haldane, J.B.S. and Jayakar, S.D.: Polymorphism due to selection of varying direction. *J Genet 58:* 237-242, 1963

Hamilton, W.D., Henderson, P.A., and Moran N.A.: Fluctuation of environment and coevolved antagonist polymorphism as factors in the maintenance of sex. *In* R.D. Alexander and D.W. Tinkle (eds.): *Natural Selection and Social Behaviour: Recent Research and Theory*, pp 363-381. Chiron, New York 1981

Hamilton, W.D., Axelrod, R., and Tanese, R.: Sexual reproduction as an adaotation to resist parasites. *Proc Natl Acad Sci USA 87:* 3566-3573, 1990

Hedrick, P.W.: Genetic variation in a heterogeneous environment. I. Temporal heterogeneity and the absolute dominance model. *Genetics 78:* 757-770, 1974

Hedrick, P.W. and Thomson, G.: Maternal-fetal interactions and the maintenance of HLA polymorphism. *Genetics 119:* 205-212, 1988

Hill, A.V.S., Allsopp, C.E.M., Kwiatkowski, D., Anstey, N.M., Twumasi, P., Rowe, P.A., Bennett, S., Brewster, D., McMichael, A.J., and Greenwood, B.M.: Common West African HLA antigens are associated with protection from severe malaria. *Nature* (in press 1991)

Hughes, A.L. and Nei, M.: Pattern of nucleotide substitution at major histocompatibility class I loci reveals overdominant selection. *Nature 335:* 167-170, 1988

Karlin, S. and Campbell, R.B.: The existence of a protected polymorphism under conditions of soft as opposed to hard selection in a multi-deme population system. *Am Nat 117:* 262-275, 1981

Karlin, S. and Lieberman, U.: Temporal fluctuations in selection intensities: case of large population size. *Theor Popul Biol 6:* 355-382, 1974

Klein, J.: Origin of major histocompatibility complex polymorphism: the trans-species hypothesis *Hum Immunol 19:* 155-162, 1987

Lawlor, D.A., Ward, F.E., Ennis, P.D., Jackson, A.P., and Parham, P.: HLA-A, and B polymorphism predate the divergence of humans and chimpanzees. *Nature 335:* 68-271, 1988

Marsh, K. and Greenwood, B.M.: The immunopathology of malaria. *Clinics Trop Med Comm Dis 1:* 91-125, 1986

Nei, M.: *Molecular Evolutionary Genetics,* pp. 335-339. Columbia University Press, New York 1987

Serjeantson, S.W.: The reasons for MHC polymorphism in man. *Transplant Proc 21:* 598-601, 1989

Snell, G.D.: The H-2 locus of the mouse: observations, and speculations concerning its comparative genetics, and its polymorphism. *Folia Biol 14:* 335-358, 1968

Takahata, N. and Nei, M.: Alleleic genealogy under overdominant, and frequency-dependent selection, and polymorphism of the major histocompatibility complex. *Genetics 124:* 967-978, 1990

Takahata, N.: Genetic variability, and rate of gene substitution in a finite population under mutation, and fluctuating selection. *Genetics 98:* 427-440, 1981

Wills, C.: Rank order selection is capable of maintaining all genetic polymorphism. *Genetics 89:* 403-417, 1978

Yamazaki, K., Boyse, E.A., Mike, V., Thaler, H.T., Mathieson, B.J., Abbott, J., Boyse, J., Zayas, Z.A., and Thomas, L.: Control of mating preferences in mice by genes in the major histocompatibility complex. *J Exp Med 144:* 1324-1335, 1976

Zavazava, N., Westphal, E., and Mueller-Ruchholtz, W.: Characterization of soluble HLA molecules in sweat and quantitative HLA differences in serum of healthy individuals. *J Immunogenet 17:* 387-394, 1990

THE EVOLUTION OF MHC-BASED MATING PREFERENCES IN *MUS*

Wayne K. Potts, C. Jo Manning, and Edward K. Wakeland
Laboratory of Molecular Genetics, Department of Pathology and Laboratory Medicine, University of Florida, Gainesville, FL 32610, USA

There is overwhelming evidence that Mhc *genetic diversity is maintained by selection. Due to the immune recognition function of* Mhc *genes, it has been generally assumed that this selection is disease-based, although* Mhc-*related reproductive mechanisms (abortion or mating preferences) have also been proposed. In a recent study designed to detect all of the proposed* Mhc *diversity-maintaining mechanisms only* Mhc-*based mating preferences were found. Here we briefly review results of* Mhc-*based mating preference studies, the associated olfactory mechanisms, and the generality and possible function of these preferences.*

The unprecedented genetic diversity found at *Mhc* loci is perhaps the most unusual feature of this genetic complex. It is therefore surprising that, despite remarkable progress in understanding the molecular details of how Mhc molecules function during immune recognition (Babbitt et al. 1985; Roy et al. 1989; Kappler et al. 1987), we still do not understand the forces that maintain *Mhc* genetic diversity. These selective forces will not simply be esoteric parameters in population genetic models, rather, they are the forces that dictate how *Mhc* genotype influences the fitness of individuals. In other words, they will define the functional significance of *Mhc* genetic diversity. The conventional assumption about what maintains *Mhc* diversity has been that *Mhc* genotype influences fitness through disease susceptibility, and the resulting pathogen-driven selection (e.g. heterozygote advantage) is responsible for the maintenance of *Mhc* polymorphisms (Doherty and Zinkernagel, 1975). But there are three problems with this assumption. First, it has not been confirmed experimentally. Second, there are negative data, primarily the general failure to find associations between *Mhc* genotype and susceptibility to infectious disease (Klein 1986; Tiwari and Terasaki 1985). Finally, there are viable alternative hypotheses which primarily include the preferential abortion of histocompatible fetuses (Hamilton and Hellstrom 1978; Thomas et al. 1985; Palm 1974) and *Mhc*-based mating preferences (Yamazaki et al. 1976; Egid and Brown 1989). These alternative hypotheses raise the intriguing possibility that the functional significance of *Mhc* genetic diversity may be non-immunological.

The idea that the functional significance of *Mhc* genetic diversity could be separate or independent of the known immune-recognition function of these genes is often confusing.

NATO ASI Series, Vol. H 59
Molecular Evolution of the
Major Histocompatibility Complex
Edited by J. Klein and D. Klein

But, consider the following example. The function of the human Hemoglobin (*Hb*) locus is to transport O_2 and CO_2. However, the functional significance of certain genetic polymorphisms found at the *Hb* locus is primarily involved in providing resistance to malaria. This secondary function of malarial resistance is what maintains the sickling allele and, despite its negative impact on fitness in the homozygous condition, is what maintains this genetic polymorphism at the *Hb* locus. Similarly, a secondary function could be involved in maintaining *Mhc* genetic diversity.

In this paper we will review the recent evidence that, at least in *Mus*, *Mhc*-based mating preferences are a major force maintaining *Mhc* genetic diversity. We will compare the competing hypotheses for what maintains these mating preferences.

Olfactory discrimination of *Mhc* genotype

The olfactory basis for *Mhc* discrimination has been well documented. Mice, rats, and humans can smell the difference between urine from Mhc congenic mice that differ genetically only at *Mhc* loci (Beauchamp et al. 1990). Both male and female mice can be trained to make these discriminations with urine from either male or female mice (Boyse et al. 1987). Either class I or class II loci can confer appropriate differences, and loci that differ by only a single amino acid are discriminable (Yamazaki et al. 1990). However, not all differences tested confer a discriminable odor difference. Under a different testing paradigm, rats have also been shown to be able to discriminate urine from both class I and class II congenic rats (Brown et al. 1990).

Investigators attempting to find the chemical basis for the Mhc-mediated odor differences in the urine of mice and rats have taken several different routes. Yamazaki et al. (Yamazaki et al. 1985) used the approach of trying to alter the Mhc-related odor profile of urine by altering the hematopoietic cells of the mice. They used radiation chimeras created by reconstituting lethally irradiated inbred mice with bone marrow transplants from *Mhc* congenic F_1 hybrid donors. (Hybrids were used to prevent graft vs host disease.) They found that the chimeras were judged, by trained mice, to have the scent typical of the *Mhc* genotypes of the donor mice. This suggests that the hematopoietic system contributes to the Mhc-related odor profile.

Singh et al. (1990) tried altering the Mhc-related odor profile by altering the bacterial fauna of rats. Rats couldn't discriminate the urine odors of Mhc congenic rats raised in a germ-free environment. They could, however, discriminate between these same Mhc congenic rats after they had been removed from the germ-free environment. Similar experiments in mice have, so far, failed to yield the same results (Gary Beauchamp, personal communication).

On a more molecular level, Singh et al. (1988) found Mhc metabolites in urine that may play a role in influencing the volatile compounds that are responsible for Mhc-related odors in urine. Svende et al. (1984) used gas chromatography to show that differences in the chemical profiles of mice correlated with *Mhc* genetic differences. While they found no volatile compounds specific to the *Mhc* genes, differences involved a large range of secondary volatile metabolites. Although the exact mechanism of the olfactory discrimination is as yet unknown, it seems clear that mice have the ability to discriminate between mice or the urine of mice that differ only at *Mhc* loci. Understanding how these Mhc-related olfactory abilities are used in nature has been a more difficult problem.

Mate choice experiments with inbred strains

Male mate choice. Male mate choice experiments involving three pairs of *Mhc* congenic strains (tested in both directions, giving six combinations) have resulted in four of the six combinations preferentially mating with the Mhc disparate females (Yamazaki et al. 1976). One strain preferred the syngeneic females while the other strain chose randomly. However, in later tests where males were raised in isolation from other strains, the strain that had previously exhibited random mating showed a preference for females of the congenic strain (Yamazaki et al. 1978). Subsequent cross-fostering experiments showed that the expected preferences of males could be reversed by cross-fostering them to parents of the congenic strain (Yamazaki et al. 1988). This experiment confirms not only that early imprinting is involved, but also that it is truly male choice, and that female preferences are not dictating the choices observed in these experiments.

Female mate choice. Theoretically, female choice is expected to be more prevalent and stringent than male mate choice because females pay a higher cost when they produce sub-optimal offspring from a given mating. Males in general are able to inseminate more females, so their strategy is more one of quantity than quality. Female mammals, however, must pay the costs of gestation and lactation, and their investment in future litters is constrained while rearing the present litter, so the number of offspring they can produce in a lifetime is limited. Consequently, theory predicts that female mate choice in *Mus* should be at least as strong as the preferences exhibited by males. Laboratory female mate choice experiments have not always supported this prediction.

Female mate choice experiments using B6(bb) and B6-$H\text{-}2k^k$ (kk) congenics have shown a non-significant trend for both types of females to prefer bb males (Beauchamp et al. 1988; C. Stahlbaum, personal communication). Cross-fostering did not effect female preferences in these experiments (Beauchamp et al. 1988). Egid and Brown (1989) have shown female mating preferences in laboratory studies using two inbred strains

(B10.CHR51 and B10.GAA37), whose *Mhc* genotypes had been recently extracted from the wild. Females of both strains preferred the male of the opposite strain.

Problems with laboratory experiments using inbred strains. There are several potential problems with drawing conclusions about natural populations from results of laboratory studies using inbred strains of mice (Partridge 1988; Nei and Hughes 1991; Potts and Wakeland 1990). The first problem involves the nature of the inbred strains themselves. They are created by breeding brothers and sisters for several generations. The animals with the strongest tendencies toward disassortative mating preferences will fail to breed under these incestuous circumstances, thereby artificially selecting against the trait under study. This, compounded by differential male sexual vigor, may account for the general "preference" for bb males and for the failure to influence female choice through cross-fostering, reported by Beauchamp et al. (1988). Data from our lab using B10 (bb) and B10.A (aa) congenics shows that B10 females do not show a preference for either male. Males get, on average, the same numbers of intromissions, with females simply going back and forth between the two males. However, in seven out of eight trials, B10 males ejaculated first, giving the illusion that B10 females preferred to mate with B10 males.

Second, individuals of *Mhc* congenic strains are monomorphic except for *Mhc*. A disassortative mating preference may, in natural populations, rely on many genetic differences other than *Mhc*. The real contribution of *Mhc*, both to individual odor differences and, consequently, to mating preferences, may be minor. But in experiments using *Mhc* congenic strains, the only genetic component of odor preferences will be *Mhc*-related, artificially inflating the importance of *Mhc*.

Third, the results of the male mate choice experiments (Yamazaki et al. 1976) have been somewhat inconsistent. Not all strain combinations show Mhc-based disassortative mating preferences. Only four (or five) out of six combinations mated disassortatively, while the sixth mated assortatively. This does not demonstrate a statistically significant disassortative mating preference, although it is usually cited as such.

Fourth, being visual animals, we have very little understanding of the *Umwelt* (perceptual world) of nocturnal, olfactory animals. What cues are mice attending to in the lab? Does a cage constitute a family unit, a shelf a deme, etc? Social context appears to influence the behavior of these animals (Hurst 1989), so we must understand these problems before we can extrapolate our laboratory findings to natural populations. Even if we understood these parameters, the artificiality of laboratory conditions makes it impossible to accurately estimate the strength of mating preferences in nature from laboratory results. Finally, the criterion most commonly used in mate choice experiments is based on the untested assumption that the preferred mate is the first mated. For males, this may be an appropriate criterion if he can only successfully inseminate one female (Dewsbury 1982;

(Dewsbury 1982; Huber et al. 1980). For females, the case may be very different. In our semi-natural population studies (Potts et al. 1991), at least half of all litters are sired by more than one male. There may be many reasons why females mate multiply. One compelling reason may be to obscure paternity and prevent infanticide (Labov et al. 1985). This does not mean that females lose control over paternity. Evidence from golden hamsters (Huck et al. 1989), rats (Spinka 1990), and deer mice (Dewsbury 1988) indicates that the outcome of sperm competition is altered by the timing of matings relative to ovulation. Female *Mus* may be able to manipulate the timing of matings to insure that some or most of her offspring are sired by the preferred male. Experiments to test this hypothesis are under way. In summary, Mhc-related mate choice experiments under laboratory conditions suffer serious problems making it difficult, or impossible, to extrapolate such findings from laboratory studies to natural populations. Laboratory results require clarification and confirmation in natural populations.

Experiments in a semi-natural enclosure

In a recent attempt to directly measure the nature and strength of selection operating on *Mhc* genes in semi-natural, captive populations of *Mus*, we found evidence of strong Mhc-based mating preferences (Potts et al. 1991). We were unable to detect any significant effect of either Mhc-related disease susceptibility or selective fertilization or abortion. This study was conducted on nine independent populations of half-wild individuals with known *Mhc* genotypes. These mice came from generations three and six of original crosses between locally caught wild mice and four different inbred strains (C57BL/10, BALB/c, B10.BR and DBA/2, carrying *Mhc* haplotypes *b, d, k,* and *q*, respectively). In the F_2 generation only mice homozygous for the inbred-derived *Mhc* were used to continue this outbred colony. Consequently, half of the genome was wild-derived and the other half was derived from one or more inbred strains, but the *Mhc* region was always derived from the inbred strains.

There were generally 24 Mhc-typed founders for each of the nine populations (8 males, 16 females) with approximately equal numbers of *Mhc* homozygotes and heterozygotes. The frequencies of the four *Mhc* haplotypes varied among the nine populations. Each animal was given a unique ear punch which allowed individual identification with the use of binoculars. This allowed us to observe and record a variety of social behaviors, including the results of aggressive encounters and identification of who was mating with whom. When offspring neared weaning age (14-21 days), they were ear punched and a 2 cm tail biopsy taken, from which DNA was extracted and used for *Mhc* genotyping by RFLP analysis. The populations were terminated after approximately 150 pups were born. At that time embryos were collected from pregnant females and *Mhc* genotyped.

Important aspects of this study are summarized in Figures 1 and 2. Figure 1 compares the departure from expected haplotypic frequencies for each haplotype across all nine populations. (All expected frequencies in Figs. 1 and 2 were calculated based on random mating between all females and only territorial males.) Although there were significant departures from expected within populations, the mean departure from expected for each haplotype was relatively small, between 2.6% and 3.1%. These results indicate that no particular haplotype experienced systematic gains or losses across the nine populations.

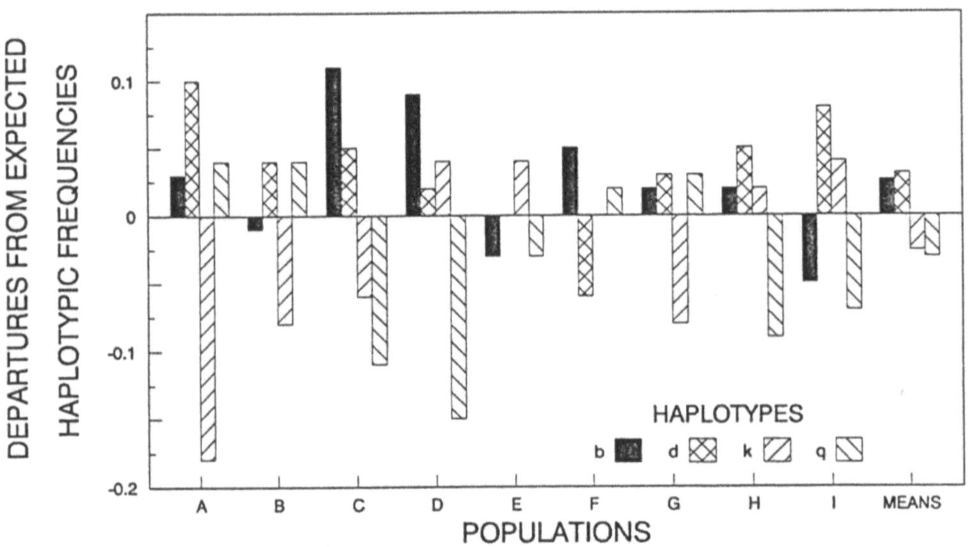

Fig. 1. Departures from expected haplotypic frequencies across nine populations of *Mus*. Departure values were calculated as observed proportions minus expected proportions for each of the four haplotypes.

The only consistent and strong departure from expected was found when the genotypic classes of heterozygotes and homozygotes were compared. All nine populations showed a deficiency of *Mhc* homozygous progeny relative to expected (Fig. 2). The mean deficiency was 27%, which was significant at the $p < 0.0001$ level. This study further demonstrated that this 27% deficiency of *Mhc* homozygous progeny was primarily due to Mhc-based mating preferences, exercised predominantly by females. This study looked for other forms of selection operating directly on *Mhc* genotype, specifically adult fitness, early pup mortality (prior to weaning) and selective abortion (or fertilization), but no significant effects were detected. These results suggest that Mhc-based mating preferences may be the primary selective force operating on *Mhc* genes in *Mus* populations. However, since the populations in this study were observed for only one generation, we could not rule out the possibility that episodic disease factors play a role, but it is unlikely that any viability selection will be as strong as the sexual selection seen in this study.

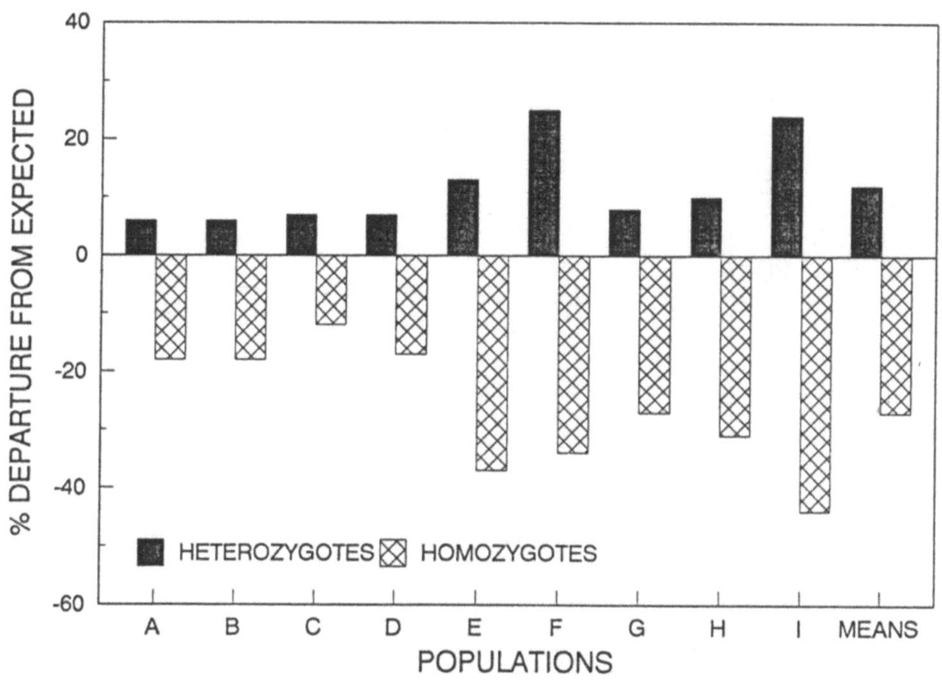

Fig. 2. Percent departure from expected genotypic frequencies of heterozygotes and homozygotes. Departure values were calculated as (observed minus expected)/expected.

Can mating preferences account for *Mhc* genetic diversity?

The precise nature of these Mhc-based mating preferences could not be deduced from this study, although the general nature must involve either a preference for rare genotypes (or haplotypes) or some form of disassortative mating preference. It has generally been assumed that Mhc-related preferences are disassortative (Beauchamp et al. 1988). Some forms of disassortative mating preferences have similar population dynamics to overdominance (heterozygote advantage; Karlin and Feldman 1968). For example, mating patterns that reduce the probability that *Mhc* homozygotes are conceived produces similar population dynamics as mortality patterns that reduce the probability that *Mhc* homozygotes survive to breed. These similarities allow us to use well developed overdominant models to take a preliminary look at how effective these mating preferences may be in contributing to the maintenance of Mhc polymorphisms.

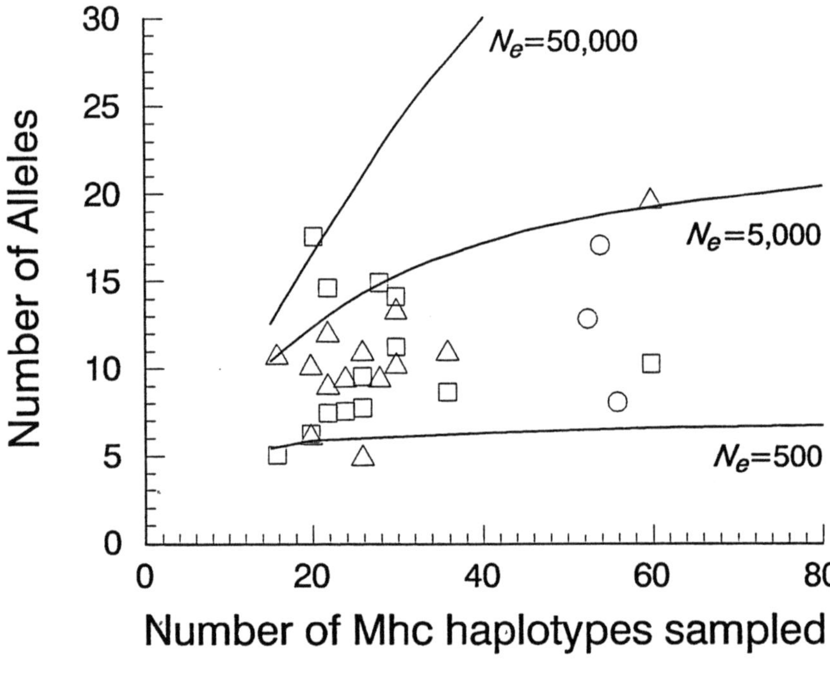

Fig. 3. The number of observed *Mhc* alleles from local *Mus* populations plotted with the expected number of alleles (lines) for three effective population sizes (N_e), under levels of disassortative mating that would yield a 27% deficiency of homozygotes. The expected values (corrected for sample size) were derived from an analytical model for symmetric overdominant selection (Yokoyama and Nei 1979), using a selection coefficient of $s = 0.27$ and a mutation rate of 10^{-6} (an empirically derived estimate for Mhc loci; Melvold and Kohn 1990). Data for class I *K* (squares) and *D* (triangles) loci are from serological analysis of 13 European populations (Nadeau et al. 1981). The total number of alleles was estimated by assuming that the allelic frequency distributions of blanks and identified alleles were the same. Data for the class II *Aβ* locus (circles) are from an RFLP study of three North American populations (Potts and Wakeland 1990).

Figure 3 plots the number of *Mhc* class I and II alleles observed in 16 natural populations (symbols), which can be compared to the expected number of alleles (lines) under levels of overdominance, or an equivalent level of disassortative mating, sufficient to produce the 27% homozygote deficiency observed in the semi-natural populations. The effective population size (N_e) for *Mus* is uncertain, and since N_e influences the number of alleles maintained, we have provided the expected number of alleles across the entire estimated range of 500 to 50 000 (Potts and Wakeland 1990). Throughout this range, disassortative mating can account for most or all of the genetic variation observed in natural *Mus* populations.

The above results suggest that Mhc-based mating preferences are currently a major force responsible for the maintenance of *Mhc* genetic diversity in *Mus*. Unlike essentially all forms of disease-based selection models, mating preferences have no associated genetic load. The genetic load associated with the overdominant models comes through the production of homozygotes which suffer lower fitness than heterozygotes. Consequently, the action of Mhc-based mating preferences would help explain the following observation that has troubled previous authors. Mhc monomorphic populations have been discovered, but no ill effects stemming from this monomorphism have been detected (Klein 1987). Klein used this observation in support of his proposal that *Mhc* genetic diversity was selectively neutral. Hughes and Nei (1988) explained the observation by postulating that disease-based selection was weak enough ($s < 0.01$) to make direct detection difficult. However, if, as suggested above, mating preferences are primarily responsible for the maintenance of *Mhc* genetic diversity, then it becomes clear why *Mhc* monomorphic populations show no ill effects, beyond dissatisfaction in love. We should point out that there are ways that strong disease-based selection could be operating on Mhc without producing visible effects in Mhc monomorphic populations (Potts and Wakeland 1990).

The generality of Mhc-based mating preferences: Who knows what the avian nose knows?

Determining the generality of Mhc-based mating preferences across vertebrates awaits experimentation, but Klein et al. (1990) has argued against the generality on the following basis. Since chickens have a poor sense of smell, but still have a polymorphic *Mhc*, then polymorphism could not be maintained by mating preferences. This argument assumes that birds cannot smell and that *Mhc* genotype only influences olfactory traits. The first assumption is demonstrably false and the second is conjecture. A variety of studies have demonstrated significant olfactory prowess in birds, including normal electrical activity of the olfactory lobe in response to odor stimuli (Tucker 1965), location of food by smell (Grubb 1972; Wenzel 1968), and odor based social preferences (Jones and Gentle 1985) and sexual preferences (Balthazart and Schoffeniels 1979). Although the sense of smell in birds may not generally be as keen as it is in mammals, it is clear that it can be keen for specific odor cues, which could certainly include Mhc-related odors.

Boyse et al. (1983) have pointed out that, from an *a priori* view, it is just as likely that *Mhc* influences visual cues as odor cues. Birds clearly use visual cues for mate choice (Bateson 1991; Zuk et al. 1990). In one of these studies (Bateson 1991) birds were able to discriminate genetic relationships among an unfamiliar panel, indicating that visual cues have a genetic basis. These results were interpreted as mating preferences for optimal outbreeding, indicating that birds are under some of the same selective pressures that we believe have been responsible for Mhc-based mating preferences in mice (see next section).

At this time very little work has been done on Mhc-based mating preferences in any species other than *Mus*. Human population studies have, as yet, yielded little evidence of human Mhc disassortative mating preferences (Tiwari and Terasaki 1985); however, high levels of heterozygosity make it difficult to detect disassortative mating preferences. Also, human behavior has been vastly complicated by cultural evolution over the past two centuries. This may render humans inappropriate animals for this kind of analysis. We conclude that any judgement about the generality of Mhc-based mating preferences across vertebrates is premature at this time.

What is the function of Mhc-based mating preferences?

It is unlikely that Mhc-based mating preferences serve no function. Two possible functions, which are not mutually exclusive, have been proposed. The first hypothesis postulates that Mhc-based mating preferences function by preferentially producing progeny with high fitness *Mhc* genotypes. These high fitness genotypes are generally thought to be Mhc heterozygotes (Doherty and Zinkernagel 1975; Nei and Hughes 1991). Mhc disassortative mating preferences would preferentially produce *Mhc* heterozygotes, thereby acting as a prezygotic filter against the production of low fitness homozygotes. If disease-based overdominant selection is operating on *Mhc* loci, Mhc-based mating preferences will be favored by natural selection because they reduce the genetic load associated with parasite-driven overdominance.

The second hypothesis postulates that Mhc disassortative mating preferences function to avoid inbreeding (Brown 1983; Partridge 1988; Uyenoyama 1988; Potts and Wakeland 1990). Many inbreeding avoidance (genetic incompatibility) systems operate by reducing the frequency of matings between individuals, or their gametes, who share alleles at one or a few polymorphic loci. The best known cases are in plants, where the growth of pollen on the stigma is inhibited if these two structures (or in some cases the parental genotype of the pollen) share an allele(s) at incompatibility (*S*) loci (Haring et al. 1990).

This enforces outbreeding at the incompatibility loci and consequently maintains high levels of polymorphism at these loci. It also increases outbreeding at the rest of the genome because allele sharing at a highly polymorphic locus will be correlated with co-ancestry. In a similar manner disassortative mating preferences based on the highly polymorphic Mhc could function primarily by reducing the impact of inbreeding depression in offspring. The correlation between Mhc sharing and co-ancestry is reflected in the way we search for Mhc compatible donors for organ transplants. Compatible donors are most common among close relatives. Reversing this process during mate choice through a preference for histoin-compatible mates, will reduce the likelihood of mating with close relatives.

The inbreeding avoidance hypothesis gains credibility from the following facts. The negative impact of inbreeding is strong enough to have been noticed for centuries (Darwin 1876), and in the only animal experiments providing reliable quantitative estimates of the inbreeding genetic load in natural populations, the fitness consequences were enormous. Enforced homozygosity for a single chromosome in *Drosophila* reduces fitness by 40% to 80% (Charlesworth and Charlesworth 1987). In contrast, the impact of *Mhc* genotype on fitness through disease susceptibility has been in general difficult to demonstrate (Tiwari and Terasaki 1985; Klein and Figueroa 1986). With a few important exceptions (e.g., Marek's Disease (Pazderka et al. 1975), malaria (Hill et al. 1991)), there have been few examples of disease resistance associating with a specific *Mhc* allele (or haplotype).

In our view it is likely that both infectious disease and genetic disease (inbreeding depression) will be involved in the evolution of Mhc-based mating preferences. Determining the relative importance of these two mechanisms will require measurements of the fitness consequences associated both with *Mhc* genotype and with inbreeding depression.

References

Babbitt, B.P., Allen, P.M., and Matsueda, G.: Binding of immunogenic peptides to Ia histocompatible molecules. *Nature 317*:359-361, 1985

Balthazart, J. and Schoffeniels, E.: Pheromones are involved in the control of sexual behaviour in birds. *Naturwissenschaften 66*:55-56, 1979

Bateson, P.: Preferences for cousins in Japanese quail. *Nature*, in press 1991

Beauchamp, G.K., Yamazaki, K., Duncan, H., Bard, J., and Boyse, E.A.: Genetic determination of individual mouse odour. In: D.W. MacDonald, D. Muller-Schwarze, and S.E. Natynczuk (eds): *Chemical Signals in Vertebrates* 5, pp. 244-254, Oxford University Press, Oxford 1990

Beauchamp, Gary K., Yamazaki, K., Bard, J. and Boyse, E.A.: Preweaning experience in the control of mating preferences by genes in the major histocompatibility complex of the mouse. *Behav Genet 18*, No.4:537-547, 1988

Boyce, E.A., Beauchamp, G.K., and Yamazaki, K.: The sensory perception of genotypic polymorphism of the major histocompatibility complex and other genes: some physiological and phylogenetic implications. *Hum Genet 6*:177-183, 1983

Boyse, E.A., Beauchamp, G.K., and Yamazaki, K.: The genetics of body scent. *Trends Genet 3*, No.4:97-102, 1987

Brown, J.L. Some paradoxical goals of cells and organisms: the role of the MHC. In: D.W. Pfaff (ed.): *Ethical Questions in Brain and Behavior*, pp. 111-124, Springer Verlag, New York 1983

Brown, R.E., Roser, B. and Singh, P.B. The MHC and individual odours in rats. In: D.W. MacDonald, D. Muller-Schwarze, and S.E. Natynczuk (eds): *Chemical Signals in Vertebrates* 5, pp. 244-254, Oxford University Press, Oxford 1990

Charlesworth, D. and Charlesworth, B.: Inbreeding depression and its evolutionary consequences. *Annu Rev Ecol Syst 18*:237-268, 1987

Darwin, C.R.: *The Effects of Cross and Self-fertilization in the Vegetable Kingdom.* John Murray, London 1876

Dewsbury, D.A.: Ejaculate cost and male choice. *Am Nat 119*:601-610, 1982

Dewsbury, D.A.: Sperm competition in deer mice (*Peromyscus maniculatus bairdi*): Effects of cycling versus postpartum estrus and delays between matings. *Behav Ecol Sociobiol 22*:251-256, 1988

Doherty, P.C. and Zinkernagel, R.M.: Enhanced immunological surveillance in mice heterozygous at the H-2 gene complex. *Nature 256*:50-52, 1975

Egid, K. and Brown, J.L.: The major histocompatibility complex and female mating preferences in mice. *Anim Behav 38*:548-549, 1989

Grubb, T.C.: Smell and foraging in shearwaters and petrels. *Nature 237*:404-405, 1972

Hamilton, M. and Hellstrom, I.: Selection for histoincompatible progeny in mice. *J Reprod Immunol 19*:267-270, 1978

Haring, V., Gray, J.E., McClure, B.A., Anderson, M.A., and Clarke, A.E.: Self-incompatibility: A self-recognition system in plants. *Science 250*:937-941, 1990

Hill, A.V.S., Allsop, C.E.M., Kwiatkowski, D., Anstey, N.M., Twumasi, P., Rowe, P.A., Bennett, S., Brewster, D., McMichael, A.J. and Greenwood, B.M.: Common West African HLA antigens are associated with protection from severe malaria. *Nature*, in press 1991

Huber, M.H.R., Bronson, F.H., and DesJardins, C.: Sexual activity of aged male mice: correlation with level of arousal, physical endurance, pathological status, and ejaculatory capacity. *Biol Reprod 23*:305-316, 1980

Huck, U.W., Tonias, B.A., and Lisk, R.D.: The effectiveness of competitive male inseminations in golden hamsters, *Mesocricetus auratus*, depends on an interaction of mating order, time delay between males, and the time of mating relative to ovulation. *Anim Behav 37*:674-680, 1989

Hughes, A.L. and Nei, M.: Pattern of nucleotide substitution at major histocompatibility complex class I loci reveals overdominant selection. *Nature 335*:167-170, 1988

Hurst, J.L.: The complex network of olfactory communication in populations of wild house mice *Mus domesticus Rutty*: urine marking and investigation within family groups. *Anim Behav 37*:705-725, 1989

Jones, R.B. and Gentle, M.J.: Olfaction and behavioral modification in domestic chicks (*Gallus domesticus*). *Physiol Behav 34*:917-924, 1985

Kappler, J.W., Roehm, N., and Marrack, P.: T cell tolerance by clonal elimination in the thymus. *Cell 49*:273-280, 1987

Karlin, S. and Feldman, M.W.: Further analysis of negative assortative mating. *Genetics 59*:117-136, 1968

Klein, J.: Origin of major histocompatibility complex polymorphism: The trans-species hypothesis. *Hum Immunol 19*:155-162, 1987

Klein, J. and Figueroa, F.: Evolution of the major histocompatibility complex. *CRC Crit Rev Immunol 6*:295-386, 1986

Klein, J. *Natural History of the Major Histocompatibility Complex.* Wiley, New York 1986

Klein, J., Kasahara, M., Gutknecht, J., and Figueroa, F.: Origin and function of Mhc polymorphism. *Chem Immunol 49*:35-50, 1990

Labov, J.B., Huck, U.W., Elwood, R.W., and Brooks, R.J.: Current problems in the study of infanticidal behavior of rodents. *Q Rev Biol 60*:1-20, 1985

Melvold, R.W. and Kohn, H.I. Spontaneous frequency of H-2 mutations. In: I.K. Egorov and C.S. David (eds): *Transgenic mice and mutants in MHC research.* Springer-Verlag, New York 1990

Nadeau, J.H., Wakeland, E.K., Gotze, D., and Klein, J.: The population genetics of the H-2 polymorphism in European and North African populations of the house mouse (*Mus musculus L.*). *Genet Res 37*:17-31, 1981

Nei, M. and Hughes, A.L.: Polymorphism and evolution of the major histocompatibility complex loci in mammals. In: R.K. Selander, A.G. Clark, and T.S. Whittam (eds): *Evolution at the Molecular Level*, pp. 222-247, Sinauer, Sunderland 1991

Palm, J.: Maternal-fetal histoincompatibility in rats: An escape from adversity. *Cancer Res 34*:2061-2065, 1974

Partridge, L.: The Rare-Male Effect: What is its Evolutionary Significance?. *Phil Trans R Soc Lond B Biol Sci 319*:525-539, 1988

Pazderka, F., Longenecker, B.M., Law, G.R.J., Stone, H.A. and Ruth, R.F.: Histocompatibility of chicken populations selected for resistance to Marek's disease. *Immunogenetics 2*:93-100, 1975

Potts, W.K., Manning, C.J., and Wakeland, E.K.: Mhc genotype influences mating patterns in semi-natural populations of *Mus. Nature*, in press 1991

Potts, W.K. and Wakeland, E.K.: Evolution of diversity at the major histocompatibility complex. *Trends Ecol Evol 5*:181-187, 1990

Roy, S., Scherer, M.T., Briner, T.J., Smith, J.A., and Gefter, M.L.: Murine MHC polymorphism and T cell specificities. *Science 244*:572-575, 1989

Singh, P.B., Brown, R.E. and Roser, B.: Soluble class I MHC antigens in solution in the body fluids. In: P. Ivanyi (ed.): *MHC+X: Complex Formation and Antibody Induction*, pp.240, Springer-Verlag, New York 1988

Singh, P.B., Herbert, J., Roser, B., Arnott, L., Tucker, D.K., and Brown, R.E.: Rearing rats in a germ-free environment eliminates their odors of individuality. *J Chem Ecol 16*:1990

Spinka, M.: The effect of time of day on sperm competition and male reproductive success in laboratory rats. *Physiol Behav 47*:483-488, 1990

Svende, F.J., Jorgenson, J.W., and Novotny, M.: Possible chemical basis for histocompatibility-related mating preferences in mice. *J Chem Ecol 10*:1603-1615, 1984

Thomas, M.L., Harger, J.H., Wagner, D.K., Rabin, B.S., and Gill, T.J.: HLA sharing and spontaneous abortion in humans. *Am J Obst G 151*:1053-1058, 1985

Tiwari, J.L. and Terasaki, P.I.: *HLA and Disease Associations.* Springer-Verlag, New York 1985

Tucker, D.: Electrophysiological evidence for olfactory function in birds. *Nature 207*:34-36, 1965

Uyenoyama, M.K.: On the evolution of genetic incompatibility systems: incompatibility as a mechanism for the regulation of outcrossing distance. In: R.E. Michod and B.R. Levin (eds): *The Evolution of Sex*, pp. 212-232, Sinauer Associates Inc., Sunderland, Mass. 1988

Wenzel, B.M.: Olfactory processes of the kiwi. *Nature 220*:1133-1134, 1968

Yamazaki, K., Boyse, E.A., Mike, V., Thaler, H.T., Mathieson, B.J., Abbott, J., Boyse, J., Zayas, Z.A., and Thomas, L.: Control of mating preferences in mice by genes in the major histocompatibility complex. *J Exp Med 144*:1324-1335, 1976

Yamazaki, K., Yamaguchi, M., Andrews, P.W., Peake, B., and Boyse, E.A.: Mating preferences of F_2 segregants of crosses between MHC-congenic mouse strains. *Immunogenetics 6*:253, 1978

Yamazaki, K., Beauchamp, G.K., Thomas, L., and Boyse, E.A.: The hematopoietic system is a source of odorants that distinguish major histocompatibility types. *J Exp Med 162*:1377-1380, 1985

Yamazaki, K., Beauchamp, G.K., Kupniewski, D., Bard, J., Thomas, L., and Boyse, E.A.: Familial imprinting determines H-2 selective mating preferences. *Science 240*:1331-1332, 1988

Yamazaki, K., Beauchamp, G.K., Bard, J., and Boyse, E.A. Single MHC gene mutations alter urine odour constitution in mice. *In* D.W. MacDonald, D. Muller-Schwarze, and S.E. Natynczuk (eds): *Chemical Signals in Vertebrates* 5, pp. 255-259, Oxford University Press, Oxford 1990

Yokoyama, S. and Nei, M.: Population dynamics of sex-determining alleles in honey bees and self-incompatibility alleles in plants. *Genetics 91*:609-626, 1979

Zuk, M., Johnson, K., Thornhill, R., and Ligon, J.D.: Mechanisms of female choice in red jungle fowl. *Evolution 44*:477-485, 1990

POSSIBLE *MHC* ASSOCIATED HETEROZYGOUS ADVANTAGE IN WILD MOUSE POPULATIONS

Uzi Ritte*, Esther Neufeld*, Uzi Motro*, Felipe Figueroa°, and Jan Klein°+
*Department of Genetics, Hebrew University of Jerusalem, Jerusalem, Israel
°Max-Planck-Institut für Biologie, Abteilung Immungenetik, 7400 Tübingen, Germany
+Department of Microbiology and Immunology, University of Miami School of Medicine, Miami, FL, 33101, USA

The variability of the major histocompatibility complex (Mhc) in natural populations is usually very extensive, but the forces that are responsible for the maintenance of this variability are still unknown. Several investigators (Hughes and Nei 1988, 1989; Jonsson et al. 1989; Takahata and Nei 1990) have recently suggested that the polymorphism in the region of the peptide-binding sites is under positive selection. A verification of this suggestion should be obtained by empirical studies of natural populations, and a good model animal for such studies is the house mouse (Mus domesticus and M. musculus), for which a great deal of information is available both as far as the structure and dynamics of natural populations (Berry 1981) and the organization of H-2, its major histocompatibility complex (Klein 1986), is concerned. A problem associated with the demonstration of the effect of selection on H-2 variability, however, is the fact that in general, natural populations of the house mouse are highly polymorphic for H-2 haplotypes (Klein and Figueroa 1986), so that all individuals are expected to be heterozygotes, even without selection (Duncan et al. 1979). The discovery of low H-2 variability in two adjacent populations of the house mouse (M. domesticus) in Jerusalem, Israel (Neufeld et al. 1986), in which some individuals were shown to be homozygotes, provided a good opportunity to find out whether H-2 heterozygotes have a selective advantage.

The two mouse populations live in the Educational Farm of the Hebrew University in Jerusalem. One ("Population I") occupies an active chicken coop, the area of which is 100 m^2. The other ("Population II") lives in an adjacent, 250 m^2 field, in which 20 permanent "mouse stations" (containing a constant supply of food and water) were established. Both populations are unconfined and no mouse was artificially introduced into them.

The dynamics of the two populations was followed by bi-weekly (each of two successive nights) or monthly (each of four successive nights) trapping series. The survey began in November 1982 and continued until October 1985. The survey in 1983 was used to characterize the different *H-2* haplotypes found in the populations. It was discovered (Neufeld et al. 1986) that there are only four haplotypes in the two populations - $H\text{-}2^{w82}$

NATO ASI Series, Vol. H 59
Molecular Evolution of the
Major Histocompatibility Complex
Edited by J. Klein and D. Klein

($K^{w16}D^{w82}$), $H\text{-}2^{w83}$ ($K^{w83}D^{w16}$), $H\text{-}2^{w84}$ ($K^{w84}D^{w84}$) and $H\text{-}2^{w85}$ ($K^{w83}D^{w84}$). $H\text{-}2^{w85}$ is a recombinant between $H\text{-}2^{w83}$ and $H\text{-}2^{w84}$. The antisera prepared against these four haplotypes were anti-K^{w16}, anti-D^{w82}, anti-K^{w83}, and anti-$H\text{-}2^{w84}$.

The analysis of all the mice that belonged to the two populations, using the four antisera, began in June 1984 (trapping series 36) and ended in October 1985 (trapping series 54). This period seems to cover, in both populations, a full cycle of population size variation.

Table 1. *H-2* genotypes deduced from the results of the cytotoxic test.

Reactivity with antisera specific for					Deduced genotype
K^{w16}	D^{w82}	K^{w83}	$H\text{-}2^{w84}$		
+	+	-	-		$H\text{-}2^{w82}/H\text{-}2^{w82}$
-	-	+	-		$H\text{-}2^{w83}/H\text{-}2^{w83}$
-	-	-	+		$H\text{-}2^{w84}/H\text{-}2^{w84}$
+	+	+	-		$H\text{-}2^{w82}/H\text{-}2^{w83}$
+	+	-	+		$H\text{-}2^{w82}/H\text{-}2^{w84}$
+	+	+	+		$H\text{-}2^{w82}/H\text{-}2^{w85}$
-	-	+	+		$H\text{-}2^{w83}/H\text{-}2^{w84}$
				or	$H\text{-}2^{w83}/H\text{-}2^{w85}$
				or	$H\text{-}2^{w84}/H\text{-}2^{w85}$
				or	$H\text{-}2^{w85}/H\text{-}2^{w85}$

+, positive; -, negative

Adult mice (15 g or more) were temporarily taken to the laboratory and one lymph node was removed from each. The *H-2* genotype of a mouse was determined by the reactivity of its lymphocytes with the four antisera, using the cytotoxic test. Table 1 gives the deduced genotypes, according to the different patterns of reactivity. One pattern (negative reactions with anti-K^{w16} and D^{w82}, positive reactions with anti-K^{w83} and anti-$H\text{-}2^{w84}$) can represent more than a single genotype. No new haplotype was discovered, and the existence of homozygotes was confirmed by matings. A sporadic survey was carried out in Population I in 1989-1990, using antisera that could identify unambiguously all genotypes.

Tables 2 and 3 give the distribution of the different *H-2* genotypes in the different series, separately for Population Is (Table 2) and II (Table 3). For each series, the tables include population size, the number of mice for which the *H-2* genotypes were known, and the number of mice that had each genotype.

Table 2. Number of mice with different *H-2* genotypes in Population I.

Series number	Population size	Number of mice with known genotype	Mice with genotype						
			82/82	*83/83*	*84/84*	*82/83*	*82/84*	*82/85*	*83/84/85*
36	12	7			3		2		2
37	14	8		1	3		2		2
38	12	9		1	3	1	3		2
39	16	12		1	3	1	2	2	3
40	15	14	1	1	3	4	3	2	3
41	20	18		2	2	3	3	5	2
42	23	20		4	1	3	3	6	3
43	23	20		4	1	3	2	8	2
44	27	23		4	1	4	1	10	4
45	36	28		4	1	5	2	10	7
46	53	40		4	1	4	4	18	8
47	49	37		4	1		3	17	8
48	38	28		1	1		2	16	8
49	39	29		1	1		2	19	7
50	36	30					1	17	12
51	29	20					1	12	7
52	13	10						8	2
53	13	11						7	4
54	7	6						3	3

At the beginning, when the size of Population I was small, *H-2* homozygotes were relatively common (Table 2). With time (as population size increased), the homozygotes became less frequent (with $H\text{-}2^{w83}H\text{-}2^{w83}$ replacing $H\text{-}2^{w84}$ ($H\text{-}2^{w84}$) and disappeared completely from Population I after series 49. They did not even reappear when population size was reduced. Another change was an increase in the proportion of genotypes $H\text{-}2^{w82}/H\text{-}2^{w85}$, or of mice with haplotype $H\text{-}2^{w85}$.

Table 3. Number of mice with different *H-2* genotypes in Population II.

Series number	Population size	Number of mice with known genotype	Mice with genotype						
			82/82	*83/83*	*84/84*	*82/83*	*82/84*	*82/85*	*83/84/85*
36	17	5			1		2	1	1
37	13	6			1		2	1	2
38	10	8			1		3	1	3
39	12	10			1		5	1	3
40	12	11			2		6	1	2
41	7	6			1		3		2
42	8	8			1		4	1	2
43	16	14		1	1		5	5	2
44	22	15		2	1		2	8	2
45	19	16		2	1		3	8	2
46	24	21		3	4		4	8	2
47	32	27		3	4	2	3	13	2
48	29	25		2	1	2	3	15	2
49	28	27		4	1	2	1	16	3
50	23	20		4	2	2	1	9	2
51	17	15		2		2	1	7	3
52	15	12		1		1		8	2
53	10	10		1		1		6	2
54	12	10		1		1		5	3

Similar changes (although less extreme) were evident in Population II (Table 3). An analysis of 35 mice from Population I in October 1989 (nine mice), February 1990 (11 mice), and May 1990 (15 mice) indicated that it continued to be composed exclusively of mice that have the same four haplotypes. The proportions of $H\text{-}2^{w82}$, $H\text{-}2^{w84}$, and $H\text{-}2^{w85}$ haplotypes were almost equal (each haplotype occurred at a frequency of about 30%), while the proportion of haplotype $H\text{-}2^{w83}$ was about 10%. As in the larger survey, the *H-2* homozygotes ($H\text{-}2^{w84}/H\text{-}2^{w84}$ and $H\text{-}2^{w85}/H\text{-}2^{w85}$) were more common (33%) when population size was small (October 1989) and less common when population size increased (17% in February 1990, 13% in May 1990).

Can the changes in the frequencies of the different genotypes be caused by random events? A major problem with the statistical analysis of our results is the overlap between the series with regard to the mice that were captured in each. To overcome this problem we decided to consider each mouse only once - either at the series of its first capture or, for a second comparison, at the series of its last capture.

For the statistical analysis we combined the mice from both populations. In each series the mice were divided into two categories - homozygotes (*H-2*^{w82}/*H-2*^{w82}, *H-2*^{w83}/*H-2*^{w83}, and *H-2*^{w84}/*H-2*^{w84}) and heterozygotes *H-2*^{w82}/*H-2*^{w83}, *H-2*^{w82}/*H-2*^{w84}, and *H-2*^{w82}/*H-2*^{w85}). Mice listed in Tables 2 and 3 in column 83/84/85 were excluded. If there is no selective difference between homozygotes and heterozygotes, they should join the population, or disappear from it, at random. To test this hypothesis we used the One-Sided Wilcoxon Rank Sum Test (Ostle 1963). All mice were ranked according to the time of their first (or last) appearance, and the positions of all members in each category were summed. The difference, T, between the two sums was used to calculate t.

For the time of first appearance, the sum of ranks was 940 for homozygotes and 7316 for heterozygotes. The value of t (2.6670) is highly significant. In the analysis of discontinuing mice we reached a figure of 1089 for homozygotes and 7617 for heterozygotes, again giving a significant value of t (-1.7083). We conclude that the average time of appearance and disappearance of *H-2* homozygotes is significantly shorter than that of heterozygotes, suggesting the operation of selection in favor of heterozygotes.

A similar calculation was carried out for mice divided into a category that included all animals with at least one copy of haplotype *H-2*^{w85}, and a category of all other animals. In contrast to the first statistical test, which excluded some mice, this test included all mice. For the time of first appearance, the sums of ranks were 11172 for mice with *H-2*^{w85} and 2689 for mice without *H-2*^{w85}. The t value (5.7551) is highly significant. For discontinuing mice the sums of ranks were 10963 (with *H-2*^{w85}) and 2898 (without *H-2*^{w85}), again giving a highly significant t value (5.0274). Selection seems to be favoring mice with haplotype *H-2*^{w85}.

A simulation study, based on an initial population with a composition similar to that of Population I at the beginning of the survey, and continued for four additional generations (in which population sizes were 5, 21, 15, and 9 mice), showed that in 2000 independent runs, the probability of increasing by chance (without selection) the proportion of heterozygotes, or of mice with *H-2*^{w85} to the level found at the end of our survey, was close to zero (Ritte et al. 1991).

H-2 homozygotes become more common when the population undergoes major declines, usually at the end of the summer. When population size increases, *H-2* homozygotes are selected against. In our particular population, the favored genotype was *H-2*^{w82}/*H-2*^{w85}.

Among homozygotes, the proportion of each genotype is often much lower than the square of the proportion of its haplotype. This is true especially for the genotype *H-2*^{w82}/*H-*

2^{w82} which was found only in one individual, although the haplotype *H*-2^{w82} was relatively common.

Acknowledgments. We thank Ms. Lynne Yakes for editorial assistance. The work was supported by grant I-39-098.11/87 from the German/Israeli Foundation for Scientific Research and Development, Jerusalem, Israel and grant AI 23667 from the National Institutes of Health, Bethesda, Maryland.

References

Berry, R.J.: Population dynamics of the house mouse. *In* R.J. Berry (ed.): *Biology of the House Mouse* (Symp Zool Soc Lond 47), pp 395-425, Academic Press, London, 1981

Duncan, W.R., Wakeland, E.K., and Klein, J.: Heterozygosity of *H-2* loci in wild mice. *Nature 281*:603-605, 1979

Hughes, A.L. and Nei, M.: Pattern of nucleotide substitutions at major histocompatibility complex class I reveals overdominant selection. *Nature 335*:167-170, 1988

Hughes, A.L. and Nei, M.: Nucleotide substitution at major histocompatibility complex class II loci: Evidence for overdominant selection. *Proc Natl Acad Sci USA 86*:958-962, 1989

Jonsson, A.-K., Andersson, L., and Rask, L.: Selection for polymorphism in the antigen recognition site of major histocmpatibility complex class II molecules. *Scand J Immunol 30*:409-417, 1989

Klein, J.: *Natural History of the Major Histocompatibility Complex*. John Wiley, New York 1986

Klein, J. and Figueroa, F.: Evolution of the major histocompatibility complex. *CRC Crit Rev Immunol 6*:296-386, 1986

Neufeld, E., Ritte, U., Figueroa, F., and Klein, J.: Low *H-2* polymorphism in some Israeli wild mouse populations. *Immunogenetics 24*:374-380, 1986

Ostle, B.: *Statistics in Research. Basic Concepts and Techniques for Research Workers*, 2nd edition, Iowa State University Press, Ames, Iowa, 1963

Ritte, U., Neufeld, E., O'hUigin, C., Motro, U., Figueroa, F., and Klein, J.: Origins of *H-2* polymorphism in the house mouse. II. Characterization of a model population and evidence for heterozygous advantage. *Immunogenetics*, in press 1991

Takahata, N. and Nei, M.: Allelic linealogy under overdominant and frequency-dependent selection and polymorphism of major histocompatibility complex loci. *Genetics 124*:967-978, 1990

ANTIGEN PRESENTATION BY NEOCLASSICAL MHC CLASS I GENE PRODUCTS IN MURINE RODENTS

Chyung-Ru Wang*, Alexandra Livingstone°[1], Geoffrey W. Butcher°, Evan Hermel*, Jonathan C. Howard°, and Kirsten Fischer Lindahl*

*Departments of Microbiology and Biochemistry and Graduate Immunology Program, Howard Hughes Medical Institute, Univ. of Texas Southwestern Medical Center, 5323 Harry Hines Blvd., Dallas, TX 75235-9050, USA

°Department of Immunology, Institute of Animal Physiology and Genetics Research, Babraham Hall, Cambridge CB2 4AT, UK

Two roads diverged in a wood, and I -
I took the one less traveled by,
And that has made all the difference.
Robert Frost: THE ROAD NOT TAKEN

A number of cytotoxic T cell responses in rats and mice are restricted by medial or non-classical class I antigens of the major histocompatibility complex (MHC). The class I heavy chain encoded by the H-2M3 *gene presents MTF, an* N*-formylated peptide derived from the amino terminus of the mitochondrially-encoded ND1 protein, which is polymorphic in the sixth residue.* H-2M3 *shows minimal polymorphism in mice, and it is more similar to its rat ortholog,* RT1.M3*, than to any other* H-2 *class I gene. In rats, a minor antigen is presented by an* RT1.C*-encoded restriction element. Other examples of antigen presentation by medial class I antigens are reviewed. Whereas the majority of medial class I genes have ill-defined, if any, function, we propose that some may serve to present special antigens, and that these neo-classical class I genes are highly conserved between species and may be localized at the distal end of the MHC.*

All mammalian major histocompatibility complexes (MHC) studied have more class I genes than they need to code for the classical molecules that present antigen to CD8+, cytotoxic T

[1]Present address: Basel Institute for Immunology, Grenzacherstraße 487, CH-4005 Basel, Switzerland

NATO ASI Series, Vol. H 59
Molecular Evolution of the
Major Histocompatibility Complex
Edited by J. Klein and D. Klein

cells (CTL). The numbers range from seven in the miniature swine (Singer et al. 1987), over ten in man (Koller et al. 1989), to over forty in mice (Stroynowski 1990; Fischer Lindahl et al. 1991) and over sixty in rats (Jameson 1988; Jameson et al. 1990). These medial, or non-classical, class I genes are expressed at lower levels and often in only a limited range of tissues, or they are pseudogenes. Cross-hybridization with probes specific for mouse *Q* and *T* class I genes failed to reveal orthologous genes in closely related rodents, leading to the suggestion that the medial class I genes have evolved separately in each species by duplication of the classical class I genes (Rogers 1985). As no biological function has been found for the medial class I genes, it was proposed that they represent the debris of an evolutionary process acting on the classical class I genes (Klein and Figueroa 1986; Howard 1987; Parham et al. 1989).

Any support for an alternative, non-immunological function for medial antigens, such as mediating cell interactions during development, was effectively crushed by the observation that mice homozygous for a deletion in the *B2m* gene, and hence unable to produce the β_2-microglobulin (β2m) light chain necessary for folding and surface expression of class I molecules, developed normally. The only detectable lesion was a lack of $CD8^+$ T lymphocytes (Koller et al. 1990; Zijlstra et al. 1990). If medial class I genes do have a biological role, it must therefore be of an immunological nature. In this paper we wish to discuss the evidence that medial class I genes can present antigen, that they may do so in a specialized manner, and that such function may have led to their conservation during the divergence of mice and rats.

The maternally transmitted antigen, Mta, of mice

Discovered as an alloantigen that induced slow skin graft rejection and a vigorous CTL response, Mta was considered exceptional because the epitope was determined by maternal inheritance and the CTL were not H-2 restricted (Fischer Lindahl et al. 1991). The molecular analysis of Mta has shown that, like all other histocompatibility antigens, it is a complex of an MHC molecule and a peptide ligand (Loveland et al. 1990) (Fig. 1); Mta is special in that the MHC molecule is a medial class I antigen, encoded by *H-2M3* (also called *Hmt*) (Wang et al. 1991), and binding of the ligand, MTF, requires that the N-terminal amino-group is formylated (Shawar et al. 1990).

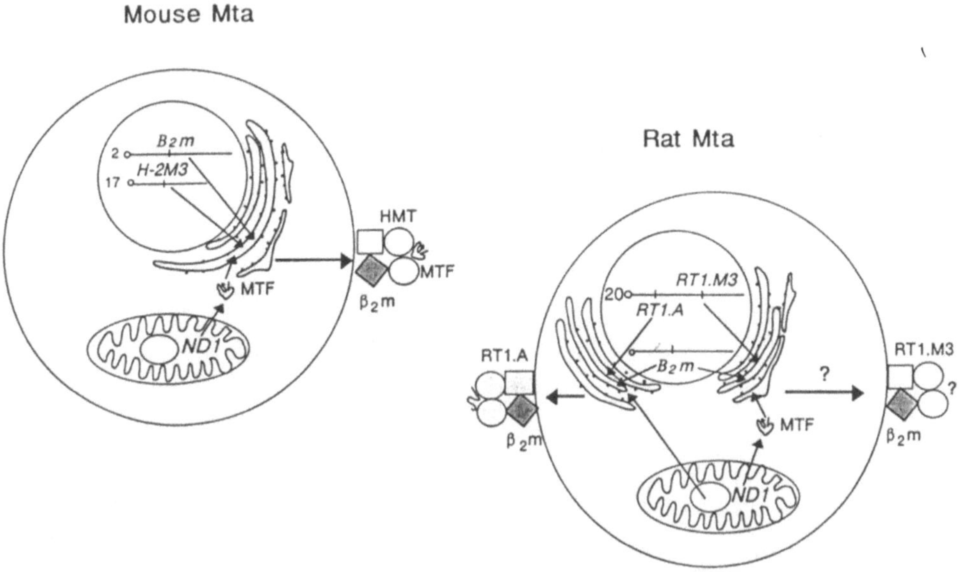

Fig. 1. The maternally transmitted antigen systems of mice and rats. In the mouse, the *H-2M3* (formerly *Hmt*) locus on Chr. 17 encodes a heavy chain, which together with β2m, encoded by the *B2m* locus on Chr. 2, forms a class I molecule that presents the MTF peptide, derived from the mitochondrially encoded ND1 protein (Loveland et al. 1990), on the cell surface as the Mta antigen (Wang et al. 1991). In the rat, the RT1.A class I molecule (complexed with β2m) presents a mitochondrially encoded peptide of unknown origin (Davies et al. 1991). In addition, a peptide cross-reacting with mouse MTF, presumably derived from the mitochondrial ND1 protein, can be presented by a suitable mouse restriction element (Loveland et al., in preparation). The medial class I gene, *RT1.M3*, which is abundantly expressed, may encode such a restriction element in the rat (C.-R. Wang et al., in preparation).

Mta is formed when the M3 class I antigen (also called HMT) binds a peptide derived from the amino-terminus of the mitochondrially encoded protein ND1, which explains the maternal transmission. ND1 is a hydrophobic, 30 kDa subunit of respiratory complex I or NADH-reductase, an enzyme complex of about 25 subunits located in the inner mitochondrial membrane. The N-terminal sequence of ND1 has been conserved throughout evolution: sequences from fungi, plants, and animals can all be aligned on the basis of a conserved Glu-Arg-Lys (ERK) motif, which is preceded by a hydrophobic stretch long enough to span a membrane (Table 1). The sixth residue of this protein is polymorphic in mice: the predominant form is Ile, and the alleles are Ala, Val, and Thr (Loveland et al. 1990); this conservative variation makes MTF an alloantigen. We do not know how MTF escapes from the mitochondria, but because the surface expression of Mta is reduced on the RMA-S cell line, which is defective in presentation of endogenous peptides (Townsend et al. 1989), we believe that MTF reaches the endoplasmic reticulum and assists in the folding of M3 (Hermel et al. 1991).

Table 1. Comparison of ND1 amino-termini through evolution

	Mtf	1 5 10 15 20 25	Reference
		* (position 6)	
Mouse C3H,ddY	[α]	M F F I N I L T L L V P I L I A M A F L T L V **E** **R** **K**	(Bibb et al. 1981)
Mouse SBB	[α]	- - - - - - - - - - L ? ? ? ? ? ? ? ? ? ? ? ? ? ? ?	(Loveland et al. 1990)
Mouse NZB	[β]	- - - - - A -	(Loveland et al. 1990)
Mouse MilP	[γ]	- - - - - V -	(Loveland et al. 1990)
Mouse SHC	[δ]	- - - - - T -	(Loveland et al. 1990)
Rat	[ρ]	- Y - - - - - - - - I - - - - - - - - - - - - - - -	(Davies et al. 1991)
Hamster		- H - - - - - M - - I - - - - - ? ? ? ? ? ? ? ? ? ?	(Kotin and Dubin, 1984)
Cow		- - M - - - - M - I I - - - L - V - - - - - - - - -	(Anderson et al. 1987)
Quagga/Horse/Zebra		? ? ? ? ? ? ? ? ? ? ? - - - L - V - - - - - - - - -	(Higuchi et al. 1987)
Fruit Bat		- Y L M - L - - T I - - V - L - V - - - - - - - - -	(C. Phillips, pers. comm.)
Man		- P M A - L - L - I - - - - - - - - - - M - T - - -	(Anderson et al. 1981)
Chicken		M T L P T L T N L L L - M T - S Y I L - - - - - V - - - - - - - - -	(Desjardins and Morais, 1990)
Xenopus		M L T I I T H L - - P - L Y M I - - - L - V - - - - - I - - -	(Roe et al. 1985)
Sea Urchin[PL]		M T Y L F T V L **E** L - S F - - - V - L - V - - - - - - - - -	(Cantatore et al. 1988)
Sea Urchin[SP]		M V Y V F S I L **E** L I S F - I - - - L S V - - - - - - - - -	(Jacobs et al. 1988)
Sea Star [PO]		M S V I L L L L - S I G F I - - V - L - V - L - - - I - - -	(J. Brown, pers.comm.)
Locust		M M Y **D** L L F M - V L N F L - I I C V - - S V - - - - - M - - -	(McCracken et al. 1987)
Fruit Fly		M **E** F I L S L - G S - L - I I C V - V S V - - - - - L - - -	(Garesse, 1988)
Mosquito		M **D** F I L S L - G S - L - V - C V M V G V - - - - - L **K** - -	(HsuChen et al. 1984)
Nematode [AS]		L I L M L V Q V I L I M I F V I Q S I - - V - - Y - - H	(D. Wolstenholme, pers.comm.)
Nematode [CR]		M I L V L L M V I L M M I F - V Q S I - - I - - Y - - H	ibid.
Nematode [MJ]		M F F F L - I I L V I F C I S L - - - L - - Y	ibid.
Liver Fluke		VVLLL **K** A F Y L G L S S F F A F - M - M V F V - - F M - G - - -	(Garey and Wolstenholme, 1989)
Trypanosome		MY**K**THLL L H L D I C - L - V I F I L V L S V L C G Y V S - C - - -	(Hensgens et al. 1984)
Paramecium		I L I Y S I V - M L V V T L - I - S I - - L - - -	(Pritchard et al. 1986)
Neurospora		M F Y S L T I I S I L **E** V - L V - - - S - L - V Y V - V A - - -	(Burger and Werner, 1985)
Oenothera		MNR **K** S **K** T Y I A V P A **E** - - G I I L - L - L G V - - - V - A - - -	(Wissinger et al. 1991)
Wheat		M Y I A V P A **E** - - C - I L - L - L G V - - - V - A - - -	(Chapdelaine and Bonen, 1991)

? ? ? = sequence unknown; * marks the polymorphic residue responsible for the mouse minor histocompatibility antigen; bold characters are charged residues.

MTF can be mimicked by a synthetic peptide as short as 8 to 12 amino acids corresponding to the N-terminal sequence of ND1, and even the N-terminal pentapeptide will compete for binding to the restriction element (Shawar et al. 1991). The first Met can be replaced by Val or Phe without significant loss of activity, but if the formyl-group is removed from the terminal amino-group or replaced by an acetyl-group, then all binding activity is lost (Shawar et al. 1990). Indeed, certain other peptides can bind to the same site, provided they carry an N-terminal formyl-methionine; however, not all *N*-formylated peptides are bound. It thus appears that the restriction element is specific for *N*-formylated peptides, which, except for the 13 mitochondrial proteins, can come only from prokaryotic sources, such as intracellular bacteria.

Structure and polymorphism of *H-2M3*

Mta has eluded all attempts to raise alloantibodies against it, and we therefore had to identify its components by classical genetic analysis and molecular cloning. *Hmt* was mapped to a region of Chr. 17, distal of *Tla*, that spans less than 0.7 cM (Richards et al. 1989). Genomic DNA digests from appropriate mice were probed for the conserved exon 4 of MHC class I genes to identify class I gene fragments from the *Hmt* region (Fischer Lindahl et al. 1987), which were then cloned. Candidate genes from Hmt^a (Mta^+) mice were transfected into an Hmt^b (Mta^-) fibroblast line, and the transfectants were assayed for Mta display. With probes isolated from the R4B2 clone from an Hmt^b mouse, known to contain exon 4 of a class I gene (Richards et al. 1989), we screened a BALB/c (Hmt^a) cosmid library and isolated a 35 kb fragment, 14RB-A, which contained the complete $H\text{-}2M3^d$ class I gene. Transfection of this cosmid, or of corresponding cDNA clones from $H\text{-}2^k$ mice, from an Hmt^a mouse into an Hmt^b cell line led to expression of Mta and conferred the ability to present a synthetic MTF peptide antigen (Wang et al. 1991).

Does the structure of M3 explain its privileged interaction with MTF, an *N*-formylated, hydrophobic, mitochondrial peptide? The *M3* gene specifies a class I molecule that would be expected to fold in the same manner as a classical class I antigen, but with only eight amino acids for a cytoplasmic anchor. The extracellular domains have the same length and conserve most residues, such as cysteines, that are considered important in the three-dimensional structure of HLA-A2, and the interactions with β2m are all conserved (Saper et al. 1991). Comparison of nucleotide (or amino acid) sequences show that *M3* is slightly more closely related to the classical class I genes, such as *H-2K*, than to other medial class I genes, such as *Q4*, *T18* and *T23*, or to the other seven class I genes of the *M* region (Table 2). The peptide-binding site of M3 is more hydrophobic than that of most class I molecules. Most importantly, the otherwise highly conserved Tyr 171 is changed to Phe in M3 (Fig. 2). Tyr 171 and Tyr 59 lie at the end of the groove, and their hydroxyl-groups are thought to coordinate the free N-terminal amino-group of common peptide ligands (D. Madden, J. Gurka, J.L. Strominger, and D.C. Wiley, in preparation).

Among over 140 MHC class I sequences, including 88 human (Bjorkman and Parham, 1990), 12 cotton top tamarin (Watkins et al. 1990), 35 mouse (Watts et al. 1989), 3 rat (Rada et al. 1990; Parker et al. 1990), 4 miniature swine (Ehrlich et al. 1987; Frels et al. 1990; Satz et al. 1985), 2 rabbit (Tykocinski et al. 1984), 1 *Xenopus* (Flajnik et al. 1991), and 1 chicken (Guillemot et al. 1988), we find Phe171 only in *H-2M1* and *M3* and in a carp class I gene (Hashimoto et al. 1990). The only other exception to Tyr171 is His in 8 human, 2 mouse and 1 tamarin sequence. Another unusual amino acid is Gln34, at the bottom of the "45" (Garrett et al. 1989) or "B" pocket (Saper et al. 1991), shared with only $Q5^k$. The consensus residue is Val, and the sequence collection has only four other exceptions: Ala in $T13^c$, Arg in Tla^{a1}, Met in *H-2M1*, and Leu in $T22^b$ (Ito et al. 1990) and *HLA-F*. Phe 171 and Gln34 may therefore contribute to the unusual peptide specificity of M3.

Table 2. Percentage DNA sequence identity of *H-2M3* to other *H-2* class I genes

Exon	K^d	D^d	L^d	*Q4*	$T18^d$	*T23*	*M1*	*M2*	*M4*	*M5*	*M6*
2	70	68	69	70	62	62	54	62	ND	56	62
3	68	69	70	66	62	65	53	64	63	59	ND
4	85	87	86	85	86	86	75	84	80	87	83
5	50	47	38	61	50	53	45	66	64	64	62

Data from Wang et al. (1991), in preparation. ND = not determined.

As defined by CTL recognition, *Hmt* shows only limited polymorphism. The *a* form is found in most laboratory and wild mice of several subspecies; the *b* form, which segregates in *Mus musculus castaneus* and is found among *t* haplotypes, behaves as a null allele, in that we have been unable to detect it with CTL; the *c* form, characteristic of *Mus spretus*, presents MTF but can be distinguished by CTL (Fischer Lindahl et al. 1986). Further alleles may be present in *t* haplotypes and *M. m. bactrianus* (Fischer Lindahl et al. 1991). The protein specified by $M3^{cas3}$, which represents an Hmt^b (Mta^-) allele, differs from $M3^k$ by only three replacement substitutions and one deletion; of these, only the Leu95 to Gln change affects the peptide-binding groove (Fig. 2). Preliminary assays of site-directed mutants confirm that this substitution abolishes recognition by standard anti-Mta CTL (C.-R. Wang, unpublished). The

M3^{sp2} gene, representing the active *Hmt*^{c} allele, has more substitutions, but only one in the peptide-binding site: Leu147 to Phe. This change will have to account for the distinct epitopes recognized by CTL (Fischer Lindahl et al. 1986). The active *spretus* form shares Gln34, Leu95 and Phe171 (Fig. 2).

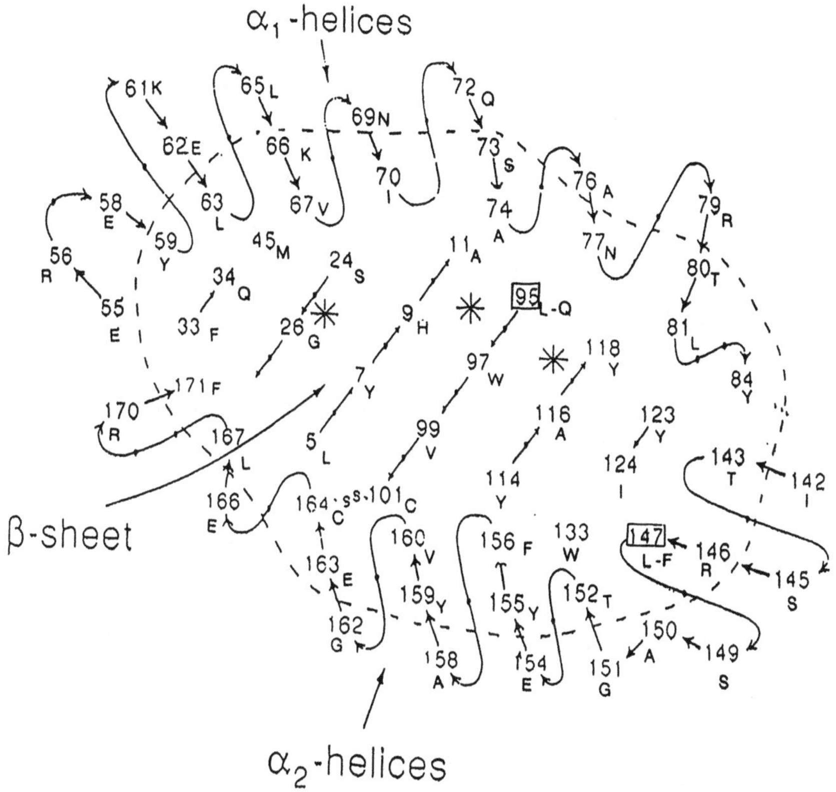

Fig. 2. Diagrammatic presentation, adapted from Garrett et al. (1989), of the peptide-binding groove of the H-2M3 molecule, showing the substitutions in the *M3*^{cas3} form, representing *Hmt*^{b} (Leu95 to Gln) (Wang et al. 1991), and in the *spretus* form, representing *Hmt*^{c} (Leu147 to Phe) (C.-R. Wang, in preparation).

We are now examining the polymorphism of *M3* at the nucleotide level. It is possible that this class I gene has not been subjected to the same selection for polymorphism because it uses a different principle to discriminate and bind foreign antigen, and that instead it has been under pressure to conserve this property. It will be of interest to see whether M3 does indeed act as a restriction element in the immune response to prokaryotic parasites.

Maternally transmitted antigens in other species

If we believe that medial class I antigens have specialized functions in antigen presentation, and that these are of selective advantage, then we should be able to find orthologous genes, or at least analogous genes, in other species. We have therefore sought a gene in the rat that cross-hybridizes with *H-2M3*.

A probe for *H-2M3* from exon 3 showed three bands in Southern blots of rat genomic DNA (Wang et al. 1991). We used this probe to screen a cDNA library from strain DA lymphoblasts and isolated seven clones, which all came from the same gene, *RT1.M3*. The sequence of the full-length transcript, which encodes a complete class I gene, is very similar to *H-2M3* (Table 3). These genes are more similar to each other than they are to any known class I gene of their own species, except in exon 4, where *RT1.M3* maintains greater similarity to rat class I genes. The *RT1.M3* gene was missed when a rat cosmid library was screened for MHC class I genes (Jameson 1988, and unpublished results), and it brings the total number to 61. *RT1.M3* encodes a peptide-binding site similar to that of *H-2M3* (Fig. 3). The signature residues Gln34, Leu95, and Phe171 are conserved.

Table 3. Percentage DNA sequence identity of *RT1.M3* to other class I MHC genes

Exon	*H-2M3*	*H-2K*^{k}	*H-2T18*^{d}	*H-2T23*^{d}	*RT1.A*^{a}	*pARI.5*	*RT.BM1*	*RT16*
1	81	ND	ND	ND	57	53	ND	ND
2	88	69	63	68	70	67	61	66
3	90	65	62	66	64	63	66	65
4	90	90	91	91	94	92	94	93
5	87	52	49	62	56	54	50	56

Data from C.-R. Wang et al., in preparation. ND = not determined.

The substitutions are found mostly on top of the α-helices, where they are more likely to affect T cell recognition than peptide selection. One exception is the Arg146 to Glu change; in most class I molecules this residue is a Lys, which is thought to form a salt bridge with the C-terminus of the peptide (D. Madden, J. Gurka, J.L. Strominger, and D.C. Wiley, in preparation); a Glu might seriously affect peptide binding. However, in H-2M3 the other end of the groove appears more important, as evidenced by the ability of an *N*-formylated pentapeptide to compete with longer peptides for binding. The possibility therefore remains that the relatively abundantly expressed *RT1.M3* gene encodes a class I molecule with specificity for *N*-formylated peptides.

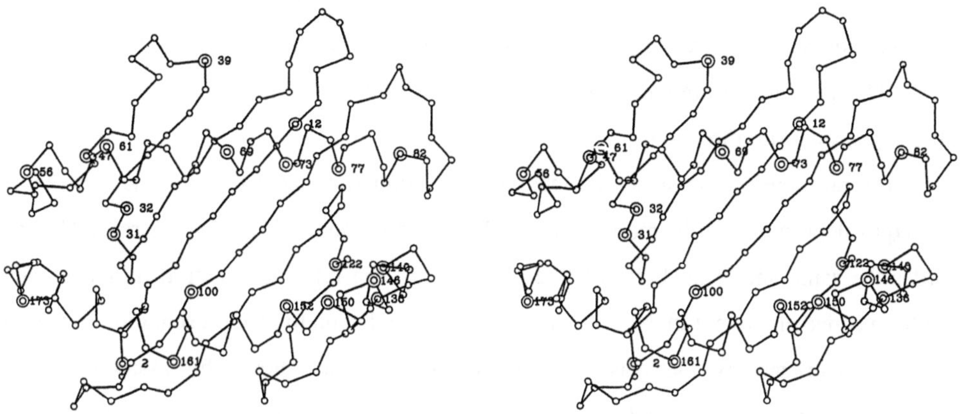

Fig. 3. Stereoview of the α-carbon backbone of the α1 and α2 domains of an MHC class I molecule (HLA-w68 from Garrett et al. [1989]). Big circles mark those residues that differ between mouse H-2M3^{k} and rat RT1.M3^{av1} (C.-R. Wang, in preparation).

This raises the question whether such peptides escape from the mitochondria in rats as they do in mice? Loveland and coworkers (manuscript in preparation) examined somatic cell hybrids, made by fusion of a rat thymoma with mouse lymphocytes with *Mtf*$^{\beta}$ mitochondria. The rat parent line is resistant to CTL specific for the complex of mouse HMT^{a} with MTF$^{\alpha}$ peptide, but the hybrids were lysed, provided they retained mouse Chr. 17 and rat mitochondria. We conclude that the amino-terminal peptide of rat ND1, which differs from MTF$^{\alpha}$ only by Phe2 to Tyr in the first ten positions (Table 1), crossreacts with mouse MTF and is equally capable of escaping from the mitochondrial source for a trip to the cell surface with HMT. Thus the components of Mta are present in the rat (Fig. 1).

Can one then identify a maternally transmitted antigen in the rat? Only serendipity uncovered mouse Mta; mitochondrial polymorphism gives little guidance, as the critical nucleotide is one among 16,300, the proverbial needle in a haystack. But serendipity works in rats, too. Davies and her colleagues (Davies et al. 1991) found that (LEWxDA)F1 rats immunized with DA T cells generated a CTL response against an antigen present on DA and (DAxLEW)F1 cells. This antigen was strictly maternally inherited, sensitive to chloramphenicol (an inhibitor of mitochondrial protein synthesis) and followed mitochondrial haplotypes (present in the DA:AUG:PVG and the BN haplotypes and absent in the LEW:WF haplotype). Earlier reports of graft-versus-host disease induced by LEW lymphocytes being more severe in (LEWxBN)F1 than in (BNxLEW)F1 hosts (Collins and Silvers 1981) and of LEW skin and sarcoma graft rejection by (BNxLEW)F1, but not (LEWxBN)F1, recipients (Kren 1990, personal communication) may be manifestations of the same antigen system acting in the opposite direction.

The rat therefore possesses a maternally transmitted antigen; the surprise was that it is not equivalent to that of the mouse (Fig. 1). Firstly, because the 5'-end of the rat *ND1* gene encodes the same amino acids in all three haplotypes, the rat mitochondrial minor antigen must be encoded elsewhere in the mitochondrial genome. Secondly, the presentation of the rat antigen is restricted by the classical RT1.A^a molecule, as shown by assay of MHC recombinants and by monoclonal antibody inhibition of lysis. This finding confirms that the specificity of *H-2M3* for MTF is based on the presence of *N*-formyl-methionine, rather than on the mitochondrial origin *per se*. The product of *RT1.M3* may well bind the amino-terminal peptide of rat ND1, but we cannot detect it because the CTL system relies on polymorphism.

A rat minor H antigen restricted by an *RT1.C* gene

Limited presentation of peptide ligands by medial class I molecules will prove to be more general than Mta. We present here evidence for presentation of an autosomal gene product by a medial class I antigen of the laboratory rat. The rat MHC is shown schematically in Figure 4. The only classical class I molecule, *i.e.* promiscuous restriction element, is specified by the *RT1.A* class I gene mapping at the left hand side of the rat MHC in the position homologous to *H-2K* (Livingstone 1983; Günther 1985). The *RT1.C* region on the right side of the rat MHC contains at least 60 class I genes; of these, several are expressed on the cell surface (Jameson 1988) and at least one specifies a typical polymorphic medial antigen, the CT antigen system (Stephenson et al. 1985; Marshak et al. 1977). No *C*-region class I antigen displays conventional classical properties (Günther 1985). However, during the course of studies intended to demonstrate the exclusive role of RT1.A as a restriction element, we found clear evidence of restriction by the product of a gene mapping in the *RT1.C* region (Livingstone 1983).

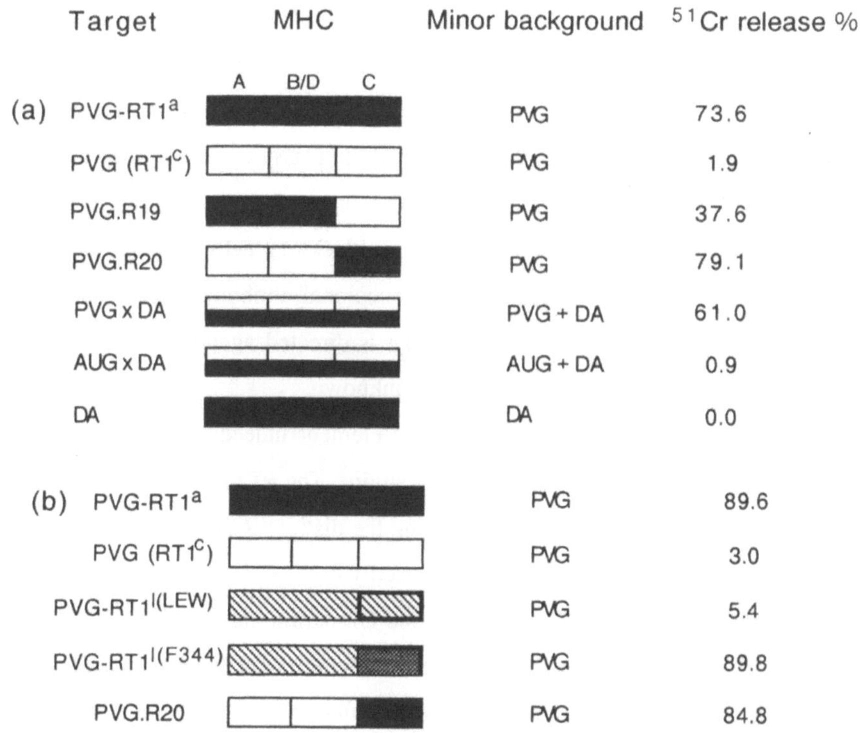

Fig. 4. Restriction of the secondary CTL response against a minor histocompatibility antigen of the rat by the product of a gene in the *RT1.C* region. In all cases, secondary CTL were obtained from (AUGxDA)F1 ($RT1^{c/a}$) rats primed in vivo and boosted in vitro with lymphoid cells from PVG.$RT1^{a}$ donors (Livingstone 1983). Concanavalin A-stimulated lymphoblasts from the indicated strains were labeled with ^{51}Cr before the assay. Effector to target ratios were titrated in each separate experiment (30:1 in *a* and 16.7:1 in *b*). Black bars indicate MHC regions derived from the $RT1^{a}$ haplotype, white bars the $RT1^{c}$ haplotype, and hatched bars the $RT1^{l(LEW)}$ haplotype, which differs from $RT1^{l(F344)}$ (shaded) in the *RT1.C* region by the polymorphic CT antigens (Stephenson et al. 1985; Marshak et al. 1977).

(AUGxDA)F1 ($RT1^{c}$ x $RT1^{a}$) rats were immunized against differential minor antigens of the PVG background carried on cells of the congenic strain PVG-$RT1^{a}$. Figure 4a shows that CTL obtained in this combination were largely specific for an antigen formed by a component from the PVG background and a component coded by the *C* region of the *RT1*^{a} haplotype. In particular, the PVG.R19 recombinant strain, carrying the RT1.A^a classical restriction element and an inappropriate *RT1.C* region, was a poor target, while the reciprocal PVG.R20 strain, carrying the inappropriate RT1.A^c classical restriction element with the *RT1.C*^{a} non-classical class I region was as good a target as the immunizing PVG-*RT1*^{a} strain itself. Importantly, the antigen could be complemented in an F1 hybrid between two negative strains, the DA responder strain (providing the *RT1*^{a} haplotype) and the PVG (*RT1*^{c}) strain providing the appropriate background. Monoclonal antibodies against the classical RT1.A^a molecule of the PVG-*RT1*^{a} targets or the RT1.A^c molecule of the recombinant PVG.R20 failed to block the strong killing. The weaker, but significant, killing of PVG.R19 cells was due to antigens restricted by RT1.A^a and could be blocked by a monoclonal antibody against this antigen. Whether this weak, classical RT1.A-restricted killing is directed against the same minor H antigen as that restricted by the *RT1.C*^{a} element is unknown.

A final piece of evidence that the restriction element indeed mapped to the *RT1.C* region came from an assay of two variant PVG.*RT1*^{l} strains. The *RT1.A* region of strains LEW and F344 are, so far as is known, the same, encoding the distinct RT1.A^l classical restriction element. However, *RT1*$^{l(LEW)}$ and *RT1*$^{l(F344)}$ differ markedly in the *C*-region in their class I gene fragments. All of *RT1.C*^{a} medial CT antigens of the DA strain are shared by *RT1*$^{l(F344)}$ but absent from *RT1*$^{l(LEW)}$ (Stephenson et al. 1985). It was therefore of interest that the RT1.C^a restriction molecule for the PVG minor antigen also appeared to be shared with PVG.*RT1*$^{l(F344)}$, but not with PVG.*RT1*$^{l(LEW)}$ (Fig. 4b).

Emphasizing the unusual nature of the restriction element, the conspicuous excess of killing restricted by an *RT1.C*^{a}-encoded element depends on the (AUGxDA)F1 hybrid responder. When parental DA rats were immunized against PVG-*RT1*^{a}, the killing restricted by RT1.A^a was at least as strong as that restricted by RT1.C^a, presumably because the response in this case involved other minor antigens not covered by the AUG parent and restricted conventionally by RT1.A^a. Similarly, (AUGxDA)F1 rats immunized against PVG (the same background antigens with *RT1*^{c}) produced CTL exclusively restricted by RT1.A^c. Preliminary genetic analysis showed that the RT1.C^a-restricted minor antigen is specified by a single autosomal locus unlinked to the MHC. Further experiments are needed to identify both interacting components in this example of restriction by a medial histocompatibility antigen. One wonders whether this system, without the special feature of a formylated peptide, might present another distinctive principle to guide the specific interaction of this minor histocompatibility antigen and its medial class I restriction element.

Other restriction systems

Qc-1 was defined as a CTL determinant of the mouse that depended on two genes: one that mapped to the *Q* region and encoded a product with Qa-2 serological epitopes, and one that could be mapped between the *Pb* and the *Ob* genes by the B10.MBR and B10.SBR recombinants (Murphy and Homer 1987). This result has been interpreted as evidence that one of the *Q* class I genes, perhaps *Q6*, presents a peptide antigen encoded in the *I-A* region; as the Qc-1 antigen is found on B lymphocytes, the peptide might be derived from one of the class II chains encoded by the *Ma* and *Mb* genes (Cho et al. 1991). Alternatively, a polymorphism affecting the specificity of the LMP proteosome or peptide transporter subunits encoded in the same region may determine whether a peptide encoded elsewhere is available for presentation (Monaco and McDevitt 1986; Monaco et al. 1990). The location and effect of this second gene is similar to *cim*, the modifier of MHC class I antigens in rats (Livingstone et al. 1989).

Spleen cells from several mouse strains of different *H-2* types all generate CTL restricted by a medial class I antigen upon coculture with a synthetic peptide corresponding to amino acids 173-190 of the influenza hemagglutinin (Milligan et al. 1991). The CTL are $CD8^+$ positive and have αβ receptors. They recognize target cells treated with either the long peptide or any of the staggered dodecapeptides that span the sequence from 179 to 197 and thus include the pentapeptide sequence Met-Leu-Ile-Ile-Trp (186-190). The restriction element is shared by the *a*, *b*, *d*, *k*, *k1*, and *s* haplotypes, but absent from the *f* haplotype, and it could be mapped distal to the *Q* region of the *H-2* complex. The restriction element must be a class I antigen, because only target cells that express β2m can be recognized. Because influenza infected cells or cells that express a minigene encoding the antigenic sequence are not lysed, Milligan and his colleagues (1991) suggest that medial class I restriction elements are "charged with peptides generated through an antigen processing and presentation pathway distinct from the pathway used by conventional MHC molecules". However, they have not ruled out the less interesting explanation that the active peptide ligand is a contaminant of their synthetic peptides (Rötzschke et al. 1990; Schumacher et al. 1991), which is not generated when the influenza peptide is made by the target cells. Another alternative, to be considered before a distinct pathway, is that the medial class I molecule must be associated with bovine, rather than mouse, β2m to be able to bind these peptides (Rock et al. 1991); the peptides synthesized inside the cell would not associate with and be protected by an MHC molecule in that case (Griem et al. 1991).

The Qa-1^a or Qa-1^b alloantigens are not fully expressed in mice that are homozygous for the *H-2D^k* region (Fischer Lindahl 1983). The difference could be quantitative, or it could be due to a qualitative lack of a determinant (Aldrich et al. 1988), perhaps a peptide ligand encoded in the *H-2D* region of other haplotypes, or it could be due to lack of a posttranslational modification (Wolf and Cook 1990). Participation of a mitochondrial peptide ligand in the Qa-1 alloantigen may explain why we can partially restore Qa-1 expression on RMA-S cells by treatment with an inhibitor of mitochondrial ATPase, which is thought to increase turnover of mitochondrial proteins (Hermel, in preparation).

Qa-1^b has also been shown to present the antigen GT (a random copolymer of Glu and Tyr) to a GT-specific T cell hybridoma with γ/δ receptors (Vidović et al. 1989). It has been proposed generally that the genes at the distal end of the MHC present antigen to T cells with γ/δ receptors (Raulet 1989; Strominger 1989). The notion is attractive because the low level of polymorphism and the limited tissue distribution of the medial class I antigens might match the peculiar distribution of the γ/δ V gene repertoire and their preference for highly conserved ligands, such as heat shock proteins (reviewed by Raulet [1989]), but it has not yet been submitted to a critical test. The response to Mta, at least, appears to be mediated by α/β T cell receptors (Fischer Lindahl et al. 1989).

Discussion

Examples of antigen presentation by medial class I antigens remain few. The low level of expression may put medial antigens at a disadvantage when competing with classical class I antigens for their ligands (Howard 1987). However, the medial class I molecules may have a higher affinity for their particular ligands than the average classical class I molecule if they serve a specialized function; this may explain why the RMA-S mutation affects Mta and Qa-1 less than H-2 alloantigens (Hermel et al. 1991). Some MHC class I genes may bind peptide antigen but fail to activate T cells because their α3 domain cannot bind CD8. Aldrich et al.

(1991) found that a hybrid L^d molecule, in which the α3 domain of the *Q7* gene is substituted for that of *L^d*, failed to positively select T cells, and a viral peptide antigen was not recognized although it was presumably bound by the L^d hybrid.

Yet another reason may be the immunologist's bias towards polymorphism. If either the restriction element or the ligand is monomorphic, then they will escape our attention unless they have a limited tissue distribution (Marrack and Kappler 1988) or are species-specific (Heath et al. 1989), as has been the case with ligands for the classical alloantigens. There is little question that the medial class I genes are less polymorphic than the classical ones (Flaherty 1988).

It is becoming clear that orthologous relationships are conserved and can be defined also among the non-classical antigens: *H-2M3* and *RT1.M3* presented in this paper, *H-2T23* (formerly "gene *37*", identified as *Qa-1^b* [Wolf and Cook 1990]) and *RT.BM1* (Parker et al. 1990; Hughes 1991), *H-2T10* and *T22* and *RT(5.8)* (Kirisits et al. 1991), and the *H-2M1* family and *Pele-C* genes of *Peromyscus leucopus* (Crew et al. 1990). A probe for *H-2M2* gives few and distinct bands in a Southern blot with rat genomic DNA, suggesting another conserved relation (Brorson et al. 1989). These relationships have all been defined by cross-hybridization or sequence gazing, and it remains to be determined whether they extend to function, *i.e.*, does RT1.M3 present formylated peptides, and does *RT.BM1* encode the CT antigen system, which is so similar to Qa-1?

At this point we must ask whether the study by Rogers (1985) was misleading? Crew and colleagues (1990) argued that he used the wrong probes: exon 5 and the 3'-end may detect conserved relationships, whereas exons 2 and 3 are strongly selected for polymorphism by antigen. However, an exon 3 probe defined the *M3* orthology. We rather believe that Rogers made the wrong choice with the *Qa-2* and *Tla* genes. The strongest mutational force moulding the MHC is unequal cross-over between members of the class I multigene family. This force has eliminated entire genes in the *H-2D* region (the *dm1* and *dm2* mutants, reviewed by Stephan et al. [1986]) and the *RT1.C* region (the *lm1* mutant [Wurst et al. 1989]) in our own scientific lifetime, rather than over evolutionary time. The same force acting on another gene family is the main cause of color blindness in humans (Nathans et al. 1986). Unequal recombination is most likely to duplicate or eliminate genes in the middle of the MHC, such as the *Qa* region, which can be entirely missing (O'Neill et al. 1986). The orientation of the *H-2T* gene clusters is still open to debate, and the conserved genes, *T22* and

T23, may be located at the distal end, closest to the *Hmt* or *M* region (Richards et al. 1989). It is possible that the field of sequence homogenization (Rada et al. 1990) is similarly confined to the center of the MHC, and that the genes at the distal end are left in peace to evolve their specialized antigen presentation. One might even propose that the transposition of the *RT1.A* and *H-2K* genes (Hanson and Trowsdale 1991), as well as the physical isolation of the *HLA-B* and *HLA-C* genes from the other class I genes of the *HLA* complex (Koller et al. 1989), serves to protect them from loss by unequal recombination. A physical map of the class I genes in the rat MHC, as well as completion of the mouse map, is urgently needed.

Conclusions

In a gene family that evolves by expansion, selection and contraction, it is not surprising that the classical class I antigens of one species are derived from genes that became the "road not taken" of another species. We have seen this with the cotton top tamarin class I genes and *HLA-G* (Watkins et al. 1990), and a similar situation has been proposed for *RT1.A* and *H-2K2* (Hughes 1991).

Ten years ago, lack of information led us to consider failure to restrict antigen recognition a defining feature of medial histocompatibility antigens (Fischer Lindahl and Langhorne 1981). Five years later, we realised that some medial class I antigens could act as restriction elements (Howard 1987). Today, our knowledge of the folding and structure of class I molecules suggests that any heavy chain that could not bind some peptide ligand would have little chance of forming a stable complex with β2m and would be unlikely to reach the cell surface at physiological temperatures (Ljunggren et al. 1990). We have here presented definitive evidence in both a rat and a mouse system for antigen presentation by restriction elements encoded in the non-classical region of the MHC; in the mouse system both the restricting molecule and the peptide ligand have been defined and found to represent a novel way of distinguishing antigen. These restriction elements might be more appropriatelynamed neo-classical class I antigens.

Acknowledgments. We thank Diana Diggs for the stereo plot, Harvey Harris for library assistance, and Lynne Yakes for printing the final version of this paper. E.H. was supported by Grant No. CA09082 from the National Cancer Institute and by the Texas Ladies' Auxiliary of Veterans of Foreign Wars.

References

Aldrich, C.J., Rodgers, J.R., and Rich, R.R.: Regulation of Qa-1 expression and determinant modification by an *H-2D*-linked gene, *Qdm*. *Immunogenetics 28*: 334-344, 1988

Aldrich, C.J., Hammer, R.E., Jones-Youngblood, S., Koszinowski, U., Hood, L., Stroynowski, I., and Forman, J.: Negative and positive selection of antigen-specific cytotoxic T lymphocytes affected by the α3 domain of MHC class I molecules. *Nature* (submitted 1991)

Anderson, S., Bankier, A.T., Barrell, B.G., deBruijn, M.H.L., Coulson, A.R., Drouin, J., Eperon, I.C., Nierlich, D.P., Roe, B.A., Sanger, F., Schreier, P.H., Smith, A.J.H., Staden, R., and Young, I.G.: Seqence and organization of the human mitochondrial genome. *Nature 290*: 457-465, 1981

Anderson, S., de Bruijn, M.H.L., Coulson, A.R., Eperon, I.C., Sanger, F., and Young, I.G.: Complete sequence of bovine mitochondrial DNA. Conserved features of the mammalian mitochondrial genome. *J Mol Biol 156*: 683-717, 1987

Bibb, M.J., Van Etten, R.A., Wright, C.T., Walberg, M.W., and Clayton, D.A.: Sequence and gene organization of mouse mitochondrial DNA. *Cell 26*: 167-170, 1981

Bjorkman, P.J. and Parham, P.: Structure, function, and diversity of class I major histocompatibility complex molecules. *Annu Rev Biochem 59*: 253-288, 1990

Brorson, K., Richards, C.S., Hunt, S.W.,III, Cheroutre, H., Fischer Lindahl, K., and Hood, L.: Analysis of a new class I gene mapping to the *Hmt* region of the mouse. *Immunogenetics 30*: 273-283, 1989

Burger, G. and Werner, S.: The mitochondrial URF1 gene in *Neurospora crassa* has an intron that contains a novel type of URF. *J Mol Biol 186*: 231-242, 1985

Cantatore, P., Roberti, M., Rainaldi, G., Saccone, C., and Gadaleta, M.N.: Clustering of tRNA genes in *Paracentrotus lividus* mitochondrial DNA. *Curr Genet 13*: 91-96, 1988

Chapdelaine, Y. and Bonen, L.: The wheat mitochondrial gene for subunit I of the NADH dehydrogenase complex: A *trans*-splicing model for this gene-in-pieces. *Cell 65*: 465-472, 1991

Cho, S., Attaya, M., Brown, M.G., and Monaco, J.J.: A cluster of transcribed sequences between the *Pb* and *Ob* genes of the murine major histocompatibility complex. *Proc Natl Acad Sci USA 88*: 5197-5201, 1991

Collins, N.H. and Silvers, W.K.: Immunologic interactions between parental and reciprocal F_1 hybrid rats. *Transplant Proc 13*: 1439-1443, 1981

Crew, M.D., Filipowsky, M.E., Zeller, E.C., Smith, G.S., and Walford, R.L.: Major histocompatibility complex class I genes of *Peromyscus leucopus*. *Immunogenetics 32*: 371-379, 1990

Davies, J.D., Wilson, D.H., Hermel, E., Fischer Lindahl, K., Butcher, G.W., and Wilson, D.B.: Generation of T cells with lytic specificity for atypical antigens. I. A mitochondrial antigen in the rat. *J Exp Med 173*: 823-832, 1991

Desjardins, P. and Morais, R.: Sequence and gene organization of the chicken mitochondrial genome. A novel gene order in higher vertebrates. *J Mol Biol 212*: 599-634, 1990

Ehrlich, R., Lifshitz, R., Pescovitz, M.D., Rudikoff, S., and Singer, D.S.: Tissue-specific expression and structure of a divergent member of a class I MHC gene family. *J Immunol 139*: 593-602, 1987

Fischer Lindahl, K.: Polymorphism and expression of Qed. *Transplant Proc 15*: 2042-2044, 1983

Fischer Lindahl, K. and Langhorne, J.: Medial histocompatibility antigens. *Scand J Immunol 14*: 643-654, 1981

Fischer Lindahl, K., Hausmann, B., Robinson, P.J., Guénet, J.-L., Wharton, D.C., and Winking, H.: Mta, the maternally transmitted antigen, is determined jointly by the chromosomal *Hmt* and the extrachromosomal *Mtf* genes. *J Exp Med 163*: 334-346, 1986

Fischer Lindahl, K., Loveland, B.E., and Richards, C.S.: The end of *H-2*. *In* C.S. David (ed.): *H-2 Antigens: genes, molecules, function*, pp. 327-338, Plenum Press, New York 1987

Fischer Lindahl, K., Hermel, E., Loveland, B.E., Richards, S., Wang, C.-R., and Yonekawa, H.: Molecular definition of a mitochondrially encoded mouse minor histocompatibility antigen. *Cold Spring Harbor Symp Quant Biol 54*: 563-569, 1989

Fischer Lindahl, K., Hermel, E., Loveland, B.E., and Wang, C.-R.: Maternally transmitted antigen of mice - a model transplantation antigen. *Annu Rev Immunol 9*: 351-372, 1991

Flaherty, L.: Major histocompatibility complex polymorphism: a nonimmune theory for selection. *Hum Immunol 21*: 3-13, 1988

Flajnik, M.F., Canel, C., Kramer, J., and Kasahara, M.: Evolution of the major histocompatibility complex: Molecular cloning of the major histocompatibility complex class I from the amphibian *Xenopus*. *Proc Natl Acad Sci USA 88*: 537-541, 1991

Frels, W.I., Bordallo, C., Golding, H., Rosenberg, A., Rudikoff, S., and Singer, D.S.: Expression of a class I MHC transgene: Regulation by a tissue-specific negative regulatory DNA sequence element. *New Biologist 2*: 1024-1033, 1990

Garesse, R.: *Drosophila melanogaster* mitochondrial DNA: gene organization and evolutionary considerations. *Genetics 118*: 649-663, 1988

Garey, J.R. and Wolstenholme, D.R.: Platyhelminth mitochondrial DNA: Evidence for early evolutionary origin of a $tRNA^{ser}AGN$ that contains a dihydrouridine arm replacement loop and of serine-specifying AGA and AGG codons. *J Mol Biol 28*: 374-387, 1989

Garrett, T.P.J., Saper, M.A., Bjorkman, P.J., Strominger, J.L., and Wiley, D.C.: Specificity pockets for the side chains of peptide antigens in HLA-Aw68. *Nature 342*: 692-696, 1989

Griem, P., Wallny, H.-J., Falk, K., Rötzschke, O., Arnold, B., Schönrich, G., Hämmerling, G., and Rammensee, H.-G.: Uneven tissue distribution of minor histocompatibility proteins versus peptides is caused by MHC expression. *Cell 66*: 633-640, 1991

Guillemot, F., Billault, A., Pourquié, O., Béhar, G., Chaussé, A.-M., Zoorob, R., Kreibich, G., and Auffray, C.: A molecular map of the chicken major histocompatibility complex: the class II β genes are closely linked to the class I genes and the nucleolar organizer. *EMBO J 7*: 2775-2785, 1988

Günther, E.: Genetic control of anti-male immune responsiveness in the rat. *Transplant Proc 17*: 1840-1841, 1985

Hanson, I.M. and Trowsdale, J.: Colinearity of novel genes in the class II regions of the MHC in mouse and man. *Immunogenetics 34*: 5-11, 1991

Hashimoto, K., Nakanishi, T., and Kurosawa, Y.: Isolation of carp genes encoding major histocompatibility complex antigens. *Proc Natl Acad Sci USA 87*: 6863-6867, 1990

Heath, W.R., Hurd, M.E., Carbone, F.R., and Sherman, L.A.: Peptide-dependent recognition of $H\text{-}2K^b$ by alloreactive cytotoxic T lymphocytes. *Nature 341*: 749-752,

1989

Hensgens, L.A.M., Brakenhoff, J., De Vries, B.F., Sloof, P., Tromp, M.C., Van Boom, J.H., and Benne, R.: The sequence of the gene for cytochrome c oxidase subunit, a frameshift containing gene for cytochrome c oxidase subunit II and seven unassigned reading frames in *Trypanosoma brucei* mitochondrial maxi-circle DNA. *Nucl Acids Res 12*: 7327-7344, 1984

Hermel, E., Grigorenko, E., and Fischer Lindahl, K.: Expression of medial class I histocompatibility antigens on RMA-S mutant cells. *Int Immunol 3*: 407-412, 1991

Higuchi, R.G., Wrischnik, L.A., Oakes, E., George, M., Tong, B., and Wilson, A.C.: Mitochondrial DNA of the extinct quagga: Relatedness and extent of postmortem change. *J Mol Biol 25*: 283-287, 1987

Howard, J.C.: MHC organization of the rat: evolutionary considerations. *In* G. Kelsoe and D.H. Schulze (eds.): *Evolution and Vertebrate Immunity. The Antigen-receptor and MHC Gene Families*, pp. 397-411, Univ. of Texas Press, Austin, TX 1987

HsuChen, C.-C., Kotin, R., and Dubin, D.T.: Sequences of the coding and flanking regions of the large ribosomal subunit RNA gene of mosquito mitochondria. *Nucl Acids Res 12*: 7771-7785, 1984

Hughes, A.L.: Independent gene duplications, not concerted evolution, explain relationships among class I MHC genes of murine rodents. *Immunogenetics 33*: 367-373, 1991

Ito, K., Van Kaer, L., Bonneville, M., Hsu, S., Murphy, D.B., and Tonegawa, S.: Recognition of the product of a novel MHC TL region gene (27^b) by a mouse $\gamma\delta$ T cell receptor. *Cell 62*: 549-561, 1990

Jacobs, H.T., Elliott, D.J., Math, V.B., and Farquharson, A.: Nucleotide sequence and gene organization of sea urchin mitochondrial DNA. *J Mol Biol 202*: 185-217, 1988

Jameson, S.C.: *Cloning and characterisation of rat MHC class I genes*. Ph.D. thesis, Cambridge University 1988

Jameson, S.C., Rada, C., Lorenzi, R., Diamond, A.G., Butcher, G.W., and Howard, J.C.: Cloning, expression, and evolution of rat classical and nonclassical class I genes. *Transplant Proc 22*: 2510-2511, 1990

Kirisits, M.J., Kunz, H.W., and Gill, T.J.,III: Genomic DNA sequence and organization of a *TL*-like gene in the *RT1.E-grc* region of the rat. *Immunogenetics* (submitted 1991)

Klein, J. and Figueroa, F.: Evolution of the major histocompatibility complex. *CRC Crit Rev Immunol 6*: 295-386, 1986

Koller, B.H., Geraghty, D.E., DeMars, R., Duvick, L., Rich, S.S., and Orr, H.T.: Chromosomal organization of the human major histocompatibility complex class I gene family. *J Exp Med 169*: 469-480, 1989

Koller, B.H., Marrack, P., Kappler, J.W., and Smithies, O.: Normal development of mice deficient in β_2M, MHC class I proteins, and $CD8^+$ T cells. *Science 248*: 1227-1230, 1990

Kotin, R.M. and Dubin, D.T.: Sequences around the 3'-end of a ribosomal RNA gene of hamster mitochondria. *Biochim Biophys Acta 782*: 106-108, 1984

Livingstone, A.M.: *Specificity of responses against rat major transplantation antigens*. Ph.D. Thesis, Cambridge University 1983

Livingstone, A.M., Powis, S.J., Diamond, A.G., Butcher, G.W., and Howard, J.C.: A *trans*-acting major histocompatibility complex-linked gene whose alleles determine gain and loss changes in the antigenic structure of a classical class I molecule. *J Exp Med 170*: 777-795, 1989

Ljunggren, H.-G., Stam, N.J., Öhlén, C., Neefjes, J.J., Höglund, P., Heemels, M.-T., Bastin, J., Schumacher, T.N.M., Townsend, A., Kärre, K., and Ploegh, H.L.: Empty MHC class I molecules come out in the cold. *Nature 346*: 476-480, 1990

Loveland, B.E., Wang, C.-R., Yonekawa, H., Hermel, E., and Fischer Lindahl, K.: Maternally transmitted histocompatibility antigen of mice: a hydrophobic peptide of a mitochondrially encoded protein. *Cell 60*: 971-980, 1990

Marrack, P. and Kappler, J.: T cells can distinguish between allogeneic major histocompatibility complex products on different cell types. *Nature 332*: 840-843, 1988

Marshak, A.D., Doherty, P.C., and Wilson, D.B.: The control of specificity of cytotoxic T lymphocytes by the major histocompatibility complex (*Ag-B*) in rats and identification of a new alloantigen system showing no Ag-B restriction. *J Exp Med 146*: 1773-1790, 1977

McCracken, A., Uhlenbusch, I., and Gellissen, G.: Structure of the cloned *Locusta migratoria* mitochondrial genome: restriction mapping and sequence of its ND-1 (URF-1) gene. *Curr Genet 11*: 625-630, 1987

Milligan, G.N., Flaherty, L., Braciale, V.L., and Braciale, T.J.: Nonconventional (TL-encoded) major histocompatibility complex molecules present processed viral antigen to cytotoxic T lymphocytes. *J Exp Med 174*: 133-138, 1991

Monaco, J.J. and McDevitt, H.O.: The LMP antigens: a stable MHC-controlled multisubunit protein complex. *Hum Immunol 15*: 416-426, 1986

Monaco, J.J., Cho, S., and Attaya, M.: Transport protein genes in the murine MHC: possible implications for antigen processing. *Science 250*: 1723-1726, 1990

Murphy, D.B. and Homer, R.J.: A Q region product as a CTL restriction element? In C S David (ed.): *H-2 Antigens Genes, Molecules, Function*, pp. 359-369, Plenum Publishing Co., New York 1987

Nathans, J., Piantanida, T.P., Eddy, R.L., Shows, T.B., and Hogness, D.S.: Molecular genetics of inherited variation in human color vision. *Science 232*: 203-210, 1986

O'Neill, A.E., Reid, K., Garberi, J.C., Karl, M., and Flaherty, L.: Extensive deletions in the *Q* region of the mouse major histocompatibility complex. *Immunogenetics 24*: 368-373, 1986

Parham, P., Benjamin, R.J., Chen, B.P., Clayberger, C., Ennis, P.D., Krensky, A.M., Lawlor, D.A., Littman, D.R., Norment, A.M., Orr, H.T., Salter, R.D., and Zemmour, J.: Diversity of class I HLA Molecules: Functional and evolutionary interactions with T cells. *Cold Spring Harbor Symp Quant Biol 54*: 529-543, 1989

Parker, K.E., Carter, C.A., and Fabre, J.W.: A rat class I cDNA clone with an Alu-like sequence and mapping to two genes in *RT1.C/E*. *Immunogenetics 31*: 211-214, 1990

Pritchard, A.E., Seilhamer, J.J., and Cummings, D.J.: Paramecium mitochondrial DNA sequences and RNA transcripts for cytochrome oxidase subunit I, URF1, and three ORFs adjacent to the replication origin. *Gene 44*: 243-253, 1986

Rada, C., Lorenzi, R., Powis, S.J., van den Bogaerde, J., Parham, P., and Howard, J.C.: Concerted evolution of class I genes in the major histocompatibility complex of murine rodents. *Proc Natl Acad Sci USA 87*: 2167-2171, 1990

Raulet, D.H.: Antigens for γ/δ T cells. *Nature 339*: 342-343, 1989

Richards, S., Bucan, M., Brorson, K., Kiefer, M.C., Hunt, S.W.,III, Lehrach, H., and Fischer Lindahl, K.: Genetic and molecular mapping of the *Hmt* region of mouse. *EMBO J 8*: 3749-3757, 1989

Rock, K.L., Gamble, S., Rothstein, L., Gramm, C., and Benacerraf, B.: Dissociationof β_2-microglobulin leads to the accumulation of a substantial pool of inactive class I

MHC heavy chains on the cell surface. *Cell 65*: 611-620, 1991

Roe, B.A., Ma, D.-P., Wilson, R.K., and Wong, J.F.-H.: The complete nucleotide sequence of the *Xenopus laevis* mitochondrial genome. *J Biol Chem 260*: 9759-9774, 1985

Rogers, J.H.: Mouse histocompatibility-related genes are not conserved in other mammals. *EMBO J 4*: 749-753, 1985

Rötzschke, O., Falk, K., Deres, K., Schild, H., Norda, M., Metzger, J., Jung, G., and Rammensee, H.-G.: Isolation and analysis of naturally processed viral peptides as recognized by cytotoxic T cells. *Nature 348*: 252-254, 1990

Saper, M.A., Bjorkman, P.J., and Wiley, D.C.: Refined structure of the human histocompatibility antigen HLA-A2 at 2.6 Å resolution. *J Mol Biol 219*: 277-319, 1991

Satz, M.L., Wang, L.-C., Singer, D.S., and Rudikoff, S.: Structure and expression of two porcine genomic clones encoding class I MHC antigens. *J Immunol 135*: 2167-2175, 1985

Schumacher, T.N.M., De Bruijn, M.L.H., Vernie, L.N., Kast, W.M., Melief, C.J.M., Neefjes, J.J., and Ploegh, H.L.: Peptide selection by MHC class I molecules. *Nature 350*: 703-706, 1991

Shawar, S.M., Cook, R.G., Rodgers, J.R., and Rich, R.R.: Specialized functions of MHC class I molecules. I. An *N*-formyl peptide receptor is required for construction of the class I antigen Mta. *J Exp Med 171*: 897-912, 1990

Shawar, S.M., Vyas, J.M., Rodgers, J.R., Cook, R.G., and Rich, R.R.: Specialized functions of MHC class I molecules. II. Hmt binds N-formylated peptides of mitochondrial and prokaryotic origin. *J Exp Med* (submitted 1991)

Singer, D.S., Ehrlich, R., Satz, L., Frels, W., Bluestone, J., Hodes, R., and Rudikoff, S.: Structure and expression of class I MHC genes in the miniature swine. *Vet Immunol Immunopathol 17*: 211-221, 1987

Stephan, D., Sun, H., Fischer Lindahl, K., Meyer, E., Hämmerling, G.J., Hood, L., and Steinmetz, M.: Organization and evolution of D region class I genes in the mouse major histocompatibility complex. *J Exp Med 163*: 1222-1244, 1986

Stephenson, S.P., Morley, R.C., and Butcher, G.W.: Genetics of the rat CT system: its apparent complexity is a consequence of cross-reactivity between the distinct MHC class I antigens RT1.C and RT1.A. *J Immunogenetics 12*: 101-114, 1985

Strominger, J.L.: The γδ T cell receptor and class Ib MHC-related proteins: enigmatic molecules of immune recognition. *Cell 57*: 895-898, 1989

Stroynowski, I.: Molecules related to class-I major histocompatibility complex antigens. *Annu Rev Immunol 8*: 501-530, 1990

Townsend, A., Öhlén, C., Bastin, J., Ljunggren, H.-G., Foster, L., and Kärre, K.: Association of class I major histocompatibility heavy and light chains induced by viral peptides. *Nature 340*: 443-448, 1989

Tykocinski, M.L., Marche, P.N., Max, E.E., and Kindt, T.J.: Rabbit class I MHC genes: cDNA clones define full-length transcripts of an expressed gene and a putative pseudogene. *J Immunol 133*: 2261-2269, 1984

Vidović, D., Roglić, M., McKune, K., Guerder, S., MacKay, C., and Dembić, Z.: Qa-1 restricted recognition of foreign antigen by a γδ T-cell hybridoma. *Nature 340*: 646-650, 1989

Wang, C.-R., Loveland, B.E., and Fischer Lindahl, K.: *H-2M3* encodes the MHC class I molecule presenting the maternally transmitted antigen of the mouse. *Cell 26*: 335-345, 1991

Watkins, D.I., Chen, Z.W., Hughes, A.L., Evans, M.G., Tedder, T.F., and Letvin, N.L.:

Evolution of the MHC class I genes of a New World primate from ancestral homologues of human non-classical genes. *Nature 346*: 60-63, 1990

Watts, S., Wheeler, C., Morse, R., and Goodenow, R.S.: Amino acid comparison of the class I antigens of mouse major histocompatibility complex. *Immunogenetics 30*: 390-392, 1989

Wissinger, B., Schuster, W., and Brennicke, A.: *Trans* splicing in Oenothera mitochondria: *nadl* mRNAs are edited in exon and *trans*-splicing group II intron sequences. *Cell 65*: 473-482, 1991

Wolf, P.R. and Cook, R.G.: The *TL* region gene *37* encodes a Qa-1 antigen. *J Exp Med 172*: 1795-1804, 1990

Wurst, W., Wonigeit, K., and Günther, E.: A mutant rat major histocompatibility haplotype showing a large deletion of class I sequences. *Immunogenetics 30*: 237-242, 1989

Zijlstra, M., Bix, M., Simister, N.E., Loring, J.M., Raulet, D.H., and Jaenisch, R.: β2-microglobulin deficient mice lack $CD4^{-}8^{+}$ cytolytic T cells. *Nature 344*: 742-746, 1990

Mls ANTIGENS (SUPERANTIGENS), CLASS II MHC, AND TCR REPERTOIRE: CO-ADAPTIVE EVOLUTION

Paul Zhou and Chella S. David

Department of Immunology, Mayo Clinic, Rochester, Minnesota 55905, USA

Mls or superantigen system differs from the conventional peptide antigens as well as mitogens and can be classified as endogenous superantigens, encoded by genes residing in the mouse genome itself, and exogenous superantigens, derived from viral, bacterial, or mycoplasma infection or immunization. In mouse the endogenous superantigens are mouse mammary tumor viral genes, which are scattered throughout the mouse genome. During thymic differentiation, T cells which respond to endogenous superantigens in the context of self MHC class II molecules are clonally deleted. Thus T cells expressing self reactive variable region receptors are eliminated. Studies on the exogenous superantigens demonstrated that superantigens bind to the region on MHC class II which is outside of the conventional peptide binding pocket and stimulate T cells through the interaction with specific Vb chain of T cell receptor. The superantigen system has not been discovered in the human.

Mls (minor lymphocyte stimulating locus) was first discovered by Festenstein twenty years ago and was defined by demonstration of a strong primary T cell proliferative response in a mixed lymphocyte culture between MHC-identical murine strains (Festenstein 1973). Soon after, the Mls locus was determined to be located on mouse chromosome 1 by demonstration of a close linkage between Mls phenotype and two mouse chromosome 1 markers *Dip-1* (Dipeptidase 1), and *ald* (adrenal lipid depletion) in AKXL and BXD recombinant inbred strains, as well as the sixth backcross generation of (BALB/c x DBA/2) x BALB/c (Festenstein et al. 1977). It was postulated at that time that Mls was controlled by four codominant alleles (Mls^a, Mls^b, Mls^c, and Mls^d) on a single Mls locus on mouse chromosome 1. Later on, the Mls locus was reassessed through more detailed segregation studies by using Mls^a and Mls^c specific cloned T cell lines (Abe et al. 1989). It was demonstrated that Mls^a and Mls^c determinants are not allelic; Mls^d phenotype results from the coexpression of Mls^a (on chromosome 1) and Mls^c (at some other chromosomal location); and Mls^b may be a null allele of these loci.

NATO ASI Series, Vol. H 59
Molecular Evolution of the
Major Histocompatibility Complex
Edited by J. Klein and D. Klein

In the early 1980's, two unique features of Mls-induced proliferative T cell responses were elucidated. It was found that (1) about one-fifth of all T cells are able to respond to Mls stimulation; (2) the MHC class II molecules on antigen-presenting cells (APC) are required, but H-2 haplotype unrestricted (Janeway et al. 1989). This intimate relationship between Mls and class II molecules in stimulating T cells lead the authors to speculate that the antigen receptor on T cells must be involved in the recognition of Mls and class II molecules on APC cells (Janeway et al. 1989). Eventually, a strong association between TCR specificity for Mls^a or Mls^c and the expression of a particular TCR Vb segment was discovered. It was shown that the Mls^a product in association with H-2E molecules or E_α/A_β hybrid molecules clonally deletes and/or stimulates the $V_\beta6$-, $V_\beta8.1$-, and $V_\beta9$-bearing T cells (Kappler et al. 1988; MacDonald et al. 1988; Happ et al. 1989), and Mls^c product(s) in association with H-2E molecules clonally delete and/or stimulate $V_\beta3$ bearing T cells (Pullen et al. 1989). In the absence of H-2E molecules, H-2A molecules could present Mls^a to varying degrees. A^b and A^k were good presenters, A^s and A^f were intermediate, while A^q and A^p were nonpresenters (Anderson et al. 1989). The clonal deletion of these Mls-reactive, specific V_β-bearing T cells takes place in the thymus during T cell development, presumably through the interaction between the TCR V_β on the CD4/CD8 double positive immature thymocytes and the Mls and class II on the bone marrow-derived APC cells (dendritic cells and macrophages) (Kappler et al. 1988; MacDonald et al. 1988). Mature thymocytes and peripheral T cells contain a reduced number of these Vb specific, Mls-reactive T cells.

Mls and mammary tumor virus

Very recent studies suggested that the Mls genes may actually be some of the mouse mammary tumor virus genomes (mmtv) and the Mls gene products may be the products of these viral genomes (Acha-Orbea et al. 1991; Choi et al. 1991; Dyson et al. 1991; Marrack et al. 1991; Frankel et al. 1991; Woodland et al. 1990; Woodland et al. 1991). Mmtv is a B-type retrovirus. It can be transmitted either exogenously, as virus particles, mainly in milk, or endogenously via the germ line as stably integrated proviral genomes. The latter was first discovered from studies of tumorigenesis in the mouse strain GR (Muhlbock 1965). It was found that females of the GR strain developed virus-associated mammary carcinomas at an early age even after foster nursing by virus-negative females. Similar phenomenon was also seen in C3H and DBA strains (Bentvelzen et al. 1970). Later on, in both cases single dominant alleles, *mtv-1* in GR strain and *mtv-2* in C3H strain, were found to be responsible for the phenotypes (Salmons and Gunzburg 1987). Among mouse inbred strains and some wild mice about 20 individual mmtv proviral loci are widely scattered throughout the mouse genome (Kozak et al. 1987). Genetically, they behave as stable Mendelian genes in mouse chromosomes, which is believed to result from rare germ line infection events. The gross structure of these endogenous mmtv genomes are the same as that of exogenous proviruses obtained by infection in the order cell-LTR-*gag-pol-env*-LTR-cell (Salmons and Gunzburg 1987).

Expression of mmtv genomes varies among these individual proviral loci (Salmons and Gunzburg 1987). Only *Mtv-1* and *Mtv-2* loci are known to be able to make infectious viral particles. Other mmtv loci are able to synthesize some viral proteins, but not the whole viral particles. Still other may be totally silent. The mmtv transcripts and/or proteins are mainly detected in the mammary glands. Salivary glands and lymphoid tissues (thymus, spleen, lymph nodes and bone marrow) may be the only other tissues, besides the mammary glands, showing the expression of mmtv. The level of mmtv expression in these two tissues is 100- to 1000-fold lower than that found in mammary glands. Mmtv expression can be greatly enhanced by the stimulation of glucocorticoid hormone (Hynes et al. 1983). This unique feature of mmtv expression is due to the existence of the hormone responsive element (HRE) within the U^3 region of LTR of mmtv genomes (Salmons and Gunzburg 1987). Some mmtv (*Mtv-9*) are found to be expressed in normal B cells and B lymphomas. The transcripts of *Mtv-9* genome are greatly enhanced in these B cells upon stimulation with lipopolysaccharide (LPS) and interleukin-4 (IL-4) (King et al. 1990).

Very recent studies from several laboratories indicated that the products of mouse Mtv genomes may serve as endogenous superantigens (cotolerogens) involving the elimination of certain TCR V_{β}-bearing T cells by MHC class II molecules during the T cell ontogeny in thymus. This notion is based upon the following observations: (1) the genetic linkage studies demonstrated that genes encoding various endogenous superantigens are genetically inseparable from endogenous mmtv genomes; for example, Mls^a is inseparable from *Mtv-7* genome on mouse chromosome 1 and Mls^c is inseparable from *Mtv-1, Mtv-6* and *Mtv-13* on mouse chromosome 4, 16 and 7, respectively (Dyson et al. 1991; Frankel et al. 1991;

Woodland et al. 1991); (2) functionally, only certain B cell lymphoma lines, which express the appropriate mmtv transcripts, were able to stimulate the corresponding V_{β}-bearing T cell lines or T hybridomas (Woodland et al. 1991); and (3) the transfectant and transgenic mouse studies demonstrated that introducing the open reading frame (ORF) of the 3' long terminal repeat (LTR) of mmtv genomes (*Mtv-2*) caused the clonal deletion of Vb14-bearing T cells in *ORF-Mtv-2* transgenics and the clonal stimulation of Vb14-bearing T cells in *ORF-Mtv-2* transfectants (Acha-Orbea et al. 1991; Choi et al. 1991). Similar deletion of the vb14-bearing T cells were also observed in the exogenous *Mtv-2* virus infection through the milk route (Marrack et al. 1991). Thus, collectively, these results strongly suggest that the ORF of mmtv LTR encodes the endogenous superantigen itself The other possibility is that the ORF may encode a trans-acting factor to induce the expression of a real endogenous superantigen leading to the clonal deletion of certain V_{β} specific T cells.

If the ORF of mmtv truly encodes the Mls and the other endogenous superantigens, this could explain several puzzling features related to Mls and its V_{β} specificity, which includes (1) its genomes behave as autosomal dominant genes; (2) the existence of both stimulatory or null alleles due to the random integration of viral genomes; (3) the intercellular transfer of Mls products in radiation bone marrow chimeras; (4) the failure to detect Mls-like loci in human and the other species; and (5) about 10% to 15% amino acid variation among ORF in these individual mmtv genomes which could account for differences in the V_{β} specificity among these Mls and other endogenous superantigens.

Exogenous superantigens

Although our understanding of the endogenous superantigens has been greatly advanced in the last three years, the chemical identity of endogenous superantigens is still unclear. So the studies of these endogenous superantigens did not tell us how the superantigens interact with the MHC class II molecules and TCR. Information along this line is mostly gathered from studies of exogenous superantigens, such as *Staphylococcal enterotoxins* (SE), whose chemical nature has already been elucidated (Kappler et al. 1989). Staphylococcal enterotoxins (SE) are a family of molecules implicated in food poisoning and toxic shock in the human and weight loss and death in the mouse (Langford et al. 1978). To date, 5 major types (A, B, C, D, and E) have been isolated (Bergdoll 1983), which comprise a group of structurally related but serologically distinct intermediately sized proteins. Three forms of SEC (C1, C2, C3) have been described (for review see Janeway et al. 1989 and Marrack and Kappler 1990). A considerable structural homology between SEB and SEC1 and between SEA and SEE was observed, while less similarity exists between SEA and SEB. SE are the most potent T cell stimulators, which stimulate human or murine T cells at the concentrations of $<10^{-9}$ M efficiency (Fleischer and Schrezenmeier 1988; Fraser 1989). This contrasts strongly with the doses of mitogenic plant lectins, which usually require about 10^{-6} M for normal T cell activation. On the other hand, the plateau level of proliferation reached by SEA and SEB are lower than those reached with lectins, which suggests that SEA and SEB, while polyclonal stimulators, act only on a subset of T cells. It has been observed that cells bearing MHC class II molecules are required for presenting these toxins to $CD4^+$ T cells (Janeway et al. 1989). Besides this SE family, several other Gram positive bacterial toxins also possess the features of superantigens, which include toxic shock syndrome toxin (TSST1), exfoliating toxins A and B from *S. aureus* and Pyrogenic exotoxins A, B and C, but fewer studies have been done on these toxins.

Interaction between superantigens and MHC class II molecules

The requirement of MHC class II to present superantigens to the T cells is established. 1) Only APC which are MHC class II positive can present Mls to the T cells; while B lymphoma cell lines which lack class II failed to elicit the T cell response to Mls. 2) In the mouse, both anti-H-2A and anti-H-2E antibodies are able to inhibit Mls-induced T cell response; in this regard, anti-H-2E antibodies exhibit more profound inhibition than anti-H-2A antibodies do. 3) Antigen processing, which is required for the conventional antigens, seems not to be required for superantigen presentation, since paraformaldehyde-fixed, or chloroquine-treated APC cells have the same capacity as the non-treated APC (Fleischer and Schrezenmeier 1988).

In search for the superantigen binding site(s) on the MHC class II molecules, various strategies have been applied. By using a panel of L cells transfected with different isotypes of human MHC class II genes as APC to present SEs to T hybrids, the hierarchy for supporting SE stimulation (DR>DQ>DP) was observed, which was similar to that seen with murine isotypes (H-2E>H-2A) (Herman et al. 1990). Furthermore, within DR alleles the difference in their capacity of presenting SEs and TSST1 to T hybrids was observed (Herman et al. 1990). Since all DR molecules have a common alpha chain, the difference in superantigen presenting capacity of these DR alleles indicates it is the DR_β chain that possesses the superantigen binding site(s). In the DQw6b single transgenic mice we found that the trans-species A_α/DQ_β hybrid molecules, but not endogenous H-2A molecules, were able to present endogenous superantigens (Mls^a and *Mtv-9*) which lead to the clonal deletion of T cells bearing $V_\beta5$, $V_\beta6$, $V_\beta7$, $V_\beta8.1$, $V_\beta9$, and $V_\beta11$ cells (Zhou et al. 1991a,b). The results also suggested that it is DQb chain that interacts with the superantigens. However, studies with murine recombinant strains and E_α transgenic mice suggested that the E_α chain is important in the deletion of Vb6 and Vb11 T cells (Anderson et al. 1989).

Although we still do not know on which chain(s) the superantigen binding site(s) reside, we do know that the binding site(s) are not in the conventional peptide binding pocket. This conclusion is based upon the following experimental observations: 1) binding SEs to the class II molecules does not compete with the presentation of the conventional peptide to the specific T cells (Brown et al. 1988); 2) mutating the amino acid residues around the peptide binding pocket of Aa^k strongly effect its presenting the conventional antigen (a hen egg lysozyme), but has no effect on its presentation of SEs to T cells (Dellabona et al. 1990); 3) the limited proteolysis of SEs results in the loss of binding of SEs to class II molecules which indicates that the intact molecules of SEs are required for their interaction to class II molecules. Since the whole molecule of SEs are too big to fit in the peptide binding pocket of the class II molecules, these superantigens are very likely binding to site(s) on the class II, which are outside of the peptide binding pocket.

Evolution of Mls binding site

The first question is, "do humans have an Mls-like system which regulates the T cell repertoire"? So far there is no evidence that such a system exists, but the studies with the transgenic mice expressing human class II genes clearly show that human HLA-class II molecules can present Mls-like antigens to mouse T cells both in vivo and in vitro. We believe that such a system must exist in the human. The reason why it has not been identified could be because all major human class II molecules, DR, DQ and DP, can effectively present Mls-type antigens in regulating the T cell repertoire. Thus, there would be no 'hole' in the system and differences in the self reactive T cells cannot be identified between individuals.

Evolutionary selection in the mouse could have caused mutations and alterations in the Mls binding sites of class II molecules. While the ancestral class II gene, Ea, has retained the ability to present Mls-type antigens, the Aa and Ab genes probably evolved where mutations altered the Mls binding site. This could have happened during the survival of mouse species in a serious episode with an infectious agent. This could have generated a polymorphic Aa gene for the Mls binding site, such that some of them can present certain Mls type antigens to partially delete certain T cells, but not to provide a complete deletion of several T cells. The loss of the Ea gene in 30-40% of the mouse population also could have been the result of such a selection process. On the other hand, the expression of endogenous retroviral genes might also be more limited in the human compared to mice. Studies with humans indicate that certain T cell receptor V_β have a very low expression in all individuals. In the context of certain autoimmune diseases, these T cells seem to expand. This could indicate an expansion of a T cell population which had been clonally deleted during thymic maturation. Studies of the T cell repertoire in patients with Bare Lymphocyte Syndrome might reveal possible self reactive T cells which are deleted in most individuals. Only extensive analysis of the human T cell repertoire in many ethnic populations would tell us whether Mls-like superantigen system exists in humans.

References

Abe, R., Foo-Phillips, M., and Hodes, R.J.: Analysis of Mls^c genetics. A novel instance of genetic redundancy. *J Exp Med 170*: 1059-1073, 1989

Acha-Orbea, H., Shakhov, A.N., Scarpellino, L., Kolb, E., Muller, V., Vessaz-Shaw, A., Fuchs, R., Blochlinger, K., Rollini, P., Billotte, J., Sarafidou, M., MacDonald, H.R., and Diggelmann, H.: Clonal deletion of Vb14-bearing T cells in mice transgenic for mammary tumour virus. *Nature 350*: 207-211, 1991

Anderson, G.A., Banerjee, S., and David, C.S.: MHC class II Aa and Ea molecules determine the clonal deletion of $Vb6^+$ T cells. *J Immunol 143*: 1757-1761, 1989

Bentvelzen, P., Daams, J.H., Hageman, P., and Calafat, J.: Genetic transmission of viruses that incite mammary tumour in mice. *Proc Natl Acad Sci 67*: 377-384, 1970

Bergdoll, M.S.: Enterotoxins. *In* C.S.F. Easman and C. Adlam (eds.): *Staphylococci and Staphylococcal Infections*, Vol II, pp. 559-598. Academic Press, New York 1983

Brown, J.H., Jardetzky, T., Saper, M.A., Barnaoui, B., Bjorkman, P.J., and Wiley, D.C.: A hypothetical model of the foreign antigen binding site of class II histocompatibility molecules. *Nature 332*: 845, 1988

Choi, Y.W., Kappler, J.W., Marrack, P.: A superantigen encoded in the open reading frame of the 3' long terminal repeat of mouse mammary tumour virus. *Nature 350*: 203-207,

1991

Dellabona, P., Peccoud, J., Kappler, J.W., Marrack, P., Benoist, C., and Mathis, D.: Superantigens interact with MHC class II molecules outside of the antigen groove. *Cell 62*: 1115-1121, 1990

Dyson, P.J., Knight, A.M., Fairchild, S., Simpson, E., and Tomonari, K.: Genes encoding ligands for deletion of Vb11 T cells cosegregate with mammary tumour virus genomes. *Nature 349*: 530-531, 1991

Festenstein, H.: Immunogenetic and biological aspects of the in vitro lymphocyte allotransformation (MLR) in the mouse. *Transplant Rev 15*: 62-88, 1973

Festenstein, H., Bishop, C., and Taylor, B.A.: Location of Mls locus on mouse chromosome 1. *Immunogenetics 5*: 357-361, 1977

Fleischer, B. and Schrezenmeier, H.: T cell stimulation by Staphylococcal enterotoxins. Clonally variable response and requirement for major histocompatibility complex class II molecules on accessory or target cells. *J Exp Med 67*: 1697-1707, 1988

Frankel, W.N., Rudy, C., Coffin, J.M., Huber, B.T.: Linkage of Mls genes to endogenous mammary tumour viruses of inbred mice. *Nature 349*: 526-528, 1991

Fraser, J.D.: High affinity binding of Staphylococcal enterotoxins A and B to HLA-DR. *Nature 339*: 221-223, 1989

Happ, M.P., Woodland, D.L., and Palmer, Ed: A third T-cell receptor b-chain variable region gene encodes reactivity to Mls-1^{a} gene products. *Proc Natl Acad Sci USA 86*: 6293-6296, 1989

Herman, A., Croteau, G., Sekaly, R.P., Kappler, J.W., and Marrack, P.: HLA-DR alleles differ in their ability to present Staphylococcal enterotoxins to T cells. *J Exp Med 172*: 709-717, 1990

Hynes, N.E., van Ooyen, A.J., Kennedy, N., Herrlich, P., Ponta, B., and Groner, B.: Subfragments of the large terminal repeat cause glucocorticoid-responsive expression of mouse mammary tumor virus and of an adjacent gene. *Proc Natl Acad Sci USA 80*: 3637-3641, 1983

Janeway, C.A., Yagi, J., Katz, M.E., Jones, B., Vroegop, S., and Buxser, S.: T cell responses to Mls and to bacterial proteins that mimics its behavior. *Immunol Rev 107*: 62-88, 1989

Kappler, J.W., Roehm, N., and Marrack, P.: T cell tolerance by clonal elimination in the thymus. *Cell 49*: 273-280, 1987

Kappler, J.W., Staerz, U.D., White, J., and Marrack, P.: Self-tolerance eliminates T cells specific for Mls-modified products of the major histocompatibility complex. *Nature 332*: 35-40, 1988

Kappler, J.W., Kotzin, B., Herron, L., Gelfand, E.W., Bigler, R.D., Boylston, A., Carrel, S., Posnett, D.N., Choi, Y., and Marrack, P.: Vb-specific stimulation of human T cells by staphylococcal toxins. *Science 244*: 811-813, 1989

King, L.B., Lund, F.E., White, D.A., Sharma, S., and Corley, R.B.: Molecular events in B lymphocyte differentiation: inducible expression of the endogenous mouse mammary tumor proviral gene *Mtv-9*. *J Immunol 144*: 3218-3227, 1990

Kozak, C., Peters, G., Pauley, R., Morris, V., Michalides, R., Dudley, J., Green, M., Davisson, M., Prakash, O., Vaidya, A., Hilgers, J., Verstraeten, A., Hynes, N., Diggelmann, H., Peterson, D., Cohen, J.C., Dickson, C., Sarkar, N., Husse, R., Varmus, H., and Callahan, R.: A standardized nomenclature for endogenous mouse mammary tumor viruses. *J Virol 61*: 1651-1654, 1987

Langford, M.P., Stanton, G.J., and Johnson, H.M.: Biological effects of staphylococcal enterotoxin A on human peripheral lymphocytes. *Infect Immun 22*: 62, 1978

MacDonald, H.R., Schneider, R., Lees, R.K., Howe, R.C., Acha-Orbea, H., Festenstein, H., Zinkernagel, R.M., and Hengartner, H.: T-cell receptor Vb use predicts reactivity and tolerance to Mls^a-encoded antigens. *Nature 332*: 40-45, 1988

Marrack, P. and Kappler, J.W.: The Staphylococcal enterotoxins and their relatives. *Science 248*: 705-711, 1990

Marrack, P., Kushnir, E., and Kappler, J.W.: A maternally inherited superantigen encoded by a mammary tumour virus. *Nature 349*: 524-526, 1991

Muhlbock, O.: Note on a new inbred strain GR/A. *Eur J Cancer 1*: 123-124, 1965

Pullen, A.M., Marrack, P., and Kappler, J.W.: Evidence that Mls-2 antigens which delete $Vb3^+$ T cells are controlled by multiple genes. *J Immunol 142*: 3033-3037, 1989

Salmons, B. and Gunzburg, W.H.: Current perspectives in the biology of mouse mammary tumour virus. *Virus Res 8*: 81-102, 1987

Woodland, D.L., Happ, M.P., Gollob, K.J., and Palmer (eds.): An endogenous retrovirus mediating deletion of ab T cells? *Nature 349*: 529-530, 1991

Zhou, P., Anderson, G.D., Savarirayan, S., Inoko, H., and David, C.S.: Thymic deletion of Vb11+, Vb5+ T cells in H-2E negative, HLA-DQb+ single transgenic mice. *J Immunol 146*: 854-859, 1991a

Zhou, P., Anderson, G.D., Savarirayan, S., Inoko, H., and David, C.S.: Human HLA-DQb chain presents Mls gene products and clonally deletes TCR Vb6+, Vb8.1+ T cells in single transgenic mice. *Human Immunol 31*: 47-56, 1991b

DIVERSITY AND EVOLUTION AT THE *Eb* RECOMBINATIONAL HOTSPOT IN THE MOUSE

D. Sant'Angelo, D. Heine, and H. Passmore

Bureau of Biological Research, Rutgers University, Piscataway, NJ 08855, USA

The second intron of the mouse Eb *gene contains a well-defined recombinational hotspot. A comparison of standard laboratory MHC haplotypes suggests a region of 650 bp in the 3' end of this intron, corresponding to the location of the breakpoint region of the recombinational hotspot, is highly conserved. This conserved segment was examined in several species and subspecies of* Mus *and revealed nucleotide diversity ranging from zero to 4.3%. Sequence analysis also resulted in the identification of a variable mononucleotide repeat within the proposed recombinational hotspot which shows at least eight different sequence configurations. In addition, a variable number tandem repeat (VNTR), previously identified in the 3' end of the intron, was examined for diversity. This analysis revealed a wide variation in the number of AGGC repeats observed ranging from as few as 5 (*Mus caroli*) to as many as 21 (*Mus musculus brevirostris*). Further, a second VNTR, consisting of the tetramer TGGA was also detected immediately preceding the AGGC tandem repeat.*

Meiotic recombination involves a complex, yet extremely precise exchange of genetic information between homologous chromosomes. Limited analysis of this process in eukaryotes suggests that crossing over may occur preferentially at specific sites in the genome (Chakravarti et al. 1986; Grimm et al. 1989; Steinmetz et al. 1986). One such hotspot of recombination has been mapped to the *Eb* gene of the *I*-region of the mouse MHC (Steinmetz et al. 1982; Kobori et al. 1984; Lafuse and David 1986; Passmore et al. 1987; Zimmerer et al. 1987). Site-specificity of this hotspot is quite dramatic since approximately 75% of the known cases of intra-*I* region meiotic recombination between standard laboratory haplotypes have occurred within this gene.

The characteristics of this hotspot suggest that a signal for the initiation and/or resolution of recombination exists in the second intron of the *Eb* gene. Several DNA sequence motifs within the intron have been proposed as putative recombinational signals. These include 1) a

NATO ASI Series, Vol. H 59
Molecular Evolution of the
Major Histocompatibility Complex
Edited by J. Klein and D. Klein

variable number tandem repeat (VNTR) in the 3' end of the intron (Steinmetz et al. 1986; Kobori et al. 1986) that has sequence similarity to minisatellite DNA (Jeffreys et al. 1985), 2) a lengthy sequence which shows significant nucleotide similarity (Shiroishi et al. 1990; Bryda et al. 1991) to the mouse transcript (MT) middle repetitive family of DNA (Heinlein et al. 1986), 3) regulatory protein binding motifs closely associated with a DNase I-hypersensitive site (Shenkar et al. 1991) and 4) nucleotide similarity to a LTR and retroviral *pol* and *env* genes in the 5' half of the intron (Zimmerer and Passmore 1991).

In the work presented in this communication, diversity at the *Eb* recombinational hotspot was studied by nucleotide sequence analysis of a number of species and subspecies of the genus *Mus*. This analysis focused on a preliminary examination of diversity and evolution of proposed signals for recombination and the identification of other sequences which could play a role in the regulation of meiotic crossing over.

Methods

Liver DNA from wild derived species and subspecies of the genus *Mus* were provided by Dr. Norman Arnheim (Dept. of Molecular Biology, University of Southern California, Los Angeles, CA). These animals were originally from the colony of Dr. Michael Potter. DNA for the *q*, *f* and *p* alleles were prepared from the inbred strains B10.G, B10.M and P/J, respectively. Genomic DNA was isolated from liver using a modification of the method of Blin and Stafford (1976).

DNA was isolated by polymerase chain reaction using primers complementary to the *b* allele of *Eb*. A 2 kb amplified fragment covering the 3' end of the intron was either directly cloned into the plasmid pCR1000 (Invitrogen Corp., San Diego, CA) or into M13mp18 and mp19. Xba I and Hind III restriction sites incorporated into the primers facilitated cloning into M13. Sequencing was done utilizing the Sequenase 2.0 system (United States Biochemical, Cleveland, OH). Sequence specific oligonucleotide primers were used to facilitate sequencing. Sequences for the *b, d, k* and *s* alleles have been published previously (Kobori et al. 1986; Bryda et al. 1991).

Results

Polymorphism Associated With the Eb Hotspot of Recombination. The segment of DNA containing the recombinational break-points for 23 intra-*Eb* recombinants have now been sequenced and mapped (Kobori et al. 1986, Zimmerer et al. 1991; Bryda et al. 1991; Passmore et al. 1991). These studies suggest that the recombinational breakpoints associated with the *Eb* hotspot can be confined to a region of approximately 800-1000 bp located in the 3' half of the large 4.3 kb second intron (Fig. 1). We have designated this segment of the intron as the breakpoint region (BPR) of the *Eb* hotspot.

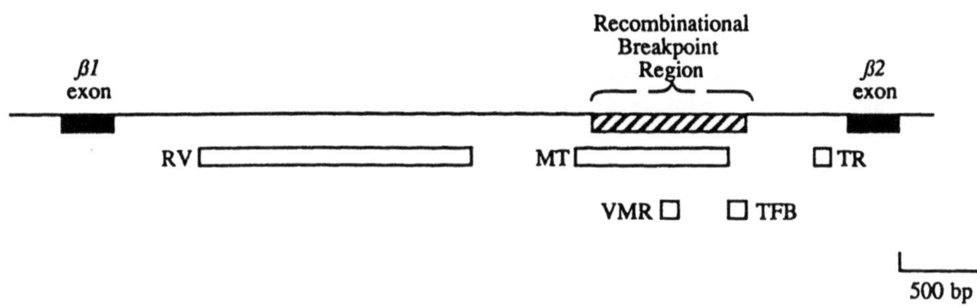

Fig. 1. Map of the second intron of the *Eb* gene showing putative recombinational signals and the proposed recombinational breakpoint region. TR = four base tandem repeat; MT = MT family of middle repetitive sequences; TFB = transcription factor binding motifs; VMR = variable mononucleotide repeat; RV = retroviral insertion.

Studies in our laboratory have shown that most of the breakpoint region is highly conserved among alleles from standard inbred strains. This conserved sequence is located between nucleotide positions 1050 and 1700 (based on the numbering of Kobori et al., 1986) and represents approximately 80% of the proposed breakpoint region. Figure 2 illustrates the nucleotide diversity seen among seven standard inbred alleles within this area of the second intron. With the exception of the *p* haplotype, these sequences show very low levels of diversity (zero to 0.8%).

b	k	s	d	q	f	p	
—	0.3	0.5	0.3	0.8	0.3	2.0	b
	—	0.2	0.3	0.5	0.0	2.0	k
		—	0.5	0.6	0.2	2.2	s
			—	0.8	0.3	2.0	d
				—	0.5	1.8	q
					—	2.0	f
						—	p

Fig. 2. A comparison of the conserved 650 bp area (positions 1050-1700) of the breakpoint region in standard inbred alleles. Numbers represent polymorphisms per 100 bp.

Polymorphism ranging from zero to 4.3% was observed when alleles from wild derived species and subspecies were compared (Fig. 3). The polymorphisms consist primarily of single nucleotide substitutions although deletions/insertions ranging in size from one to seven bp were also detected. The lowest level of polymorphism was seen in comparisons of three of the four *Mus musculus* subspecies. In this group, which includes *Mus musculus domesticus*, *Mus musculus molossinus* and *Mus musculus brevirostris*, the amount of polymorphism never exceeded 0.3%. The fourth subspecies studied, *Mus musculus musculus* (Czech) showed a difference of 1.8% when compared to the other three. However, due to shared polymorphisms, this number is reduced to only 0.9% when the *Mus musculus musculus* allele is compared to the *p* haplotype suggesting a common ancestry between the *p* haplotype and *Mus musculus musculus*.

Based on shared polymorphisms (data not shown) the alleles from *Mus caroli*, *Mus cervicolor* and *Mus cervicolor popaeus* appear to share a common lineage. *Mus spretus* also shares a number of similar polymorphisms with this group. Despite these similarities, significant levels of nucleotide polymorphism exist among these species. Most remarkable are two alleles of *Mus caroli* (derived from a single heterozygous mouse) which showed as high a level of polymorphism when compared to each other than when compared to other species.

Mus caroli (Allele 1)	*Mus caroli* (Allele 2)	*Mus cervicolor*	*Mus cervicolor popaeus*	*Mus spretus*	*Mus musculus musculus*	*Mus musculus domesticus*	*Mus musculus brevirostris*	*Mus musculus molossinus*	*b*-haplotype	
—	3.8	2.3	3.5	3.5	3.8	3.2	3.2	3.2	3.2	*Mus caroli* (Allele 1)
	—	3.4	3.4	4.0	4.2	3.7	3.7	3.7	3.4	*Mus caroli* (Allele 2)
		—	1.4	3.8	4.0	3.7	3.7	3.7	3.7	*Mus cervicolor*
			—	4.0	4.3	4.3	4.3	4.3	4.3	*Mus cervicolor popaeus*
				—	2.9	2.9	2.9	2.9	2.9	*Mus spretus*
					—	1.8	1.8	1.8	1.8	*Mus musculus musculus*
						—	0.3	0.3	0.3	*Mus musculus domesticus*
							—	0.0	0.3	*Mus musculus brevirostris*
								—	0.3	*Mus musculus molossinus*
									—	*b*-haplotype

Fig. 3. A comparison of the conserved 650 bp area of the breakpoint region from wild *Mus* species and subspecies. The *b* haplotype (allele) is included for comparison.

Tandem Repeat Creates Diversity. A four base repeat (AGGC) has been previously reported in the 3' end of the *Eb* second intron. This repeat has been suggested as a possible recombination signal (Steinmetz et al. 1986; Kobori et al. 1986). In our sequence analysis, variation in the AGGC repeat among standard laboratory alleles ranged from 18 in the *d* haplotype to only six repeats in the *p*-haplotype. This variation is even more pronounced in the wild derived *Mus* alleles (Fig. 4). where the number of repeats is as high as 21 in an allele from *Mus musculus brevirostris* and as low as only five in one of the alleles from *Mus caroli*. Although most of the repeats are perfect in nucleotide sequence, a small number are imperfect. For example, in the *Mus musculus brevirostris* allele, four of the 21 repeats differ from the AGGC sequence by one base.

CCTGGATCCAGTC	$(TGGA)_{4}$	$(AGGC)_{6}$	TACAAGG	*p*-haplotype
-------------	$(TGGA)_{4}$	$(AGGC)_{10}$	-------	*b*-haplotype
-------------	$(TGGA)_{4}$	$(AGGC)_{18}$	-------	*d*-haplotype
-------------	$(TGGA)_{4}$	$(AGGC)_{5}$	-------	*Mus caroli*
-------------	$(TGGA)_{4}$	$(AGGC)_{21}$	-------	*Mus musc.brevirostris*
-------------	$(TGGA)_{3}$	$(AGGC)_{5}$	-------	*Mus cervicolor*
-------------	$(TGGA)_{13}$	$(AGGC)_{8}$	-------	*Mus spretus*

Fig. 4. Variation in the tandem repeats associated with the 3' end of the intron. The bold numbers represent the total number of both the perfect and imperfect repeats. The last base of the TGGA repeat is identical to the first base of the AGGC repeat.

In our studies we have detected a second tetrameric VNTR located 5' to the AGGC repeat (Fig. 4). This TGGA repeat appears four times in all the alleles from standard inbred strains although some of the repeats are imperfect. The wild alleles sequenced for this report, however, show that this tetramer is repeated from as few as three times in a *Mus cervicolor* allele up to 13 times in a *Mus spretus* allele. The repeats in these two species are all perfect TGGA tetramers and they immediately precede the AGGC repeat.

Variable Mononucleotide Repeat Marks the Center of the Breakpoint Segment. As previously discussed, there is a 650 bp region of the recombination breakpoint region (BPR) that shows little or no polymorphism when alleles from standard and some subspecies of *Mus musculus* are compared. Even when more distantly related species are compared the amount of polymorphism does not exceed 5% and usually involves single nucleotide substitutions. However, a remarkable exception to this pattern is apparent near the center of the BPR (positions 1281-1295). This region is marked by a run of pyrimidines that is variable in both length and, to a lesser extent, nucleotide composition. From a total 16 alleles (seven standard and nine wild-derived), eight different sequence configurations have been determined at this site (Fig. 5). Importantly, no polymorphism is seen for at least 20 nucleotides in either direction from this variable mononucleotide repeat (VMR). The repeat consists primarily of cytosine bases, however in five of the configurations one or two thymine bases are seen. Also, in *Mus musculus molossinus* a purine (guanine) was present.

CCCACAGTGGGTGAGT	CCCTCCCCCCCCC	ATCAAT	*b,p*-haplotypes
----------------	CCCCCTTCCCCCC	------	*Mus cervicolor*
----------------	CCCTCCCCC	------	*Mus caroli*
----------------	CCCCCCCCC	------	*q*-haplotype
----------------	CCCCCCCC	------	*d,f,k,s*-haplotypes
----------------	CTCCCCCC	------	*Mus spretus*
----------------	CCTCCCCC	------	*Mus musc. domesticus*
----------------	CCCCCGCC	------	*Mus musc. molossinus*

Fig. 5. Variations in the run of pyrimidines seen at the midpoint (position 1281-1295) of the proposed hotspot. Both variation in length and base composition are noted. Boxed sequence is identical to the immunoglobulin V(D)J heptamer signal sequence.

Discussion

Considerable data is now available concerning the location of crossover events within the *Eb* recombinational hotspot. Little information, however, exists on the physical or molecular properties that result in this chromosomal region being an attractive site for recombination. Molecular attributes that may confer this site-specificity include specific DNA motifs which could act as recognition sites for proteins involved in recombination, and/or more general sequences that influence the configuration of the DNA or its accessibility within the chromatin structure. It also seems likely that a requirement for a high level of homology between the DNA strands may be necessary at sites of recombination.

In an effort to address the possibility of a nucleotide homology requirement, we have examined the level of sequence diversity within the breakpoint region (BPR) of the *Eb* recombinational hotspot. Most of this segment (positions 1050 to 1700) is highly conserved among alleles from standard inbred strains (*b*, *d*, *k*, *f*, *q* and *s* haplotypes). These alleles have all been involved in recombination at the *Eb* recombination hotspot. Therefore, the low level of diversity, ranging from zero to 0.8%, is clearly permissive for recombination among these alleles. A similar level of limited polymorphism would suggest that recombination between the *Mus* subspecies *Mus musculus domesticus* (including standard alleles), *Mus musculus brevirostris*, and *Mus musculus molossinus* should also occur freely at the *Eb* hotspot. In contrast, if an increased level of diversity inhibits crossing over, the *Eb* recombinational hotspot may be inactive in crosses between these three subspecies and *Mus musculus musculus* or the standard *p* haplotype. Alleles derived from the same species that have significant amounts of intraspecies polymorphism may also predict an inactive hotspot of recombination. This situation would be suggested, for example, by the diversity seen between the two alleles of *Mus caroli* sequenced for this study.

The relatively low level of polymorphism seen in the conserved area of the breakpoint region is interrupted by a variable number mononucleotide repeat (VMR). This run of pyrimidines varies in length from 8 to 13 nucleotides and also shows some variation in nucleotide composition. The location of the VMR, near the center of the breakpoint region, suggests a potential relationship between this sequence and the hotspot of recombination. It has been noted that oligopurine and oligopyrimidine sequences can form transient regions of single stranded DNA and have also been implicated as binding sites for nuclear proteins from meiotic cells (Edelmann et al. 1989).

Interestingly, a perfect match to the heptamer signal sequence associated with V(D)J recombination in immunoglobulins is present upstream from the VMR (see Figure 4). Normal V(D)J recombination relies on both a heptamer and a nonamer sequence. However, isolated heptamers such as seen in this segment from *Eb*, have been associated with aberrant events such as somatic translocations and inversions.

Other potential signals of recombination associated with the Eb breakpoint region include protein binding motifs for $NF_{\kappa}B$ and H2TF1/KBF1 that are associated with a DNase I-hypersensitivity site in the breakpoint region (Shenkar et al. 1991). These sites are conserved in the various species of *Mus* with the exception of a single polymorphism seen in the $NF_{\kappa}B$ motif in *Mus cervicolor popaeus*.

Steinmetz and colleagues (1986) have suggested that the molecular signal for this recombinational hotspot may be a variable number tandem repeat (VNTR). This AGGC repeat shows sequence similarity to minisatellite DNA (Jeffreys et al. 1985) and the Chi recombination sequence (Smith 1983). In the 16 inbred and wild *Mus* alleles that have been sequenced there are eight variations in the number of AGGC repeats. The allele from *Mus musculus brevirostris* with 21 repeats has the distinction of having expanded the most, while in contrast, one of the *Mus caroli* alleles stands at only 5 repeats. The TGGA repeat immediately 5' to the AGGC repeat has not been previously reported. This VNTR displays a range in number of repeats from three in *Mus cervicolor* to 13 in *Mus spretus*.

Acknowledgments. This work was supported by NIH grant HG00188 and the Anne B. & James H. Leathem Scholarship Fund.

References

Bryda, E.C., Depari, J.A., Sant'Angelo, D.B., Murphy, D.B., and Passmore, H.C.: Multiple sites of crossing over within the *Eb* recombinational hotspot in the mouse. 1991 (Submitted)

Blin, N. and Stafford, D.W.: A general method for isolation of high molecular weight DNA from eukaryotes. *Nucleic Acids Research 3*: 2303-2308, 1976

Chakravarti, A., Elbein, S.C., Permutt, M.A.: Evidence for increased recombination near the human insulin gene: implication for disease association studies. *Proc Natl Acad Sci 83*: 1045-1049, 1986

Edelmann, W., Kroger, B., Goller, M., and Horak, I.: A recombination hotspot in the LTR of a mouse retrotransposon identified in an in vitro system. *Cell 57*: 937-946, 1989

Grimm, T., Muller, B., Dreier, M., Kind, E., Bettecken, T., Meng, G., and Muller, C.R.: Hot spot of recombination within XS164 in the Duchenne Muscular compatibility complex of the mouse: mapping of crossover sites within the I region. *Biochem Genet 25*: 513-526, 1987

Heinlein, U.A.O., Lange-Sablitzky, R., Schaal, H., and Wille, E.: Molecular characterization of the MT-family of dispersed middle-repetitive DNA in rodent genomes. *Nucleic Acids Res 14*:6403-6416, 1986

Jeffreys, A.J., Wilson, V., and Thein, S.L.: Hypervariable minisatellite regions in human DNA. *Nature 314*:67-73, 1985

Kobori, J.A., Winton, A., McNicholas, J., and Hood, L.: Molecular characterization of the recombination region of six murine major histocompatibility complex (MHC) I-region recombinants. *J Mol Cell Immunol 1*:125-131, 1984

Kobori, J.A., Strauss, E., Minard, K., and Hood, L.: Molecular analysis of the hotspot of recombination in the murine major histocompatibility complex. *Science 234*:173-179, 1986

Lafuse, W.P. and David, C.S.: Recombination hotspots within the I region of the mouse H-2 complex map to the Eß and Eα genes. *Immunogenetics 24*:352-360, 1986

Passmore, H.C., Kobori, J.A., Zimmerer, E.J., Spinella, D.G., and Hood, L.: Molecular characterization of meiotic recombination within the major histocompatibility complex of the mouse: mapping of crossover sites within the I region. *Biochem Genet 25*:513-526, 1987

Passmore, H.C., Sant'Angelo, D.B., and Bryda, E.C.: Molecular dissection of the Eb recombinational hotspot. (These proceedings) 1991

Shenkar, R., Shen, M., and Arnheim, N.: DNase I-hypersensitive sites and transcription factor-binding motifs within the mouse Eß meiotic recombinational hot spot. *Molec Cell Biol 11*: 1813-1819, 1991

Shiroishi, T., Hanzawa, N., Sagai, T., Ishiura, M., Gojobori, T., Steinmetz, M., and Moriwaki, K.: Recombinational hotspot specific to female meiosis in the mouse major histocompatibility complex. *Immunogenetics 31*: 79-88, 1990

histocompatibility complex. *Immunogenetics 31*: 79-88, 1990
Smith, G.R.: Chi hotspots of generalized recombination. *Cell 34*: 709-710, 1983
Steinmetz, M., Minard, K., Horvath, S., McNicholas, J., Frelinger, J., Wake, C., Long, E., Mach, B., and Hood, L.: A molecular map of the immune response region from the major histocompatibility complex of the mouse. *Nature 300*: 35-42, 982
Steinmetz, M., Stephan, D., and Fischer Lindahl, K.: Gene organization and recombination hotspots in the murine major histocompatibility complex. *Cell 44*: 895-904, 1986
Zimmerer, E.J. and Passmore, H.C.: Structural and genetic properties of the E_β recombinational hot spot in the mouse. *Immunogenetics 33*: 132-140, 1991
Zimmerer, E.J., Ogin, E., Shreffler, D.C., and Passmore, H.C.: Molecular mapping of crossover sites within the I region of the mouse MHC: analysis of ten recombinant chromosomes. *Immunogenetics 25*: 274- 276, 1987

MOLECULAR DISSECTION OF THE *Eb* RECOMBINATIONAL HOTSPOT IN THE MOUSE

H. Passmore, D. Sant'Angelo and E. Bryda
Department of Biological Sciences, Rutgers University, Piscataway, NJ 08855, USA

Meiotic crossing over within the I region of the mouse major histocompatibility complex (MHC) has been shown to be limited to a number of well-defined recombinational hotspots. Twenty cases of recombination mapping to the Eb gene of the I region have been sequenced to more precisely define the segment of DNA carrying meiotic crossover sites. This analysis was based on positioning recombination breakpoints between nucleotide polymorphisms found in the nucleotide sequences of parental haplotypes. Each of the twenty cases of recombination mapped to the second intron of the Eb gene. The exact sites of crossing over could not be determined in these recombinants since the nucleotide sequences of the parental alleles are identical between the polymorphic sites, and because no deletion or duplication of nucleotides was observed in recombinant sequences. These studies indicate that the second intron of the Eb gene contains at least two closely linked clusters of recombinational breakpoints or, alternatively, a broad area of approximately 800-1000 bp in which recombination can occur at multiple sites.

Over the past ten years, studies using restriction fragment length polymorphisms (RFLP) and nucleotide sequence analysis have led to the discovery of four well-defined recombinational hotspots associated with the *I* region of the mouse major histocompatibility complex (Steinmetz et al. 1982, 1986; Lafuse et al. 1986; Uematsu et al. 1986; Shiroishi et al. 1990, 1991). The site specificity of meiotic crossing over observed in these studies appears to be strictly regulated by the MHC haplotypes involved in the recombination event. For example, recombination between haplotypes found in standard inbred strains (with the exception of the *p* haplotype) occurs primarily at a recombinational hotspot located within the *Eb* gene (Steinmetz et al., 1982; Lafuse and David 1986; Saha and Cullen 1986; Passmore et al. 1987; Zimmerer et al. 1987). In this report, we present a nucleotide sequence analysis of twenty cases of crossing over associated with the *Eb* recombinational hotspot. This analysis more precisely identifies the recombinational breakpoint sites associated with this hotspot and defines a segment of DNA that may be involved in the regulation of site-restricted recombination.

NATO ASI Series, Vol. H 59
Molecular Evolution of the
Major Histocompatibility Complex
Edited by J. Klein and D. Klein

Methods

Intra-*I* region recombinant strains were obtained from our own animal colony or were provided from the colonies of Dr. Donald Shreffler, Washington University School of Medicine, St Louis, Missouri, USA or from Dr. Donal Murphy, Wadsworth Center, New York State Department of Health, Albany, New York, USA. Genomic DNA from strains A.TFR3, A.TFR4 and B10.QSR2 was isolated from liver tissue by using a modification of the method of Blin and Stafford (1976).

The analysis in this report is based on nucleotide sequences and polymorphism in the second intron of the mouse *Eb* gene for six different alleles. A more detailed account of the nucleotide sequence of this intron can be found in Kobori et al. (1986) and in Zimmerer and Passmore (1991). Segments of DNA carrying recombination breakpoints were cloned from partial genomic libraries constructed with the vector lambda gt7lac5 (Passmore et al. 1987) or by polymerase chain reaction (PCR) using primers complementary to the *b* allele of *Eb* (Bryda et al. 1991). Fragments amplified by PCR were cloned into M13mp18 and mp19. Single-stranded DNA obtained from M13 clones was used as a template for dideoxy sequencing utilizing the Sequenase Version 2.0 system (United States Biochemical) with ^{35}S-dATP as label.

Results and Discussion

Substantial information has accumulated detailing the predominant role of the *Eb* hotspot in *I* region recombination between MHC haplotypes derived from standard inbred strains. Table 1 provides a summary of the haplotype combinations and general location of previously published *I* region recombinants involving standard haplotypes. Most of these initial map assignments are based on the analysis of restriction fragment length polymorphisms. The site and haplotype specificity of this recombinational hotspot is dramatically illustrated by the observation that 42 of 45 *I* region recombinants between the *b, d, k, f, q* and *s* haplotypes have occurred in close association with the *Eb* gene. In contrast, all nine cases of recombination involving the *p* parental haplotype have occurred in association with the *Ea* gene located approximately 25 kb distal to *Eb*.

Table 1. Location of *I* region recombinants derived from standard inbred strains

Heterozygous Combination	Eb Hotspot	Ea Hotspot	Other	References
k,b	14	0	1 (B10.MBR)	1,2,4,6,7,11
k,s	9	0	1 (A.TL)	1,2,5,7
k,f	5	0	0	4,7,8
k,q	4	0	1 (B10.AQR)	1,4
d,s	2	0	0	11
d,b	2	0	0	1,2,9
s,b	2	0	0	1
s,f	2	0	0	12
s,q	1	0	0	12
b,f	1	0	0	4
p,k	0	7	0	3,4
p,f	0	2	0	10
TOTALS	42	9	3	

References: 1, Steinmetz et al. 1982, 1986; 2, Kobori et al. 1984; 3, Lafuse et al. 1986; 4, Lafuse and David 1986; 5, Saha and Cullen 1986; 6, Passmore et al. 1987; 7, Zimmerer et al. 1987; 8, Begovich and Jones 1985; 9, Epstein et al. 1990; 10, Lafuse et al. 1990; 11, Bryda et al. 1991; 12, Passmore et al. (this report).

The *p* haplotype, carried by the standard inbred strain P/J, may be unusual among standard haplotypes. Preliminary data from our laboratory suggests that this MHC haplotype may have been derived at least in part from *Mus musculus musculus* rather than *Mus musculus domesticus* (the likely origin of most inbred strains). Thus the site-specificity of the *p* haplotype for *I* region recombination may be considerably different from other standard haplotypes that were derived from *Mus musculus domesticus*.

A more precise picture of the location of the recombination events associated with the *Eb* gene can be obtained by cloning and nucleotide sequence analysis of DNA fragments carrying the recombinational breakpoints. This approach is based on positioning recombination sites between nucleotide polymorphisms that distinguish the alleles used in the original heterozygote. Figure 1 illustrates this kind of analysis for two cases of recombination (strains A.TFR3 and A.TFR4) involving the *f* and *s* haplotypes. In both recombinants, the recombination events can be assigned to a 746 bp segment in the 3' half of the second intron of the *Eb* gene. Since recombination was accomplished without the addition or deletion of nucleotides within the 746 bp fragment, the precise location of the recombination events could not be determined.

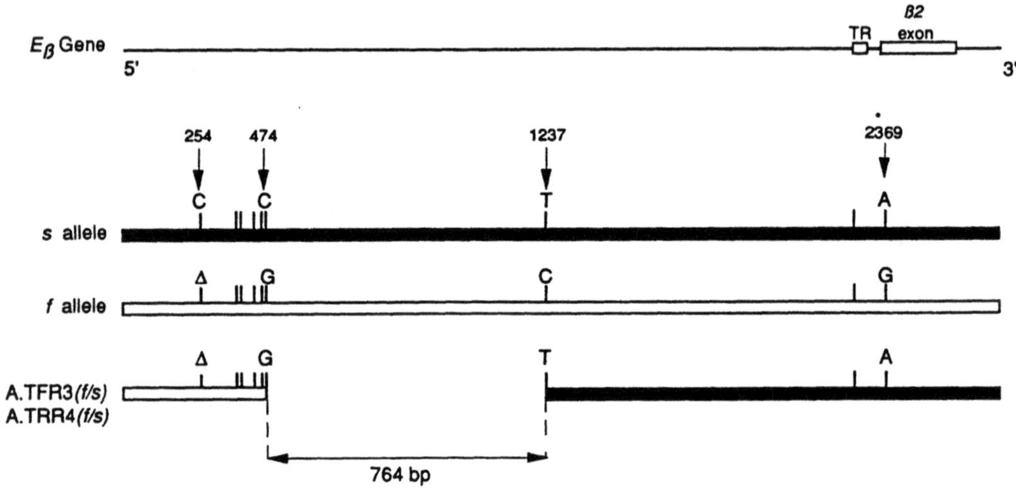

Fig. 1. Map of the 3' half of the second intron of the *Eb* gene indicating sites of recombination in two recombinant strains derived from crosses between the *f* and *s* haplotypes. Positions of the nucleotide polymorphisms which distinguish between these two haplotypes are indicated by short vertical lines.

We have now completed the sequencing of twenty cases of intra-*Eb* recombination, some of which have been reported previously (Zimmerer and Passmore 1991). Figure 2 provides a summary of the breakpoint segments carrying each of these recombinants. Three cases of recombination sequenced by Kobori et al. (1986) have also been included. A number of important characteristics of intra-*Eb* recombination become immediately evident from this summary diagram. First, all cases of chromosomal exchange fall into the second intron of the *Eb* gene. Thus, although these crossover sites are closely associated with a functional class II MHC gene, no case of intraexonic exchange was documented. Second, most of the recombinational breakpoint segments do not include the four base tandem repeat (designated TR in Figure 2), which has been proposed as a possible signal for recombination (Kobori et al. 1986; Steinmetz et al. 1986). This observation, however, does not eliminate a potential role for this repeat sequence, and may be consistent with a model in which the repeat acts as an enhancer or initiator of recombination. In addition, analysis of the segments carrying the recombination breakpoints reveals no change or errors in nucleotide sequences covering the breakpoint segments indicating that the recombination process is extremely precise.

One possible interpretation of the mapping data in Figure 2 suggests that all the sites of recombination may fall within the region of the intron bounded by positions 843 and 1698. A number of the recombinant breakpoints map fully within this area, while all others have a significant portion of their arrow bar within the region. It remains unclear whether this breakpoint region represents the recombination hotspot in its entirety, or alternatively, represents only a portion of the hotspot.

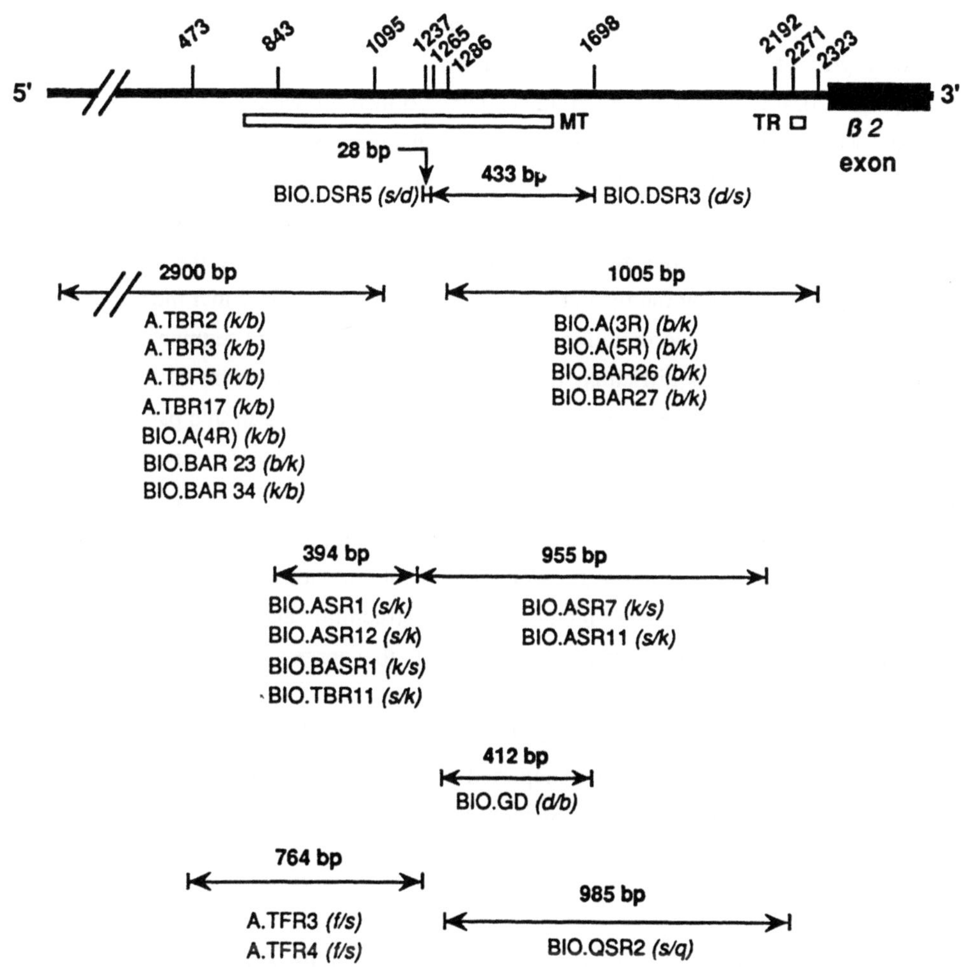

Fig. 2. Summary of twenty-three cases of recombination within the second intron of the *Eb* gene. Informative polymorphic sites are indicated on the composite map at the top of the diagram.

The distribution of recombination breakpoints seen in Figure 2 suggests there are at least two recombinationally active sites within the intron. This division into two sites is particularly apparent in the series of eleven recombinant between the *b* and *k* haplotypes as well as in the six cases involving *k* and *s*. Although it is possible that the placement of these recombinants into two distinct segments is a consequence of two closely linked recombinational hotspots, it seems more likely that the multiple location of breakpoints is a reflection of the variable manner in which transient heteroduplex DNA, formed during recombination, is resolved or repaired. A similar type of mechanism has been proposed for homologous recombination in prokaryotes (Radman 1988), and is based on the random mismatch repair of nucleotide polymorphisms located within the heteroduplex DNA formed immediately after recombination. This model would suggest that the breakpoints segments mapped in Figure 1 could contain, in some cases, the site at which recombination was initiated, while in other cases they would contain the site of resolution of a Holliday-type intermediate. Furthermore, differences in the length of the branch migration step would likely create variation in the location of the breakpoint segments carrying the resolution site. In any case, it is clear that the recombination process is complex and the interpretation of the locations of breakpoint segments is not always straightforward. Despite this difficulty, the overall pattern of the location of the breakpoints seen in the 23 independent cases of intra-*Eb* recombination presented in this paper does permit the identification of a 800 to 1000 bp region which is likely to carry both initiation and resolution sites of recombination occurring at the *Eb* hotspot. It is of particular interest that Shiroishi and colleagues (1990, 1991) have also defined multiple breakpoints in a segment of similar length (about 1 kb) in another I region recombinational hotspot located between the *Pb* and *Ob* genes. Thus, a clearer picture of the structural and molecular features of recombinational hotspots in the mouse is now beginning to emerge.

Acknowledgments. This work was supported by NIH grants GM-40399 and HG-00188 from the National Institutes of Health. E.B. was supported by a Charles and Johanna Busch Predoctoral Fellowship and a grant from the Carl N. Steinetz Memorial Fund.

References

Begovich, A.B. and Jones, P.P.: Free Ia Eα chain expression in the E^-_α:E^-_β recombinant strain A.TFR5. *Immunogenetics 22*: 523-532, 1985

Bryda, E.C., Depari, J.A., Sant'Angelo, D.B., Murphy, D.B. and Passmore, H.C.: Multiple sites of crossing over within the *Eb* recombinational hotspot in the mouse. 1991 (Submitted)

Blin, N. and Stafford, D.W.: A general method for isolation of high molecular weight DNA from eukaryotes. *Nucleic Acids Research 3*:2303-2308, 1976

Epstein, S.L., Arn, J.S., Sachs, D.H.: Seven new MHC recombinant strains defining new *H-2* haplotypes. *Immunogenetics 32*:142-145, 1990

Kobori, J.A., Winto, A., McNicholas, J. and Hood, L.: Molecular characterization of the recombination region of six murine major histocompatibility complex (MHC) I-region recombinants. *J Mol Cell Immunol 1*: 125-131, 1984

Kobori, J.A., Strauss, E., Minard, K., and Hood, L.: Molecular analysis of the hotspot of recombination in the murine major histocompatibility complex. *Science 234*: 173-179, 1986

Lafuse, W.P. and David, C.S.: Recombination hot spots within the *I* region of the mouse *H-2* complex map to the E_β and E_α genes. *Immunogenetics 24*: 352-360, 1986

Lafuse, W.P., Berg, N., Savarirayan, S., and David, C.S.: Mapping of a second recombination hot spot within the I-E region of the mouse *H-2* gene complex. *J Exp. Med. 163*: 1518-1528, 1986

Lafuse, W.P., Lee, S.T., and David, C.S.: Molecular analysis of two recombinant mouse strains with crossovers in the E_α recombination hot spot. *J Immunogenetics 17*: 169-176, 1990

Passmore, H.C., Kobori, J.A., Zimmerer, E.J., Spinella, D.G. and Hood, L.: Molecular characterization of meiotic recombination within the major histocompatibility complex of the mouse: mapping of crossover sites within the *I* region. *Biochemical Genetics 25*: 513-526, 1987

Radman, M.: Mismatch repair and genetic recombination. In R. Kucherlapati and G.R. Smith (eds.). *Genetic Recombination*, pp. 169-191, American Society of Microbiology, Washington DC 1988

Saha, B.K. and Cullen, S.E.: Molecular mapping of murine *I* region recombinants: crossing over in the *E* gene. *J Immunol 136*:1112-1116, 1986

Shiroishi, T., Hanzawa, N., Sagai, T., Ishiura, M., Gojobori, T., Steinmetz, M., and Moriwaki, K.: Recombinational hotspot specific to female meiosis in the mouse major histocompatibility complex. *Immunogenetics 31*: 79-88, 1990

Shiroishi, T., Sagai, T., Hanzawa, N., Gotoh, H., and Moriwaki, K.: Genetic control of sex-dependent meiotic recombination in the major histocompatibility complex of the mouse. *EMBO J 10*: 681-686, 1991

Steinmetz, M., Minard, K., Horvath, S., McNicholas, J., Frelinger, J., Wake, C., Long, E., Mach, B., and Hood, L.: A molecular map of the immune response region from the major histocompatibility complex of the mouse. *Nature 300*: 35-42, 1982

Steinmetz, M., Stephan, D., and Fischer Lindahl, K.: Gene organization and recombination hotspots in the murine major histocompatibility complex. *Cell 44*: 895-904, 1986

Uematsu, Y., Kiefer, H., Schulze, R., Fischer-Lindahl, K., and Steinmetz, M.: Molecular characterization of a meiotic recombinational hotspot enhancing homologous equal crossing-over. *EMBO J 5*: 2123-2129, 1986

Zimmerer, E.J. and Passmore, H.C.: Structural and genetic properties of the $\underline{E}_\beta$ recombinational hot spot in the mouse. *Immunogenetics 33*: 132-140, 1991

Zimmerer, E.J., Ogin, E., Shreffler, D.C. and Passmore, H.C.: Molecular mapping of crossover sites within the *I* region of the mouse MHC: analysis of ten recombinant chromosomes. *Immunogenetics 25*: 274- 276, 1987

MOLECULAR CLONING OF NURSE SHARK cDNAs WITH HIGH SEQUENCE SIMILARITY TO NUCLEOSIDE DIPHOSPHATE KINASE GENES[1]

Masanori Kasahara, Camilo Canel, E. Churchill McKinney, and Martin F. Flajnik

Department of Microbiology and Immunology, University of Miami School of Medicine, P.O. Box 016960 (R-138), Miami, FL 33101, USA

In the course of our study aimed at isolating class II genes of the major histocompatibility complex from primitive vertebrates, nurse shark cDNA clones assumed to encode a nucleoside diphosphate kinase were isolated. The deduced amino acid sequence of the putative nurse shark nucleoside diphosphate kinase protein has two cysteinyl residues approximately 60 amino acid residues apart from each other. Weak amino acid sequence similarity was found between the putative nurse shark nucleoside diphosphate kinase and C1-SET members of the immunoglobulin superfamily.

Nucleoside diphosphate (NDP) kinases (EC 2.7.4.6) are a group of ubiquitous enzymes that catalyze the transfer of the terminal phosphate group of nucleoside 5'-triphosphates to nucleoside 5'-diphosphates by a ping-pong mechanism involving a high energy phosphorylated enzyme intermediate (Parks and Agarwal 1973). The action of these enzymes is non-specific in the sense that they show no preference either for purines or pyrimidines or for ribose or deoxyribose. It is believed that NDP kinases play a major role in the biosynthesis of nucleoside triphosphates other than ATP. Nucleoside triphosphates are required for the synthesis of nucleic acids, CTP for the synthesis of lipids, UTP for the synthesis of polysaccharides, and GTP for protein elongation, microtubule polymerization, and signal transduction. Thus, biochemical processes dependent on NDP kinases are quite diverse.

Recently, two groups of investigators (Biggs et al. 1988; Rosengard et al. 1989) have reported the isolation of human, mouse, and insect genes, the predicted protein products of which show a high degree of sequence similarity to the NDP kinase proteins of the slime mold *Dictyostelium discoideum* (Lacombe et al. 1990), the Gram-negative eubacterium *Myxococcus*

[1]The nucleotide sequence reported here has been submitted to the GenBank and has been assigned the accession number M63964

NATO ASI Series, Vol. H 59
Molecular Evolution of the
Major Histocompatibility Complex
Edited by J. Klein and D. Klein

xanthus (Muñoz-Dorado et al. 1990a,b), and the rat (Kimura et al. 1990). The mouse gene, designated *nm23*, was identified on the basis of its differential expression in mouse melanoma cell lines of varying metastatic potentials (Steeg et al. 1988); the expression of *nm23* mRNA was significantly reduced in highly metastatic tumor cell lines in comparison to those with no or low metastatic potential. Subsequent studies have shown that such inverse correlation between the metastatic potential of the tumor and the level of *nm23* transcripts is also found in other rodent and possibly human tumors (Bevilacqua et al. 1989). The insect gene, termed *awd* (*abnormal wing discs*), was isolated from a fruit fly *Drosophila melanogaster* on the basis of the fact that its mutations result in gross developmental anomalies, some features of which appeared to resemble those of malignant transformation (Biggs et al. 1988). The deduced amino acid sequences of the human and mouse Nm23 proteins (Rosengard et al. 1989) and the *D. melanogaster* Awd protein (Biggs et al. 1988) are 89%, 88%, and 77% identical to the deduced amino acid sequence of the rat NDP kinase, respectively. This observation has led to the proposal that the Nm23 and Awd proteins are NDP kinases and that their aberrant expression can influence metastatic potentials of tumors as well as normal development (Liotta and Steeg, 1990). The evidence that the *D. melanogaster* Awd protein is indeed an NDP kinase was recently obtained (Biggs et al. 1990).

We describe here the isolation and characterization of nurse shark (*Ginglymostoma cirratum*) cDNA clones assumed to encode an NDP kinase. These cDNA clones were isolated in the course of our study aimed at cloning major histocompatibility complex (*MHC*) class II genes from primitive vertebrates. We discuss the possible significance of weak amino acid sequence similarity found between the putative nurse shark NDP kinase and C1-SET members of the immunoglobulin (Ig) superfamily (Williams and Barclay 1988; Hunkapiller and Hood 1989).

Materials and methods

Oligonucleotides. Three oligonucleotides used in this study are as follows: Primer 21 (5'-CAAAGCTTGTGACNGRYTTCTAYCC-3'; where N is any one of the four bases, Y is a pyrimidine, and R is a purine), an adapter primer (5'-GACTCGAGTCGACATCG-3'), and a $(dT)_{17}$-adapter primer (5'-GACTCGAGTCGACATCGATTTTTTTTTTTTTTTTT). The 5' eight bases of Primer 21 contain an artificial *Hin*dIII site aimed at facilitating subsequent subcloning; the sequence of the remaining 17 bases was designed on the basis of the fact that most MHC class II ß chains have a conserved amino acid stretch, Val-Thr-(Gly or Asp)-Phe-Tyr-Pro, in the ß strand B region of the Ig-like domain. The sequences of the adapter and $(dT)_{17}$-adapter primer are as described by Frohman and his co-workers (1988).

RNA isolation. RNA was isolated from the spleen of a nurse shark as described by Chomczynski and Sacchi (1987). $Poly(A)^+$ RNA was isolated by using an mRNA isolation kit purchased from Pharmacia (Piscataway, NJ).

Polymerase chain reaction (PCR). Total cellular RNA (7 µg) isolated from the nurse shark spleen was reverse transcribed in a total volume of 50 µl containing 50 m*M* Tris-HCl (pH 8.3), 6 m*M* $MgCl_2$, 40 m*M* KCl, 1 m*M* DTT, 1 m*M* dNTPs, 300 ng of the $(dT)_{17}$-adapter primer, and 20 units of Moloney murine leukemia virus reverse transcriptase (Stratagene, La Jolla, CA) at 37°C for 1 h. The cDNA thus synthesized was then subjected to PCR as described by Frohman and his co-workers (1988). Briefly, 100 µl of PCR cocktail contained 3 µl of the cDNA, 200 µ*M* each dNTP, 1 µ*M* of Primer 21, 1 µ*M* of the adapter primer, and 2.5 units of *Thermus aquaticus* DNA polymerase (Perkin Elmer Cetus, Norwalk, CT) in 10 m*M* Tris-HCl (pH 8.4), 50 m*M* KCl, 2.5 m*M* $MgCl_2$ and 0.2 mg/ml of gelatin. This cocktail was denatured at 91°C for 5 min and cooled to 54°C for 5 min. It was then amplified using an automatic thermocycler under the following condition: three cycles of 1 min at 94°C, 2 min at 37°C, and 3 min at 72°C, followed by 40 cycles of 1 min at 94°C, 2 min at 48°C, and 3 min at 72°C, and a final extension of 20 min at 72°C. After gel fractionation, the amplified DNA was subcloned into the plasmid vector pBluescript SKII(+) (Stratagene, La Jolla, CA).

Construction of the shark cDNA library and screening. Double-stranded cDNA was synthesized from the $poly(A)^+$ fraction of nurse shark spleen cell RNAs using the Uni-Zap XR cloning system (Stratagene, La Jolla, CA). The cDNA ligated to the Uni-Zap vector was *in vitro* packaged and plated out on *E. coli PLK-F'*. Screening of the library was performed according to the standard method in a buffer containing 50% (vol./vol.) formamide, 1 *M* NaCl, 10 X Denhardt's solution, 50 m*M* Tris-HCl (pH 7.5), 1% $Na_4P_2O_7$, 1% sodium dodecyl sulfate (SDS), and 100 µg/ml of denatured salmon sperm DNA at 42°C for ~ 24 h. The filters were washed twice with 2 X SSC (1 X SSC is 150 m*M* NaCl, 15 m*M* sodium citrate, pH 7.0), 0.1%

SDS at 60°C, and then twice with 0.1 X SSC, 0.1% SDS at 60°C. After plaque-purification of positive clones, the inserts were subcloned into the plasmid vector pBluescript SK(-).

DNA sequencing. Double-stranded cDNA inserts were sequenced by the chain-termination method (Sanger et al. 1977) using the T7 polymerase sequencing kit purchased from Pharmacia (Piscataway, NJ).

DNA sequence analysis. Protein and DNA database searches as well as sequence analyses were performed using the software of IntelliGenetics, Inc. (Mountain View, CA). The following databases were searched using the FASTDB program (Brutlag et al. 1990): GenBank Nucleotide Sequence Data Library, release 65.0; EMBL Genetic Sequence Data Bank, release 24; NBRF-PIR Protein Sequence Database, release 25.0; and Swiss Protein Database, release 14.0.

Results and Discussion

Isolation and sequence determination of nurse shark cDNA clones assumed to encode an NDP kinase

With the aim of isolating *MHC* class II genes from a cartilaginous fish, cDNAs synthesized from the nurse shark splenic RNA were subjected to PCR using Primer 21 and the adapter primer. One of the cDNA clones thus obtained was found to encode a IgM heavy chain (Kokubu et al. 1987; data not shown). Another cDNA clone we sequenced predicted a polypeptide with a cysteinyl residue about 50 residues apart from the conserved Phe-Tyr-Pro stretch. This clone, designated S1-3, was used as a probe to isolate cDNA clones from the nurse shark splenic cDNA library. Screening of 1 X 10^5 plaques resulted in the identification of approximately 50 positive cDNA clones, indicating that the mRNA species detected by S1-3 is abundant. Two cDNA clones (designated 6-2-2 and 2-2-1), which appeared to have the longest insert, were chosen and sequenced in their entirety. The nucleotide sequence of 6-2-2 (Fig. 1) is made up of 610 base pairs (bp) and ends with a 18-bp poly(A) tail. A putative polyadenylation signal AATAAA is located 14-bp upstream from the poly(A) tail. The sequences of the two cDNA clones were identical to each other, with the exception of the fact that the 5'-end of 6-2-2 has 5-bp absent in 2-2-1. The nucleotide sequence of 6-2-2 contains an open reading frame capable of encoding a polypeptide made up of 151 amino acids (Fig.1).

```
         *         *         *         *         *         *         *         *
GGTAACAAGGAACGAACCTTCATTGCTGTGAAGCCAGATGGAGTCCAGAGGTGCATTGTGGGTGAAGTCATCAAACGCTT   80
GlyAsnLysGluArgThrPheIleAlaValLysProAspGlyValGlnArgCysIleValGlyGluValIleLysArgPh   27

TGAACAGAAAGGCTTTAAACTGGTTGCCATGAAATTTCTTCAGGCTCCTAAGGATCTGCTTGAAAAGCATTACTGCGAGC  160
eGluGlnLysGlyPheLysLeuValAlaMetLysPheLeuGlnAlaProLysAspLeuLeuGluLysHisTyrCysGluL   54

TTTCAGACAAACCTTTCTACCCCAAGTTGATCAAGTACATGAGTTCAGGACCTGTCGTTGCCATGGTGTGGGAGGGCCTG  240
euSerAspLysProPheTyrProLysLeuIleLysTyrMetSerSerGlyProValValAlaMetValTrpGluGlyLeu   80

AATGTGGTGAAGACTGGTAGAGTGATGCTGGGTGAGACCAATCCAGCAGATTCCAAGCCTGGCACCATTCGTGGTGACTT  320
AsnValValLysThrGlyArgValMetLeuGlyGluThrAsnProAlaAspSerLysProGlyThrIleArgGlyAspPh  107

CTGTATTCAGGTTGGCAGGAACATCATCCACGGCAGTGACTCTGTAGAAAGTGCAAAGAAGGAGATTAGCCTTTGGTTCA  400
eCysIleGlnValGlyArgAsnIleIleHisGlySerAspSerValGluSerAlaLysLysGluIleSerLeuTrpPheL  134

AACGCGAGGAGCTGGTGGAGTACCAGAACTGTGCCCAAGACTGGATCTATGAGTGAACCATTCCAAAATTGCATGGACGC  480
ysArgGluGluLeuValGluTyrGlnAsnCysAlaGlnAspTrpIleTyrGluEnd                         151

ATCCTAATTAAATCATTCCAAATTGCCTTCGGAAGTTGATTTTTTTAAAACTGACTTCATATCATGTCTTGGTCCTGAAT  560
GTAAAAGCAAGTAACTACAATAAACTAACTTTAAAAAAAAAAAAAAAAAA                               610
```

Fig. 1. Nucleotide sequence of the nurse shark cDNA clone 6-2-2. A putative polyadenylation signal is underlined.

A computer search of nucleotide and protein databases revealed the existence of six sequences quite similar to the nurse shark sequence shown in Figure 1. The three most similar sequences (~ 80% amino acid sequence identity) were the NDP kinase of rats (Shimada et al. 1990), and the Nm23 proteins of humans and mice that are assumed to encode NDP kinases (Rosengard et al. 1989). The other three sequences with significant sequence similarity were the NDP kinases of *D. melanogaster* (the Awd protein; Biggs et al. 1988; ~ 75% amino acid sequence identity), *D. discoideum* (Lacombe et al. 1990; ~ 58% amino acid sequence identity), and *M. xanthus* (Muñoz-Dorado et al. 1990a,b; ~ 41% amino acid sequence identity). The extent of sequence similarity as well as the distribution of shared residues (Fig. 2) suggests strongly that the nurse shark protein we isolated is an NDP kinase. The cDNA clone 6-2-2 appears to contain the entire coding region except for the start codon. It is notable that, like typical Ig superfamily members, the nurse shark sequence has two cysteinyl residues approximately 60 amino acids apart from each other (indicated by # in Figure 2). The top four, vertebrate-derived sequences shown in Figure 2 are completely identical in the region spanning amino acid residues 70 and 123, suggesting that this region is under strong functional constraint.

It has been suggested that the NDP kinase sequence might contain two sequence motifs involved in nucleotide binding (Lacombe et al. 1990; Kimura et al. 1990): a glycine-rich loop which is found in many nucleotide-binding proteins (Wierenga and Hol 1983), and a conserved sequence of the ATP binding region of cyclic nucleotide-dependent protein kinases, Ala-Met-Lys-X-Leu (Hanks et al. 1988). Like the rat NDP kinase (Shimada et al. 1990), the putative nurse shark NDP kinase has the latter sequence motif at residues 36 - 40 (Fig. 2). However, the nurse shark sequence does not appear to have the former sequence motif since it has cysteine instead of glycine at residue 18 (Fig. 2).

```
                               *         *         *         *         *      #  *
Nm23     (Human)         M/ANCERTFIAIKPDGVQRGLVGEIIKRFEQKGFRLVGLKFMQASEDLLKEHYVDLK  56
Nm23     (Mouse)         -/--S------------------------------------L-----------T---  56
NDPK     (Rat)           -/--L--------------------------------AM--LR---EH--Q--I---  56
CL6-2-2  (Shark)           G-K------V-------CI---V---------K--AM--L--PK---EK--CE-S  55
Awd      (Fruit fly)     -A--K-----MV-----------K--E-------K--A---TW---E--EK--A--S  57
NDPK     (Slime mold) MSTNKV-K----L-V-----A--------A-Y-K---V-----QLVPTK--AES--AEH-  60
NDPK     (Myxococcus)    -/-/I---LSI-----LEK-VI-K--S---E--LKP-AIRLQHL-QAQAEGF-AVH-  55

                               *         *         *         *         *  #      *
Nm23     (Human)      DRPFFAGLVKYMHSGPVVAMVWEGLNVVKTGRVMLGETNPADSKPGTIRGDFCIQVGRNI  116
Nm23     (Mouse)      -----T------------------------------------------------------  116
NDPK     (Rat)        -----P------N-----------------------------------------------  116
CL6-2-2  (Shark)      -K--YPK-I---S-----------------------------------------------  115
Awd      (Fruit fly)  A----P---N--N-----P-------------Q---A------L----------------  117
NDPK     (Slime mold) E----G---SFIT--------F--KG--ASA-L-I-V---LA-A--S-----GVD-----  120
NDPK     (Myxococcus) A----KD--QF-I-----L--L--E-A-LAN-DIM-A----QA-E----K--ATSIDK-T  115

                               *         *         *
Nm23     (Human)      IHGSDSVESAEKEIGLWFHPEELVDYTSCAQNWIYE                          152
Nm23     (Mouse)      -------K------S---Q-----E-K---------                          152
NDPK     (Rat)        ------------------K----I--K---HD-V--                          152
CL6-2-2  (Shark)      ----------K---S---KR----E-QN---D----                          151
Awd      (Fruit fly)  -----A--------A---NEK---TW-PA-KD----                          153
NDPK     (Slime mold) ----------NR--A---K----LTEVKPNP-/L--                          155
NDPK     (Myxococcus) V-----L-N-KI--AYF-RET-IHS-PYQK//////                          145
```

Fig. 2. Comparison of the nurse shark sequence to NDP kinases (NDPK) and Nm23 proteins. "-" indicates identity to the top sequence. "/", absence of residues. The sources of the sequences are given in the text. The locations of the putative glycine-rich loop and the conserved Ala-Met-Lys-X-Leu sequence motif are inicated by bold lines.

Sequence comparison between NDP kinases and members of the Ig superfamily

A computer analysis showed that our nurse shark sequence was not significantly related to those of Ig superfamily members. Nevertheless, manual sequence alignment depicted in Figure 3 reveals weak sequence similarity between the nurse shark protein and representative C1-SET members of the Ig superfamily. The highest sequence similarity was found in the region surrounding the Phe-Tyr-Pro stretch (which was used to design the PCR amplification primer); the region surrounding the second cysteinyl residue showed the second highest sequence similarity (Fig. 3). The secondary structure of the putative nurse shark NDP kinase predicted by the algorithm of Chou and Fasman (Fig. 3) is generally consistent with the known structure of the Ig-like domain. It is notable that most of the residues, which are shared or conservatively substituted between the C1-SET members and the nurse shark sequence, are located in the region of the NDP kinase molecule that is virtually invariant across the species (compare Figures 2 and 3).

```
                        B            C               D            E                 F
                    ---------    ---------       -----------  ----------         ---------

MHC Class I         CWALG--FYPAEITLTWQRDGEDQTQDTELVETRPAG---DGTFQKWAAVVV-PS--GEEQRYTC-HVQH
MHC Class II        CSVSG--FYPGSIEVRWFRNGQ---EEKTGVVSTGLIH--NG-DWTFQTLVMLETVPRSGEVYTC-QVEH
IgC Lambda          CLISD--FYPGAVTVAWKADSSP---VKAGVETTTPSKQSNNKYAASSYLSLTPEQWKSHRSYSC-QVTH
                    *      ***  *   *   *        *                       *        * *  * *
CL6-2-2 (Shark)     CELSDKPFYPKLIK--YMSSGPVVAMVWEGLNVVKTGR--VMLGETNPADSK-PGT--IRGDF-CIQVGR

                                           AAAAAA        A--AAAAAA
                    BBB        BBB--BB   BBBBBBB BBBBB BBB--BBBB                 BB-BBBB
                      TTTTTTTTTT      TTTT                         TTTTT-TTT            TT
```

Fig. 3. Sequence similarity of the putative nurse shark NDP kinase to representative C1-SET members of the Ig superfamily. "*" indicates residues which are characteristic fo the C1-SET members of the Ig superfamily as identified by Williams and Barclay (1988) and which are shared or conservatively substituted between the nurse shark sequence and the C1-SET sequences. The top three sequences were taken from Fig. 2 in Williams and Barklay (1988). The predicted secondary structure of the nurse shark protein is indicated by A, for the alpha helix; B, for the beta sheet; and T, for the turn. The locations of the beta strands in the Ig-like domain are indicated by B-F above the sequences. "-", absence of residues.

The putative nurse shark NDP kinase has, however, at least two features that do not appear to conform to the known characteristics of the Ig superfamily molecules. First, the alignment shown in Figure 3 requires the insertion of a few gaps in the predicted ß-strand regions. Changes in the length of the ß-strand may significantly alter the tertiary structure of the molecule. Second, among the seven sequences shown in Figure 2, the cysteinyl residue adjacent to the conserved Phe-Tyr-Pro stretch is found only in the nurse shark sequence. In addition, the cysteinyl residue occurs four residues upstream from the Phe-Tyr-Pro stretch in the C1-SET member of the Ig superfamily, whereas this residue is located seven residues upstream from the Phe-Tyr-Pro stretch in the putative nurse shark NDP kinase.

At present, it is not clear whether the weak sequence similarity observed between the NDP kinases and representative C1-SET members of the Ig superfamily has any evolutionary significance. Determination of the exon-intron structure of the NDP kinase gene and of the crystal structure of the NDP kinase protein should provide important information in resolving this issue. It should be pointed out, however, that highly divergent members of the Ig superfamily, to which the NDP kinase might belong, are not likely to fulfill the strict membership criteria proposed by Williams and Barclay (1988).

Acknowledgments. This work was supported by Grant R01 RR06603-01 from the National Institutes of Health, Bethesda, Maryland.

References

Bevilacqua, G., Sobel, M.E., Liotta, L.A., and Steeg, P.S.: Association of low *nm23* RNA levels in human primary infiltrating ductal breast carcinomas with lymph node involvement and other histopathological indicators of high metastatic potential. *Cancer Res 49:* 5185-5190, 1989

Biggs, J., Tripoulas, N., Hersperger, E., Dearolf, C., and Shearn, A.: Analysis of the lethal interaction between the *prune* and *Killer of prune* mutations of *Drosophila. Genes & Dev 2:* 1333-1343, 1988

Biggs, J., Hersperger, E., Steeg, P.S., Liotta, L.A., and Shearn, A.: A Drosophila gene that is homologous to a mammalian gene associated with tumor metastasis codes for a nucleoside diphosphokinase. *Cell 63:* 933-940, 1990

Brutlag, D.L., Dautricourt, J.-P., Maulik, S., and Relph, J.: Improved sensitivity of biological sequence database searches. *CABIOS 6:* 237-245, 1990

Chomczynski, P. and Sacchi, N.: Single-step method of RNA isolation by acid guanidinium thiocyanate-phenol-chloroform extraction. *Anal Biochem 162:* 156-159, 1987

Frohman, M.A., Dush, M.K., and Martin, G.R.: Rapid production of full-length cDNAs from rare transcripts: amplification using a single gene-specific oligonucleotide primer. *Proc Natl Acad Sci USA 85:* 8998-9002, 1988

Hanks, S.K., Quinn, A.M., and Hunter, T.: The protein kinase family: Conserved feature and deduced phylogeny of the catalytic domains. *Science 241:* 42-52, 1988

Hunkapiller, T. and Hood, L.: Diversity of the immunoglobulin gene superfamily. *Adv Immunol 44:* 1-63, 1989

Kimura, N., Shimada, N., Nomura, K., and Watanabe, K.: Isolation and characterization of a cDNA clone encoding rat nucleoside diphosphate kinase. *J Biol Chem 265:* 15744-15749, 1990

Kokubu, F., Hinds, K., Litman, R., Shamblott, M.J., and Litman, G.W.: Extensive families of constant region genes in a phylogenetically primitive vertebrate indicate an additional level of immunoglobulin complexity. *Proc Natl Acad Sci USA 84:* 5868-5872, 1987

Lacombe, M.-L., Wallet, V., Troll, H., and Véron, M.: Functional cloning of a nucleoside diphosphate kinase from *Dictyostelium discoideum. J Biol Chem 265:* 10012-10018, 1990

Liotta, L.A. and Steeg, P.S.: Clues to the function of Nm23 and Awd proteins in development, signal transduction, and tumor metastasis provided by studies of *Dictyostelium discoideum. J Natl Cancer Inst 82:* 1170-1172, 1990

Muñoz-Dorado, J., Inouye, M., and Inouye, S.: Nucleoside diphosphate kinase from

Myxococcus xanthus. I. Cloning and sequencing of the gene. *J Biol Chem 265:* 2702-2706, 1990a

Muñoz-Dorado, J., Inouye, S., and Inouye, M.: Nucleoside diphosphate kinase from *Myxococcus xanthus*. II. Biochemical characterization. *J Biol Chem 265:* 2707-2712, 1990b

Parks, R.E., Jr. and Agarwal, R.P.: Nucleoside diphosphokinases. *In* P.D. Boyer (ed.): *The Enzymes*, Vol. 8, pp. 307-334. Academic Press, New York 1973

Rosengard, A.M., Krutzsch, H.C., Shearn, A., Biggs, J.R., Barker, E., Margulies, I.M.K., King, C.R., Liotta, L.A., and Steeg, P.S.: Reduced Nm23/Awd protein in tumour metastasis and aberrant *Drosophila* development. *Nature 342:* 177-180, 1989

Sanger, F., Nicklen, S., and Coulson, A.R.: DNA sequencing with chain-terminating inhibitors. *Proc Natl Acad Sci USA 74:* 5463-5467, 1977

Steeg, P.S., Bevilacqua, G., Kopper, L., Thorgeirsson, U.P., Talmadge, J.E., Liotta, L.A., and Sobel, M.E.: Evidence for a novel gene associated with low tumor metastatic potential. *J Natl Cancer Inst 80:* 200-204, 1988

Wallet, V., Mutzel, R., Troll, H., Barzu, O., Wurster, B., Véron, M., and Lacombe, M.-L.: *Dictyosterium* nucleoside diphosphate kinase highly homologous to Nm23 and Awd proteins involved in mammalian tumor metastasis and *Drosophila* development. *J Natl Cancer Inst 82:* 1199-1202, 1990

Wierenga, R.K. and Hol, W.G.J.: Predicted nucleotide-binding properties of p21 protein and its cancer-associated variant. *Nature 302:* 842-844, 1983

Williams, A.F. and Barclay, A.N.: The immunoglobulin superfamily - Domains for cell surface recognition. *Annu Rev Immunol 6:* 381-405, 1988

Author Index

Subject Index

NATO ASI Series H

Vol. 1: **Biology and Molecular Biology of Plant-Pathogen Interactions.** Edited by J. A. Bailey. 415 pages. 1986.

Vol. 2: **Glial-Neuronal Communication in Development and Regeneration.** Edited by H. H. Althaus and W. Seifert. 865 pages. 1987.

Vol. 3: **Nicotinic Acetylcholine Receptor: Structure and Function.** Edited by A. Maelicke. 489 pages. 1986.

Vol. 4: **Recognition in Microbe-Plant Symbiotic and Pathogenic Interactions.** Edited by B. Lugtenberg. 449 pages. 1986.

Vol. 5: **Mesenchymal-Epithelial Interactions in Neural Development.** Edited by J. R. Wolff, J. Sievers, and M. Berry. 428 pages. 1987.

Vol. 6: **Molecular Mechanisms of Desensitization to Signal Molecules.** Edited by T. M. Konijn, P. J. M. Van Haastert, H. Van der Starre, H. Van der Wel, and M. D. Houslay. 336 pages. 1987.

Vol. 7: **Gangliosides and Modulation of Neuronal Functions.** Edited by H. Rahmann. 647 pages. 1987.

Vol. 8: **Molecular and Cellular Aspects of Erythropoietin and Erythropoiesis.** Edited by I. N. Rich. 460 pages. 1987.

Vol. 9: **Modification of Cell to Cell Signals During Normal and Pathological Aging.** Edited by S. Govoni and F. Battaini. 297 pages. 1987.

Vol. 10: **Plant Hormone Receptors.** Edited by D. Klämbt. 319 pages. 1987.

Vol. 11: **Host-Parasite Cellular and Molecular Interactions in Protozoal Infections.** Edited by K.-P. Chang and D. Snary. 425 pages. 1987.

Vol. 12: **The Cell Surface in Signal Transduction.** Edited by E. Wagner, H. Greppin, and B. Millet. 243 pages. 1987.

Vol. 13: **Toxicology of Pesticides: Experimental, Clinical and Regulatory Perspectives.** Edited by L. G. Costa, C. L. Galli, and S. D. Murphy. 320 pages. 1987.

Vol. 14: **Genetics of Translation. New Approaches.** Edited by M. F. Tuite, M. Picard, and M. Bolotin-Fukuhara. 524 pages. 1988.

Vol. 15: **Photosensitisation. Molecular, Cellular and Medical Aspects.** Edited by G. Moreno, R. H. Pottier, and T. G. Truscott. 521 pages. 1988.

Vol. 16: **Membrane Biogenesis.** Edited by J. A. F. Op den Kamp. 477 pages. 1988.

Vol. 17: **Cell to Cell Signals in Plant, Animal and Microbial Symbiosis.** Edited by S. Scannerini, D. Smith, P. Bonfante-Fasolo, and V. Gianinazzi-Pearson. 414 pages. 1988.

Vol. 18: **Plant Cell Biotechnology.** Edited by M. S. S. Pais, F. Mavituna, and J. M. Novais. 500 pages. 1988.

Vol. 19: **Modulation of Synaptic Transmission and Plasticity in Nervous Systems.** Edited by G. Hertting and H.-C. Spatz. 457 pages. 1988.

Vol. 20: **Amino Acid Availability and Brain Function in Health and Disease.** Edited by G. Huether. 487 pages. 1988.

NATO ASI Series H

Vol. 21: **Cellular and Molecular Basis of Synaptic Transmission.**
Edited by H. Zimmermann. 547 pages. 1988.

Vol. 22: **Neural Development and Regeneration. Cellular and Molecular Aspects.**
Edited by A. Gorio, J.R. Perez-Polo, J. de Vellis, and B. Haber. 711 pages. 1988.

Vol. 23: **The Semiotics of Cellular Communication in the Immune System.**
Edited by E.E. Sercarz, F. Celada, N.A. Mitchison, and T. Tada. 326 pages. 1988.

Vol. 24: **Bacteria, Complement and the Phagocytic Cell.**
Edited by F.C. Cabello und C. Pruzzo. 372 pages. 1988.

Vol. 25: **Nicotinic Acetylcholine Receptors in the Nervous System.**
Edited by F. Clementi, C. Gotti, and E. Sher. 424 pages. 1988.

Vol. 26: **Cell to Cell Signals in Mammalian Development.**
Edited by S.W. de Laat, J.G. Bluemink, and C.L. Mummery. 322 pages. 1989.

Vol. 27: **Phytotoxins and Plant Pathogenesis.**
Edited by A. Graniti, R.D. Durbin, and A. Ballio. 508 pages. 1989.

Vol. 28: **Vascular Wilt Diseases of Plants. Basic Studies and Control.**
Edited by E.C. Tjamos and C.H. Beckman. 590 pages. 1989.

Vol. 29: **Receptors, Membrane Transport and Signal Transduction.**
Edited by A.E. Evangelopoulos, J.P. Changeux, L. Packer, T.G. Sotiroudis, and K.W.A. Wirtz. 387 pages. 1989.

Vol. 30: **Effects of Mineral Dusts on Cells.**
Edited by B.T. Mossman and R.O. Bégin. 470 pages. 1989.

Vol. 31: **Neurobiology of the Inner Retina.**
Edited by R. Weiler and N.N. Osborne. 529 pages. 1989.

Vol. 32: **Molecular Biology of Neuroreceptors and Ion Channels.**
Edited by A. Maelicke. 675 pages. 1989.

Vol. 33: **Regulatory Mechanisms of Neuron to Vessel Communication in Brain.**
Edited by F. Battaini, S. Govoni, M.S. Magnoni, and M. Trabucchi. 416 pages. 1989.

Vol. 34: **Vectors as Tools for the Study of Normal and Abnormal Growth and Differentiation.**
Edited by H. Lother, R. Dernick, and W. Ostertag. 477 pages. 1989.

Vol. 35: **Cell Separation in Plants: Physiology, Biochemistry and Molecular Biology.**
Edited by D.J. Osborne and M.B. Jackson. 449 pages. 1989.

Vol. 36: **Signal Molecules in Plants and Plant-Microbe Interactions.**
Edited by B.J.J. Lugtenberg. 425 pages. 1989.

Vol. 37: **Tin-Based Antitumour Drugs.** Edited by M. Gielen. 226 pages. 1990.

Vol. 38: **The Molecular Biology of Autoimmune Disease.**
Edited by A.G. Demaine, J-P. Banga, and A.M. McGregor. 404 pages. 1990.

Vol. 39: **Chemosensory Information Processing.** Edited by D. Schild. 403 pages. 1990.

Vol. 40: **Dynamics and Biogenesis of Membranes.**
Edited by J.A.F. Op den Kamp. 367 pages. 1990.

Vol. 41: **Recognition and Response in Plant-Virus Interactions.**
Edited by R.S.S. Fraser. 467 pages. 1990.

NATO ASI Series H

Vol. 42: **Biomechanics of Active Movement and Deformation of Cells.**
Edited by N. Akkaş. 524 pages. 1990.

Vol. 43: **Cellular and Molecular Biology of Myelination.**
Edited by G. Jeserich, H. H. Althaus, and T. V. Waehneldt. 565 pages. 1990.

Vol. 44: **Activation and Desensitization of Transducing Pathways.**
Edited by T. M. Konijn, M. D. Houslay, and P. J. M. Van Haastert. 336 pages. 1990.

Vol. 45: **Mechanism of Fertilization: Plants to Humans.**
Edited by B. Dale. 710 pages. 1990.

Vol. 46: **Parallels in Cell to Cell Junctions in Plants and Animals.**
Edited by A. W. Robards, W. J. Lucas, J. D. Pitts, H. J. Jongsma, and D. C. Spray. 296 pages. 1990.

Vol. 47: **Signal Perception and Transduction in Higher Plants.**
Edited by R. Ranjeva and A. M. Boudet. 357 pages. 1990.

Vol. 48: **Calcium Transport and Intracellular Calcium Homeostasis.**
Edited by D. Pansu and F. Bronner. 456 pages. 1990.

Vol. 49: **Post-Transcriptional Control of Gene Expression.**
Edited by J. E. G. McCarthy and M. F. Tuite. 671 pages. 1990.

Vol. 50: **Phytochrome Properties and Biological Action.**
Edited by B. Thomas and C. B. Johnson. 337 pages. 1991.

Vol. 51: **Cell to Cell Signals in Plants and Animals.**
Edited by V. Neuhoff and J. Friend. 404 pages. 1991.

Vol. 52: **Biological Signal Transduction.**
Edited by E. M. Ross and K. W. A. Wirtz. 560 pages. 1991.

Vol. 53: **Fungal Cell Wall and Immune Response.**
Edited by J. P. Latgé and D. Boucias. 472 pages. 1991.

Vol. 54: **The Early Effects of Radiation on DNA.**
Edited by E. M. Fielden and P. O'Neill. 448 pages. 1991.

Vol. 55: **The Translational Apparatus of Photosynthetic Organelles.**
Edited by R. Mache, E. Stutz, and A. R. Subramanian. 260 pages. 1991.

Vol. 56: **Cellular Regulation by Protein Phosphorylation.**
Edited by L. M. G. Heilmeyer, Jr. 520 pages. 1991.

Vol. 57: **Molecular Techniques in Taxonomy.**
Edited by G. M. Hewitt, A. W. B. Johnston, and J. P. W. Young. 420 pages. 1991.

Vol. 58: **Neurocytochemical Methods.**
Edited by A. Calas and D. Eugène. 352 pages. 1991.

Vol. 59: **Molecular Evolution of the Major Histocompatibility Complex.**
Edited by J. Klein and D. Klein. 528 pages. 1991.

MIX
Papier aus verantwortungsvollen Quellen
Paper from responsible sources
FSC® C105338

If you have any concerns about our products,
you can contact us on
ProductSafety@springernature.com

In case Publisher is established outside the EU,
the EU authorized representative is:
Springer Nature Customer Service Center GmbH
Europaplatz 3, 69115 Heidelberg, Germany

Printed by Libri Plureos GmbH
in Hamburg, Germany